Principles
of
Farm
Irrigation
System
Design

Larry G. James
Washington State University

John Wiley & Sons
New York   Chichester   Brisbane   Toronto   Singapore

Library of Congress Cataloging in Publication Data:

James, Larry G.
  Principles of farm irrigation.
  Includes index.
  1. Irrigation engineering. I. Title.
TC805.J36 1988   627'.52   87-6289
ISBN 0-471-83954-X

Printed in the United States of America

10 9 8 7 6 5 4 3 2 1

About
the
Author

Larry G. James is an associate professor of agricultural engineering at Washington State University, Pullman, Washington. Dr. James received a B.S.Ag.E. degree from Washington State University in 1970 and a Ph.D. in agricultural engineering from the University of Minnesota in 1975. He was an assistant professor of agricultural engineering at Cornell University in Ithaca, New York, before joining the Washington State faculty in 1977. At Washington State he has twice been a Featured Teacher in the College of Agriculture and Home Economics. Dr. James has received the R. M. Wade Award for Excellence in Teaching, has been a Mortar Board Outstanding WSU Faculty Member of the Month, and has received the College of Engineering and Architecture Research Excellence Award. In 1983 he was named the Engineer of the Year by the Inland Empire Section of the American Society of Agricultural Engineers (ASAE); in 1984 he received the Dow Outstanding Young Faculty Award for the Pacific Northwest Region of the American Society for Engineering Education, and in 1986 received an ASAE Paper Award. Dr. James's research interests include crop water requirements, irrigating with limited water, energy requirements for irrigation, mathematical modeling of irrigation systems, and water infiltration into sprinkle irrigated soils. He has authored or coauthored more than 50 papers on irrigation engineering, crop water requirements, and water infiltration into soils.

Preface

This book is a text for beginning students in irrigation. It is designed to guide students from a basic knowledge of soils, botany, mathematics, and hydraulics to state-of-the-art irrigation system design and management. Although based on current design practices and the latest research, this book is not intended to be a design manual or a review of irrigation research. Instead, the method and order of presentation have been carefully developed and classroom tested to make this book an effective and useful teaching tool. My goal is to present the principles and concepts of farm irrigation in a manner that maximizes student learning, understanding, and motivation.

The book contains 60 solved example problems, FORTRAN code for 7 computer programs (disks with these programs can be obtained from either the author or publisher), 197 homework problems (a solutions manual with code for an additional 23 computer programs is available), and more than 100 references for student use and enrichment. The introduction section of each chapter includes learning objectives and an overview of the material to be covered in the chapter. Most of the 177 equations in the book have a unit constant, K, which allows the use of either SI (the International System of Units) or English units.

The title and a summary of the contents of each chapter follow.

Chapter 1

Irrigation Requirements and Scheduling. This chapter presents an overview of how plants use water and how they interact with the soil and atmosphere to obtain it. Detailed discussions of such topics as soil water-holding capacity and evapotranspiration are included. Several methods for quantifying irrigation requirements and schedules are set forth.

Chapter 2

Irrigation Systems and System Design Fundamentals. The types and functions of farm irrigation systems, the major steps in the design process, and the role of the designer are explained in this chapter. Methods are presented for determining the design daily irrigation requirement, the effectiveness of an

v

irrigation system, and the annual ownership and operating costs of a farm irrigation system.

Chapter 3

Water for Irrigation. The characteristics of various surface and ground water sources are spelled out. Procedures for evaluating water quality and the hydrologic suitability of surface and ground water sources, and the basic principles of water law are presented.

Chapter 4

Pumps. In this chapter, the operation, performance, and selection of pumps for farm irrigation systems are considered. System curves, affinity laws, and series and parallel hookups are among the topics examined.

Chapter 5

Sprinkle Irrigation Systems. The operating characteristics of several types of sprinkle irrigation systems, sprinkler selection, the hydraulics of pipelines with constant and diminishing flow, water hammer, air entrainment, valving, and the design of set-move, traveler, center-pivot, and linear-move systems are the primary topics covered in this chapter.

Chapter 6

Trickle Irrigation. The operation, management, and design of drip, subsurface, bubbler, and spray-type trickle systems are described. Clogging and chemical injection are considered.

Chapter 7

Surface Irrigation. In this chapter, surface irrigation processes, systems, and components are discussed. Detailed descriptions of low head pipelines, lined and unlined open channel conveyance systems, and design of level basin, graded border, and furrow systems are important parts of this chapter

Chapter 8

Flow Measurement. The principles and methods of measuring flow in open channels and pipelines are defined in this chapter.

Larry G. James

Contents

Irrigation Requirements and Scheduling

1.1 Introduction

The primary reason for irrigating crops is to supplement water available from natural sources of water, such as rainfall, dew, floods, and groundwater which seep into the root zone. Irrigation is needed in areas where water from natural sources is adequate for crop production during only a part of the year or is sufficient in some years and not in others. The amount and timing of irrigations depends on several climatic, soils, and crop factors.

This chapter focuses on quantifying crop irrigation requirements and schedules (i.e., the amount and timing of individual irrigations). It begins with a description of how plants use water and how they interact with the soil and atmosphere to obtain it. These fundamental principles form the basis for the presentation of several methods of measuring and estimating crop water requirements from atmospheric and soil parameters. Procedures for estimating irrigation requirements and schedules from crop water requirements are then presented and demonstrated. After completing this chapter the reader should be able to use a variety of techniques to compute irrigation requirements and to schedule irrigations considering differences in crops and locations. Information in this chapter is fundamental to the design of farm irrigation systems.

1.2 Plant-Soil-Atmosphere (Climatic) Relationships

Plant-soil-atmosphere relationships can be summarized as follows: The plant needs water, the soil stores the water needed by the plant, and the atmosphere

1

provides the energy needed by the plant to withdraw water from the soil. The role of the plant, the soil, and the atmosphere are described in the following sections.

1.2.1 The Plant

Sixty to 95 percent of a physiologically active plant is water. Water is required for such plant processes as:

1. digestion,
2. photosynthesis,
3. transport of minerals and photosynthates,
4. structural support,
5. growth, and
6. transpiration.

The plant uses water primarily for transpiration. The process of *transpiration*, defined as evaporation from a living surface, usually accounts for about 99 percent of the water used by plants (Wilson et al., 1962). Transpiration involves the conversion of water from the liquid to vapor phase within the leaf and its transport through stomata of the leaf into the atmosphere.

Transpiration occurs when the vapor pressure within the leaf exceeds that of the surrounding air and stomata are open to allow carbon dioxide into the plant for photosynthesis. The rate at which water vapor escapes the leaf, that is, the transpiration rate is given by:

$$T = \frac{e_{\text{leaf}} - e_{\text{air}}}{r_{\text{leaf}} + r_{\text{air}}} \tag{1.1}$$

where

T = transpiration rate;
e_{leaf} = vapor pressure within the leaf;
e_{air} = vapor pressure of air;
r_{leaf} = resistance to vapor flow through the stomata;
r_{air} = resistance to vapor flow through the air boundary layer around the leaf.

Transpiration generally occurs whenever stomata are open, since e_{leaf} usually exceeds e_{air}. It is commonly assumed that e_{leaf} equals the saturation vapor pressure for the temperature within the leaf, since there is usually at least some water in the leaf even if it is wilted.

The resistance term r_{air} depends primarily on air movement. Higher winds tend to break up the boundary layer surrounding the leaves and reduce r_{air}. The resistance term r_{leaf} for a given plant is mainly related to the degree of stomatal closure, that is, r_{leaf} increases as stomata close.

The plant extracts water from the soil to replenish water lost by transpiration. Water moves through the soil into the roots, up the xylem and into the leaves due to a water potential gradient (difference) between the leaf and the soil. This process is called passive absorption. The rate of water flow is given by

$$Q = \frac{\psi_{leaf} - \psi_{soil}}{r_{plant} + r_{soil}} \tag{1.2}$$

$$\psi_{leaf} = \psi_T + \psi_\pi \tag{1.2a}$$

where

$\quad Q$ = rate of flow;
$\;\psi_{leaf}$ = total water potential in the leaf;
$\;\psi_{soil}$ = total water potential in the soil;
$\quad \psi_T$ = turgor pressure within the leaf;
$\quad \psi_\pi$ = osmotic pressure within the plant;
r_{plant} = resistance to water movement into the roots, up the xylem,
$\qquad\quad$ and into the leaf;
$\;r_{soil}$ = resistance to water movement in the soil.

$\qquad$ The term r_{soil} can also be written as:

$$r_{soil} = \frac{1}{K} \tag{1.2b}$$

where, K is the hydraulic conductivity of the soil.

$\qquad$ As the plant removes water from the soil, the water content of soil decreases and ψ_{soil} decreases (becomes more negative). At the same time the conductivity of the soil decreases (hence r_{soil} increases) according to Figure 1.1a. Decreasing ψ_{soil} and increasing r_{soil} tends to decrease water flow into the plant. As this situation continues, the plant begins to dehydrate and Ψ_T, hence Ψ_{leaf}, decreases in order to maintain the transpiration rate, T given by Eq. 1.1. When ψ_{leaf} is sufficiently low, the leaf will be dehydrated to the extent that stomata begin to close. This increases the resistance to vapor flow from the leaf, r_{leaf}, and results in a lower rate of transpiration as predicted by Eq. 1.1. Since CO_2 enters the leaves through the same

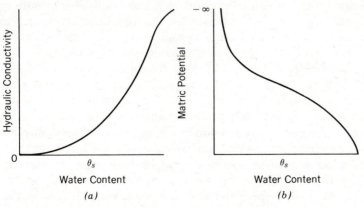

Water Content
(a)

Water Content
(b)

Figure 1.1   Example soil water characteristic curves.

pathway as water vapor escapes, stomatal closure will also result in a decreased photosynthetic rate, causing a reduction in growth, yield, and/or quality.

1.2.2   The Soil

The soil stores water needed by plants. Adsorptive and capillary forces, called *matric forces*, hold (i.e., store) significant amounts of water, which can be removed and used by plants, in the voids between individual soil particles. Adsorptive forces result because of the attraction between negatively charged clay particles and the positive end of dipole water molecules. Capillary forces are due to adhesion between soil particles and water, and the surface tension of water.

Matric forces must be overcome to remove water from the soil. The minimum force required to remove water from a soil varies with the amount of water in the soil, as illustrated by the soil characteristic curve in Figure 1.1b. As the voids between soil particles are filled with water and the soil approaches saturation, the matric forces holding water in the soil approach zero. Conversely, as the water content of the soil approaches zero the matric forces approach negative infinity. Thus, it is much easier for plants to obtain water when the soil is moist than when it is dry.

Between saturation and absolutely dry are two important soil water contents relative to the plant. These water contents, *field capacity* (fc) and the *permanent wilting point* (pwp), define, respectively, the upper and lower limits of soil water that is available to plants. The water content when the soil is at field capacity is less than saturation while the soil is not absolutely dry at the permanent wilting point. Neither field capacity nor permanent wilting point is a sharply defined quantity. Permanent wilting point, for example, is a function of crop and stage of growth.

The following illustrates the concepts of field capacity and permanent wilting point. Consider a completely saturated root zone of soil (all voids within root zone are full of water) that is receiving as much water as it is losing. Suddenly inflow to the soil ends. Flow from the soil will continue as long as gravitational forces exceed the matric forces tending to hold water in the soil. As the soil dewaters, the matric forces steadily increase (negatively) to contribute to a steadily declining outflow rate. This continues until the rate of water movement through the soil is negligible. The water content when this occurs is field capacity. Because many soils are dewatered from saturation to field capacity in less than 48 hours, water in excess of field capacity is not considered to be available to plants. Field capacity is sometimes defined as the soil water content corresponding to a matric potential ranging from 0.1 bar for sands to 0.5 bar or more for very fine-textured soils. A matric potential of 1/3 bar is used to define field capacity for most soils.

Evaporation and transpiration, often combined and called evapotranspiration, are necessary to dewater the soil from field capacity. As water is removed and soil matric forces increase, it becomes increasingly more difficult for the plant to remove water. When the plant is unable to remove any more water and it permanently wilts, the soil is at its permanent wilting point. Water left in the soil when the permanent wilting point is reached is not available to plants. Most of this water is held by adsorptive forces and is called *hygroscopic water*.

Because water between field capacity and the permanent wilting point is available to plants, it is called *available water* (AW). The following equation is used to compute available water:

$$AW = D_{rz}(fc - pwp)/100 \qquad\qquad (1.3)$$

where

AW = available water (cm, in)
D_{rz} = depth of the root zone (cm, in) (D_{rz} is the lesser of the normal effective rooting depth of the plant and the depth of the soil, i.e., the depth to a soil layer that restricts water movement/root penetration);
  fc = field capacity in percent by volume;
pwp = permanent wilting point in percent by volume.

Although plants are theoretically able to obtain water from the soil whenever water contents exceed pwp, the actual rate at which they transpire decreases as stomata close in response to declining soil water contents (as explained in Section 1.2.1). Figure 1.2 shows the variation of a typical plant's actual transpiration rate with soil water content and defines a critical soil water content, θ_c. The relatively small declines in actual transpiration associated with soil water content reductions between fc and θ_c indicate that water is more readily available and that higher crop yield and/or quality should be expected in this range than between θ_c and pwp. Irrigations are normally scheduled to maintain soil water contents above θ_c. The volume per unit surface area of soil water between fc and θ_c is called *readily available water*, (RAW). The following equation can be used to compute RAW.

$$RAW = D_{rz}(fc - \theta_c)/100 \qquad\qquad (1.3a)$$

where, θ_c is water content in percent by volume.

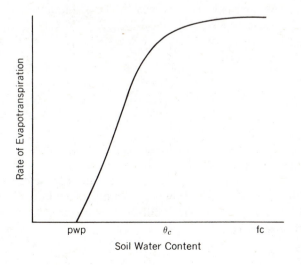

Figure 1.2   Variation of actual transpiration rate with soil water content for a typical crop.

The concept of *maximum allowable deficiency*, (MAD), is also used to estimate the amount of water that can be used without adversely affecting the plant. MAD is defined as

$$MAD = \frac{RAW}{AW} \tag{1.4}$$

where RAW and AW are as previously defined. MAD for most crops is about 0.65. Table 1.1 lists MADs and effective rooting depths for several crops.

When the MAD concept is used, RAW is computed using the following:

$$RAW = (MAD)(AW)$$
$$= (MAD)(D_{rz})(fc - pwp)/100 \tag{1.5}$$

Table 1.2 lists values of fc and pwp for different soils.

Although θ_c and MAD are normally assumed constant, they can vary considerably with the potential rate of transpiration (i.e., the transpiration rate that occurs when stomata are fully open). Figure 1.3 shows that θ_c ranges from fc for high potential transpiration rates to pwp for low potential transpiration rates. The MAD values in Table 1.1 provide guidance for selecting MAD. These MAD values may be adjusted according to potential transpiration rates, soils, and economics. For example, lower MAD values (higher θ_c values) may be desirable where extremely high potential rates of transpiration are expected or when market conditions require the highest possible yield per unit land area. Conversely, when the cost of water and/or energy are high relative to crop revenues, highest economic returns may be obtained by allowing soil moisture to drop below θ_c. This practice often called *deficit irrigation*, involves sacrificing crop revenues (by sacrificing crop yield and/or quality) to achieve reductions in water and/or energy costs that exceed the sacrificed crop revenue (water and energy reductions occur, since lower soil water contents result in increased stomatal resistance and reduced transpiration). Deficit irrigation normally involves maximizing crop production per unit of water and/or energy used. Deficit irrigation is also used when the ability of the irrigation system to deliver the full irrigation requirement is restricted by the availability of capital and other limitations.

1.2.3 The Atmosphere

The atmosphere provides the energy needed by the plant to withdraw water from the soil. If soil water is not limiting and stomata are fully open, conditions in the atmosphere control the rate of transpiration.

The most important atmospheric factors affecting transpiration are the humidity of the air surrounding the plant, the temperature and humidity of air carried to the plant by wind, and the net radiation available to the plant. Increasing the humidity of the air surrounding the leaf, other things remaining constant, will decrease the vapor pressure difference between the leaf and the surrounding air. A reduced rate of transpiration will result, as indicated by Eq. 1.1.

Table 1.1 Values of Maximum Allowable Depletion (MAD) and Maximum Rooting Depth for 39 Crops

Crop	Maximum Allowable Depletion[a]	Maximum Root Depth not Limited by Soil Depth (ft)	(cm)
Alfalfa	0.65	6	180
Apples (with/without cover)	0.65	6	180
Apricots (with/without cover)	0.65	6	180
Beans, dry	0.50	3	90
Beans, green	0.50	3	90
Carrots	0.50	3	90
Cherries (with/without cover)	0.65	6	180
Clover	0.65	2	60
Corn, grain	0.65	4	120
Corn, sweet	0.65	4	120
Crucifers	0.50	2	60
Cucumbers	0.50	4	120
Grapes	0.65	6	180
Hops	0.65	6	180
Mint	0.35	2	60
Onions, dry	0.50	2	60
Onions, green	0.50	2	60
Pasture/turf	0.65	2	60
Peaches (with/without cover)	0.65	6	180
Peas	0.65	2	60
Pears and Plums (with/without cover)	0.65	6	180
Potatoes	0.30	2	60
Radishes	0.50	2	60
Raspberries	0.65	4	120
Safflower	0.65	6	180
Sorghum	0.65	3	90
Soybeans	0.65	3	90
Spinach	0.50	2	60
Spring grain	0.65	3	90
Strawberries	0.65	1	30
Sugarbeets	0.65	3.5	105
Sunflowers	0.65	6	180
Tomatoes	0.50	6	180
Winter Wheat	0.65	3	90

Source: L. G. James, J. M. Erpeneck, D. L. Bassett, and J. E. Middleton, "Irrigation Requirements for Washington—Estimates and Methodology," Research Bulletin XB0925 (1982), Agricultural Research Center, Washington State University, Pullman, 37 pp.

[a] Values vary on different soils. Values listed are maximum for ordinary irrigated soils. There may be an advantage to quality and yield from using lesser values between irrigations, especially on heavy soils.

Table 1.2 Representative Physical Properties of Soils

Soil Texture	Saturated Hydraulic Conductivity,[a] k_s (mm/h) (ins/h)		Total Pore Space (% by vol)	Apparent Specific Gravity (A_s)	Field Capacity fc (% by vol)	Permanent Wilting pwp (% by vol)	Percent by Volume (v)	Total Available Water[b] AW = (0.10)(θ_v) (mm/cm)	AW = (0.12)(θ_v) (in/ft)
Sandy	50 (25–250)	2 (1–10)	38 (32–42)	1.65 (1.55–1.80)	15 (10–20)	7 (3–10)	8 (6–10)	0.8 (0.7–1.0)	1.0 (0.8–1.2)
Sandy Loam	25 (12–75)	1 (0.5–3)	43 (40–47)	1.50 (1.40–1.60)	21 (15–27)	9 (6–12)	12 (9–15)	1.2 (0.9–1.5)	1.4 (1.1–1.8)
Loam	12 (8–20)	0.5 (0.3–0.8)	47 (43–49)	1.40 (1.35–1.50)	31 (25–36)	14 (11–17)	17 (14–20)	1.7 (1.4–1.9)	2.0 (1.7–2.3)
Clay Loam	8 (3–5)	0.3 (0.1–0.6)	49 (47–51)	1.35 (1.30–1.40)	36 (31–42)	18 (15–20)	18 (16–22)	1.9 (1.7–2.2)	2.3 (2.0–2.6)
Silty Clay	3 (0.25–5)	0.1 (0.01–0.2)	51 (49–53)	1.30 (1.25–1.35)	40 (35–46)	20 (17–22)	20 (18–23)	2.1 (1.8–2.3)	2.5 (2.2–2.8)
Clay	5 (1–10)	0.2 (0.05–0.4)	53 (51–55)	1.25 (1.20–1.30)	44 (39–49)	21 (19–24)	23 (20–25)	2.3 (2.0–2.5)	2.7 (2.4–3.0)

Source: V. E. Hansen, O. A. Israelson, G. Stingham, *Irrigation Principles and Practice*, Copyright © 1979 by John Wiley & Sons, Inc., New York, 52 pp. Reprinted by permission of John Wiley & Sons, Inc.
Note: Normal ranges are shown in parentheses.
[a] Saturated hydraulic conductivities vary greatly with soil structure and structural stability, even beyond the normal ranges shown.
[b] Readily available water is computed using Eq. 1.5.

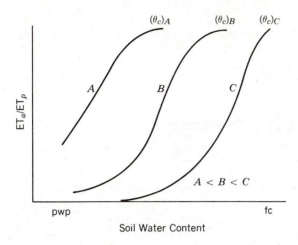

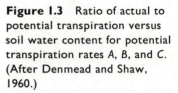

Figure 1.3 Ratio of actual to potential transpiration versus soil water content for potential transpiration rates A, B, and C. (After Denmead and Shaw, 1960.)

Wind sweeps away any layer of water vapor accumulated around the leaf and either increases or decreases the transpiration rate. If the air around the leaf is replaced with warmer and/or dryer air, the transpiration rate increases. Conversely, if the wind brings cooler and/or more humid air, the transpiration rate decreases, whether in shade or in the sun.

Radiation influences the rate of transpiration in two ways. First, radiation raises leaf temperatures above that of the surrounding air and hence increases T (since the saturation vapor pressure within the leaf increases as leaf temperature rises). The temperature of leaves in the shade is approximately equal to the air temperature, while leaf temperatures in the sunlight are 5° to 10°C (10° to 20°F) above air temperature. Second, the presence of light (shortwave radiation) triggers the opening or closing of stomata. The stomata of most plants are open during the day and closed at night. The pineapple plant is one notable exception to this, however. Its stomata are closed during the day and open at night.

1.3 Consumptive Use and Evapotranspiration

Water is transferred to the atmosphere by direct evaporation of solid and liquid water from soil and plant surfaces as well as by transpiration. Since these processes each involve evaporation and are not easily separated, they are combined and called *evapotranspiration* (ET).

Consumptive use (CU) includes water used in all of the plant processes discussed in Section 1.2 (rather than just transpiration) as well as direct evaporation from soil and plant surfaces. Thus, CU exceeds ET by the amount of water used for digestion, photosynthesis, transport of minerals and photosynthates, structural support, and growth. Since this difference is usually less than 1 percent, ET and CU are normally assumed to be equal.

1.4   Determining Evapotranspiration

Crop ET is determined by direct measurement or calculated from crop and climate data. Direct measurement techniques involve isolating a portion of the crop from its surroundings and determining ET by measurement. Several theoretical and empirical equations have been developed for computing crop ET. These equations are used to estimate ET for crops and locations where measured ET data are not available.

1.4.1   Direct Measurement of Evapotranspiration

The most widely used direct measurement techniques are based on the conservation of mass principle. Equation 1.6 defines the conservation of mass principle for the control volume in Figure 1.4. Equation 1.6 is:

$$\Delta S = D_{rz}(\theta_f - \theta_i) = \text{inflow} - \text{outflow} \tag{1.6}$$

where

$$
\begin{aligned}
\text{inflow, outflow} &= \text{total flow into and out of the control volume during the} \\
&\quad \text{time interval being considered, respectively (cm, in)} \\
\Delta S &= \text{change in soil moisture within the control volume during} \\
&\quad \text{the time interval being considered (cm, in)} \\
D_{rz} &= \text{depth of root zone (below soil surface) (cm, in)} \\
\theta_f, \theta_i &= \text{soil moisture contents by volume at end (final) and} \\
&\quad \text{beginning (initial) of the time interval being considered,} \\
&\quad \text{respectively (decimal).}
\end{aligned}
$$

From Figure 1.4

$$\text{Inflow} = I + P + SFI + LI + GW \tag{1.6a}$$

and

$$\text{Outflow} = ET + RO + LO + L + DP \tag{1.6b}$$

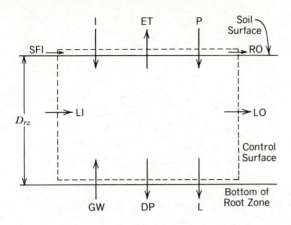

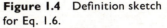

Figure 1.4 Definition sketch for Eq. 1.6.

where

I = irrigation (cm, in);

P = precipitation (cm, in);

SFI = surface flow into the control volume (cm, in);

LI = subsurface lateral flow into the control volume (cm, in);

GW = groundwater seepage into the control volume (cm, in);

ET = evapotranspiration (cm, in);

RO = surface flow out of the control volume (cm, in);

LO = subsurface lateral flow out of the control volume (cm, in);

L = leaching requirement (i.e., amount of water that must flow from the root zone to maintain a favorable salt balance in the root zone, see Section 3.3.1c) (cm, in);

DP = deep percolation (i.e., downward movement of water from the control volume in excess of that needed for leaching) (cm, in).

All terms in Eqs. 1.6, 1.6a, and 1.6b are depths and have dimensions of length (i.e., mass/density/surface area).

Substitution of Eqs. 1.6a and 1.6b into Eq. 1.6 and solving for ET yields Eq. 1.7.

$$ET = I + P + SFI + LI + GW - RO - LO - L - DP - D_{rz}(\theta_f - \theta_i) \qquad (1.7)$$

1.4.1a Lysimeters

Crops are often grown in buried soil-filled tanks called *lysimeters*, to facilitate application of Eq. 1.7. Lysimeters hydrologically isolate soil within them from surrounding soil and make it possible to eliminate SFI, Li, and LO, while GW, RO, L, and DP are either eliminated or measured. ET can be calculated when I, P, D, θ_i and θ_f have been measured.

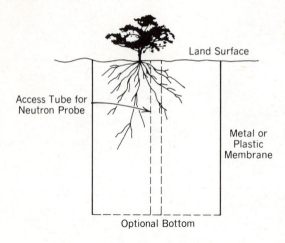

Figure I.5 A nonweighing lysimeter.

Lysimeters differ in the way in which ΔS is determined. Weighing lysimeters are constructed so that ΔS is determined by weighing. Various techniques for measuring or inferring changes in soil moisture described in Section 1.6.2b are used to determine ΔS in nonweighing lysimeters. Figures 1.5 and 1.6 show the primary differences in the construction of weighing and nonweighing lysimeters. Weighing lysimeters have a second tank that retains surrounding soil so that the inside container is free for weighing. They also usually have a means for removing and measuring DP and L. Nonweighing lysimeters may or may not have this capability.

The reliability of ET data collected with lysimeters depends on how well conditions within the lysimeter (i.e., soil structure and density, drainage characteristics, temperature, and density, height, etc., of the crop) match conditions surrounding the lysimeter. Lysimeters must be large enough to minimize boundary effects and to avoid restricting root development. High installation costs and the

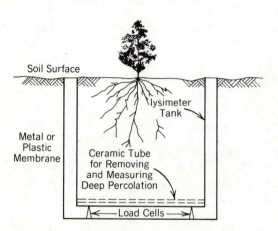

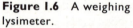

Figure I.6 A weighing lysimeter.

immobility of lysimeters preclude their use as routine field instruments. Lysimeters are used primarily as research tools for checking the accuracy of other methods of determining ET.

1.4.1b   Field Water Balances

Equation 1.7 can also be used to determine ET in irrigated fields (without the use of lysimeters). The accuracy of ET estimates obtained in this manner is usually reduced because it is normally more difficult to control and/or measure one or more of the terms in Eq. 1.7. The terms θ_f and θ_i are measured using techniques described in Section 1.6.2b.

1.4.1c   Evapotranspiration Chambers

Another direct measurement technique uses an aboveground chamber to enclose a vegetated area. The chamber is transparent to radiation and prevents water exchange with the atmosphere. Though useful for many studies, the space inside the chamber is not representative of conditions outside the chamber, since radiation exchange and turbulent transfer within the enclosure are altered. Reicosky and Peters (1977) have described a portable chamber for the rapid measurement of ET on field plots.

1.4.2   Calculating Evapotranspiration

All methods for computing crop ET involve the following equation

$$ET = K_c ET_0 \tag{1.8}$$

where

   ET = evapotranspiration for a specific crop;
   ET_0 = potential ET or reference crop ET;
   K_c = crop coefficient.

ET$_0$ may be either potential ET or reference crop ET. *Potential ET* is the maximum rate at which water, if available, can be removed from soil and plant surfaces. Potential ET depends on the amount of energy available for evaporation and varies from day to day. *Reference crop ET* is the potential ET for a specific crop (usually either grass or alfalfa) and set of surrounding (advective) conditions. Doorenbos and Pruitt (1977) define reference crop ET as the "ET from an extensive surface of 8 to 15 cm (3 to 6 ins) tall, green grass cover of uniform height, actively growing, completely shading the ground and not short of water." Wright (1981) define reference crop ET as being "equal to daily alfalfa ET when the crop occupies an extensive surface, is actively growing, standing erect and at least 20 cm (8″) tall, and is well watered so that soil water availability does not limit ET." Reference crop ET is preferred over potential ET, since potential ET can vary from crop to crop due to differences in aerodynamic roughness and surface reflectance (albedo), and from location to location because of differences in the amount of sensible and latent heat transferred into the area. Reference crop ET is, on the other hand, defined for a specific crop and set of advective conditions.

Many methods with differing data requirements and levels of sophistication have been developed for computing ET_0. Some of these methods require daily relative humidity, solar radiation, wind and air temperature data, while others need only mean monthly temperatures. Some are physically based while others were determined empirically. These methods may be classified as

1. aerodynamic methods,
2. energy balance methods,
3. combination methods, and
4. empirical methods (including temperature, solar radiation, and pan evaporation based methods).

These methods are described in the following sections.

The crop coefficient, K_c, relates the actual rate at which a crop uses water (ET) to ET_0. Coefficient K_c for a crop is determined experimentally and reflects the physiology of the crop, the degree of crop cover, the location where data were collected, and the method used to compute ET_0. The variation of K_c with location and ET_0 method is minimized when ET_0 is reference crop ET.

Values of K_c for field and vegetable crops generally increase from an initial plateau to a peak plateau and then decline as the plant progresses through its growth stages. Figure 1.7 shows the typical variation of K_c with growth stage and defines four growth stages for field and vegetable crops. Table 1.3 lists K_c values (based on a grass reference crop) and the duration (in days) of each of the growth stages defined in Figure 1.7. Coefficient K_c and growth stage information for several other field and vegetable crops is available in Doorenbos and Pruitt (1977). Alfalfa-based reference crop coefficients for several crops are given in Burman et al. (1980).

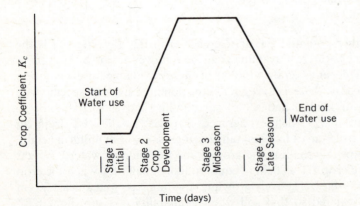

Figure 1.7   Seasonal variation of the crop coefficient, K_c, in Eq. 1.8. (After Dorrenbos and Pruitt, 1977.)

Table 1.3   Equations to Compute K_c for Growth Stage 1 (see Figure 1.7) for Annual Crops Listed in Table 1.4 and in Doorenbos and Pruitt (1977).

Average Interval of Irrigation or Rainfall (days)	Units of ET			
	mm/day		in/day	
	a	b	a	b
2	1.049	−0.119	0.714	−0.119
4	0.904	−0.216	0.450	−0.216
7	0.742	−0.319	0.264	−0.319
10	0.580	−0.408	0.155	−0.408
20	0.438	−0.455	0.101	−0.455

A procedure for determining K_c values based on the information in Table 1.3 and Doorenbos and Pruitt is illustrated in Example 1.1. The K_c for growth stage 1 for annual crops is estimated with the following equation:

$$K_c = a\mathrm{ET}_0^b \tag{1.9}$$

where

$K_c = K_c$ for growth stage 1;
a = coefficient from Table 1.3;
ET_0 = average daily reference crop ET during growth stage 1;
b = exponent from Table 1.3.

EXAMPLE 1.1   Determining K_c Values for Field and Vegetable Crops

Given:
• average daily level of ET_0 during stage 1 = 4 mm/day
• the interval between irrigation is expected to be approximately 7 days
• dry desert location ($\mathrm{RH}_{min} < 20\%$) when the average wind speed is 3 m/s

Required:
Seasonal variation of K_c values for 140 day corn and 135 day wheat.

Solution:
   For Corn

K_c for stage 1 = 0.742 (4 mm/day)$^{-0.319}$
   = 0.48 $\tag{1.9}$

K_c for stage 2 ranges between 0.48 and 1.15 (from Table 1.4)
K_c for stage 3 = 1.15 (from Table 1.4)
K_c for stage 4 ranges between 1.15 and 0.60 (from Table 1.4)
   For Wheat
K_c for stages 1, 2, and 3 same as corn
K_c for stage 4 ranges between 1.15 and 0.20

Table 1.4 K_c Values Based on a Grass Reference Crop and the Duration of Each of the Growth Stages in Figure 1.7 for Selected Field and Vegetable Crops

Crop	Growth stage	min RH > 70% 0–5[a]	min RH > 70% 5–8[a]	min RH < 20% 0–5[a]	min RH < 20% 5–8[a]	Days 1	Days 2	Days 3	Days 4	
Barley (also wheat and oats)	3	1.05	1.10	1.15	1.20	15	25	50	30 (120)	Varies widely with variety; wheat central India Nov. planting
	4	0.25	0.25	0.20	0.20	20	25	60	30 (135)	Early spring sowing, semiarid, 35–45° latitudes and Nov. planting Republic of Korea
						15	30	65	40 (150)	Wheat sown in July in East Africa highlands at 2500 m altitude and Republic of Korea
Beans (dry)	3	1.05	1.10	1.15	1.20	20	30	40	20 (110)	Continental climates late spring planting
	4	0.30	0.30	0.25	0.25	15	25	35	20 (95)	June planting Central California and West Pakistan
Corn (grain)	3	1.05	1.10	1.15	1.20	30	50	60	40 (180)	Spring planting East African highlands
	4	0.55	0.55	0.60	0.60	25	40	45	30 (140)	Late cool season planting, warm desert climates
						20	35	40	30 (125)	June planting subhumid Nigeria, early October India
						30	40	50	30 (150)	Early April planting Southern Spain

Crop	Wind[a] (m/s)	Kc (crop development stages)				Length of growing stages[b] (days)				(total)	Remarks
Cotton	3	1.05	1.15	1.20	1.25	30	50	60	55	(195)	March planting Egypt, April–May planting Pakistan, September Planting South Arabia
	4	0.65	0.65	0.65	0.70	30	50	55	45	(180)	Spring planting, machine harvest Texas
Potato (Irish)[c]	3	1.05	1.10	1.15	1.20	25	30	30	20	(105)	Fall planting warm winter desert climates
	4	0.70	0.70	0.75	0.75	25	30	45	30	(130)	Late winter planting arid and semiarid climate and late spring–early summer planting continental climate
						30	35	50	30	(145)	Early–mid spring planting, central Europe

Source: J. Doorenbos and W. O. Pruitt, "Guidelines for Predicting Crop Water Requirements," Irrigation and Drainage Paper 24 (1977), FAO. United Nations, 144 pp.
[a] Wind speed in m/s.
[b] Numbers in parenthesis are total number of days in growing season.
[c] Slow emergence may increase length of initial period by 15 days during cold spring.

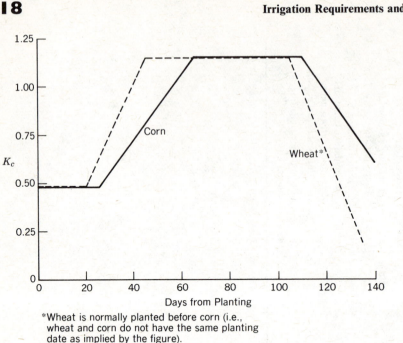

*Wheat is normally planted before corn (i.e., wheat and corn do not have the same planting date as implied by the figure).

Coefficient K_c values for most other crops also vary during the irrigation season. Tables 1.5, 1.6, 1.7 and 1.8 list K_c values (based on a grass reference crop) for several other crops. All of the K_c values in Tables 1.4 through 1.8 are based on a grass reference crop and should be used only when ET_0 has been calibrated to a grass reference crop. Doorenbos and Pruitt (1977) present K_c values for many additional crops.

1.4.2a   Aerodynamic Methods

Water vapor moves through the thin layer of air next to the leaf by the process of molecular diffusion. Outside of this layer the unevenness of the surface and differential heating lead to turbulent motion and a region where turbulent diffusivity far exceeds molecular diffusivity. In the aerodynamic methods vapor flux is proportional to mean wind speed and the vapor pressure difference between the evaporating surface and the surrounding air.

The Dalton equation is one of the earliest aerodynamic equations for estimating evaporation from a water surface. This equation is:

$$ET_0 = (e_s - e)f(u) \tag{1.10}$$

where

    e_s = vapor pressure at the plant surface (within the boundary layer surrounding the leaf);

    e = vapor pressures at some height above the plant;

  $f(u)$ = function of the horizontal wind velocity.

Table 1.5 $K_c^{a,b}$ **Values Based on a Grass Reference Crop for Citrus**

	Jan.[d]	Feb.	Mar.	Apr.	May	June	July	Aug.	Sept.	Oct.	Nov.	Dec.
Large mature trees providing $\cong 70\%$ tree ground cover												
Clean cultivated	0.75	0.75	0.70	0.70	0.70	0.65	0.65	0.65	0.65	0.70	0.70	0.70
No weed control[c]	0.90	0.90	0.85	0.85	0.85	0.85	0.85	0.85	0.85	0.85	0.85	0.85
Trees providing $\cong 50\%$ tree ground cover												
Clean cultivated	0.65	0.65	0.60	0.60	0.60	0.55	0.55	0.55	0.55	0.55	0.60	0.60
No weed control[c]	0.90	0.90	0.85	0.85	0.85	0.85	0.85	0.85	0.85	0.85	0.85	0.85
Trees providing $\cong 20\%$ tree ground cover												
Clean cultivated	0.55	0.55	0.50	0.50	0.50	0.45	0.45	0.45	0.45	0.45	0.50	0.50
No weed control[c]	1.00	1.00	0.95	0.95	0.95	0.95	0.95	0.95	0.95	0.95	0.95	0.95

Source: J. Doorenbos and W. O. Pruitt, "Guidelines for Predicting Crop Water Requirements," Irrigation and Drainage Paper 24 (1977), FAO, United Nations, 144 pp.
[a] Values of K_c for citrus are for dry, Mediterranean climates. The effect of wind stronger than moderate is negligible. Values of K_c may need to be increased by 15 to 20 percent during midsummer in humid and cooler climates (since the stomatal resistance of citrus is higher under dry and hot conditions and lower under humid and cooler conditions).
[b] Some studies indicate somewhat higher K_c values, up to 10–15 percent for grapefruit and lemons compared to those given.
[c] With frequent rain or irrigation, K_c values for clean cultivation will approach those of no weed control.
[d] Months refer to northern hemisphere; for southern hemisphere add 6 months.

The Dalton equation is not widely used to estimate ET because of the difficulty of determining e_s within the boundary layer surrounding the plant. This is due primarily to the very large vapor pressure and temperature gradients that often exist within the boundary layer.

A more common aerodynamic method of estimating ET involves determining the vapor pressure and wind speed differences between measurement heights Z_1 and Z_2 (Z_1 and Z_2 are usually both in the turbulent zone outside of the boundary layer that surrounds the plant).

$$\text{ET}_0 \propto \frac{K^2(\bar{u}_2 - \bar{u}_1)(\rho v_1 - \rho v_2)}{\ln(Z_1/Z_2)^2} \tag{1.11}$$

where

$$K = \text{von Karman constant;}$$
$$\bar{u}_1, \bar{u}_2 = \text{mean wind velocity at heights } Z_1 \text{ and } Z_2;$$
$$\rho v_1, \rho v_2 = \text{mean density of water vapor at heights } Z_1 \text{ and } Z_2.$$

The use of Eq. 1.11 and other similar aerodynamic equations is limited to research situations because of the high accuracy necessary in the measurement of u and the vapor pressure difference. Short measurement intervals of an hour or less as well as expensive equipment and well-trained people are required to obtain acceptable accuracy using these methods.

Table 1.6 K_c Values Based on Grass Reference Crop for Apples and Cherries

	Jan.	Feb.	Mar.	Apr.	May	June	July	Aug.	Sept.	Oct.	Nov.	Dec.
With ground cover[a,b]												
(a) Humid, light to moderate wind	—	—	—	0.50	0.75	1.00	1.10	1.10	1.10	0.85	—	—
(b) Humid, strong wind	—	—	—	0.50	0.75	1.10	1.20	1.20	1.15	0.90	—	—
(c) Dry, light to medium wind	—	—	—	0.45	0.85	1.15	1.25	1.25	1.20	0.95	—	—
(d) Dry, strong wind	—	—	—	0.45	0.85	1.20	1.35	1.35	1.25	1.00	—	—
Clean, cultivated, weed free[a,c]												
(a) Humid, light to moderate wind	—	—	—	0.45	0.55	0.75	0.85	0.85	0.80	0.60	—	—
(b) Humid, strong wind	—	—	—	0.45	0.55	0.80	0.90	0.90	0.85	0.65	—	—
(c) Dry, light to medium, wind	—	—	—	0.40	0.60	0.85	1.00	1.00	0.95	0.70	—	—
(d) Dry, strong wind	—	—	—	0.40	0.65	0.90	1.05	1.05	1.00	0.75	—	—

Source: J. Doorenbos and W. O. Pruitt, "Guidelines for Predicting Crop Water Requirements," Irrigation and Drainage Paper 24 (1977), FAO, United Nations, 144 pp.

[a] Values of K_c are for climates with cold winters with killing frost. Ground cover starts in April.

[b] Values of K_c need to be increased if frequent rain occurs (use Eq. 1.9 to determine K_c for growth stage 1). For young orchards with tree ground cover of 20 to 50 percent, reduce midseason K_c values by 10 percent to 15 percent and 5 percent to 10 percent, respectively.

[c] Values of K_c are for infrequent wetting by irrigation or rain (every 2 to 4 weeks). For frequent irrigation for March, April, and November use Eq. 1.9 to determine K_c for growth stage 1; for May to October use K_c values of table "with ground cover crop." For young orchards with tree ground cover of 20 percent and 50 percent reduce midseason K_c values by 25 percent to 35 percent and 10 percent to 15 percent, respectively.

Table 1.7 K_c^a **Values Based on a Grass Reference Crop for Alfalfa, Clover, Grass–Legume and Pasture**

Climate		Alfalfa	Grass Hay	Clover Grass Legumes	Pasture
Humid, light to moderate wind	K_c (mean)^c	0.85	0.80	1.00	0.95
	K_c (peak)^d	1.05	1.05	1.05	1.05
	K_c (low)^e	0.50	0.60	0.55	0.55
Dry, light to moderate wind	K_c (mean)^c	0.95	0.90	1.05	1.00
	K_c (peak)^d	1.15	1.10	1.15	1.10
	K_c (low)^e	0.40	0.55	0.55	0.50
Strong wind	K_c (mean)^c	1.05	1.00	1.10	1.05
	K_c (peak)^d	1.25	1.15	1.20	1.15
	K_c (low)^e	0.30	0.50	0.55	0.50

Source: J. Doorenbos and W. O. Pruitt, "Guidelines for Predicting Crop Water Requirements," Irrigation and Drainage Paper 24 (1977), FAO, United Nations, 144 pp.
[a] The variation of K_c between initial and harvest stages for crops in this table is similar to field crops except it is repeated each cutting (i.e., the cycle is repeated 2 to 8 times per year).
[b] Alfalfa grown for seed production will have a K_c value equal to K_c (peak) during full cover until the middle of full bloom.
[c] The term K_c (mean) is the mean value between cuttings.
[d] The term K_c (peak) is the value just before harvest (cutting)
[e] The term K_c (low) is the value just after cutting for dry soil conditions. For wet conditions increase table values by 30 percent.

1.4.2b   Energy Balance Methods

When a vapor pressure gradient exists and water is readily available, ET is controlled by the availability of energy for vaporizing water. The energy available for ET can be computed using the following equation.

$$ET = Q_n + AD - S - A - C - P \qquad (1.12)$$

where

Q_n = net ratiation
AD = advection
S = heat flux to the soil
A = heat flux to the air
C = heat storage in crop
P = photosynthesis

The term Q_n is the amount of solar radiation that reaches the earth's surface minus reflected and reradiated energy. Terms C and P are usually neglected, since over the normal growing season, $C + P$ account for less than 2 percent of Q_n. For short periods of a few hours, P and C can be significant, however. Heat flux S is also commonly neglected even though it can be as much as 15 percent of Q_n.

Table 1.8 K_c Values Based on a Grass Reference Crop for Rice[a,b]

	Planting	Harvest	First and Second Month	Mid-season	Last 4 Weeks
Humid Asia					
Wet season (monsoon)					
light to moderate wind	June–July	Nov–Dec.	1.10	1.05	0.95
strong wind			1.15	1.10	1.00
Dry season[c]					
light to moderate wind	Dec.–Jan,	Mid–May	1.10	1.25	1.00
strong wind			1.15	1.35	1.05
North Australia					
Wet season					
light to moderate wind	Dec.–Jan.	Apr.–May	1.10	1.05	0.95
strong wind			1.15	1.10	1.00
South Australia					
Dry summer					
light to moderate wind	Oct.	Mar.	1.10	1.25	1.00
strong wind			1.15	1.35	1.05
Humid S. America					
Wet season					
light to moderate wind	Nov.–Dec.	Apr.–May	1.10	1.05	0.95
strong wind			1.15	1.10	1.00
Europe (Spain, S. France and Italy)					
Dry season					
light to moderate wind	May–June	Sept.–Oct.	1.10	1.20	0.95
strong wind			1.15	1.30	1.00
United States					
Wet summer (south)					
light to moderate wind	May	Sept.–Oct.	1.10	1.10	0.95
strong wind			1.15	1.15	1.0
Dry summer (Calif.)					
light to moderate wind	Early May	Early Oct.	1.10	1.25	1.00
strong wind			1.15	1.35	1.05

Source: J. Doorenbos and W. O. Pruitt, "Guidelines for Predicting Crop Water Requirements," Irrigation and Drainage Paper 24 (1977), FAO, United Nations, 144 pp.
[a] Values of K_c for broadcast or sown, and transplanted rice are assumed to be identical, since the percent cover during the first month after transplant/seeding is approximately equal.
[b] For upland rice, the same coefficients given for paddy rice will apply, since recommended practices involve the maintenance of top soil layers very close to saturation. The term K_c should be reduced during the initial crop stage.
[c] Use wet season values when minimum relative humidity exceeds 70 percent.

   Advection is the transfer of sensible and latent heat from adjacent areas into the area being considered. Even though advection may equal Q_n in arid climates, advection usually is omitted, largely because no simple way has been devised to evaluate it. Thus, the energy budget approach is most accurate when AD is small.
   When C, P, S, and AD are neglected, Eq. 1.12 becomes

$$Q_n = ET + A$$

(1.12a)

A method of partitioning energy used in evaporation and in heating of the air is often used to determine the relative magnitude of A and ET and to solve Eq. 1.12a. In 1926 Bowen proposed the following relationship between energy used in evaporation and in heating the air.

$$\beta = \frac{A}{ET} = \gamma \frac{K_n}{K_w}\left(\frac{T_s - T_a}{e_s - e_a}\right) \tag{1.13}$$

where

γ = psychrometric constant;
K_n, K_w = eddy diffusivities for heat and water vapor, respectively;
T_s, e_s = temperature and vapor pressure of the surface, respectively;
T_a, e_a = temperature and vapor pressure of the air, respectively.

The Bowen ratio, β, is negative when heat is transferred from air to crop and positive when heat transfer is from crop to air. Accurate determination of β is difficult because of problems in measuring leaf temperature and the vapor pressure of the evaporating surface. As with aerodynamic methods, the most accurate determination of β requires instantaneous measurements. Using mean daily values to compute β can be misleading.

1.4.2c   Combination (Penman) Methods

In 1948, Penman combined the aerodynamic and energy budget methods to obtain an equation for computing ET. Combination equations have the form

$$ET_p = \frac{\Delta Q_n + \gamma E_a}{\Delta + \gamma} \tag{1.14}$$

where

$$\Delta = \frac{4098 e_{sa}}{(T_a + 237.3)^2} \tag{1.14a}$$

$$e_{sa} = \exp\left(\frac{19.08 T_a + 429.4}{T_a + 237.3}\right) \tag{1.14b}$$

$$\gamma = \frac{1615 P_a}{2.49(10)^6 - 2.13(10)^3 T_a} \tag{1.14c}$$

$$P_a = 1013 - 0.1152h + 5.44(10)^{-6}h^2 \tag{1.14d}$$

Δ = slope of the saturation vapor pressure versus temperature
     curve at air temperature T_a (mbar/°C);
Q_n = net radiation (mm/day);
γ = psychrometric constant (mbar/°C);
E_a = aerodynamic term = $f(e_{sa}, e_a, u_1)$(mm/day);
e_{sa} = saturation vapor pressure at air temperature T_a (mbar);
e_a = actual vapor pressure of the air (mbar);
P_a = air pressure (mbar);
h = elevation above mean sea level (m).

Table 1.9 Common E_a Relationships Used in the Penman Equation

E_a	Reference
$(0.2625 + 0.1409\,u)(e_{sa} - e_a)$	Penman (1948)
$212.6\,\rho\,u/P_a\,(e_{sa} - e_a)$	Van Bavel (1966), Businger (1956)
$(0.197 + 0.261\,u)(e_{sa} - \bar{e}_s)$	Wright and Jensen (1978)
$(0.27 + 0.2333\,u)\,(e_{sa} - e_a)$	Doorenbos and Pruitt (1977)
$(0.2738 + 0.147\,u)(e_{sa} - e_a)$	Thom and Oliver (1977)

Note: u = wind velocity (m/s); P_a = air pressure (mbar) (compute using Eq. 1.14d); e_{sa} = saturation vapor pressure at mean air temperature (the saturation vapor pressure at a temperature can be computed using Eq. 1.14b); e_a = vapor pressure at mean air temperature = e_{sa}(RH/100)(RH in percent); $\bar{e}_s = (e_s$ at maximum air temperature + e_s at minimum air temperature)/2.

The primary attributes of the Penman equation are that it is based on reasonable physical principles and the need for measurements of leaf temperature and e_s at the leaf surface or within the thin boundary layer surrounding the leaf are eliminated.

Variations of the Penman equation are widely used today to estimate ET. The primary difference between the variations is how Q_n and E_a are evaluated. Table 1.9 includes several methods of determining E_a, and Table 1.10 lists several different equations that can be used when direct measurements of Q_n are not available. The selection of an equation for Q_n depends upon data availability.

EXAMPLE 1.2 Calculating ET_0 Using a Penman-type Equation

Given:
The following daily climate data from Prosser, WA (latitude = 46.25° N; elevation = 275 m)

Date	Day from Jan. 1	Mean Temp (°C)	Mean RH(%)	Solar Rad (cal/cm²)	Mean Wind Speed (m/s)
7/1	182	21.5	39.5	701	2.10
7/2	183	18.0	52.5	713	3.00
7/3	184	14.5	55.0	762	0.75
7/4	185	16.5	56.0	796	0.30
7/5	186	18.0	58.0	510	2.37
7/6	187	14.0	60.0	815	2.01
7/7	188	14.5	61.0	773	1.30
7/8	189	16.5	48.5	536	0.78
7/9	190	22.0	46.0	733	1.08
7/10	191	23.5	44.0	796	0.71
10 day average		17.9	52.1	714	1.44

Table 1.10 Common Q_n Relationships Used in the Penman Equation

$$Q_n = 0.77 R_s - 1.00(10)^{-9}\left((T_{max} + 273.16)^4 + (T_{min} + 273.16)^4\right)\left(a_1 - 0.044\sqrt{e_a}\right)\left(-0.18 + 1.22\frac{R_s}{R_{s0}}\right)$$ Wright and Jensen (1978)

$$a_1 = 0.325 + 0.045\sin(0.01745J - 0.26175)$$

$$Q_n = 0.75R_s - 2.00(10)^{-9}(T_a + 273.16)^4\left(0.34 - 0.044\sqrt{e_a}\right)\left(-0.35 + 1.8\frac{R_s}{R_a}\right)$$ Doorenbos and Pruitt (1977)

$$Q_n = 0.75R_s - 2.00(10)^{-9}(T_a + 273.16)^4\left(0.31 - 0.044\sqrt{e_{sd}}\right)\left(-0.35 + 1.35\frac{R_s}{R_{s0}}\right)$$ Pair et al. (1983)

$$Q_n = 0.80R_a(0.18 + 0.55n/N) - 2.01(10)^{-9}(T_a + 273.16)^4(0.56 - 0.092\sqrt{E_{sd}})(0.10 + 0.90n/N)$$ Brunt (1944, 1952) (used by Penman)

Note: R_a = extraterrestrial solar radiation (mm/day) (see Appendix A); R_s = observed solar radiation in evaporation equivalents (mm/day) (see Eq. 1.19) when R_s is not available, $R_s = (0.25 + 0.50n/N)R_a$; R_{s0} = solar radiation on a clear day (for the same day as R_s except under cloudless conditions) (mm/day) (R_{s0} is evaluated using several years of daily R_s data. R_s is plotted versus date and a curve fit through the larger R_s values. Values of R_{s0} for each day are estimated from this curve.); T_a = average air temperature (°C); T_{max} = maximum air temperature (°C); T_{min} = minimum air temperature (°C); e_a = vapor pressure of air (mbar); $e_a = e_{sa}(RH/100)$; RH = relative humidity (%); e_{sd} = saturation vapor pressure at dew point temperature in °C (T_d) of air (mbar), $T_d = \dfrac{429.4 - 237.3\ln(e_a)}{\ln(e_a) - 19.08}$; n/N = ratio between actual and possible hours of sunshine; J = day of the year.

25

Required:

Daily ET_0 for grass reference crop using Doorenbos and Pruitt version of Penman equation

Solution:

The following is tabular solution of the Doorenbos and Pruitt version of the Penman equation

Day	e_{sa} (mbar)	e_a (mbar)	Δ (mbar/°C)	γ (mbar/°C)	Ra (mm/day)	Qn (mm/day)	Ea (mm/day)	ET_0 (mm/day)
182	25.6	10.1	1.57	0.65	17.2	6.3	11.8	7.9
183	20.6	10.8	1.30	0.65	17.2	6.5	9.5	7.5
184	16.5	9.1	1.07	0.64	17.1	6.9	3.3	5.5
185	18.8	10.5	1.19	0.65	17.1	7.2	2.8	5.7
186	20.6	12.0	1.30	0.65	17.1	5.0	7.1	5.7
187	16.0	9.6	1.04	0.64	17.1	7.3	4.7	6.3
188	16.5	10.1	1.07	0.64	17.0	7.0	3.7	5.8
189	18.8	9.1	1.19	0.65	17.0	5.1	4.4	4.8
190	26.4	12.2	1.61	0.65	17.0	6.6	7.4	6.9
191	29.0	12.7	1.74	0.65	16.9	7.1	7.1	7.1

Total $ET_0 = 63.2$ mm

Sample calculations for day 182

$$P_a = 1013 - 0.1152(275) + 5.44(10)^{-6}(275)^2 = 981.7 \text{ mbar} \tag{1.14d}$$

$$e_{sa} = \exp\left(\frac{19.08(21.5) + 429.4}{21.5 + 237.3}\right) = 25.6 \text{ mbar} \tag{1.14b}$$

$$\Delta = \frac{(4098)(25.6)}{(21.5 + 237.3)^2} = 1.57 \tag{1.14a}$$

$$\gamma = \frac{1615(981.7)}{2.49(10)^6 - 2.13(10)^3(2.15)} = 0.65 \tag{1.14c}$$

Computing extraterrestrial radiation using Appendix A

$$\theta = 0.986(182 - 1) = 178.5° \tag{A6}$$

$$\delta = \frac{180}{\pi}(0.006918 - 0.399912 \cos(178.5°) + 0.070257 \sin(178.5°)$$
$$- 0.06758 \cos(356.9°) + 0.000907 \sin(356.9°)$$
$$- 0.002697 \cos(535.4°) + 0.001480 \sin(535.4°)) \tag{A5}$$

$$= 23.2°$$

$$h_s = \cos^{-1}(-\tan(46.25°)\tan(23.2°)) = 116.6° \tag{A4}$$

$$r_{ve} = 0.98387 - 1.11403(10)^{-4}(182) + 5.27747(10)^{-6}(182)^2$$
$$- 2.68285(10)^{-8}(182)^3 + 3.61634(10)^{-11}(182)^4 \tag{A3}$$

$$= 1.02$$

$$h_{do} = 12.126 - 1.85191(10)^{-3} ABS(46.25°) + 7.61048(10)^{-5}(46.25°)^2 \qquad \text{(A2)}$$

$$= 12.20°$$

$$R_a = 1.26714 \frac{12.20}{(1.02)^2} \left[\left(116.6 \frac{\pi}{180} \sin(46.25°)\sin(23.2°) \right) \right.$$

$$\left. + \cos(46.25°)\cos(23.2°)\sin(116.6°) \right] \qquad \text{(A1)}$$

$$= 17.2 \text{ mm/day}$$

$$R_s = \frac{701}{58.48} = 12.0 \text{ mm/day} \qquad \text{(1.19)}$$

$$e_a = 25.6 \left(\frac{39.5}{100} \right) = 10.1 \text{ mbar}$$

$$Q_n = 0.75(12.0) - 2.00(10)^{-9}(21.5 + 273.16)^4(0.34 - 0.044\sqrt{10.1})$$

$$\times \left[-0.35 + 1.8 \left(\frac{12.0}{17.2} \right) \right] \qquad \text{(Table 1.9)}$$

$$= 6.3 \text{ mm/day}$$

$$E_a = (0.27 + 0.2333(2.10))(25.6 - 10.1) = 11.8 \text{ mm/day} \qquad \text{(Table 1.8)}$$

$$ET_0 = \frac{(1.57)(6.3) + (0.65)(11.8)}{1.57 + 0.65} = 7.9 \text{ mm/day} \qquad \text{(1.14)}$$

1.4.2d   Empirical Methods

Many simpler methods of estimating ET based on one or more of the basic parameters controlling ET have been developed. Air temperature, solar radiation, and pan evaporation are the most commonly used parameters. In general, these methods are more convenient to use but are not regarded as being as accurate as the Penman-type equations for periods of less than 5 days. Empirical methods are used when all the data needed for Penman-type equations are not available, adequate accuracy can be obtained using simple empirical equations that require less time and effort to apply, and sufficient equipment and/or technical expertise and experience to apply the more complete methods is not available or economically attainable. Three different empirical methods are presented and described. These are the Jensen–Haise, pan evaporation, and Blaney–Criddle methods.

(i) Jensen–Haise Method   The Jensen–Haise method is based on the energy balance equation and approximately 1000 measurements of evapotranspiration (ET) made over a 35-year period at 20 locations in the western U.S. Climatic data needed for this method include solar radiation expressed in evaporation equivalent of inches or millimeters/day, mean daily temperature, and the long-term mean maximum and minimum temperatures for the month of highest mean air temperatures. Elevation above sea level of the location being considered is also needed. The basic Jensen–Haise equation is

$$ET_0 = C_T(T - T_x)R_s \tag{1.15}$$

where

C_T = air temperature coefficient for the location being considered;
 T = mean daily air temperatures;
T_x = constant for the location being considered;
R_s = total solar radiation for period in inches or millimeters.

ET$_0$ for each day of the period can be calculated with Eq. 1.15 by using T and R_s for each day (rather than mean daily temperature and total R_s for the period, respectively). Mean daily ET$_0$ for the period is computed by substituting mean daily T and R_s for the period into Eq. 1.15.

The coefficients C_T and T_x are determined using Eqs. 1.16, 1.17 and 1.18

$$C_T = \frac{1}{K_1 - (h/K_2) + (K_3/(e_{s\,max} - e_{s\,min}))} \tag{1.16}$$

$$T_x = K_4 - K_5(e_{s\,max} - e_{s\,min}) - \frac{h}{K_6} \tag{1.17}$$

$$e_{si} = \exp\left(\frac{K_7 T_i + K_8}{K_9 T_i + K_{10}}\right) \tag{1.18}$$

where

T_i = average daily air temperature for period i (°C, °F);
$K_1, K_2, K_3, K_4, K_5, K_6,$ = constants from Table 1.11 which depend on the
K_7, K_8, K_9, K_{10} units of h and T_i and the reference crop;
h = elevation above sea level (m, ft);
$e_{s\,max}, e_{s\,min}$ = saturation vapor pressure at mean maximum and minimum air temperature during the warmest month of the year, respectively (mbar).

The units of ET calculated with Eq. 1.15 are the same as the units of R_s. Daily solar radiation is normally reported in cal/cm^2 or langleys. These are converted to equivalent depths of evaporation assuming a heat of vaporization of 585 cal/g using the following:

$$R_s = \frac{R_s^*}{K} \tag{1.19}$$

where

R_s = solar radiation in equivalent depths of evaporation (mm, in);
R_s^* = solar radiation (cal/cm^2);
K = unit constant ($K = 58.48$ when R_s is in mm, $K = 1486$ for R_s in in);

Table 1.11   Values of K_1, K_2, K_3, K_4, K_5, K_6, K_7, K_8, K_9, and K_{10} in Equations 1.16, 1.17, and 1.18 for Different Units of h and T_i, and Alfalfa and Grass Reference Crops.

Units of T_i	°C	°F	°C	°F
Units of h	m	ft	m	ft
Reference Crop	Alfalfa	Alfalfa	Grass	Grass
K_1	38	68	45	81
K_2	152	278	137	250
K_3	365	650	365	650
K_4	−2.5	27.5	−2.5	27.5
K_5	0.14	0.44	0.14	0.44
K_6	550	1000	500	1000
K_7	19.08	10.60	19.08	10.60
K_8	429.41	90.20	429.41	90.20
K_9	1.00	0.56	1.00	0.56
K_{10}	237.30	219.52	237.30	219.52

EXAMPLE 1.3   Calculating ET_0 Using the Jensen–Haise Equation

Given:
- data from Example 1.2
- warmest month of year is July
- average high daily temperature during July is 31.3°C
- average low daily temperature during July is 11.8°C

Required:
ET_0 for grass reference crop for the 10-day period in Example 1.2 for Prosser, WA using the Jensen–Haise equation

Solution:
average temperature during period is 17.9°C (from Example 1.2)
total R_s^* for period is 7135 cal/cm^2 (from Example 1.2 data)

$$e_{smin} = \exp\left(\frac{19.08(11.8) + 429.41}{(1.0)(11.8) + 237.3}\right) = 13.8 \text{ mbar} \tag{1.18}$$

$$e_{smax} = \exp\left(\frac{19.08(31.3) + 429.41}{(1.0)(31.3) + 237.3}\right) = 45.7 \text{ mbar} \tag{1.18}$$

$$T_x = -2.5 - 0.14(45.7 - 13.8) - \frac{275}{550} = -7.47 \tag{1.17}$$

$$C_T = \left(\frac{1}{45 - (275/137) + (365/(45.7 - 13.8))}\right) = 0.0184 \tag{1.16}$$

$$R_s = \frac{7135}{58.48} \doteq 122.0 \text{ mm} \tag{1.19}$$

$$ET_0 = 0.0184(17.9 + 7.47)(122.0) = 61.9 \text{ mm} \tag{1.15}$$

(ii) *Pan Evaporation*   Measuring the loss of water from an open faced pan of water is a relatively inexpensive and simple way of assessing the evaporative capability of the atmosphere. The amount of water evaporating from a pan (i.e., pan evaporation) is determined by measuring the change in water level in the pan and correcting for precipitation (assuming that water loss due to wind action, animals, birds, etc., has been prevented or is negligible). The frequency of pan evaporation measurements normally ranges from hourly to weekly with daily observations being most common. Daily pan evaporation is routinely measured at weather stations throughout the United States.

There are many different types of evaporation pans in use. Table 1.12 describes the U.S. Class A type pan used at most U.S. weather stations and the Colorado sunken pan. The Colorado sunken pan is sometimes preferred for crop water requirements studies, since it gives a better direct prediction of potential ET of grass than Class A pans. Ten gallon washtubs have also been successfully used as evaporation pans (Westesen and Hanson, 1981).

Reference crop evapotranspiration, ET_0 is related to pan evaporation, E_p, by the following

$$ET_0 = K_p E_p \tag{1.20}$$

where, K_p, is a pan coefficient that accounts for differences in pan type and conditions upwind of the pan, and for dissimilarities between plants and evaporation pans. The shape and color of evaporation pans significantly influences K_p. For example, water loss from circular pans is independent of wind direction, while evaporation from square pans depends on wind direction (since the amount of evaporation depends on the length of the water surface). Color differences between pans affect the reflection of radiation and hence evaporation. Screens mounted above pans to prevent birds and animals from drinking from evaporation pans reduce pan evaporation by as much as 10 percent (Doorenbos and Pruitt, 1977).

Conditions upwind of the evaporation pans also significantly affect K_p. Table 1.13, lists K_p values for different winds, relative humidities, and windward groundcover conditions (illustrated in Figure 1.8) for Class A evaporation pans.

Table 1.12   **Description of U.S. Class A and Colorado Sunken Evaporation Pans.**

Pan	Dimensions	Situation
U.S. Class A	121 cm diameter 255 mm deep	Made of 22 guage galvenized iron; mounted level on an open wooden frame with its bottom 15 cm above ground; water level maintained between 5 and 7.5 cm below rim; water is changed as needed to eliminate extreme turbidity; is painted annually with aluminum paint
Sunken Colorado	92 cm square 56 cm deep	Made of galvanized iron; is set in the ground with the rim 5 cm above ground level; water level inside the pan is maintained at or slightly below ground level

Table 1.13 K_p for Class A Pan for Different Groundcover and Levels of Mean Relative Humidity and 24-hour wind.

Class A Pan	Case A[a] Pan Surrounded by Short Green Crop				Case B[ab] Pan Surrounded by Dry-Fallow Land			
RH Mean Percent	Upwind Distance of Green Crop (m)	Low <40	Medium 40–70	High >70	Upwind Distance of Dry Fallow (m)	Low <40	Medium 40–70	High >70
Average Daily Wind Speed (m/s)								
Light <2	0	0.55	0.65	0.75	0	0.7	0.8	0.85
	10	0.65	0.75	0.85	10	0.6	0.7	0.8
	100	0.7	0.8	0.85	100	0.55	0.65	0.75
	1000	0.75	0.85	0.85	1000	0.5	0.6	0.7
Moderate 2–5	0	0.5	0.6	0.65	0	0.65	0.75	0.8
	10	0.6	0.7	0.75	10	0.55	0.65	0.7
	100	0.65	0.75	0.8	100	0.5	0.6	0.65
	1000	0.7	0.8	0.8	1000	0.45	0.55	0.6
Strong 5–8	0	0.45	0.5	0.60	0	0.6	0.65	0.7
	10	0.55	0.6	0.65	10	0.5	0.55	0.65
	100	0.6	0.65	0.7	100	0.45	0.50	0.6
	1000	0.65	0.7	0.75	1000	0.4	0.45	0.55
Very strong >8	0	0.4	0.45	0.5	0	0.5	0.6	0.65
	10	0.45	0.55	0.6	10	0.45	0.5	0.55
	100	0.5	0.6	0.65	100	0.4	0.45	0.5
	1000	0.55	0.6	0.65	1000	0.35	0.4	0.45

Source: J. Doorenbos and W. O. Pruitt, "Guidelines for Predicting Crop Water Requirements," Irrigation and Drainage Paper 24 (1977), FAO, United Nations, 144 pp.
[a] Cases A and B are defined in Figure 1.8.
[b] For extensive areas of bare-fallow soils and areas without agricultural development, reduce K_p values by 20 percent under hot windy conditions; by 5 to 10 percent for moderate wind, temperature, and humidity conditions.

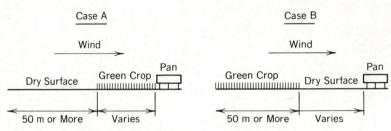

Figure 1.8 Definition sketch for Cases A and B in Table 1.13. (From Doorenbos and Pruitt, 1977.)

Similar tables are available for Colorado sunken pans in Doorenbos and Pruitt (1977). In general, K_p rises as relative humidity increases and wind speed decreases.

There are several differences between plants and evaporation pans that cause their responses to differ and to make K_p vary from unity (Doorenbos and Pruitt, 1977). These include the following.

1. The reflectance of solar radiation from a water surface is only 5 to 8 percent, while the reflectance of plant surfaces is 20 to 25 percent.
2. Heat storage within an evaporation pan can be significant and may cause almost equal evaporation during night and day. Most crops transpire only during the daytime.
3. The wind profile over a crop surface differs from that over a pan. In crops that allow sun to pass through them rather than merely over them, water use may exceed pan evaporation because lower leaves may transpire large amounts of water.
4. Evaporation pans offer no stomatal resistance to water loss.

EXAMPLE 1.4 Calculating ET_0 Using Pan Evaporation Data

Given:

The following daily pan evaporation data

Date	Ep (mm)
7/1	12.2
7/2	11.2
7/3	8.4
7/4	6.6
7/5	9.7
7/6	9.7
7/7	8.1
7/8	6.6
7/9	8.6
7/10	5.8
Total	86.9 mm

Data are for a USWB Class A pan with 100 m of clipped grass on upwind side of the pan (Case A, Table 1.13)

Required:
ET_0 for a grass reference crop for the 10-day period

Solution:
$K_p = 0.8$ (from Table 1.13) since RH is medium and wind is light
$ET_0 = 0.8 (86.9) = 69.5$ mm

(iii) Blaney–Criddle Approach   One of the most used temperature-based methods of estimating evapotranspiration is the Blaney–Criddle equation as modified by the Soil Conservation Service (SCS). Equation 1.21 is the SCS-modified Blaney–Criddle equation.

$$ET = K_{scs}K_t NP\left(\frac{\bar{T}}{K_1} + K_2\right) \tag{1.21}$$

$$K_t = K_3\bar{T} + K_4 \tag{1.21a}$$

where

ET = evapotranspiration for specified crop (mm, in);
K_{scs} = crop coefficient from Appendix C (crop coefficients in Tables, 1.4, 1.5, 1.6, 1.7 and 1.8 do not apply to this equation. Similarly, K_{scs} values from Appendix C apply only to Eq. 1.21);
K_1, K_2, K_3, K_4 = constants from Table 1.14 which depend on the units of $\bar{T}$ and ET_0;
N = number of days in the time period (N should not be less than 10 days or longer than 1 month);
P = mean daily percentage of annual daytime hours for the time period (from Table 1.15);
$\bar{T}$ = average daily temperature during the time period (°C, °F).

Since, Eq. 1.21 does not compute to reference crop ET, crop coefficients in Tables 1.4, 1.5, 1.6, 1.7, and 1.8 can *not* be used with this equation. Instead, the crop coefficients in Appendix C must be used.

Doorenbos and Pruitt (1977) further modified the Blaney–Criddle approach to include long-term average values of minimum relative humidity, daytime wind speed, and the ratio of actual to possible sunshine hours. This equation is

$$ET_0 = N\left[\frac{a}{K_5} + bP\left(\frac{\bar{T}}{K_1} + K_2\right)\right] \tag{1.22}$$

$$a = 0.0043(RH_{min}) - n/N - 1.41 \tag{1.22a}$$

$$b = 0.81917 - 0.0040922\,(RH_{min}) + 1.0705\,(n/N)$$
$$+ 0.065649\,(U_{day}) - 0.0059684\,(RH_{min})\,(n/N)$$
$$- 0.0005967\,(RH_{min})\,(U_{day}) \tag{1.22b}$$

Table 1.14 Values of K_1, K_2, K_3, K_4, and K_5 Used in Equations 1.21, 1.21a and 1.22.

Units of	ET (mm) $\bar{T}$(°C)	inches (°F)
K_1	2.19	100
K_2	8.13	0
K_3	$3.11(10)^{-2}$	$1.73(10)^{-2}$
K_4	$2.40(10)^{-1}$	$-3.14(10)^{-1}$
K_5	1.00	25.4

where

a, b = factors that depend on long-term average minimum relative humidity, daytime wind speed, and the ratio of actual to possible sunshine hours;

RH_{min} = minimum relative humidity, (percent);

n/N = ratio of actual to maximum sunshine hours, (decimal);

U_{day} = daytime wind speed at 2 m height (m/s).

Equation 1.22b for b was developed by Frevert et al. (1983).

EXAMPLE 1.5 Calculating ET_0 Using the Blaney–Criddle Approach

Given:
• data from Example 1.2
• average minimum RH during July = 25 percent at Prosser
• average ratio of actual to possible sunshine hours (n/N) during July = 0.86 at Prosser

Required:
ET_0 for a grass reference crop for the 10-day period

Solution:
$T = 17.9°C$ from Example 1.3
$P = 0.34$ from Table 1.15
$U_{day} = 1.4$ m/s   from Example 1.3
SCS method K_{scs} (from Appendix C for July) = 0.92

$$K_t = 0.0311(17.9) + 0.240 = 0.80$$

$$ET_0 = (0.92)(0.80)(10)(0.34)\left(\frac{17.9}{2.19} + 8.13\right) = 40.8 \text{ mm}$$

Table 1.15 Mean Daily Percentage of Annual Daytime Hours, *P*, by Month for Different Northern and Southern Latitudes

Latitude

North South[a]	Jan. July	Feb. Aug.	Mar. Sept.	Apr. Oct.	May Nov.	June Dec.	July Jan.	Aug. Feb.	Sept. Mar.	Oct. Apr.	Nov. May	Dec. June
60°	0.15	0.20	0.26	0.32	0.38	0.41	0.40	0.34	0.28	0.22	0.17	0.13
58°	0.16	0.21	0.26	0.32	0.37	0.40	0.39	0.34	0.28	0.23	0.18	0.15
56°	0.17	0.21	0.26	0.32	0.36	0.39	0.38	0.33	0.28	0.23	0.18	0.16
54°	0.18	0.22	0.26	0.31	0.36	0.38	0.37	0.33	0.28	0.23	0.19	0.17
52°	0.19	0.22	0.27	0.31	0.35	0.37	0.36	0.33	0.28	0.24	0.20	0.17
50°	0.19	0.23	0.27	0.31	0.34	0.36	0.35	0.32	0.28	0.24	0.20	0.18
48°	0.20	0.23	0.27	0.31	0.34	0.36	0.35	0.32	0.28	0.24	0.21	0.19
46°	0.20	0.23	0.27	0.30	0.34	0.35	0.34	0.32	0.28	0.24	0.21	0.20
44°	0.21	0.24	0.27	0.30	0.33	0.35	0.34	0.31	0.28	0.25	0.22	0.20
42°	0.21	0.24	0.27	0.30	0.33	0.34	0.33	0.31	0.28	0.25	0.22	0.21
40°	0.22	0.24	0.27	0.30	0.32	0.34	0.33	0.31	0.28	0.25	0.22	0.21
35°	0.23	0.25	0.27	0.29	0.31	0.32	0.32	0.30	0.28	0.25	0.23	0.22
30°	0.24	0.25	0.27	0.29	0.31	0.32	0.31	0.30	0.28	0.26	0.24	0.23
25°	0.24	0.26	0.27	0.29	0.30	0.31	0.31	0.29	0.28	0.26	0.25	0.24
20°	0.25	0.26	0.27	0.28	0.29	0.30	0.30	0.29	0.28	0.26	0.25	0.25
15°	0.26	0.26	0.27	0.28	0.29	0.29	0.29	0.28	0.28	0.27	0.26	0.25
10°	0.26	0.27	0.27	0.28	0.28	0.29	0.29	0.28	0.28	0.27	0.26	0.26
5°	0.27	0.27	0.27	0.28	0.28	0.28	0.28	0.28	0.28	0.27	0.27	0.27
0°	0.27	0.27	0.27	0.27	0.27	0.27	0.27	0.27	0.27	0.27	0.27	0.27

Note: Appendix B details methodology and equations for computing daytime hours and *P* to facilitate computer application of this table.

[a] Southern latitudes: apply 6 month difference as shown.

Doorenbos and Pruitt method (using Eq. 1.22)

$$a = 0.0043(25\%) - 0.86 - 1.41 = -2.16 \tag{1.22a}$$

$$b = 0.81917 - 0.0040922(25\%) + 1.0705(0.86) + 0.065649(1.4)$$
$$\quad\; - 0.0059684(25\%)(0.86) - 0.005967(25\%)(1.4) \tag{1.22b}$$

$$= 1.58$$

$$ET_0 = 10\left[\frac{-2.16}{1.00} + 1.58(0.34)\left(\frac{17.9}{2.19} + 8.13\right)\right]$$

$$= 66.0 \text{ mm}$$

1.5   Irrigation Requirements

The *irrigation requirement* of a crop is the total amount of water that must be supplied by irrigation to a disease-free crop, growing in a large field with adequate soil water and fertility, and achieving full production potential under the given growing environment (Doorenbos and Pruitt, 1977). The irrigation requirement includes water used for crop consumptive use, maintaining a favorable salt balance within the root zone (see Chapter 3), and overcoming nonuniformity and inefficiencies of irrigation. The irrigation requirement does not include water from natural sources (such as precipitation) that crops can effectively use.

      The irrigation requirement (I) can be computed when ET is known by solving Eq. 1.7 for I. Equation 1.23 is a simpler and more commonly used equation for computing I.

$$I = ET - P + RO + DP + L + D_{rz}(\theta_f - \theta_i) \tag{1.23}$$

Equation 1.23 applies when SFI, LI, SG, and LO are negligible. Equation 1.7 should be rearranged and used to compute I when one or more of these quantities cannot be neglected.

      Because DP and RO can result from either irrigation or precipitation, Eq. 1.23 can be rewritten as:

$$I = ET - P_e + RO_i + DP_i + L + D_{rz}(\theta_f - \theta_i) \tag{1.24}$$

where

$$P_e = \text{effective precipitation} = P - RO_p - DP_p;$$
$$RO_i, RO_p = \text{runoff due to irrigation and precipitation, respectively (cm, in);}$$
$$DP_i, DP_p = \text{deep percolation due to irrigation and precipitation, respectively}$$
$$\text{(cm, in).}$$

Table 1.16   Typical Overall On-farm Efficiencies for Various Types of Irrigation Systems

System	Overall Irrigation Efficiency (%)
Surface	
Average system, no treatment	50
Partial treatment, i.e., land leveling or irrigation pipelines	60
Land leveling, delivery pipeline, and drainage system meeting design standards	70
Tailwater recovery system with proper land leveling, delivery pipeline, and drainage system	80
Sprinkle	60–75
Trickle	80–90

Source: Bureau of Reclamation and Bureau of Indian Affairs, 1978.

Equation 1.25 is an equivalent form of Eq. 1.24.

$$I = 100\left(\frac{D_{rz}(\theta_f - \theta_i) + ET + L - P_e}{E_i}\right) \tag{1.25}$$

where

E_i = overall efficiency of irrigation in percent.

In addition to accounting for deep percolation and runoff caused primarily by nonuniformity of irrigation, E_i also includes losses due to seepage, evaporation, and spillage from open canals and pipes (i.e., conveyance losses) as well as evaporation and wind drift losses from sprinkle irrigation. Typical overall efficiencies for various types of irrigation systems are listed in Table 1.16.

Mean monthly effective precipitation is estimated from mean monthly ET and rainfall using Table 1.17. For example, mean monthly effective precipitation equals 3.16 inches for a location where mean monthly ET and rainfall are 5.00 and 4.50 inches, respectively.

1.6   Irrigation Scheduling

Irrigation scheduling is the process of determining when to irrigate and how much water to apply per irrigation. Proper scheduling is essential for the efficient use of water, energy, and other production inputs, such as fertilizer. It allows irrigations

Table 1.17 Average Monthly Effective Rainfall as Related to Mean Monthly Rainfall and Mean Monthly ET

Monthly Mean Rainfall (in)	Mean Monthly Consumptive Use													
	1.0	2.0	3.0	4.0	5.0	6.0	7.0	8.0	9.0	10.0	11.0	12.0	13.0	14.0
0.0	0.0	0.0	0.0	0.0	0.0	0.0	0.0	0.0	0.0	0.0	0.0	0.0	0.0	0.0
0.50	0.30	0.31	0.34	0.35	0.39	0.41	0.44	0.46	0.49	0.49	0.49	0.49	0.49	0.49
1.00	0.59	0.64	0.69	0.71	0.73	0.78	0.81	0.87	0.96	0.98	0.98	0.98	0.98	0.98
1.50	0.89	0.94	1.03	1.08	1.11	1.11	1.20	1.30	1.43	1.48	1.48	1.48	1.48	1.48
2.00	0.98	1.27	1.36	1.41	1.44	1.54	1.59	1.72	1.85	1.97	1.97	1.97	1.97	1.97
2.50	0.98	1.56	1.67	1.81	1.91	1.91	1.99	2.11	2.26	2.46	2.46	2.46	2.46	2.46
3.00	0.98	1.82	1.96	2.07	2.17	2.26	2.37	2.51	2.66	2.90	2.95	2.95	2.95	2.95
3.50	0.98	1.97	2.23	2.37	2.51	2.60	2.74	2.90	3.06	3.33	3.44	3.44	3.44	3.44
4.00	0.98	1.97	2.51	2.83	2.83	2.92	3.10	3.27	3.45	3.74	3.94	3.94	3.94	3.94
4.50	0.98	1.97	2.78	2.95	3.16	3.25	3.43	3.65	3.86	4.13	4.37	4.41	4.41	4.41
5.00	0.98	1.97	2.95	3.21	3.45	3.56	3.77	4.02	4.25	4.53	4.53	4.92	4.92	4.92
5.50	0.98	1.97	2.95	3.49	3.75	3.89	4.09	4.37	4.65	4.96	5.20	5.39	5.39	5.39
6.00	0.98	1.97	2.95	3.75	4.02	4.17	4.41	4.72	5.00	5.35	5.63	5.91	5.91	5.91
6.50	0.98	1.97	2.95	3.94	4.29	4.45	4.72	5.04	5.31	5.71	6.02	6.30	6.38	6.38
7.00	0.98	1.97	2.95	3.94	4.53	4.72	5.00	5.31	5.63	6.06	6.46	6.69	6.89	6.89
7.50	0.98	1.97	2.95	3.94	4.76	4.96	5.28	5.63	5.94	6.34	6.69	7.05	7.28	7.36
8.00	0.98	1.97	2.95	3.94	4.92	5.24	5.51	6.22	6.61	7.01	7.40	7.40	7.72	7.87
8.50	0.98	1.97	2.95	3.94	4.92	5.67	5.94	6.30	6.73	7.17				
9.50	0.98	1.97	2.95	3.94	4.92	5.91	6.34	6.69	7.20	7.64				
10.50	0.98	1.97	2.95	3.94	4.92	5.91	6.73	7.13	7.64	8.07				
11.50	0.98	1.97	2.95	3.94	4.92	5.91	6.89	7.48	7.99	8.46				
12.50	0.98	1.97	2.95	3.94	4.92	5.91	6.89	7.80	8.39	8.82				
13.50	0.98	1.97	2.95	3.94	4.92	5.91	6.89	7.87	8.66	9.13				
14.50	0.98	1.97	2.95	3.94	4.92	5.91	6.89	7.87	8.66	9.45				
15.50	0.98	1.97	2.95	3.94	4.92	5.91	6.89	7.87	8.86	9.72				
16.50	0.98	1.97	2.95	3.94	4.92	5.91	6.89	7.87	8.86	9.84				
17.50	0.98	1.97	2.95	3.94	4.92	5.91	6.89	7.87	8.86	9.84				

Source: USDA-SCS as presented in R. D. Burman, P. R. Nixon, J. L. Wright, and W. O. Pruitt, "Water Requirements," in *Design and Operation of Farm Irrigation Systems* (1980), M. E. Jensen (Ed.), ASAE Monograph 3, St. Joseph, MI, p. 189. Copyright © 1980 ASAE, pp. 580. Reprinted by permission of ASAE.

to be coordinated with other farming activities including cultivation and chemical applications. Among the benefits of proper irrigation scheduling are: improved crop yield and/or quality, water and energy conservation, and lower production costs.

1.6.1 Scheduling Strategies

Irrigation schedules are designed to either fully or partially provide the irrigation requirement. These strategies are disussed in the following sections.

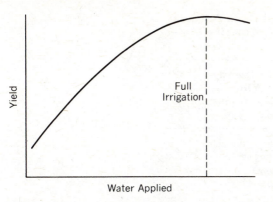
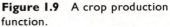

Figure 1.9   A crop production function.

1.6.1a   Full Irrigation

Full irrigation involves providing the entire irrigation requirement and results in maximum production as shown for the production function in Figure 1.9. Exceeding full irrigation reduces crop yields by reducing soil aeration and restricting gas exchange between the soil and atmosphere. Full irrigation is economically justified when water is readily available and irrigation costs are low. It is accomplished by irrigating to minimize the occurrence of plant stress (i.e., irrigating so that actual transpiration rates do not drop below potential rates).

1.6.1b   Deficit Irrigation

Partially supplying the irrigation requirement, a practice that has been called *deficit irrigation*, reduces yield as smaller amounts of water, energy, and other production inputs are used to irrigate the crop. Deficit irrigation is economically justified when reducing water applications below full irrigation causes production costs to decrease faster than revenues decline (because of high production costs and the relatively flat slope of production functions in the vicinity of full irrigation). Application levels can be reduced below full irrigation until the slope of the production function is such that the decrease in revenue due to an incremental reduction in water application equals the accompanying decline in production costs. Irrigations should be scheduled to apply the seasonal application at this point on the production function, since net benefits of irrigation are maximized.

Deficit irrigation is also used when the water supply or the irrigation system limit water availability. In these situations the level of irrigation, the amount of land to be irrigated, and the crop mix that maximize the benefits of irrigation must be determined.

Deficit irrigation is accomplished by allowing planned plant stress during one or more periods of the growing season. Adequate water is supplied during critical growth stages to maximize water use efficiency (i.e., maximizing crop production per unit of water applied). Critical growth stages for several crops are listed in Table 1.18.

Table 1.18 Growth Stages in Which an Adequate Water Supply is Most Critical for Managing Water Use Efficiency, WUE

Crop	Growth Period Most Sensitive to Water Stress	Growth Interval in which Irrigation Produces Greatest Benefits
Sorghum	Boot—heading	Boot—soft dough
Wheat	Boot—flowering	Jointing-soft dough
Corn	Tassel—pollination	12 leaf—blister kernal
Cotton	First bloom—peak bloom	First bloom—boils well—formed
Dry beans	Flowering—early podfill	Axillary bud—podfill
Potatoes	Tuberization	Tuberization—maturity
Soybean	Flowering—early podfill	Axillary bud—podfill
Sugarbeets	No critical stages	WUE[a] is maximized when water depletion is limited to about 50% available water depletion

Source: E. C. Stegman, J. T. Musick, J. I. Stewart, "Irrigation water management," in *Design and Operation of Farm Systems* (1980), M. E. Jensen (Ed.), ASAE Monograph 3, St. Joseph, MI, p. 779. Copyright © 1980 by ASAE. Reprinted by permission of ASAE.
[a] WUE = water use efficiency

1.6.2 When to Irrigate

Several different methods are used to determine when to irrigate. They may be classed as:

1. plant indicators,
2. soil indicators, and
3. water budget technique.

Plant and soil indicators involve monitoring the plant and soil, respectively, to determine when to irrigate. The water budget technique is based on Eq. 1.24 and normally uses Penman-type equations and Eq. 1.8 to compute crop ET.

1.6.2a Plant Indicators

Monitoring plants is the most direct method of determining when to irrigate, since the primary objective of irrigation is to supply plants with the water they need when it is needed. Normally, it is necessary to relate plant parameters to soil water content to determine the amount of irrigation. Required instruments and/or procedures and the principal advantages and disadvantages of several plant indicators are listed in Table 1.19.

   (i) *Appearance and Growth*   Visual indicators (appearance) of the need for water include leaf and shoot wilting and leaf color. Measurements of stem diameter and height can be made regularly to determine the growth rate. The need to irrigate is indicated by a low growth rate. Appearance and growth often are not effective parameters for scheduling full irrigations, since plants have been short of water long enough to adversely affect production when changes in appearance and/ or growth can be detected. Crop appearance must be carefully interpreted, since disease and improper nutrient levels may produce changes in appearance similar to those associated with water stress. The primary advantage of appearance and growth as an indicator of when to irrigate is simplicity.

   (ii) *Leaf Temperature*   Rises in leaf temperature, in addition to those related to increases in air temperature, are associated with reduced transpiration rates resulting from partial or full stomatal closure. Air and leaf temperatures may be remotely sensed from the ground, from aircraft, and possibly from satellites. One popular method is to use a hand-held infrared thermometer to measure the difference between plant canopy and ambient air temperatures each day near the time of maximum surface temperatures (usually about 1 to 1.5 hours after solar noon). The number of degrees by which canopy temperature exceeds air temperature for each day are accumulated until a certain critical level is reached. When this level, which depends on the crop and soil, is reached, it is time to irrigate. Days when canopy temperature is less than the air temperature are neglected.

   (iii) *Leaf Water Potential*   The measurement of leaf water potential, ψ_{leaf} in Eq. 1.2, is another indicator of the plant's need for water. Lower (more negative) potentials indicate a greater need for water. Leaf water potential measurement is a destructive procedure that involves removing a leaf and placing it in a pressurized chamber (called a pressure bomb). The pressure in the chamber is slowly increased until fluid is forced from the leaf stem. The pressure required is a measure of the leaf's moisture potential. Care must be exercised in making these measurements, as leaf age, leaf exposure to solar radiation, and time of day can significantly affect the results. Usually, mature leaves are selected from a specific, standardized location on the plant and measurements are made at a particular time of day. Although commercial instruments for measuring leaf water potential are available, the use of leaf water potential by irrigators is not widespread, since considerable time, care, and training are required to obtain reliable results.

   (iv) *Stomatal Resistance*   Stomatal resistance is an index to the need for water, since it is related to the degree of stomatal opening and the rate of transpiration. In general, high resistances indicate significant stomatal closure, reduced transpiration rates, and the need for water. Commercially available leaf or diffusion porometers are used to measure stomatal resistance. The large amount of time and the high level of skill required to make and interpret stomatal resistance measurements limit their use to research purposes.

Table 1.19 Plant-Based Indicators of When to Irrigate

Observed or Measured Parameters	Required Instruments or Procedures	Principal Advantages	Principal Disadvantages
Appearance	Eye	Simple	Yield potential is usually affected before color and other changes are observed
Leaf temperature	Noncontacting thermometers	Can be sensed remotely	Application methods not well developed
Leaf water potential	Pressure chamber or thermocouple psychrometer	Indicates integrated effect of aerial and soil environment on degree of plant hydration; is correlated with metabolic processes; a fundamental parameter affecting water flux	Subject to large durinal variation; method is time-consuming; requires samping skill; data not easily interpreted
Stomatal resistance	Diffusion porometer	Measures stomatal opening	Same as for leaf water potential

Source: E. C. Stegman, J. T. Musick, J. I. Stewart, "Irrigation water management," in *Design and Operation of Farm Systems* (1980), M. E. Jensen (Ed.), ASAE Monograph 3, St. Joseph, MI, p. 789. Copyright © 1980 ASAE. Reprinted by permission of ASAE.

1.6.2b Soil Indicators

Soil-based irrigation scheduling involves determining the current water content of the soil, comparing it to a predetermined minimum water content (such as θ_c) and irrigating to maintain soil water contents above the minimum level. The minimum water content is often varied according to growth stage, especially for deficit irrigation schedules (see Section 1.6.1b). Soil indicators of when to irrigate also provide data for estimating the amount of water to apply per irrigation. The following example illustrates the use of soil indicators to determine when to irrigate.

EXAMPLE 1.6   Using Soil Indicators to Determine When to Irrigate

Given:
• the measured soil water contents in the table below
• soil water content must be 16 percent by volume or greater

Required:
date of next irrigation

Solution:

Date (June)	θ (%)
1	21.8
2	20.7
3	19.6
4	18.4
5	17.1
6	15.6

Conclusion:
Irrigate the morning of June 6 (if one waits until the evening of June 6, θ would equal 15.6 percent, which would be less than 16 percent).

The soil water contents in Example 1.6 are determined either by direct measurement or inference from measurements of other soil parameters such as soil water potential or electrical conductivity. Several common methods of estimating soil water contents are listed in Table 1.20, along with the principal advantages and disadvantages of various soil indicators.

(i) *Appearance and Feel*   With experience, irrigators can judge soil water content by the appearance and feel of the soil. A soil probe is needed to obtain samples from lower portions of the root zone for examination. Table 1.21 presents guides for judging how much available water has been removed from the soil.

(ii) *Gravimetric Sampling*   Gravimetric sampling is a direct method of measuring the water content of soil samples taken from a field. Samples are weighed, dried at 105 to 110°C, and reweighed after drying. Usually, 24 hours is

Table 1.20 Soil-Based Indicators of When to Irrigate

Observed or Measured Parameters	Required Instruments or Procedures	Principal Advantages	Principal Disadvantages
Appearance and feel	Hand probe	Simple	Time-consuming, approximate; requires interpretative skills
Gravimetric sampling	Sample cans, soil agar, scale, and oven	Simple and accurate	Destructive technique (successive samples at same location and depth are impossible)
Electrical resistance	Porous blocks	Provides indirect measure of soil water content	Requires careful installation, calibration, and frequent readings; not sufficiently sensitive in coarse-textured soils; short block life; need multiple sites
Soil matric potential	Tensiometer	Measures fundamental parameter affecting soil water flux	Requires preinstallation preparation; careful installation, frequent readings and service; need multiple sites
Soil matric potential	Porous (ceramic) blocks	Accurate over a wide range of soil contents; facilitates automatic scheduling of irrigations	Limited experience in use; calibration can deteriorate with time
Neutron scattering	Neutron probe and access tubes	Successive measurements at same location and soil depth are possible; quick and accurate	Relatively expensive equipment that requires special handling, and storage precautions; calibration affected by changes in soil organic matter

Source: E. C. Stegman, J. T. Musick, and J. I. Stewart, "Irrigation Water Management," in *Design and Operation of Farm Systems* (1980), M. E. Jensen (Ed.), ASAE Monograph 3, St. Joseph, MI, p. 789. Copyright © 1980 by ASAE. Reprinted by permission of ASAE.

Table 1.21 Guide for Judging How Much of the Available Moisture has been Removed from the Soil

Soil Moisture Deficiency (%)	Feel or Appearance of Soil and Moisture Deficiency in Centimeters of Water Per Meter of Soil			
	Coarse Texture	Moderately Coarse Texture	Medium Texture	Fine and Very Fine Texture
0 (field capacity)	Upon squeezing, no free water appears on soil but wet outline of ball is left on hand 0.0	Upon squeezing, no free water appears on soil but wet outline of ball is left on hand 0.0	Upon squeezing, no free water appears on soil but wet outline of ball is left on hand 0.0	Upon squeezing, no free water appears on soil but wet outline of ball is left on hand 0.0
0–25	Tends to stick together slightly, sometimes forms a very weak ball under pressure 0.0 to 1.7	Forms weak ball, breaks easily, will not slick 0.0 to 3.4	Forms a ball, is very pliable, slicks readily if relatively high in clay 0.0 to 4.2	Easily ribbons out between fingers, has slick feeling 0.0 to 5.0
25–50	Appears to be dry, will not form a ball with pressure 1.7 to 4.2	Tends to ball under pressure but seldom holds together 3.4 to 6.7	Forms a ball somewhat plastic, will slick slightly with pressure 4.2 to 8.3	Forms a ball, ribbons out between thumb and forefinger 5.0 to 10.0
50–75	Appears to be dry, will not form a ball with pressure[a] 4.2 to 6.7	Appears to be dry, will not form a ball[a] 6.7 to 10.0	Somewhat crumbly but holds together from pressure 8.3 to 12.5	Somewhat pliable, will ball under pressure[a] 10.0 to 15.8
75–100	Dry, loose single-grained, flows through fingers 6.7 to 8.3	Dry, loose, flows through fingers 10.0 to 12.5	Powdery, dry, sometimes slightly crusted but easily broken down into powdery condition 12.5 to 16.7	Hard, baked, cracked, sometimes has loose crumbs on surface 15.8 to 20.8

Source: V. E. Hansen, O. A. Israelson, G. E. Stringham, *Irrigation Principles and Practices.* Copyright © (1979) by John Wiley & Sons, Inc. New York. Reprinted by permission.

[a] Ball is formed by squeezing a handful of soil very firmly.

required to dry samples to a constant weight. Equation 1.26 is used to compute the percent water content on a dry weight basis.

$$\theta_w = \frac{W_w - W_d}{W_d} 100 \tag{1.26}$$

where

θ_w = soil water content on a dry weight basis in percent;
W_w = wet weight (g, lbs);
W_d = dry weight (after drying) (g, lbs).

The percent water content on a volume basis is computed using Eq. 1.27.

$$\theta_v = \frac{\forall_w}{\forall_T} 100 \tag{1.27}$$

where

θ_v = soil water content on a volume basis in percent;
$\forall_T$ = total volume of soil and voids in cm^3 or ft^3;
$\forall_w$ = volume of water in soil (cm^3, ft^3).

A relationship between θ_w and θ_v is obtained by solving Eq. 1.26 for the weight of water ($W_w - W_d$) and observing that

$$\forall_w = \frac{W_w - W_d}{\rho\, g} = \frac{\theta_w W_d}{100\rho\, g}$$

where ρ = density of water (g/cm^3, slugs/ft^3). Equation 1.28 results when this result is substituted into Eq. 1.27:

$$\theta_v = \frac{\theta_w W_d}{(100)\rho\, g\, \forall_T} (100) = \frac{\rho_b}{\rho} \theta_w = A_s \theta_w \tag{1.28}$$

where

$$\rho_b = \frac{W_d}{g\, \forall_T} = \text{bulk density of soil (g/cm}^3\text{, slugs/ft}^3\text{);}$$

$$A_s = \frac{\rho_b}{\rho} = \text{apparent specific gravity of soil (see Table 1.2).}$$

The following example illustrates the use of Eq. 1.26 and 1.27.

EXAMPLE I.7   Gravimetric Analysis to Determine the Water Content of Soil Samples

Given:
• 100 cm^3 of wet soil that weighs 131 g
• the sample weighs 121 g after drying

Required:
θ_w and θ_v

Solution:

$$\theta_w = \frac{W_w - W_d}{W_d}(100) = \left(\frac{131 - 121}{121}\right)100 = 8.26\%$$

$$\theta_v = \frac{\theta_w W_d}{\rho\forall_T} = \frac{(8.26)(121)}{(1)(100)} = 10.00\%$$

Although simple and reliable, gravimetric sampling is a destructive (i.e., a sample cannot be reused) and time-consuming technique. In addition, data are at least one day old when they become available for scheduling (because of the 24-hour drying time).

   (iii) Tensiometers   There are many commercially available (see Figure 1.10) tensiometers for measuring soil water potential. A tensiometer is a ceramic cup filled with water and connected through a water-filled tube to either a vacuum gauge (for normal field use) or a mercury manometer (when more precise measurements are required). Water moves in and out of the cup in response to changes in soil water content. Tensiometers provide measurements of soil water potential that are related to soil water content by the characteristic curve of the soil (see Figure 1.1*b*). Tensiometers may interfere with cultivation and can require considerable time for installation and maintenance. In addition, they have a relatively limited range of operation (0 to −80 centibars). Table 1.22 provides guidelines for interpreting tensiometer readings.

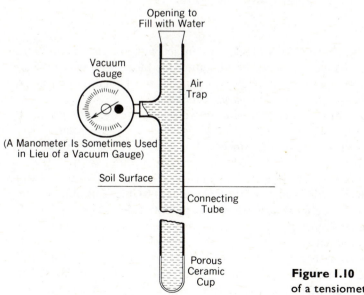

Figure 1.10   The essential parts of a tensiometer.

Table 1.22   Interpretation of Tensiometer Readings

	Dial Reading in Centibars	Interpretation
Nearly saturated	0 10	Near saturated soil often occurs for a day or two following irrigation. Danger of waterlogged soils, a high water table, poor soil aeration, or the tensiometer may have broken tension, if readings persist.
Field capacity	11 20 30	Field capacity. Irrigations discontinued in this range to prevent waste by deep percolation and leaching of nutrients below the root zone. Sandy soils will be at field capacity, in the lower range, with clay soil at field capacity in the upper range.
	40	Usual range for starting irrigations. Soil aeration is assured in this range. In general, irrigations start at readings of *30–40* in *sandy textured soils* (loamy sands and sandy loams).
Irrigation range	50	Irrigations usually start from *40 to 50* on *loamy-soils* (very fine sandy loam and silt loams). On *clay soils* (silty clay loams, silty clays, etc.) irrigations usually start from *50 to 60*.
	60	Starting irrigations in this range insures maintaining readily available soil moisture at all times
Dry	70	This is the stress range. However, crop not necessarily damaged or yield reduced. Some soil moisture is readily available to the plant but is getting dangerously low for maximum production.
	80	Top range of accuracy of tensiometer readings above this are possible but the tensiometer will break tension between 80 to 85 centibars

Source: P. E. Fischbach and R. E. Schleusener, "Tensiometers a Tool to Help Control Irrigation Water" (1961), Univ. of Nebraska Agricultural Experimental Station, Lincoln, Ext. Ctr. 61-716, 8 pp.

(iv) *Porous Blocks*   Porous blocks in contact with moist soil equilibrate with the soil water. Changes in the water content of the block (and hence the soil) cause various properties of the block to change. The two most commonly measured properties are *electrical resistance* and *thermal conductivity*.

Embedded pairs of stainless-steel wires or wire grids are used to measure the electrical resistance of gypsum (plaster of paris), nylon, or fiberglass blocks. Gypsum blocks are most common, since nylon and fiberglass are extremely sensitive to the concentration of dissolved solids in the soil water solution (i.e., soil salinity). Calibration curves relating electrical resistance of the block to soil water potential and the characteristic curve of the soil are used to infer soil water content. Because gypsum blocks are most accurate at soil potentials less than -1 bar, they are often used with tensiometers to obtain the entire range of soil water.

The thermal conductivity of porous ceramic blocks is also an index to soil water. Ceramic blocks with a heater and several temperature sensors are used to measure the rate at which heat is conducted through the block. The thermal conductivity of the ceramic block is directly related to the water content of the block and the soil water potential. The characteristic curve of the soil is needed to determine soil water content. These sensors are not affected by soil salinity, operate over a large range of soil potential, are commercially manufactured, and have been used with microcomputers to automatically schedule irrigations (Phene et al., 1981).

(v) Neutron Scattering   Neutron probes are widely used to measure the volumetric water content of soils. The technique involves an access tube, a source of high energy or fast neutrons (usually americium), and a detector. The access tube is an aluminum or steel tube installed in a hole drilled to the desired depth (usually the root zone depth). The source and detector are contained in a single unit that is lowered into the access tube to a particular depth. The fast neutrons are emitted into the soil and gradually lose energy by collision with various atomic nuclei. Hydrogen, present in water and organic matter, is the most effective element in the soil in slowing down the neutrons. The slowed or thermalized neutrons form a cloud around the source and some of them randomly return to the detector. The detector consists of a charged wire and a ratemeter located above the ground surface. The returning thermalized neutrons cause an electrical pulse on the charged wire that is counted by the ratemeter. The number of pulses counted over an interval of time is linearly related to volumetric water content (when the amount of soil organic matter is constant) by a calibration curve. It is normally recommended that locally determined curves be used in lieu of factory supplied curves.

When the count at one depth has been determined, the probe can be positioned at another depth to determine the water content at that depth. Sometimes the average water content of the root zone is obtained by moving the probe through the root zone and recording the counts. A single probe can be moved from one access tube to another around the farm. Repeated measurements may be made at the same location and depth, thus minimizing the effect of soil variability on successive measurements. Measurements near the surface are not accurate because neutrons are lost into the atmosphere. Because the source of high-energy neutrons is radioactive, operators must receive radiation safety training and follow strict handling, shipping, and storage procedures.

1.6.2c   Water budget Technique

The *water budget technique* of determining when to irrigate is similar to soil indicator methods. Instead of measuring soil water content, it is computed using Eq. 1.29.

$$\theta_i = \theta_{i-1} - 100\left(\frac{\text{ET} - P_e}{D_{rz}}\right)i \tag{1.29}$$

where

θ_i, θ_{i-1} = the soil water content in percent by volume at the end of day i and day $i - 1$, respectively.

Equation 1.29 is derived from Eq. 1.25 with $I = L = 0$. ET is computed with Eq. 1.8 and ET_0 is normally determined using Penman-type equations. The following example illustrates the water budget technique.

EXAMPLE 1.8   Water Budget Technique of Determining When to Irrigate

Given:
- daily ET and P_e in the following table
- $D_{rz} = 24$ in
- soil water content on the morning of July 1 is 23 percent by volume
- $\theta_c = 14\%$

Required:
date of next irrigation

Solution:

Date (July)	ET	P_e	θ^a (%)
1	0.33	0	21.63
2	0.32	0.05	20.51
3	0.26	0	19.43
4	0.29	0	18.22
5	0.30	0	16.97
6	0.33	0	15.59
7	0.23	0.10	15.05
8	0.30	0	13.80

[a] water content by volume at end of day

Sample calculation (for day 1)

$$\theta_i = 23 - 100\left(\frac{0.33 - 0}{24}\right) = 21.63\%$$

Conclusion:
Irrigate on the morning of July 8.

1.6.3   Amount to Apply per Irrigation

Once it has been determined that it is time to irrigate, the usual practice is to fill the root zone to field capacity. The amount of water required to do this is given by the following equation.

$$\text{IRRI} = \frac{D_{rz}(f_c - \theta)}{E_i} \tag{1.30}$$

where, θ = percent water content of the soil prior to irrigation. Plant indicators of when to irrigate do not provide information about θ. In such cases it is necessary to measure or estimate θ.

Irrigators do not always fill the soil to field capacity. In locations with significant amounts of precipitation during the irrigation season, irrigators often choose to only partially fill the root zone, so that at least some precipitation can be stored. The recent management concept of deficit irrigation, that is, not replacing all the water that is used by the plant, also involves partial replacement and eventually depleting soil water below θ_c. It has been found, for example, that the yield and fruit quality of mature apple trees is not affected by partial replenishment of the water used during the time prior to fruiting. Full replenishment is required during and following fruiting. Tree growth is significantly reduced, however. Other research indicates that in the future, it may be more economical to irrigate to achieve maximum production per unit of water and/or energy rather than the current practice of irrigating for maximum production per unit of land. This involves partial replenishment and depleting the root zone water below θ_c at least during a portion of the growing season. Equation 1.30 cannot be used to estimate the amount to apply per irrigation in these situations.

EXAMPLE 1.9   Determining How Much Water to Apply Per Irrigation

Given:
• Example Problem 1.8
• $f_c = 26\%$
• $D_{rz} = 24$ in
• $E_i = 80\%$

Required:
amount of water to apply per irrigation

Solution:

$$I = \frac{24(26 - 15.05)}{80} = 3.29 \text{ in}$$

Thus, 3.29 in of water should be applied the morning of July 8.

Homework Problems

A homework problem preceded by an asterisk indicates that a computer program will facilitate the solution of the problem.

1.1 Determine the depth (volume per unit area) of water in a 75-cm-deep soil. The average water content of the soil is 20 percent by volume.

1.2 Determine the depth of water in a 100-cm-deep:
a. sand,
b. loam, and
c. clay soil

for each of the following conditions.
a. saturation,
b. field capacity, and
c. permanent wilting point.

1.3 Compare the amount of water held by the soils considered in Problem 1.2 by plotting soil texture (on the horizontal axis) versus the depths of water at saturation, field capacity, and permanent wilting point.

1.4 Determine the available water holding capacity of a 100-cm-deep:
a. sand,
b. loam, and
c. clay soil.

1.5 Compare the available water holding capacity of the soils considered in Problem 1.4 by plotting soil texture (on the horizontal axis) versus available water holding capacity.

1.6 Determine the depth (volume per unit area) of gravimetric and hygroscopic water for the soils in Problem 1.2. Gravimetric water is water that can be drained from the soil by gravity. Hygroscopic water is water that is held too tightly in the soil to be removed by plants.

1.7 Compare the depth (volume per unit area) of gravimetric and hygroscopic water for the soils in Problem 1.2 by plotting soil texture (on the horizontal axis) versus the depth of gravimetric and hygroscopic water.

1.8 Compute the amount of readily available water for
a. corn,
b. clover, and
c. potatoes

grown in a 75-cm-deep loam soil.

1.9 Determine the total amount of available water that can be held by the following layered soil profile.

Soil Texture	fc Percent by Volume	pwp Percent by Volume	Depth (cm)
Sandy loam	14	6	15
Loam	22	10	13
Clay loam	27	13	75

1.10 Determine the available water holding capacity of the layered soil in Problem 1.9 for potatoes.

1.11 Determine the amount of readily available water for potatoes that can be stored in the soil in Problems 1.9 and 1.10.

1.12 Determine θ_c for potatoes for the soil in Problems 1.9, 1.10, and 1.11.

1.13 Determine the water content of a 100-cm-deep soil after an irrigation in which 12 cm of water was applied. The initial water content of the soil was 15 percent by volume. Assume that no runoff or deep percolation occurred.

1.14 Determine the amount of deep percolation that would occur if the water content at field capacity for the soil in Problem 1.13 was 23 percent by volume. Compute the depth (volume per unit area) of irrigation water that can be applied without deep percolation. Assume that there is no surface runoff.

1.15 A 4-ha field is being irrigated with a sprinkle irrigation system that applies water at a rate of 3000 l/min. The water content of the 75-cm-deep loam soil when the irrigation began was 13 percent by volume. Determine
a. The depth of water that can be applied without deep percolation, and
b. The duration of time that the irrigation system can be operated without causing deep percolation.

Assume that the entire field is being irrigated simultaneously and that there are no sprinkler spray, wind drift, or runoff losses.

1.16 Given the following information
- 1-m^2 by 3-m-deep lysimeter with corn
- no drainage or runoff from the lysimeter
- 10 cm of rainfall fell during the growing season
- 66.2 cm of irrigation water was applied to the corn
- the water content of soil in the lysimeter was 20 percent by volume initially and 15 percent by volume at the end of the growing season
- measured ET_0 for grass for the growing season was 100 cm

compute seasonal ET and the seasonal K_c for corn.

1.17 Determine monthly and seasonal ET_0 using the following data collected in a 100-cm-deep lysimeter.

Month	Precipitation (mm)	Irrigation (mm)	Drainage (mm)	SW[a] Content by Volume[b]
Apr.	16.0	136.0	25.0	20
May	39.6	112.6	10.3	25
June	37.8	133.1	10.6	24
July	4.6	217.1	5.4	15
Aug.	1.5	200.3	1.2	18
Sep.	13.0	98.9	0	25
Oct.	13.2	24.5	10.2	20

[a] SW = soil water
[b] End of the month soil water contents are given in the table. The soil water content on April 1 was 20 percent by volume.

No runoff or other losses were observed.

1.18 Using ET_0 data computed in Problem 1.17, determine monthly and seasonal ET for apples with a cover crop. The apple orchard is located in a relatively dry climate (the average relative humidity during the irrigation season is about 20 percent). There are light to moderate winds of about 2 m/s during the irrigation season.

1.19 Using the ET_0 data computed in Problem 1.17 and the climatic data from Problem 1.18, determine monthly and seasonal ET for potatoes. The planting date is May 6.

1.20 Using the solar radiation data in Appendix D, plot solar radiation versus date and use the resulting plot to estimate R_{so} (see Table 1.10).

*1.21 Utilizing data from Appendix D and Problem 1.20, compute grass ET_0 for the period June 3 through 10 (154 through 161 days from Jan. 1) using:
a. The Wright–Jensen Penman equation, and
b. The Doorenbos–Pruitt Penman equation (use Table A1).

The station is 100 m above mean sea level and located at 46°N latitude. R_{so} for the Wright-Jensen method is given by

$$R_{so} = -17.01 + 0.43 \, J - 1.73(10)^{-3} \, J^2 + 1.85(10)^{-6} \, J^3$$

where, J is the number of days from January 1.

1.22 Utilizing data from Appendix D, compute daily ET_0 for grass during the period June 3 through 10 using the pan evaporation method. A Class A evaporation pan located in the middle of a 1000 m^2 grassed field was used to collect the pan evaporation data in Appendix D.

*1.23 Utilizing data from Appendix D, compute ET_0 for grass for the period June 3 through 10 (154 through 161 days from Jan. 1) using the Jensen–Haise method. The location is 700 m above mean sea level.

1.24 Utilizing data from Appendix D, compute ET_0 for grass for June using the SCS version of the Blaney–Criddle equation. The latitude of the location is 46°N.

1.25 Utilizing data from Appendix D, compute ET_0 for grass for June using the Doorenbos and Pruitt version of the Blaney–Criddle method. The value of n/N for June is 0.74 at this 46°N location.

*1.26 Using the following cloud cover data for 30°S latitude, estimate extra-terrestrial and solar radiation

Date (July)	Days from Jan. 1	n/N
10	191	0.80
11	192	0.75
12	193	0.20
13	194	0.55
14	195	1.00

*1.27 Utilizing data from Appendix D, compute monthly and seasonal ET for apples with a cover crop at the location described in Problem 1.18.

*1.28 Utilizing data from Appendix D and information from Problem 1.19, compute monthly and seasonal ET for potatoes for the location in Problem 1.18.

1.29 Given the following information
- RAW = 30 mm
- depth (volume/area) of water in the soil at saturation = 10 cm
- depth (volume/area) of water in the soil at field capacity = 7 cm
- the soil is at field capacity the morning of day 1 (in the following table)
- the following table of climatic data

Day	ET (mm)	Precipitation (mm)
1	7.9	0
2	6.4	0
3	5.3	0.
4	10.4	0
5	3.0	15.0
6	8.9	0
7	10.2	0
8	7.9	0
9	5.8	0
10	4.6	0

- no runoff was observed

Determine:

a. when irrigations are required during the 10-day period,

b. the depth of each irrigation,

c. the total depths of irrigation and deep percolation for the 10-day period, and

d. the depth of water remaining in the soil at the end of the 10-day period.

1.30 Given the following information:

- grain corn is being grown in a 90-cm-deep sandy loam soil (fc = 20 percent)
- the available water holding capacity of soil is 0.1 cm per centimeter of soil depth
- the field is located in an area with an average relative humidity and wind speed of 75 percent and 2 m/s, respectively
- the soil water content on the morning of day 81 is 18 percent by volume
- data in the following table

Days from start of ET	ET_o (grass) (mm)	Precipitation (mm)
81	7.9	0
82	8.9	0
83	9.1	0
84	10.2	2.5
85	10.4	0
86	9.9	0
87	8.9	38.1
88	7.6	0
89	7.4	0
90	8.4	0

Determine:
a. the total effective precipitation,
b. the date and amount of each irrigation,
c. the total depth of irrigation water applied during the 10-day period, and
d. the total depth of water used by the corn during the 10-day period.

*1.31 Given the following information
- corn is being grown at 46°N latitude
- corn is normally planted during the first week of May
- the water contents of the 75-cm-deep loam soil at field capacity and permanent wilting point are 31 and 14 percent by volume, respectively
- the following long-term average monthly climatic data

Month	Temperature (°C)	Precipitation (mm)	n/N	U (m/s)	RH (%)
May	16.0	15.0	0.72	2.6	31
June	21.0	33.2	0.74	2.5	31
July	27.0	5.8	0.86	2.4	25
Aug.	24.0	9.4	0.86	2.2	28
Sept.	21.0	10.3	0.74	2.3	31

Estimate the number of irrigations required during the growing season. Assume the soil is at field capacity on May 6.

1.32 Repeat Problem 1.31 for the following soils:

Soil Type	fc[a]	pwp[a]
Sand	15	7
Clay	44	21

[a] Water content in percent by volume.

Compare the number of irrigations and the depths of irrigation for the three soils by plotting soil texture (on the horizontal axis) versus the depth and number of irrigations.

*1.33 Using daily pan evaporation data from Appendix D, schedule irrigations for apples with a cover crop. The soil is a sandy loam. Use a pan coefficient of 0.80. Plot the days since the beginning of the growing season (on the horizontal axis) versus the interval between successive irrigations and describe how the irrigation interval varies during the irrigation season. The soil is 150 cm deep.

1.34 Each week during the growing season an irrigator obtains soil samples for 10 locations around an irrigated field. Rain gauge data are also collected to determine weekly depths of irrigation and rain for the field. The approximate volume of each soil sample is 100 cm³. The depth of the root zone is 100 cm. Determine crop water use for the following set of data:

Date	Wet Weight[a] (g)	Dry Weight[a] (g)	Irrigation and Rain (mm)
July 1	1555	1354	40
July 8	1514	1346	10

[a] Total weight of the 10 samples.

1.35 Determine the soil water tension at which irrigations should be initiated for corn grown in a very deep soil with the following water content versus matric potential relationship.

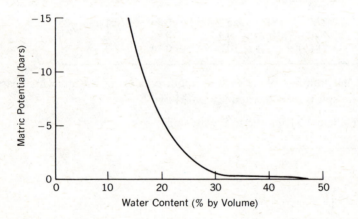

*1.36 Write a computer program that uses the Wright–Jensen Penman-type equation to schedule irrigations. Using the data from Appendix D and the equation given in Problem 1.21 for R_{so}, schedule irrigations for the following crops:

a. grass,
b. apples with a cover crop, and
c. corn

grown in 200-cm-deep
a. sand,
b. loam, and
c. clay soils.

Assume that the growing season begins May 1.

*1.37 Modify the computer program developed in Problem 1.36 to schedule deficit irrigations using the following relationship between the ratio of actual to potential ET and soil water content.

$$ET_a = ET_p\left(1 - \frac{\theta_c - \theta}{\theta_c - \text{pwp}}\right) \qquad \text{for } \theta \geq \theta_c$$

Use this program to determine seasonal water use for corn planted on May 1 and grown in a 200-cm-deep

a. sand,
b. loam, and
c. clay soil.

Irrigate the crop fully until August 1. Do not irrigate after this date.

References

Anon. (1978). Water conservation opportunities study. U.S. Dept. of Interior, Bureau of Reclamation and Bureau of Indian Affairs, Washington, D.C.

Burman, R. D., P. R. Nixon, J. L. Wright, and W. O. Pruitt (1980). Water requirements. In *Design and Operation of Farm Irrigation Systems*, M. E. Jensen (Ed.), *ASAE Monograph 3*. St. Joseph, MI, pp. 189–232.

Brunt, D. (1952). *Physical and Dynamic Meteorology* (2nd ed.), Cambridge Univ. Press, Cambridge, England, 428 pp.

Businger, J. A. (1956). Some remarks on Penman's Equation for the evapotranspiration. *Meth. J. Agri. Sci.*, **4**, pp. 77–80.

Denmead, O. T., and R. H. Shaw (1960). The effects of soil moisture stress at different stages of growth on the development and yield of corn. *Agronomy J.* **52**(5), pp. 272–274.

Doorenbos, J., and W. O. Pruitt (1977). Guidelines for predicting crop water requirements. Irrigation and Drainage Paper 24, FAO, United Nations, 144 pp.

Fischbach, P. E., and R. E. Schleusener (1961). Tensiometers a tool to help control irrigation water. Univ. of Nebraska Agricultural Experiment Station, Lincoln, Ext. Ctr. 61–716, 8 pp.

Frevert, D. K., R. W. Hill, and B. C. Braaten (1983). Estimation of FAO evapotranspiration coefficients. *J. Irrigation and Drainage Eng.*, **109**(2), pp. 265–270.

Hansen, V. E., O. A. Israelson, and G. Stingham (1979). *Irrigation Principles and Practices*. Wiley, New York.

James, L. G., J. M. Erpenbeck, D. L. Bassett, and J. E. Middleton (1982). Irrigation requirements for Washington—Estimates and methodology. *Research Bulletin XBO925*, Agricultural Research Center, Washington State Univ. Pullman, 37 pp.

Jensen, M. C. (1970). Irrigation system design capacity functions for Washington. *Circular 525*, Agricultural Experiment Station, Washington State Univ., Pullman, 30 pp.

Jensen, M. C., S. T. Lin, A. J. Anderson, C. W. Ek, G. N. Sundstrom, and J. W. Trull (1969). Irrigation water requirement estimates for Washington. *Circular 512*, Agricultural Experiment Station, Washington State Univ., Pullman, 11 pp.

Jensen, M. E. (Ed.) (1974). Consumptive use of water and irrigation water requirements. Tech. Commun. on Irrigation Water Requirements, American Society of Civil Engineers, New York. 215 pp.

Jensen, M. E., and H. R. Haise (1963). Estimating evapotranspiration from solar radiation. *J. Irrig. and Drainage Div.*, **ASCE 89**(IR4), pp. 15–41.

Pair, C. H., W. H. Hinz, K. R. Frost, R. E. Sneed, and T. J. Schiltz (Eds.) (1983). *Irrigation.* The Irrigation Assoc., Silver Spring, Md., 686 pp.

Penman, H. L. (1948). Natural evaporation from open water, bare soil, and grass. *Proc. Royal Society of London, Ser. A.*, **193**, pp. 120–145.

Phene, C. J., J. L. Fouss, and T. A. Howell (1981). Scheduling and monitoring irrigation with the new soil matric potential sensor. *Proc. American Society of Agricultural Engineers Irrigation Scheduling Conference*, St. Joseph, Mo, pp. 91–105.

Reicosky, D. C., and D. B. Peters (1977). A portable chamber for rapid evapotranspiration measurements on field plots. *Agron. J.* **69**, pp. 729–732.

Soil Conservation Service (SCS) (1970). Irrigation water requirements. *Tech. Release 21*, USDA–SCS, 88 pp.

Stegman, E. C., J. T. Musick, and J. I. Stewart (1980). Irrigation water management. In *Design and Operation of Farm Systems*, M. E. Jensen (Ed.), *ASAE Monograph 3*. St. Joseph, MI, pp. 763–816.

Thom, A. S., and H. R. Oliver (1977). On Penman's equation for estimating regional evaporation. *Quart. J. R. Meteor Soc.*, **103**, pp. 345–357.

Van Bavel, C. H. M. (1966). Potential evaporation: The combination concept and its experimental verification. *Water Resources Research*, **2**(3), pp. 455–467.

Westesen, G. L., and T. L. Hanson (1981). Irrigation scheduling using wash tub evaporation pans. *Proc. American Society of Agricultural Engineers Irrigation Scheduling Conference*, St. Joseph, MI, pp. 144–149.

Wilson, C. L., and W. E. Loomis (1962). *Botany*. Holt, Rinehart & Winston, New York, 573 pp.

Wright, J. L. (1981). Crop coefficients for estimates of daily crop evapotranspiration. *Proc. American Society of Agricultural Engineers Irrigation Scheduling Conference*, St. Joseph, MI, pp. 18–26.

Wright, J. L., and M. E. Jensen (1978). Development and evaluation of evapotranspiration models for irrigation scheduling. *Trans. ASAE* **21**(1), pp. 88–91 and 96.

2

Farm Irrigation Systems and System Design Fundamentals

2.1 Introduction

Farm irrigation systems must supply water at rates, in quantities, and at times needed to meet farm irrigation requirements and schedules. They divert water from a water source, convey it to cropped areas of the farm, and distribute it over the area being irrigated. In addition, it is essential that the farm irrigation system facilitate management by providing a means of measuring and controlling flow.

This chapter describes farm irrigation systems and their design. The major steps of the design process and the role of the designer are presented and discussed. Several parameters and procedures for establishing the technical and economic feasibility of alternate system designs are described and demonstrated. After reading and studying this chapter the reader should be able to: describe the functions and types of farm irrigation systems, understand the design process and the role of the designer, estimate the design daily irrigation requirement for a farm, evaluate the uniformity, efficiency, adequacy, and effectiveness of irrigation, and compute the total cost of owning and operating a farm irrigation system.

2.2 Functions of Farm Irrigation Systems

The primary function of farm irrigation systems is to supply crops with irrigation water in the quantities and at the time it is needed. Specific functions include:

1. diverting water from the water source,
2. conveying it to individual fields within the farm,
3. distributing it within each field, and
4. providing a means for measuring and regulating flow.

61

Other functions of farm irrigation systems include crop and soil cooling, protecting crops from frost damage, delaying fruit and bud development, controlling wind erosion, providing water for seed germination, application of chemicals, and land application of wastes.

2.2.1 Crop and Soil Cooling

Sprinkling normally tends to cool the air, soil, and crop. During periods when the potential transpiration rate is especially high (and θ_c approaches fc), cooling due to sprinkler spray evaporation reduces plant water deficits and associated stomatal closure. Cooling has been found to improve the yield and/or quality of several crops including almonds, apples, beans, cherries, cotton, cranberries, cucumbers, flowers, grapes, potatoes, prunes, strawberries, sugar beets, tomatoes, and walnuts.

2.2.2 Frost Protection

Sprinkle irrigation systems have long been used to protect plants from freezing damage during radiation frosts. Radiation frosts occur on clear, calm nights when plants are cooled 0.5° to 2.0°C (1° to 4°F) below the ambient air temperature as they radiate energy to a cold sky. Heat to maintain plant temperatures above lethal levels is obtained from sprinkled water as it cools and freezes on plant surfaces. For low-growing crops in zero wind conditions sprinkling at a rate of 2 to 3 mm/h (about 0.1 in/h) should start before the temperature reaches 0°C (32°F) at the plant level and continue until the plant is free of ice. Higher application rates will be required if wind is present, when air temperatures are low, or for taller growing crops. Strawberries have been protected from temperatures as low as -6°C (22°F). Alfalfa, tomatoes, peppers, cranberries, apples, cherries, and citrus have also been successfully protected.

Fruit trees have been protected from freezing damage with undertree sprinklers that do not wet tree foliage. Since plant protection depends primarily on the convective transfer of heat from the sprinkled water to plant surfaces, undertree sprinkling provides less frost protection than overtree sprinkling. Undertree sprinkling does not, however, have the limb breakage (due to ice accumulation) problem associated with overtree sprinkling.

2.2.3 Delaying Fruit and Bud Development

The date of deciduous fruit-tree blossoming can be controlled by sprinkling during the bud growth period in the spring. Cooling of buds resulting from the evaporation of sprinkler spray slows bud growth and delays blossoming. Freezing damage is prevented by sprinkling in a manner that delays blossoming until the danger of frost has past.

2.2.4   Controlling Wind Erosion

Wind erosion of bare soil surfaces that are in a loosened condition because of tillage or freezing can be controlled with irrigation. Wet soils are more resistant to wind erosion. Upon drying, however, sandy soils resume their erodibility, while surface crusts that are largely resistant to wind erosion develop in medium-to-fine-textured soils. Tilling fine sandy loams and other similar textured soils after irrigation produces clods that are erosion resistant until broken down by water or additional tillage. Sandy soils must be kept moist to control wind erosion.

2.2.5   Germinating Seeds

Sprinkling can be used to control the temperature and salt content of seed beds. When soil temperatures are high due to high incoming radiation, sprinkling can reduce soil surface temperatures and prevent "burning off" of young seedlings. Where salts contained in irrigation water accumulate on the surface of furrow irrigated beds, sprinkle irrigation is applied during seed germination to move salts below the seed bed.

2.2.6   Chemical Application

Fertilizers, pesticides, herbicides, desicants, and defoliants are applied with irrigation systems. Some of the advantages of the conjunctive application of chemicals with irrigation water include savings in labor and equipment, better timing, ease of split and multiple controlled application, greater flexibility of farm operations, and enhanced crop production. Only irrigation systems that distribute water uniformly should be used to apply chemicals.

2.2.7   Land Application of Liquid Wastes

Irrigation systems are used to apply liquid wastes from cities, towns, farms, and industrial plants to the land. The wastes, which may contain suspended and/or dissolved materials, are spread over the land where the water is cleaned by filtration as it percolates through the soil. Water treatment prior to land application is necessary when the wastewater contains materials, either organic or inorganic, that are harmful to plants, fungi, bacteria, and other beneficial flora.

2.3   Types of Farm Irrigation Systems

There are several ways of diverting, conveying, and applying irrigation water on farms. The following sections describe alternate types of on-farm diversion, conveyance, and application systems. Facilities for flow measurement and regulation are also discussed.

2.3.1   Diversion Methods

There are two primary ways of diverting surface and ground waters: gravity diversions and pumping plants. When water surface elevations or heads at the water source are sufficient, gravity diversions are used. Otherwise, a pumping plant is utilized to lift and/or provide pressure for conveying and/or applying irrigation water.

2.3.1a   Gravity Diversions

The most common type of gravity diversion uses a turnout to admit water from an open water source into farm canals and pipelines. A turnout consists of an inlet, a conduit or other means of conveying water through the bank of the supply canal, and where required, an outlet transition. Turnouts normally include a means of regulating and measuring inflow to the farm such as weirs, sluice gates, or valves. Typical turnouts are diagrammed in Figure 2.1.

On farms that obtain water from pressurized pipelines, a valve is used in lieu of a turnout to admit water into the farm pipeline. A pumping plant is necessary only when the delivery pressure (from the off-farm pipeline) is not sufficient to provide the head needed to operate the farm irrigation system. The inflow rate to the farm is controlled by regulating the delivery pressure and valve opening.

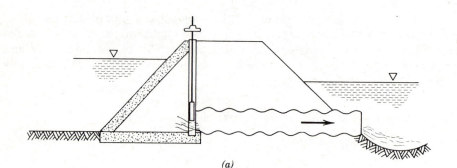

(a)

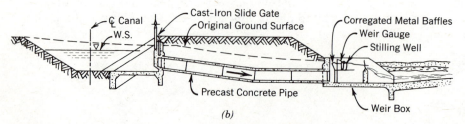

(b)

Figure 2.1   Some canal turnout structures. (After Skogerboe et al., 1971.) (*a*) Pipe turnout into a corrugated pipe section without provision for water measurement. (*b*) Pipe turnout into a concrete pipe with downstream flow measurement and energy dissipation.

2.3.1b   Pumping Plants

Pumping plants are used when water must be lifted from the water source and/or when sufficient head (pressure) is not available to operate the farm irrigation system. Pumping plants normally have one or more horizontal or vertical centrifugal pumps powered by either electric motors or internal combustion engines.

(i) Pumps   Horizontal centrifugal pumps are normally used with surface sources of water and springs. The pump and power unit can be positioned above the water surface or in a dry pit below the water surface (as in Figure 2.2). Positioning the pump above the water surface facilitates access to the pump for maintenance but makes it necessary to prime the pump (i.e., fill the pump with water) prior to starting the pump. Prime can be maintained when the pump is not operating with foot (check) valves.

Vertical centrifugal pumps similar to the one in Figure 2.3 can be used with either surface (Figure 2.3*a*) or ground water (Figure 2.3*b*) sources. Vertical

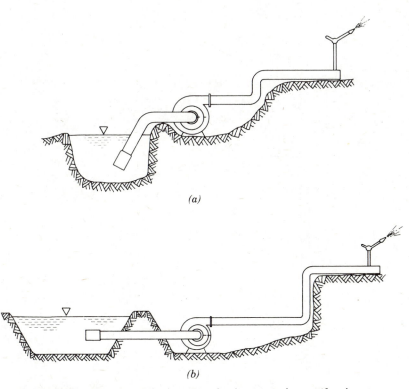

(a)

(b)

Figure 2.2   Typical mounting positions for horizontal centrifugal pumps.
(*a*) Horizontal centrifugal pump located above the water surface.
(*b*) Horizontal centrifugal pump located below the water surface (the pump does not require "priming" prior to starting).

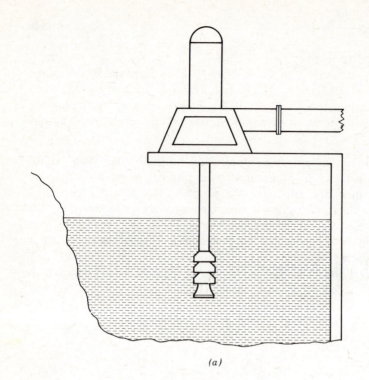

(a)

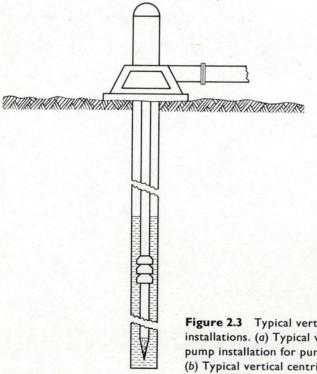

(b)

Figure 2.3 Typical vertical centrifugal pump installations. (a) Typical vertical centrifugal (turbine) pump installation for pumping surface waters. (b) Typical vertical centrifugal (turbine) pump installation for pumping ground waters.

centrifugal pumps are more difficult to maintain but do not require priming since the pump is submerged. The power unit may be located above the water surface or submerged beneath the pump. Long drive shafts are required when water is being pumped from deep wells and power units are above ground.

(ii) *Power Units* Electric motors and internal combustion engines are the most common power units for irrigation pumps. Animal- and human-powered pumps are, however, utilized in locations where electricity and fuel are either not available or too expensive. Gasoline, diesel, liquified petroleum gas, and natural gas are the most commonly used fuels for internal combustion engines.

2.3.2 Conveyance Methods

Water is conveyed from the water source to cropped areas of the farm in networks of open channels and/or pipelines. Open channels may be lined or unlined and pipelines partially open to the atmosphere or pressurized.

2.3.2a Open Channels

Open channels are usually graded (have a slope) in the direction of flow and may be either lined or unlined. They are used with all surface and ground water sources and farm application systems. Open channels are lined with hard surface linings such as concrete and asphalt, exposed or covered membranes, and soil sealants to reduce maintenance costs, channel size, and seepage losses through the channel bed and walls. Unlined ditches are used because of their low capital costs and ease of construction and relocation.

2.3.2b Pipelines

On-farm pipe networks are classified as open (low head) or closed (pressurized) depending on whether the system is open to the atmosphere. Both types of pipelines can be laid on the ground surface or buried. Surface pipes have the advantage (over buried pipes) of portability, while buried lines can be placed below the tillage zone where damage by farm machinery, vehicles, vandals, etc., is minimized. When buried above the frost depth, provisions for draining buried pipelines are required.

Pipelines have several advantages over open channels. Well-constructed pipelines eliminate seepage and evaporative losses, avoid weed problems, and are normally safer than open channels (since humans and equipment cannot fall into the water stream). Pipelines also permit the conveyance of water uphill against the normal slope of the land and, unlike open channels, can be installed on nonuniform grades. The use of buried pipe eliminates pad construction for open channels, allows use of the most direct routes from the water source to fields, and minimizes the loss of productive land (since crops can be planted up to or over the pipelines). Portable pipe systems laid on the soil surface can be removed while cultural operations are in progress. Open channels may, however, be more economical than pipelines when land is flat and flows are large.

(i) Open (Low Head) Pipelines   Because low head pipelines are open to the atmosphere, heads in these pipelines seldom exceed 15 m (50 ft). A profile view of a typical low head pipeline is shown in Figure 2.4. The stand pipe provides head regulation, water release, vacuum relief, and air release. Open pipelines are not generally used with application systems that require more than 6 m (20 ft) of head.

(ii) Closed (Pressurized) Pipelines   Pressurized pipelines normally supply application systems that require more than 6 m (20 ft) of head. Pressurized pipelines are not open to the atmosphere and do not contain structures such as standpipes. Pressure regulating, check, air release, and vacuum relief valves are used instead of standpipes and pump stands to provide flow and pressure control as well as to protect the pipeline.

2.3.3   Application Systems

Sprinkle-, trickle-, and surface- (gravity-) type application systems are used to distribute water to crops. Each of these sytems and the conditions that favor their use are discussed in the following sections.

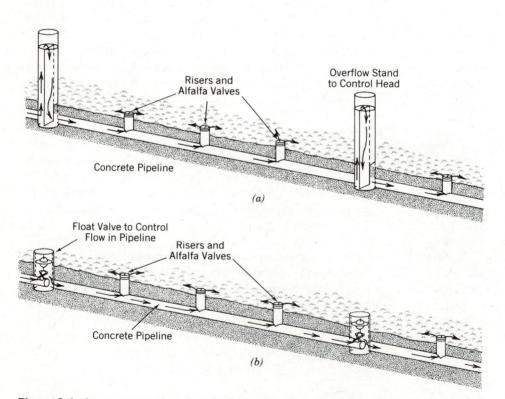

Figure 2.4   Low-pressure (open) pipelines with two types of overflow stands. (*a*) Overflow stands with baffles. (*b*) Overflow stands with float valve.

2.3.3a Sprinkle Systems

Sprinkle irrigation systems similar to the ones in Figure 2.5 use sprinklers operating at pressures ranging from 70 to over 700 kPa (10 to over 100 psi) to form and distribute "rainlike" droplets over the land surface. Sprinkle systems apply water efficiently, have relatively high capital costs and low labor requirements, and use more energy than other application methods. Sprinkle irrigation is adaptable to many soils and terrains. It can be successfully used to irrigate

1. permeable soils that are difficult to irrigate using other application systems,
2. lands with combinations of shallow soils and terrain that prevent proper land smoothing needed for other application systems,
3. land having steep slopes and easily erodible soils, and
4. undulating terrain that would be too costly to smooth for use of other application systems.

Sprinkle systems can be used for frost protection, fertilizer and chemical application, wind erosion control, crop and soil cooling, delaying fruit and bud development, germinating seeds, and land application of wastes.

2.3.3b Trickle Systems

Trickle irrigation is the frequent, slow application of water either directly onto the land surface or into the root zone of the crop. It is based on the fundamental concepts of irrigating only the root zone of the crop (rather than the entire land surface) and maintaining the water content of the root zone at near optimum levels. Trickle irrigation is accomplished using pressures ranging from 15 to 200 kPa (2 to 30 psi) to drip water one-drop-at-a-time onto the land or into the root zone, spray

Figure 2.5 Different types of sprinkle irrigation systems.
(*a*) Center pivot. (*b*) Traveler (soft hose). (*c*) Hand move.
(*d*) Side roll. (*e*) Solid set.

(b)

(c)

(d)

70

it as a fine mist over portions of the land surface, or bubble it onto the land surface in small streamlets. Different types of trickle irrigation systems are shown in Figure 2.6.

Conditions that favor the use of trickle irrigation include

1. high-value row crops,
2. a limited, expensive or saline water supply,

Figure 2.6 Different types of trickle irrigation systems. (*a*) Drip system on grapes. (*b*) In-line drip emitter. (*c*) Spray system. (*d*) Microspray sprinkler.

(b)

(c)

(d)

72

3. water application must be precise in location and amount to minimize drainage or for precise salinity management, and
4. above ground portions of plants must be dry to control bacteria, fungi, and other pests and diseases.

Waters with high concentrations of particulate, chemical, and/or biological materials that clog trickle system components make trickle irrigation difficult and expensive. Trickle irrigation is adaptable to most soils and terrains.

2.3.3c   Surface (Gravity) Irrigation

Most irrigation throughout the world is accomplished via *surface (gravity) techniques*. Types of surface irrigation systems are shown in Figure 2.7. Surface irrigation systems generally require a smaller initial investment (except when extensive land smoothing is needed), are more labor intensive, and apply water less efficiently than other types of irrigation systems. Surface irrigation systems are best suited to soils with moderate to low infiltration capacities and land with relatively uniform terrain and slopes less than 2 percent.

2.3.4   Flow Measurement and Regulating Methods

Flow data are invaluable in adjusting (regulating) farm irrigation system operation to properly irrigate all parts of the farm and in detecting maintenance problems such as clogged or plugged screens, worn pumps and sprinklers, pipeline and ditch leaks etc. Since flow measurement and regulation are essential prerequisites to effective system management and maintenance, they must be carefully and

Figure 2.7   Different types of surface irrigation systems.
(a) Water advancing across a graded border. (b) Level basin with a corner inlet. (c) Basins in an orchard. (d) Siphon tubes delivering water from an unlined head ditch into two furrows.

thoroughly considered in the design process. Flow measurement in open channels and pipelines is discussed in Chapter 8.

Weirs, flumes, and orifice plates are the most commonly used devices to measure flow in open channels. In pipelines, venturi meters, orifices, elbow meters, rotating mechanical flowmeters, ultrasonic devices, and various types of Pitot tubes are used to measure flow. Flow is regulated with checks and check structures, division boxes, and gates in open channels and various types of valves in pipelines.

Flow measuring and regulating devices should be located in the diversion facilities to measure and control system inflow, where flows are divided between and within fields, and at other criticial points throughout the system. Drainage water leaving each field can also be measured.

2.4   Designing Farm Irrigation Systems

Farm irrigation systems are designed to match the physical and economic setting in which they are to operate. The physical setting is determined by climate, soils, topography, and the location of buildings, rivers and streams, roads, etc. Such factors as the cost of land, the existence of markets, production costs (for fertilizers, seed, herbicides, etc.), the cost of energy and water for irrigation, the landowner's personal debt situation, and crop prices establish the economic setting of the farm.

The job of the designer is to match the system to the physical and economic setting in which it is to operate. The primary steps in farm irrigation system design include

1. assembling data needed for design,
2. identifying and evaluating a water source,
3. determining the design daily irrigation requirement (DDIR),
4. designing alternative systems for the farm,
5. evaluating the performance of alternative system designs,
6. determining the annual cost of alternative system designs, and
7. selecting the most suitable system design.

These design steps are described in the following sections.

2.4.1   Data Requirements for Design

An essential part of the design process is quantifying the physical and economic setting of the farm. The success of the project may depend on this important and often time-consuming task. The time and effort spent gathering design data should not be underestimated, since it can exceed the time spent designing the system.

Table 2.1 lists the principal data required to design a farm irrigation system. In the United States many of these data are routinely collected and published.

Table 2.1   Principal Data Needed for Farm Irrigation System Design

Data	Specific Requirements
Climate	Several years of temperature, relative humidity, wind, or solar radiation data for estimating daily irrigation requirements for each crop (Precise climatic requirements depend on the ET method used)
Crop	Areal distribution and amount (area) of each crop to be grown; suitability of each crop to climate, soils, farming practices, markets, etc.; K_c values, planting dates, etc., for each crop to be grown over the expected life of the project
Soils	Areal distribution of soils; water holding and infiltration characteristics, depth, drainage requirements, salinity, erodibility of each soil.
Water supply (Chapter 3 details water supply data and analysis)	Location of water source; water surface elevation; hydrologic and water quality information for assessing the availability and suitability of the water for irrigation; water right information
Energy source	Location, availability, and type of source(s); cost information
Capital and labor	Capital available for system development and availability, level of technical skill, and cost of labor
Other	Topographic map showing location of roads, buildings, drainways, and other physical features that influence design; financial situation of farmer, farmer preferences.

Climatic data can be obtained from the National Oceanic and Atmospheric Administration (NOAA) for thousands of locations throughout the United States. For some locations all the data needed for Penman-type ET calculations are recorded, while only daily temperature and precipitation data are available for other locations. Soils data are published by the U.S. Department of Agriculture Soil Conservation Service (SCS) for most U.S. locations and are available from the SCS, universities, and libraries. Irrigation requirement information and system design recommendations are also available from the SCS. The U.S. Department of Interior Geological Survey monitors and publishes water supply and quality data for the principal surface and ground water sources of the United States.

      Other important sources of data include agricultural scientists such as agricultural engineers, agronomists, horticulturalists, economists, and soil scientists. One of the most important sources of data is field measurements. Legal assistance may also be required to resolve water right conflicts and other legal questions.

2.4.2 Water Source Evaluation

Identifying a reliable water source is a prerequisite to successful irrigation. Irrigation water is obtained from a variety of surface and ground water sources. Surface sources include streams, lakes, and canals, while wells and springs are the principal ground water sources.

A detailed analysis of legal, hydrologic, and water quality factors is required to determine the total volume and volumetric flowrate (volume per unit time) that a water source can reliably supply year after year. The suitability of the water source is determined by comparing these values to the irrigation requirements of the farm. In situations where the expense of obtaining water or its limited availability does not allow full irrigation of the farm, strategies for partially irrigating the farm should be considered. Partial irrigation strategies include restricting the area of the farm that is fully irrigated and/or limiting the depth of water applied (deficit irrigation).

Surface and ground water sources of water are discussed in Chapter 3. Other topics covered in Chapter 3 include the hydrologic, water quality, economic and legal factors that determine the suitability of a water source for irrigation, procedures for determining the total volume and volumetric flowrate that surface and ground waters can reliably supply, and a discussion of water quality and its effect on the suitability of water for irrigation.

2.4.3 Determining the Design Daily Irrigation Requirement

The design daily irrigation requirement (DDIR) is usually the rate at which an irrigation system must supply water to achieve the desired level of irrigation. In some situations, however, the largest daily irrigation requirement is associated with land preparation (like rice paddy formation) and not evapotranspiration (ET). The remainder of this section assumes that DDIR is determined by ET.

DDIR has dimensions of length per unit of time. Typical units for DDIR are inches per day or gallons per minute per acre (gpm/ac) in the English system and millimeters per day or liters per minute per hectare (l/min/ha) in the SI system.

The DDIR for an irrigation system varies with the crops, climate, and soils of the farm. DDIR values are largest for crops that have relatively shallow rooting depths, are sensitive to water stress, and/or use water rapidly. Farms located in climates with high daily ET rates and low precipitation have the largest DDIR. Several years of climatic data are required to quantify year-to-year variations in daily ET and precipitation and to properly evaluate DDIR.

Generally, DDIR's for crops grown in soils with low water holding capacities, such as sands, are higher than those for crops grown in finer textured soils with higher water holding capacities. This is because the interval between irrigations (i.e., the irrigation interval) increases with water holding capacity and the average daily irrigation requirement is smallest for longer irrigation intervals.

The DDIR value for a farm is determined from several years of daily irrigation requirement (DIR) data. DIR's for each year of climatic record are usually computed with one of the Penman-type equations or pan evaporation and Eqs. 1.8 and 1.25 with $E_i = 100$ percent. DDIR is normally less than the peak DIR, since some of the water needed to meet the peak DIR can normally be obtained from the soil. In situations where no water can be obtained from the soil, DDIR equals the peak DIR.

DDIR is determined using Eq. 2.1.

$$DDIR = \frac{AD}{II_{min}}$$

(2.1)

where

AD = allowed depletion of soil water between irrigations (mm, in);
II_{min} = minimum irrigation interval during the irrigation season (days).

Although AD normally equals RAW, AD may exceed RAW for deficit irrigation strategies. The use of Eq. 2.1 is illustrated in Example 2.1

EXAMPLE 2.1   Determining DDIR When RAW Exceeds the Peak DIR

Given:
• daily irrigation requirement data in Figure 2.8
• RAW = 150 mm
Required:
DDIR
Solution:
from Figure 2.8 II_{min} = 14 days

$$DDIR = \frac{RAW}{II_{min}} = \frac{150 \text{ mm}}{14 \text{ days}} = 10.7 \text{ mm/day}$$

The design daily irrigation requirement for a farm, $DDIR_f$, is determined by computing the cumulative irrigation requirement for the farm using the procedure illustrated in Example 2.1. The farm's cumulative irrigation requirement is computed by summing the daily irrigation requirement of the crops grown on the farm. The farm's daily irrigation requirement is computed with

$$(DIR_f)_j = \frac{\sum_{i=1}^{n} (A_i)(DIR_i)_j}{\sum_{i=1}^{n} A_i}$$

(2.2)

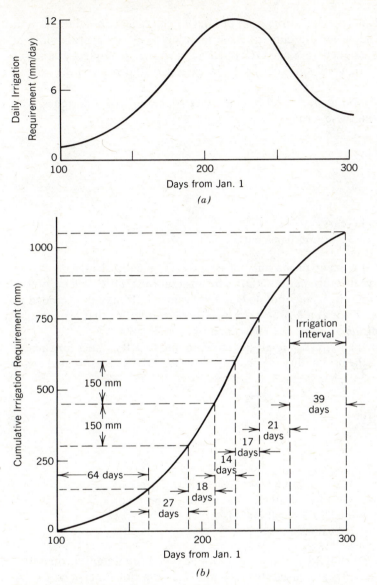

Figure 2.8   (*a*) Daily and (*b*) cumulative irrigation requirements.

where

DIR_f = daily irrigation requirement for the farm (mm/day, in/day);
DIR_i = daily irrigation requirement for crop i (mm/day, in/day);
  A_i = area of crop i (ha, acres);
   n = number of crops grown on farm;
   j = day of growing season.

A frequency analysis of several years of $DDIR_f$ values is required to account for year-to-year fluctuations in climate. Such an analysis allows a probability of occurrence to be assigned to each $DDIR_f$. For example, a frequency analysis enables the $DDIR_f$ that will, on the average, be exceeded 10 percent of the time to be determined.

The return period is often used in lieu of the probability of occurrence. The relationship between these terms is given in Eq. 2.3.

$$RP = \frac{100}{P} \tag{2.3}$$

where

PR = return period (years);
 P = probability of occurrence (percent).

Using Eq. 2.3, a 20 percent probability of occurrence is equivalent to a 5-year return period. A 5-year return period $DDIR_f$ means that the $DDIR_f$ will be, on the average, exceeded once in 5 years but does not guarantee it. It may be exceeded in each of the 5 years or not at all. A 5-year return period indicates that historically, the $DDIR_f$ has, on the average, been exceeded once in 5 years.

The first step in a frequency analysis is to compute $DDIR_f$ values for each of several years. Next, the probability of occurrence of each $DDIR_f$ is estimated using Eq. 2.4.

$$P = \left(1 - \frac{R}{M+1}\right)100 \tag{2.4}$$

where

 P = probability that a given value will be exceeded in percent;
 R = rank of $DDIR_f$ on a list of $DDIR_f$ values in ascending order
     (R for the smallest $DDIR_f$ value $= 1$);
 M = number of $DDIR_f$ values.

A plot of P versus $DDIR_f$ or an extreme value type I (minimum) probability distribution (Haan, 1977) is used to smooth the data for interpolation. A probability distribution transforms P so that a linear relationship between the transform of P and $DDIR_f$ results. The Weibull transformation of P is

$$W = \log\left[-\log\left(\frac{P}{100}\right)\right] \tag{2.5}$$

where W is the Weibull transform of P.

The following example illustrates a frequency analysis of $DDIR_f$ data using Eqs. 2.4 and 2.5.

EXAMPLE 2.2   A Frequency Analysis to Determine Design Daily Irrigation Requirements for a Farm (DDIR$_f$) for Various Return Periods

Given:
• 22 years of DDIR$_f$ values

Required:
DDIR$_f$ values that will be exceeded 50, 20, 10, and 5 percent of the time (i.e., for return periods of 2, 5, 10, and 20 years, respectively).

Solution:

Solution Steps
1. arrange DDIR$_f$ data in ascending order,
2. compute P for each DDIR$_f$ using Eq. 2.4.
3. compute W for each DDIR$_f$ using Eq. 2.5,
4. plot W versus DDIR$_f$,
5. compute W values for P values 50, 20, 10, and 5 percent,
6. read DDIR$_f$ values from plot for W values corresponding to P values of 50, 20, 10, and 5 percent (2, 5, 10, and 20 year return periods).

The following table summarizes solution steps 1–3.

DDIR$_f$ (mm)	Rank (R)	P	RP (years)	W
7.1	1	95.65	1.04	−1.71
7.4	2	91.30	1.10	−1.40
7.9	3	86.96	1.15	−1.22
8.1	4	86.61	1.21	−1.08
8.4	5	78.26	1.28	−0.97
8.4	6	73.91	1.35	−0.88
8.6	7	69.56	1.44	−0.80
8.9	8	65.22	1.53	−0.73
8.9	9	60.87	1.64	−0.67
8.9	10	56.52	1.77	−0.61
9.1	11	52.17	1.92	−0.55
9.1	12	47.83	2.09	−0.49
9.1	13	43.48	2.30	−0.44
9.4	14	39.13	2.56	−0.39
9.7	15	34.78	2.88	−0.34
9.7	16	30.44	3.29	−0.29
9.7	17	26.09	3.83	−0.23
9.9	18	21.74	4.60	−0.18
9.9	19	17.39	5.75	−0.12
10.2	20	13.04	7.67	−0.05
10.2	21	8.70	11.50	0.03
10.9	22	4.35	23.00	0.13

Table 2.2 Design Daily Irrigation Requirement Values at Selected U.S. Locations for Various Crops Grown on Deep, Medium-Textured, Moderately Permeable Soils

	Washington (Columbia Basin)		California (San Joaquin Valley)		Texas (Southern High Plains)		Arkansas (Mississippi Bottoms)		Nebraska (Eastern Part)		Colorado (Western Part)	
	mm/day	in/day	mm/day	in/day	mm/day	in/day	mm/day	in/day	mm/day	in/day	mm/day	in/day
Corn	6.9	0.27	6.6	0.26	7.6	0.30	5.8	0.23	7.1	0.28	5.8	0.23
Alfalfa	6.4	0.25	6.4	0.25	7.6	0.30	6.1	0.24	6.9	0.27	5.8	0.23
Pasture	7.4	0.29	8.1	0.32	6.4[b] 7.6[c]	0.25[b] 0.30[c]	3.3[b] 5.6[c]	0.13[b] 0.22[c]	7.4	0.29	5.8	0.23
Grain	5.3	0.21	4.3	0.17	3.8	0.15	3.8	0.15	6.6	0.26	5.6	0.22
Sugar beets	6.6	0.26	5.6	0.22					6.6	0.26	5.1	0.20
Cotton			5.6	0.22	6.4	0.25	4.6	0.18				
Potatoes	7.4	0.29	6.1	0.24					6.6	0.26	5.6	0.22
Deciduous orchards			5.3	0.21							4.6	0.18
Citrus orchards			4.8	0.19								
Grapes			4.6	0.18								
Annual legumes					4.6	0.18	7.1	0.28				
Soybeans							4.8	0.19	6.9	0.27		
Shallow truck							5.1[d]	0.20[d]				
Medium truck							3.0[e]	0.12[e]				
Deep truck												
Tomatoes											5.6	0.22
Tobacco												
Rice							4.3	0.17				

	Wisconsin (State)		Indiana (State)		Piedmont Plateau[a]		Virginia (Coastal Plain)		New York (State)	
	mm/day	in/day	mm/day	in/day	mm/day	in/day	mm/day	in/day	mm/day	in/day
Corn	7.6	0.30	7.6	0.30	5.6	0.22	4.6	0.18	5.1	0.20
Alfalfa	7.6	0.30	7.6	0.30	6.4	0.25	5.6	0.22	5.1	0.20
Pasture	5.1	0.20	7.6	0.30	6.4	0.25	5.6	0.22		
Grain	6.4	0.25			4.1	0.16				
Sugar beets	6.4	0.25								
Cotton					5.3	0.21				
Potatoes	5.1	0.20	6.4	0.25	4.6	0.18	4.6	0.18	4.6	0.18
Deciduous orchards	7.6	0.30			6.4	0.25	5.6	0.22	5.1	0.20
Citrus orchards										
Grapes			6.4	0.25	5.1	0.20				
Annual legumes										
Soybeans	6.4	0.25	7.6	0.30	4.6	0.18			4.6	0.18
Shallow truck	5.1	0.20	5.1	0.20	3.6	0.14			4.6	0.18
Medium truck	5.1	0.20	5.1	0.20	3.6	0.14	4.1	0.16	4.6	0.18
Deep truck	5.1	0.20	5.1	0.20	4.6	0.18			4.6	0.18
Tomatoes	5.1	0.20	5.1	0.20	5.3	0.21	4.6	0.18	4.6	0.18
Tobacco			6.4	0.25	4.6	0.18	4.3	0.17		
Rice										

Source: Adopted from Chapter 1, Section 15 of *SCS National Engineering Handbook.*

[a] Parts of Georgia, Alabama, North Carolina, and South Carolina.
[b] Cool-season pasture.
[c] Warm-season pasture.
[d] Summer.
[e] Fall

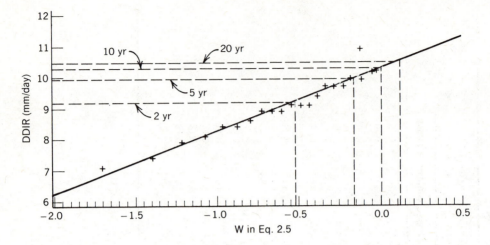

Solution step 5:

W for P = 50% = −0.52
    P = 20% = −0.16
    P = 10% = 0
    P = 5% = 0.11

Solution step 6:

$DDIR_f$ for P = 50% (RP = 2 yr) = 9.2 mm/day
      P = 20% (RP = 5 yr) = 9.9 mm/day
      P = 10% (RP = 10 yr) = 10.2 mm/day
      P = 5% (RP = 20 yr) = 10.4 mm/day

DDIR values for various crops and location throughout the United States have been developed and published by the Soil Conservation Service (SCS). Table 2.2 summarizes some of these data. The SCS has also developed an equation for estimating DDIR values from peak monthly evapotranspiration for various values of AD. This equation is

$$DDIR = (0.034)\frac{ET_m^{1.09}}{AD^{0.09}}$$ (2.6)

where

DDIR = design daily irrigation requirement (mm/day, inches/day)
  ET_m = average total evapotranspiration for the peak month (mm, in)
  AD = soil water depletion allowed between irrigations (mm, in)

2.4.4   Alternative Designs

Normally, there are several alternative system types and configurations that will satisfactorily irrigate a farm. The designer identifies these alternatives and develops detailed designs for the most feasible systems. Each system should provide flexibility for future expansion or changes in management objectives. Some alternatives may be based on full irrigation, while others may involve deficit irrigation strategies.

Identification of alternative systems begins with the selection of an application method. This choice is influenced by landowner preferences, and the physical and economic setting of the farm. Table 2.3 lists major factors that affect application method selection. In situations where this choice is not obvious, each potentially feasible aplication method is considered in subsequent steps of the design process.

Next, application and conveyance subsystems are located (i.e., laid out) according to farm geometry and terrain. Several alternative layouts may be possible. After eliminating some alternatives by inspection, detailed designs are developed for the remaining layouts.

The hydraulic design of the system begins with the application system, progresses through the conveyance facilities, and ends with the diversion subsystem. During design, sprinklers, emission devices, or furrow shape, spacing, slope, length, and streamsize selection is followed by the determination of pipeline or canal specifications. Pumping plant components and diversion structure specifications are considered last.

2.4.5   Performance of Farm Irrigation Systems

Farm irrigation systems are designed and operated to supply the individual irrigation requirements of each field on the farm while controlling deep percolation, runoff, evaporation, and operational losses. The performance of a farm irrigation system is determined by the efficiency with which water is diverted, conveyed, and applied, and by the adequacy and uniformity of application in each field on the farm. Each of these performance parameters (efficiency, uniformity, and adequacy) is considered in one of the following sections.

2.4.5a   Efficiency

The overall *efficiency* of a farm irrigation system is defined as the percent of water supplied to the farm that is beneficially used for irrigation on the farm. Overall system efficiency, also known as the *irrigation efficiency*, is defined mathematically by Eqs. 2.7 and 2.8.

$$E_i = 100\left(\frac{I + L}{S}\right) \tag{2.7}$$

or

$$E_i = 100\left(\frac{S - DP - RO - O}{S}\right) \tag{2.8}$$

Table 2.3 Comparison of Irrigation Systems in Relation to Site and Situation Factors

Site and Situation Factors	Redesigned Surface Systems	Level Basins	Intermittent Mechanical Move	Continuous Mechanical Move	Solid Set and Permanent	Emitters and Porous Tubes
Infiltration rate	Moderate to low	Moderate	All	Medium to high	All	All
Topography	Moderate slopes	Small slopes	Level to rolling	Level to rolling	Level to rolling	All
Crops	All	All	Generally shorter crops	All but trees and vineyards	All	High value required
Water supply	Large streams	Very large streams	Small streams nearly continuous	Small streams nearly continuous	Small streams	Small streams, continuous and clean
Water quality	All but very high salts	All	Salty water may harm plants	Salty water may harm plants	Salty water may harm plants	All can potentially use high salt waters
Efficiency	Average 60–70%	Average 80%	Average 70–80%	Average 80%	Average 70–80%	Average 80–90%
Labor requirement	High, training required	Low, some training	Moderate, some training	Low, some training	Low to seasonal high, little training	Low to high, some training
Capital requirement	Low to moderate	Moderate	Moderate	Moderate	High	High
Energy requirement	Low	Low	Moderate to high	Moderate to high	Moderate	Low to moderate
Management skill	Moderate	Moderate	Moderate	Moderate to high	Moderate	High
Machinery operations	Medium to long fields	Short fields	Medium field length, small interference	Some interference circular fields	Some interference	May have considerable interference
Duration of use	Short to long	Long	Short to medium	Short to medium	Long term	Long term, but durability unknown
Weather	All	All	Poor in windy conditions	Better in windy conditions than other sprinklers	Windy conditions reduce performance; good for cooling	All
Chemical application	Fair	Good	Good	Good	Good	Very good

Source: G. O. Schwab, R. K. Frevert, T. W. Edminster, and K. K. Barnes, *Soil and Water Conservation Engineering*, copyright © 1981 by John Wiley & Sons, Inc., New York, pp. 430–431. Reprinted by permission of John Wiley & Sons, Inc.

where

E_i = irrigation efficiency (percent);
  I = irrigation requirement;
  L = leaching requirement (see Section 3.3.1c);
  S = amount of water supplied to the farm;
DP = total deep percolation on farm;
RO = total runoff from farm;
  O = operational losses due to planned and accidental spillage
       from open channels and pipelines

The following example illustrates the use of Eqs. 2.8 to compute the efficiency of an irrigation.

EXAMPLE 2.3 Computing the Efficiency of an Individual Irrigation

Given:
- 1900 l/min is diverted onto farm each day (24 h)
- each day 0.6 ha of corn and 1.0 ha of alfalfa are irrigated
- RAW for corn is 8 cm
- RAW for alfalfa is 15 cm
- assume that water is uniformly applied over each field (i.e., that the entire corn field receives 8 cm and the entire alfalfa field receives 15 cm of water)

Required:
irrigation efficiency for the farm

Solution:
  use Eq. 2.7 with $L = 0$

$$S = (1900 \text{ L/min})\left(\frac{60 \text{ min}}{h}\right)(24 \text{ h})\left(\frac{m^3}{10000 \text{ l}}\right) = 2736 \text{ m}^3$$

$$I = ((8 \text{ cm})(0.6 \text{ ha}) + (15 \text{ cm})(1.0 \text{ ha}))\left(\frac{1 \text{ m}}{100 \text{ cm}}\right)\left(\frac{10000 \text{ m}^2}{ha}\right) = 1980 \text{ m}^3$$

$$E_i = 100\left(\frac{1980}{2736}\right) = 72.4\%$$

Thus, 72.4 percent of the water supplied to the farm is used beneficially for irrigation and 27.6 percent is lost as deep percolation, runoff, evaporation, and/or spillage. The actual E_i for the farm will probably not be 72.4 percent, however, since it is unlikely that water application would be perfectly uniform as was assumed in the example. Nonuniform application will affect the amount of losses by increasing deep percolation and runoff in overirrigated areas of each field and reducing these losses in under irrigated areas.

When evaluating the performance of a farm irrigation system it is often useful to examine the efficiency of each system component. This allows components that are not performing well to be identified. The following sections define reservoir storage, conveyance, and application efficiencies. The overall system efficiency is the product of these efficiencies as in Eq. 2.9.

$$E_i = \left(\frac{E_r}{100}\right)\left(\frac{E_c}{100}\right)\left(\frac{E_a}{100}\right)(100) \tag{2.9}$$

where

E_i = irrigation efficiency in percent;
E_r = reservoir storage efficiency in percent;
E_c = conveyance efficiency in percent;
E_a = application efficiency in percent.

     (i) Reservior Storage Efficiency   The efficiency with which water is stored in a reservoir is reduced by evaporation and seepage losses. Equation 2.10 defines reservoir storage efficiency.

$$E_r = 100\left(1 - \frac{V_s + V_e}{V_i}\right) = 100\left(\frac{V_o + \Delta S}{V_i}\right) \tag{2.10}$$

E_r = reservoir storage efficiency in percent;
V_e = evaporation volume from the reservoir;
V_s = seepage volume from the reservoir;
V_i = inflow to the reservoir during a time interval;
V_o = outflow volume from the reservoir during a time interval;
ΔS = change in reservoir storage during the time interval, that is, amount of water needed to maintain the water surface in the reservoir at the level that existed at the beginning of the time interval. (ΔS is negative when water must be added to the reservoir, and positive when water must be removed.)

     The ΔS term is often neglected when long time periods are considered. This term should not, however, be neglected for short time periods.

EXAMPLE 2.4   Computing Reservoir Storage Efficiency

Given:
3220 l/min are being turned into a reservoir
$\Delta S = 380$ m^3 (water must be removed to restore initial water level in reservoir)

Required:
reservoir storage efficiency for a 24 h period during which 2650 l/min are being diverted from the reservoir

Solution:

$$V_i = (3220 \text{ l/min})(24 \text{ h})(60 \text{ min/h})\left(\frac{1 \text{ m}^3}{1000 \text{ l}}\right) = 4637 \text{ m}^3$$

$$V_o = (2650 \text{ l/min})(24 \text{ h})(60 \text{ min/h})\left(\frac{1 \text{ m}^3}{1000 \text{ l}}\right) = 3816 \text{ m}^3$$

$$E_r = 100\left(\frac{3816 + 380}{4637}\right) = 90.5\%$$

(ii) Conveyance Efficiency   Water conveyance efficiency (E_c) is the ratio, in percent, of the amount of water delivered by a canal or pipeline to the amount of water delivered to the conveyance system. Efficiency E_c is computed using Eq. 2.11.

$$E_c = 100\left(\frac{V_{co}}{V_{ci}}\right) \tag{2.11}$$

where

E_c = conveyance efficiency in percent;
V_{co} = volume of water delivered by conveyance system (i.e., outflow);
V_{ci} = volume of water delivered to the conveyance system (i.e., inflow).

EXAMPLE 2.5   Computing Conveyance Efficiency

Given:
• 2650 l/min of water is being turned into an unlined canal from the reservoir in Example 2.4
• 96 furrows are required to irrigate a field
• the inflow rate to 26 of the furrows is 19 l/min
• the inflow rate to 70 of furrows is 27 l/min

Required:
conveyance efficiency

Solution:

$$E_c = 100\left(\frac{(26)(19)(t) + (70)(27)(t)}{(2650)(t)}\right)$$

where t = duration of each irrigation

$$E_c = 100\left(\frac{2384}{2650}\right) = 90.0\%$$

(iii) Application Efficiency   Water application efficiency for an irrigated area (E_a) is the ratio, expressed in percent, of the volume of water beneficially used by the crop to the volume of water delivered to the area. Application efficiency can be computed for each field of the farm or for the entire farm. Efficiency E_a is computed using Eq. 2.12:

$$E_a = 100\left(\frac{V_{bu}}{V_a}\right) = 100\left(\frac{I + L}{V_a}\right) \tag{2.12}$$

where

E_a = application efficiency in percent;
$\forall_{bu}$ = volume of water beneficially used by crop(s) in an area;
$\forall_a$ = volume of water applied in an area;
  I = irrigation requirement for the area;
  L = leaching requirement for the area.

EXAMPLE 2.6   Computing Application Efficiency

Given:
- each day 0.6 ha of corn and 1.0 ha of alfalfa are irrigated
- RAW for corn is 8 cm
- RAW for alfalfa is 15 cm
- corn is irrigated with 26 furrows each discharging 19 l/min
- alfalfa is irrigated with 70 furrows each discharging 27 l/min
- neglect leaching (L = 0)
- assume that water is uniformly applied over each field (see Example 2.3)

Required:
Application efficiency for

a. corn
b. alfalfa
c. farm

Solution:
a. E_a for corn

$$\forall_{bu} = (0.6 \text{ ha})(8 \text{ cm})\left(\frac{1 \text{ m}}{100 \text{ cm}}\right)\left(\frac{10000 \text{ m}^2}{\text{ha}}\right) = 480 \text{ m}^3$$
$$\forall_a = (26 \text{ furrows})(19 \text{ l/min/furrow})(24 \text{ h})(60 \text{ min/h})(1 \text{ m}^3/1000 \text{ l})$$
$$= 711 \text{ m}^3$$
$$E_a = 100\left(\frac{480}{711}\right) = 67.5\%$$

b. E_a for alfalfa

$$\forall_{bu} = (1.0 \text{ ha})(15 \text{ cm})\left(\frac{1 \text{ m}}{100 \text{ cm}}\right)\left(\frac{10000 \text{ m}^3}{\text{ha}}\right) = 1500 \text{ m}^3$$
$$\forall_a = (70 \text{ furrows})(27 \text{ l/min})(24 \text{ h})(\text{ha})(60 \text{ min/h})(1 \text{ m}^3/1000 \text{ l})$$
$$= 2722 \text{ m}^3$$
$$E_a = 100\left(\frac{1500}{2722}\right) = 55.1\%$$

c. E_a for farm

$$E_a = 100\left(\frac{480 + 1500}{711 + 2722}\right) = 57.7\%$$

EXAMPLE 2.7   Overall Irrigation Efficiency for the Farm Irrigation System in Examples 2.4, 2.5 and 2.6

Given:
E_r, E_i and E_a values from Examples 2.4, 2.5, and 2.6

Required:
E_i for farm

Solution:
Using Eq. 2.9

$$E_i = \left(\frac{90.5}{100}\right)\left(\frac{90.0}{100}\right)\left(\frac{57.7}{100}\right)(100) = 47.0\%$$

Thus, 47.0 percent of the water delivered to the reservoir is beneficially used by the crop.

2.4.5b   Application Uniformity

The uniformity of application describes how evenly an application system distributes water over a field. The uniformity of application is evaluated using the Christiansen uniformity coefficient (C_u). C_u is computed using the following equation:

$$C_u = 100\left(1.00 - \frac{\sum\limits_{i=1}^{n} |d_i|}{\forall_T}\right) \tag{2.13a}$$

$$d_i = X_i A_i - \bar{\forall}_i$$

$$\forall = \frac{\forall_T}{n}$$

$$\forall_T = \sum_{i=1}^{n} (A_i)(X_i)$$

where

X_i = depth/volume caught/infiltrated at observation point i;
A_i = field area represented by observation point i;
n = number of observation points.

When application is perfectly uniform, C_u equals 100 percent.

When the areas represented by each observation point are equal, Eq. 2.13a becomes:

$$C_u = 100\left(1.00 - \frac{\sum |d|}{n\bar{X}}\right) \tag{2.13b}$$

$$d = X_i - \bar{X}$$

where

n = number of observations
$\bar{X}$ = average depth/volume amount caught/infiltrated.

The coefficient C_u for sprinkle systems is often evaluated using a grid of catch cans. The volume caught in each can is divided by the area of the can opening to calculate the depth of catch. When catch cans are not used or when the uniformity of surface application methods is being considered, the amount of infiltration at each observation point is used (rather than cup catch) to compute C_u. For trickle systems, the volume of water discharged in a specified interval of time at several emission device locations is used.

When numerous observation points are being utilized to evaluate sprinkle or trickle system uniformity and the distribution pattern is nearly normal, C_u can be estimated using Equation 2.14:

$$C_u = 100 - 80.0 \frac{S}{\bar{X}} \qquad (2.14)$$

where

S = standard deviation of the observations;
$\bar{X}$ = average depth/volume caught/infiltrated.

Equation 2.14 is not recommended for use with surface systems, since their distribution patterns are seldom normally distributed.

Distribution uniformity (DU) is another index of application uniformity. DU is the ratio, expressed in percent, of the average low-quarter amount caught/infiltrated to the average amount caught/infiltrated. DU is defined by Eq. 2.15.

$$DU = 100 \frac{\bar{X}_{LQ}}{\bar{X}} \qquad (2.15)$$

where

$\bar{X}_{LQ}$ = low-quarter average-depth/volume amount caught/infiltrated;
$\bar{X}$ = average amount depth/volume caught/infiltrated.

EXAMPLE 2.8   Computing Application Uniformity

Given:
depths of infiltration in centimeters around a field

4.0	3.5	3.4	3.7
3.9	3.3	3.4	3.5
2.6	2.8	2.7	3.2
3.7	3.0	2.8	2.6
4.0	3.5	3.2	4.3

Required:

a. C_u

b. DU

Solution:

a. C_u

$$\bar{X} = 3.36 \text{ cm}$$
$$S = 0.50 \text{ cm}$$
$$\sum|d| = 7.30 \text{ cm}$$

C_u using Eq. 2.13

$$C_u = 100 \left(1.00 - \frac{7.30}{20(3.36)} \right) = 89.3\%$$

C_u using Eq. 2.14

$$C_u = 100 - 80.0 \left(\frac{0.47}{3.36} \right) = 88.8\%$$

b. DU
lowest 5 of catches: 2.6, 2.6, 2.7, 2.8, 2.8 cm

$$\bar{X}_{LQ} = 2.70$$

$$DU = 100 \left(\frac{2.70}{3.36} \right) = 80.4\%$$

2.4.5c   Adequacy of Irrigation

The adequacy of irrigation is the percent of the field receiving sufficient water to maintain the quantity and quality of crop production at a "profitable" level. Since this definition requires crop, soil, and market conditions to be specified, adequacy is normally defined to be the percent of the field (farm) receiving the desired amount of water or more.

The adequacy of irrigation is evaluated using a cumulative frequency distribution like the one in Figure 2.9. This figure shows the percent of the field (farm) receiving a given amount of water or more. The dashed line in the figure is the desired depth of application. The adequacy of irrigation for the field (farm) in Figure 2.9 is 50 percent, since 50 percent of the field receives the desired depth of application or more.

Cumulative frequency distribution patterns like the one in Figure 2.9 are constructed by determining the amount of water caught/infiltrated at locations

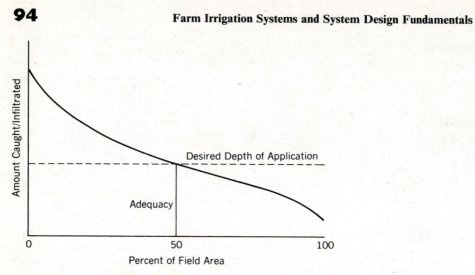

Figure 2.9 Cumulative frequency distribution pattern for determining the adequacy of irrigation.

around the field (farm) and the percent of the total area represented by each location. The amounts are then arranged in descending order and the percent of the field (farm) receiving each amount or more computed. These values are plotted as in Figure 2.9. Example 2.9 illustrates this procedure for determining the adequacy of irrigation for the field in Example 2.8.

EXAMPLE 2.9   Determining the Adequacy of Irrigation for the Field in Example 2.8.

Given:
• infiltrated depths from Example 2.8
• full irrigation = 3.25 cm

Required:
adequacy of irrigation

Solution:

Solution Steps
1. arrange depths in descending order,
2. compute percent of field represented by each depth,
3. compute cumulative area for each depth,
4. plot cumulative area versus depth,
5. determine adequacy from plot.

Steps 1, 2 and 3 calculations are summarized in the following table and plotted in the graph below.

Infiltrated Depth (cm)	Percentage of Field	Cumulative Percentage of Field
4.3	5	5
4.0	5	—
4.0	5	15
3.9	5	20
3.7	5	—
3.7	5	30
3.6	5	35
3.5	5	—
3.5	5	—
3.5	5	50
3.4	5	—
3.4	5	60
3.3	5	65
3.2	5	—
3.2	5	75
3.0	5	80
2.8	5	—
2.8	5	90
2.7	5	95
2.6	5	100

From the graph, adequacy = 67 percent.

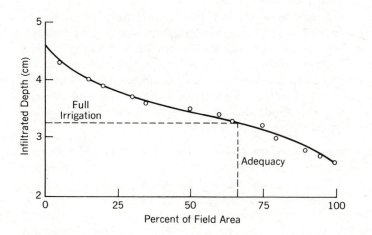

When the desired depth of irrigation fills the soil to field capacity, a term called the storage efficiency is often used as an index to adequacy. The term E_s is computed using Eq. 2.16:

$$E_s = 100 \left(\frac{S_{rz}}{S_{fc}}\right) \qquad (2.16)$$

where

S_{rz} = amount of water stored in the root zone during the irrigation;
S_{fc} = amount of water to fill the root zone to field capacity

EXAMPLE 2.10   Computing the Storage Efficiency of an Irrigation

Given:
- infiltration depths from Example 2.8
- average preirrigation water content for the field is 18 percent by volume
- root zone of crop is 60 cm deep
- field capacity is 25 percent by volume

Required:
storage efficiency

Solution:
S_{rz} = average depth infiltrated = 3.36 cm
Compute S_{fc} using Eq. 1.30 with $S_{fc} = IRRI$

$$S_{fc} = 60 \left(\frac{25 - 18}{100}\right) = 4.20 \text{ cm}$$

$$E_s = 100 \left(\frac{3.36}{4.20}\right) = 80.0\%$$

2.4.5d   Effectiveness of Irrigation

The *effectiveness of irrigation* is a term that qualitatively describes the application efficiency, uniformity, and adequacy of irrigation. The desired effectiveness of irrigation (i.e., the desired combination of efficiency, uniformity, and adequacy) maximizes net farm profit. Irrigations with the highest application efficiencies, uniformities, and adequacies are not always desirable, since they do not always maximize net farm profit. An understanding of the relationship between application efficiency, uniformity, and adequacy is needed to identify irrigation systems and strategies that maximize net farm profit.

(i) Relation Between Uniformity and Application Efficiency   The relation between uniformity and application efficiency is demonstrated using Figure 2.10. Curves A and B are the cumulative frequency distributions for application systems A and B, which are designed and managed to fill the soil to field capacity and to have equal adequacies (of 50 percent). The flatter slope and smaller range of infiltrated depths for curve A indicates that system A applies water more uniformly

than system B. The areas a_1 and a_2 are the amounts of over and under irrigation for system A, respectively, while system B amounts of over and under irrigation are given by areas $a_1 + b_1$ and $a_2 + b_2$, respectively. Since a full irrigation (with either system) brings the soil to field capacity, all overirrigation is lost as deep percolation and/or runoff, (in most situations, runoff does not occur when the ordinate of Figure 2.10 is infiltrated depth). Because these losses are largest for system B, it has the lowest application efficiency. Thus, irrigation system designs and management strategies that improve uniformity can be expected to increase application efficiency when the irrigation fills the soil to field capacity. Improved uniformity will not necessarily increase application efficiency when maximum amounts of catch/infiltration are less than the amount needed to fill the soil to field capacity.

Achieving maximum efficiency does not always maximize net farm profit, since increased initial and operating costs are usually associated with improving system uniformities. The benefits of high application efficiency must therefore be carefully balanced against the higher costs associated with higher uniformities. Maximum net farm profit can be achieved with less than maximum attainable uniformities when water, energy, and fertilizer are plentiful and/or inexpensive or when the amount and quality of irrigation water leaving the farm is not a concern.

(ii) *Relation Between Adequacy and Application Efficiency*   The relation between adequacy and application efficiency is demonstrated using the cumulative frequency distributions in Figure 2.11. The adequacy was decreased from 52 to 16 percent between curves A and B by reducing the depth applied during an irrigation from 3.0 to 2.5 cm while the uniformity remained constant. The area $a + b$ is the amount of deep percolation and runoff resulting from a full irrigation of 3 cm that fills the soil to field capacity, while area b is the loss associated with the 2.5 cm irrigation. Thus, the reduction in adequacy improved the application efficiency. This will be true as long as there are runoff and deep percolation losses.

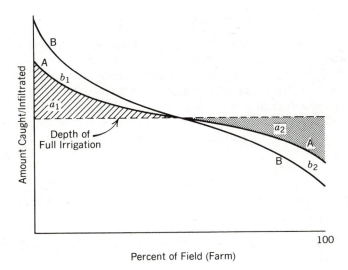

Figure 2.10
Cumulative frequency distributions for two different irrigation systems with different uniformities of application.

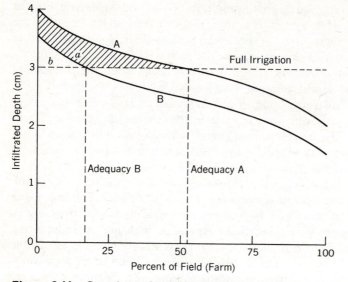

Figure 2.11 Cumulative frequency distributions for irrigations with identical uniformities of application and different application depths.

Improving application efficiency by decreasing the adequacy, however, increases the amount of the field (farm) that is underirrigated and thus, reduces the amount and/or quality of crop produced. Achieving maximum net farm profit in this situation requires balancing the benefits associated with higher efficiency and the losses associated with reduced crop yield and/or quality.

2.4.6 Farm Irrigation System Costs

Another important part of irrigation system design is determining the expected annual cost of owning and operating each feasible alternative design. These data are included in the designer's report and used by landowners to assess the feasibility of irrigating the farm, for selecting the most suitable irrigation system, and in determining the optimum crop mix for the farm. It is utilized by banks, government agencies, and other sources of capital to evaluate the economic soundness of the project and to develop suitable repayment arrangements.

2.4.6a Ownership Costs

Annual ownership costs are often called fixed costs, since they are generally independent of the level of system use. Fixed costs include annual depreciation and interest costs and yearly expenditures for taxes and insurance.

(i) Depreciation Depreciation is the decrease in system value due to age and obsolesensce. Investments that have an indefinite useful life such as water rights and land are not depreciated. The depreciation of a system component that has a finite useful life is the difference between the item's initial cost and its salvage value.

The initial cost of an item is best determined from actual price quotations. In many situations, however, initial costs are estimated by adjusting the initial costs of identical or similar components of previously designed systems to the current date. Cost trend data such as published by Engineering News-Record or the U.S. Bureau of Reclamation are especially useful in making these adjustments.

A component's salvage value is its value at the end of its useful life and may be positive, zero, or negative. Salvage values are negative when additional expenditures are required to inactivate the component at the end of its useful life.

Table 2.4 gives the expected useful life of several irrigation system components. Ranges are listed since useful life can vary significantly depending on the level of repair, operation, and maintenance practices, and the length of time the system is used each year. The smaller values apply to small units and normal operation and maintenance practices. The larger values are suggested for vigorously engineered, carefully constructed and installed items that are thoroughly and deligently maintained. The useful life values in Table 2.4 are based on an average 2000 hours of use per year.

(ii) Interest Costs   Interest is the return from productively invested capital. When money is borrowed to finance the initial cost of the irrigation system, interest is the money paid for the use of the borrowed money. For landowner financed systems, interest costs reflect returns that could be earned if the capital expended for the irrigation system were invested elsewhere.

Interest costs depend on the minimum attractive rate of return (i.e., the interest rate) and the total initial cost of the irrigation system. A systems cost includes the initial cost of all depreciable components and items such as water rights and land, that are not depreciated. Depreciable items in addition to those in Table 2.4 include fuel storage facilities (for internal combustion engine driven pumps), buildings for housing or storing pumps and other equipment, farm road and drainage facility construction, etc.

(iii) Computing Annual Depreciation and Interest Costs   Equation 2.18 is used to compute annual depreciation and interest costs for an irrigation system.

$$\text{ADIC} = \text{CRF} \sum_{j=1}^{\text{NC}} \text{PW}_j \tag{2.18}$$

$$\text{CRF} = \frac{(i)(1+i)^{\text{AP}}}{(1+i)^{\text{AP}} - 1} \tag{2.17a}$$

where

ADIC = annual depreciation and interest costs;
  CRF = capital recovery factor;
    NC = number of system components;
  PW_j = present worth of component j;
      i = annual interest rate (decimal);
    AP = analysis period (years).

Table 2.4 Annual Maintenance and Repairs, and Depreciation Guidelines for Irrigation System Components.

Component	Depreciation (h)	Period (yr)	Annual Maintenance and Repairs Percent[a]
Wells and casings	—	20–30	0.5–1.5
Pumping plant structure	—	20–40	0.5–1.5
Pump, vertical turbine			
Bowls	16,000–20,000	8–10	5–7
Column, etc.	32,000–40,000	16–20	3–5
Pump, centrifugal	32,000–50,000	16–25	3–5
Power transmission			
Gear head	30,000–36,000		5–7
V-belt	6,000	3	5–7
Flat belt, rubber and fabric	10,000	5	5–7
Flat belt, leather	20,000	10	5–7
Prime movers			
Electric motor	50,000–70,000	25–35	1.5–2.5
Diesel engine	28,000	14	5–8
Gasoline engine			
Air cooled	8,000	4	6–9
Water cooled	18,000	9	5–8
Propane engine	28,000	14	4–7
Open farm ditches (permanent)		20–25	1–2
Concrete structure		20–40	0.5–1.0
Pipe, asbestos—cement and PVC buried		40	0.25–0.75
Pipe, aluminum, gated surface		10–12	1.5–2.5
Pipe, steel, waterworks class, buried		40	0.25–0.50
Pipe, steel, coated and lines, buried		40	0.25–0.50
Pipe, steel, coated, buried		20–25	0.50–0.75
Pipe, steel coated, surface		10–12	1.5–2.5
Pipe, steel, galvanized, surface		15	1.0–2.0
Pipe, steel, coated and lined, surface		20–25	1.0–2.0
Pipe, wood, buried		20	0.75–1.25
Pipe, aluminum, sprinkler use, surface		15	1.5–2.5
Pipe, reinforced plastic mortar, buried		40	0.25–0.50
Pipe, plastic, trickle, surface		10	1.5–2.5
Sprinkler heads		8	5–8
Trickle emitters		8	5–8
Trickle filters		12–15	6–9
Landgrazing[b]		none	1.5–2.5
Reservoirs[b]		none	2.0–2.0
Mechanical move sprinklers		12–16	5–8
Continuously moving sprinklers		10–15	5–8

Source: G. T. Thompson, L. B. Spiess, and J. N. Krider, "Farm Resources and System Selection." In *Design and Opteration of Farm Irrigation, Systems*, (1980), M. E. Jensen (Ed.), ASAE Monograph 3, St. Joseph, MI, p. 45. Copyright © 1980 ASAE, pp. 58. Reprinted by permission of ASAE.

[a] Annual maintenance and costs are expressed as a percentage of the initial cost.

[b] Various stages of expected life, from 7–50 years have been applied to land grading and reservoir costs. If adequate maintenance is practiced, these items will remain unaffected by depreciation. For economic analysis, interest on the investment will cover the costs involved. Life may be limited for reservoirs if watershed sedimentation will reduce its usefulness. Costs associated with water rights can also be handled by an interest charge.

The analysis period used for economy studies of on-farm irrigation systems is typically 20, 25, or 30 years. For large complex projects, periods of 40, 50, and 100 years are commonly used (Thompson et al., 1980).

Present worth (PW) is the amount that must be invested at the beginning of the analysis period to return the equivalent of a component's initial cost plus interest by the end of the analysis period. When the analysis period equals the component's useful life, PW is computed using

$$PW = IC - SV \left(\frac{1 + r}{1 + i} \right)^{AP} \tag{2.18}$$

where

PW = present worth of component (dollars);
IC = initial cost of component (dollars);
SV = salvage value of component (dollars);
 r = expected annual rate of cost escalation (decimal);
AP = analysis period (years).

The second term in Eq. 2.18 is the present worth of the salvage value considering the effect of cost escalation.

When the analysis period is shorter than the components useful life, the component will not be fully depreciated at the end of the analysis period. In this case, the final salvage value (at the end of the analysis period) is the sum of the undepreciated and salvage values. Equation 2.19 is used to compute the final salvage value SV_f:

$$SV_f = IC - (IC - SV) \frac{AP}{UL} \tag{2.19}$$

where UL is the useful life of the component in years. Equation 2.19 uses straight line depreciation over the useful life to estimate the undepreciated value at the end of the analysis period. When AP is less than UL, PW is computed using Eq. 2.18, with $SV = SV_f$.

In situations where AP exceeds UL, the component will need to be replaced one or more times during the analysis period. Equation 2.20 is used in such situations.

$$PW = IC + (IC - SV) \left[\sum_{j=1}^{N} \left(\frac{1 + r}{1 + i} \right)^{(j)(UL)} \right] - Z \left(\frac{1 + r}{1 + i} \right)^{AP} \tag{2.20}$$

$$N = \text{integer portion of } \frac{AP - 1}{UL}$$

$$Z = IC - (IC - SV) \left(\frac{AP - (N)(UL)}{UL} \right)$$

The second and third terms are, respectively, the present worth of the component replacement costs and the final salvage value (including any undepreciated value).

Example 2.11 demonstrates the use of Eqs. 2.18, 2.19, and 2.20 to compute annual depreciation and interest costs for a pumping plant.

EXAMPLE 2.11 Computing Annual Depreciation and Interest Costs.

Given:
centrifugal pump electric motor, and steel pipeline

The following information.

Component	Initial Cost ($)	Salvage Value ($)
Pump	2,000	200
Motor	1,000	150
Pipe	6,500	650

Required:
annual depreciation and interest costs for an annual interest rate of 10 percent, an annual cost escalation rate of 5 percent, and a 30 year analysis period

Solution:
useful lives (from Table 2.4) for

Centrifugal pump = 20 years
Electric motor = 30 years
Steel pipe = 40 years

Present worth calculations:

For Pump
(use Eq. 2.20, since AP > UL)

$$N = \text{integer portion of } \frac{30-1}{20} = 1$$

$$Z = \$2000 - (\$2000 - \$200)\frac{30 - (1)(20)}{20} = \$1100$$

$$PW = \$2000 + (\$2000 - \$200)\left(\frac{1.05}{1.10}\right)^{20} - \$1100\left(\frac{1.05}{1.10}\right)^{30}$$

$$= \$2437$$

For Motor
(use Eq. 2.18, since AP = UL)

$$PW = \$1000 - \$150 \left(\frac{1.05}{1.10}\right)^{30}$$

$$= \$963$$

For Pipe
(use Eqs. 2.18 and 2.19, since AP < UL)

$$SV_f = \$6500 - (\$6500 - \$650)\frac{30}{40} = \$2113$$

$$PW = \$6500 - \$2113 \left(\frac{1.05}{1.10}\right)^{30} = \$5977$$

$$ADIC = \left(\frac{(0.1)(1.10)^{30}}{(1.10)^{30} - 1}\right)(\$2437 + \$963 + \$5977)$$

$$= \$995$$

(iii) *Taxes and Insurance*   The annual costs of taxes and insurance are normally obtained from the appropriate taxing entity and insurance companies, respectively. The combined cost for taxes and insurance normally range from 1.5 to 2.5 percent of the initial value of the irrigation system.

2.4.6b   Operating Costs

Annual operating costs include the cost of water, energy, maintenance and repair, and labor. The cost of professional services for such things as irrigation scheduling and fertilizer recommendations should also be included in annual operating costs. The effect of escalating costs can be included by multiplying estimated annual operating costs for the initial year of operation by the equivalent annual cost factor (EACF). EACF is defined by the following:

$$EACF = \left(\frac{(1 + r)^{AP} - (1 + i)^{AP}}{(r - i)}\right)\left(\frac{i}{(1 + i)^{AP} - 1}\right) \tag{2.21}$$

(i) *Annual Water Costs*   In many locations (especially in those served by irrigation districts), irrigators are charged for the water they use. These charges are normally assessed on a volume basis.

(ii) *Annual Energy Costs*   The annual energy cost includes the cost of all energy used to operate the irrigation system. Energy used for pumping, moving equipment within and between fields, injecting fertilizers and other chemicals into the system, etc., must be considered. Energy costs are estimated by calculating the quantity of energy used annually to irrigate the farm and applying the appropriate prices. Procedures for estimating the energy used for pumping, normally the primary use of energy in irrigation, are presented in Section 4.3.6.

(iii) *Annual Maintenance and Repair Costs* Maintenance and repair costs depend on the number of hours the irrigation system operates, the operating environment, and the quality of maintenance. In addition, there is substantial variation in the prices paid for parts and supplies, and in the wages paid repair and maintenance personnel. Annual maintenance and repair costs should therefore be based on local data whenever possible. When local data are not available, annual maintenance and repair costs for an irrigation system component can be approximated as a percentage of the components initial cost. Table 2.4 lists ranges of percent of initial costs that can be used to estimate annual maintenance and repair costs for several irrigation system components. The total annual maintenance and repair costs for the system is the sum of the component costs.

(iv) *Annual Labor Costs* The labor required to operate an irrigation system depends on many factors, including the type of application system, the degree of automation, the crop, the frequency and number of irrigations, and the terrain. Labor requirements are estimated by careful analysis of operations or obtained from actual irrigation with similar conditions and systems.

EXAMPLE 2.12 Computing Total Annual Ownership and Operating Costs for a Pumping Plant

Given:
- the pumping plant in Example 2.11
- information from Example 2.11
- current energy costs of $0.05/kWh are expected to escalate at a rate of 12 percent per year during the analysis period
- total seasonal energy use is expected to be 17,500 kWh·
- neglect water and labor costs

Required:
total annual ownership and operating costs

Solution:

annual depreciation and interest costs (from Example 2.12)	$995
taxes and insurance (2 percent of initial value)	
(0.02) ($2000 + $1000 + $6500)	$190
energy costs ($0.05/kWh) (17,500 kWh) (3.80)[1]	$3325
maintenance and repair	
pump: (6 percent of initial cost) = (0.06) ($2000)	$120
motor: (2 percent of initial cost) = (0.02) ($1000)	$20
pipe: (0.5 percent of initial cost) = (0.005) ($6500)	$33

Total annual ownership and operating costs $4683

$$^1EACF = \left(\frac{(1 + 0.12)^{30} - (1 + 0.10)^{30}}{(1 + 0.12) - (1 + 0.10)}\right)\left(\frac{0.10}{(1 + 0.10)^{30} - 1}\right)$$

$$= 3.80$$

2.4.7 Selecting the Most Suitable System Design

The designer should provide the landowner with specifications and a technical and economic analysis for several alternative systems. Each alternative should be thoroughly explained and discussed with the landowner. The landowner then selects, from the alternative designs prepared by the designer, the one that best satisfies landowners needs, desires, and financial situation. The landowner may also decide that none of the alternatives is acceptable or that irrigation of the farm is not justifiable.

Homework Problems

2.1 Using data from Appendix D determine the design daily irrigation requirement (DDIR) for the following.
   a. grass and
   b. apples with a cover crop

   grown in
   a. sand,
   b. loam, and
   c. clay soil.

   Compare the DDIR values using a bar graph. Plot DDIR on the vertical axis and crop and soil on the horizontal axis.

*2.2 Use the following air temperature data for July (the warmest month) of each year to determine
   a. the range of DDIR values,
   b. the average DDIR value, and
   c. the DDIR value that will, on the average, be exceeded only 10 percent of the time

   for apples with a cover crop grown in a 180-cm-deep sandy loam soil.

---

* Indicates that a computer program will facilitate the solution of the problems so marked.

Year	Mean Temperature[a] (°C)	Year	Mean Temperature[a] (°C)
1956	24.9	1971	22.4
1957	22.2	1972	21.6
1958	25.4	1973	21.0
1959	24.0	1974	20.2
1960	26.1	1975	23.2
1961	25.8	1976	20.2
1962	22.4	1977	22.2
1963	22.6	1978	21.1
1964	22.7	1979	20.8
1965	23.6	1980	19.7
1966	20.2	1981	21.9
1967	22.8	1982	20.3
1968	21.8	1983	20.9
1969	19.3	1984	20.7
1970	21.6	1985	23.2

[a] Mean temperature for July.

The latitude, average minimum relative humidity, wind speed, and ratio of actual to possible sunshine during the irrigation season are 48°N, 25 percent, 2.5 m/s, and 0.85, respectively.

*2.3 Use the data in Appendix D and the computer program developed for Problem 1.36 to determine DDIR for the following.
a. grass,
b. apples with a cover crop, and
c corn

grown in a very deep
a. sand,
b. loam, and
c. clay soil.

Compare these DDIR values with those obtained in Problem 2.1.

2.4 An irrigator plans to deficit irrigate corn. Deficit irrigation will be accomplished by allowing 80 percent (rather than 65 percent) of the available water to be depleted between irrigations. The soil is a 150-cm-deep loam. Use data from Appendix D to determine DDIR for the following.
a. full irrigation, and
b. deficit irrigation.

Compare the values of DDIR for the full irrigation and deficit irrigation strategies. Assume that corn is in growth stage 3 (see Figure 1.7) during July.

*2.5 Use pan evaporation data from Appendix D and K_c information from Example 1.1 to estimate the DDIR for a farm with 50 ha of pasture, 30 ha of corn, and 45 ha of wheat. The predominant soil on the farm is a 100-cm-deep sandy loam.

2.6 A 0.5-ha portion of a corn field is irrigated once a week with a sprinkle irrigation system for 12 hours. Water is applied at a rate of 1000 l/min. There is no runoff. The readily available water holding capacity of the soil for corn is 10 cm. Determine the overall irrigation efficiency.

2.7 6500 l/min is diverted from a stream to irrigate a 25-ha hay field. It takes a week to irrigate the entire field. The readily available water holding capacity of the soil for hay is 15 cm. Estimate the overall irrigation efficiency.

2.8 During periods of peak water use, water is diverted from an irrigation canal into a storage reservoir at a rate of 0.8 m^3/s one day/week (water is distributed to other irrigators along the canal during the remaining days of the week). Water is conveyed from the reservoir to a 50-ha field in a 2000-m long unlined ditch. The field is irrigated continuously during peak water use periods. Determine
a. the reservoir storage efficiency,
b. the conveyance efficiency,
c. the application efficiency, and
d. the overall irrigation efficiency

for the following conditions
• the average daily irrigation requirement during peak water use periods is 10 mm/day
• seepage and evaporation losses in the unlined ditch total 1.0 l/min/m
• total seepage and evaporation losses from the reservoir are 100 l/min.

2.9 In order to evaluate irrigation system performance, an irrigator used a neutron probe to measure the water content of the soil before and after an irrigation. Sampling sites were located in a 100-m-square grid throughout the field. The data in the following table are the average water contents of the top 100 cm of soil in percent by volume. The water content when the soil is at field capacity is 30 percent by volume.

soil water contents prior to irrigation

14.3	16.1	15.2	13.3	14.8	15.5
15.2	15.4	13.6	15.8	14.3	15.5
16.2	14.9	15.4	13.8	14.5	15.0
12.9	14.2	15.0	16.4	17.1	16.2
14.9	15.3	14.8	15.9	14.2	15.3

soil water contents after irrigation

30.2	29.8	31.5	32.0	31.5	29.8
30.5	30.4	31.2	31.6	31.8	32.1
29.4	28.5	31.0	31.2	29.9	30.5
30.6	31.2	31.5	30.1	29.5	30.8
31.0	31.4	30.6	29.8	32.5	32.0

Determine:
a. the uniformity of application,
b. the distribution uniformity, and
c. the storage efficiency.

2.10 Determine the adequacy of the irrigation in Problem 2.9 for the following desired depths of irrigation:
a. 16.0 cm,
b. 14.0 cm, and
c. 18.0 cm.

*2.11 Using the cumulative frequency distribution curve from Problem 2.10, estimate the application efficiency for a desired depth of irrigation of 16.0 cm.

2.12 Using the cumulative frequency distribution curve from Problem 2.10, estimate the depth of application that is required to obtain 100 percent adequacy. Estimate the application efficiency for this irrigation. The desired irrigation depth is 16 cm.

2.13 An irrigator is considering purchasing one of the following irrigation systems.

	System A	System B
Initial cost	$50,000	$40,000
Salvage value	$5000	0
Expected life	15 years	20 years
Annual energy costs	$2500	$2000
Annual labor costs	$250	$1000
Annual taxes and insurance	$1000	$800
Maintenance and repair	$2500	$2000
Annual water costs	0	0

For an annual interest rate of 12.0 percent, determine
a. annual interest and depreciation costs,
b. total annual fixed costs,
c. total annual operating costs, and
d. total annual ownership and operation costs.

Which system has the lowest total annual costs? Use a 20 year analysis period.

2.14 Determine the total annual ownership and operation costs of system A in Problem 2.13 if the irrigator plans to retire in 10 years.

2.15 Repeat Problem 2.13 using an estimated annual rate of cost escalation of 3.8 percent.

References

Anon. (1983). Sprinkler irrigation. Chapter 11, Section 15 (Irrigation) of SCS National Engineering Handbook, 121 pp.

Anon. (1964). Soil-plant-water relationships. Chapter 1, Section 15 (Irrigation) of SCS National Engineering Handbook, 72 pp.

Fangmeier, D. D. (1977). Alternative irrigation systems. Agricultural Engineering and Soil Science Series (Lithographed), University of Arizona, Tucson.

Haan, C. T. (1977). *Statistical Methods in Hydrology.* Iowa University Press, Ames, 378 pp.

James, L. G., J. M. Erpenbeck, and D. L. Bassett (1983). Estimating seasonal irrigation requirements for Washington. *Trans. ASAE*, **26**(5), St. Joseph, MI, pp. 1380–1385.

Jensen, M. E., D. S. Harrison, H. C. Karven, and F. E. Robinson (1980). The role of irrigation in food and fiber production. In *Design and Operation of Farm Irrigation Systems*. M. E. Jensen (Ed.), ASAE Monograph 3, St. Joseph, MI, pp. 15–41.

Merriam, J. L., M. N. Shearer, and C. M. Burt (1980). Evaluating irrigation systems and practices. In *Design and Operation of Farm Irrigation Systems*, M. E. Jensen (Ed.), ASAE Monograph 3, St. Joseph, MI, pp. 721–760.

Pair, C. H., W. H. Hinz, K. R. Frost, R. E. Sneed, and T. J. Schiltz (Eds.). (1983). *Irrigation.* The Irrigation Association, Silver Springs, Md., 686 pp.

Skogerboe, G. V., V. T. Someray, and W. R. Walker (1971). Check-drop energy dissipation structures in irrigation systems. Water Management *Tech. Rep. 9*, Colorado State University, Fort Collins.

Stegman, E. C., J. T. Musick, and J. I. Stewart (1980). Irrigation water management. In *Design and Operation of Farm Irrigation Systems*, M. E. Jensen (Ed.), ASAE Monograph 3, St. Joseph, MI, pp. 763–816.

Thompson, G. T., L. B. Spiess, and J. N. Krider (1980). Farm resources and system selection. In *Design and Operation of Farm Irrigation Systems*, M. E. Jensen (Ed.), ASAE Monograph 3, St. Joseph, MI, pp. 45–73.

3

Water for Irrigation

3.1 Introduction

Water for irrigation is obtained from surface and ground water sources. Surface sources include lakes, reservoirs, streams, water-user association distribution facilities, and waste waters, while the primary ground water sources are wells and springs. The suitability of a water source for irrigation depends on several factors including legal constraints, the quality of the water (i.e., the amount and identity of suspended and dissolved materials in the water), as well as the ability of the source to supply the total irrigation requirement and seasonally varying irrigation requirements year after year.

This chapter describes the characteristics of several surface and ground water sources of water. The effect of water quality (i.e., salinity, exchangeable sodium, and toxicity) on the suitability of a water source for irrigation is discussed and several salinity control measures described. Procedures for evaluating the hydrologic suitability of surface and ground water sources are included. These procedures involve methods of estimating the volume and rate that a source can dependably supply and for computing the volume and rate of water supply required to irrigate a farm. Legal constraints that may influence water source suitability are also presented.

The intent of this chapter is to make the reader aware of the various water quality, hydrologic and legal factors affecting the suitability of a water source for irrigation. After reading and studying material in this chapter, the reader should be able to evaluate the hydrologic suitability of surface and ground water sources of irrigation water.

3.2 Sources of Water

3.2.1 Surface Sources of Irrigation Water

Surface sources of farm irrigation water include streams, lakes and water-user organization distribution facilities. Another surface source of irrigation water that is becoming increasingly important is industrial and agricultural waste waters and sewage.

3.2.1a Streams

Flowing streams are an important source of irrigation water. Where stream-flows are large enough to meet irrigation demands throughout both wet and dry years, water needed for irrigation may be directly withdrawn from the stream. Gravity diversion structures, like those in Figure 2.1, divert streamflow into open canals and pipelines. Pumping plants are used when it is necessary to "lift" streamflow to a higher elevation and/or to pressurize pipelines.

On many streams, however, flow does not always exceed irrigation demand. On some of these streams, annual streamflow is not adequate every year, while other streams have sufficient annual flow but daily flows are inadequate during at least a part of the irrigation season. Such streams require storage reservoirs to be useful for irrigation. These storage reservoirs regulate the stream so that the natural flow is adjusted to meet, as nearly as possible, the rate of demand (i.e., storage reservoirs store water during periods of high flow for use during periods of inadequate flow).

Several years of daily and annual streamflow data are needed to evaluate the suitability of a stream for irrigation. These data are also needed to design diversion structures, pumping plants, and storage reservoirs.

3.2.1b Lakes

Lakes are another surface source of water for irrigation. As with streams, gravity diversions and pumping plants are used to withdraw water. Several years of water surface elevation data are needed to establish the suitability of a lake for irrigation and to design diversion structures and pumping plants. There are often legal barriers to the use of lakes for irrigation, such as prior water rights, laws mandating minimum water levels, and recreational rights.

3.2.1c Water-User Organization Distribution Facilities

Water delivery facilities owned and operated by water-user organizations (irrigation districts, corporations, or companies) are a major source of irrigation water. Irrigators belonging to these organizations share construction, operating, and maintenance costs of sometimes extremely large diversion dams, reservoirs, pumping plants, canals, and pipelines to have access to a more reliable and/or larger source of water than would otherwise be available. Diversion structures, called *turnouts*, are used to remove water from the delivery system.

Table 3.1 Operational Characteristics of Several Irrigation Company Delivery Schedules

		Operational Characteristics	
Delivery Schedule	**Description**	**Delivery System**	**On-Farm System**
Rotation Schedules Fixed amount– fixed frequency	The amount of water delivered is determined by farm size and crop. The time between successive deliveries is constant. For example, 1000 l/min is delivered for 24 hr every 2 weeks	Since delivery schedules are not changed once they have been established, the delivery system is designed for a constant flow rate that is maintained almost continuously. This minimizes capital and operating costs	The rigid schedule contributes to low on-farm water use efficiencies, since excess water is delivered in the spring and late summer. On-farm reservoirs may be required to effectively use water
Fixed amount– variable frequency	An amount of water determined by farm size and crop is delivered with the time between successive irrigations changing from irrigation to irrigation. For example, 1000 l/min each 2 weeks in the spring and 1000 l/min is delivered for 24 hr each week during peak periods	Delivery systems have relatively low capacities and require considerable management to accommodate seasonal variations within the serviced areas	The imposed frequency may not match variable soil and crop needs. Some crops and soils may receive too much water while others receive too little
Variable amount– fixed frequency	The amount of water delivered is varied (by changing the flow rate) according to the irrigation requirement. The time between irrigations is constant	Delivery systems have relatively low capacities. Delivery flow rates are changed according to seasonal variations in the irrigation requirement.	This method may not suit a particular soil–crop combination

Demand	Each irrigator is able to have water as desired (i.e., there is not restriction on the rate, duration, or frequency of irrigation)	The delivery system capacity is excessively high	This schedule provides maximum flexibility to the irrigator
Frequency demand	Any desired rate of delivery can be obtained if ordered at least 24 hr prior to the date of delivery. The duration of delivery is usually in 24 hr increments	Larger lateral capacities are required near the downstream end of the system than for a rigid schedule delivery system. Reservoirs at the downstream end of the system may reduce required capacities in the downstream end of the system	On-farm reservoirs may be required to achieve reasonable on-farm water and labor use efficiencies
Limited rate demand	Any desired rate up to a maximum rate can be obtained if ordered at least 24 hr prior to the desired delivery date. Some water supplies require 3 or 4 days notice and delivery is usually made within 1 day of the requested time	Spillage, reservoir storage, or automation are required, since flows are varied by the irrigator. Limited storage capacity and incomplete automation may limit the number of changes farmers can make	This system limits only the maximum rate of water delivery to the farm
Continuous flow	Water is delivered continuously to a farm during the irrigation season	Minimizes delivery system capacity and management	Low on-farm water-use efficiencies are associated with this schedule

Source: J. A. Replogle, J. L. Merrium, L. R. Swarner, and J. T. Phelen, "Farm Water Delivery Systems." In *Design and Operation of Farm Irrigation Systems* (1980), M. E. Jensen (Ed.), ASAE Monograph 3, p. 317. Copyright © 1980 by ASAE. Reprinted by permission of ASAE.

The design and operation of farm irrigation systems is influenced by the quantity and timing of water delivery to the farm (i.e., the delivery schedule). Delivery schedules are either demand, rotational, or continuous flow. The demand schedule is the ideal strategy from the irrigator's point of view, since water is delivered to the farm at the time and in the quantity requested by the irrigator. The rotation and continuous-flow schedules do not provide as much flexibility to the irrigator, but result in smaller delivery systems with lower capital and operating costs. Delivery systems operating under rotation schedules deliver water for a fixed duration of time according to a prearranged schedule. Under a continuous-flow schedule, each irrigator receives water as a continuous flow. The operational characteristics of several typical delivery systems using rotation and demand schedules are given in Table 3.1. Both the water-supplier's and water-user's point of view are considered.

3.2.1d Waste-Water Sources

Sewage and industrial waste waters are being used more and more for irrigation, especially in areas where the availability of other water sources is limited. The quality of waste waters must be closely evaluated and frequently checked to protect human health, and to prevent damage to crops and soils. Laws in some locations prohibit the use of untreated sewage, while in other places, sewage effluent that has received the required treatment may be used the same as any other source. Benefits occur when waste waters contain quantities of nitrogen, potassium, and phosphates that reduce crop fertilizer requirements. State and local laws and regulations should be checked when the suitability of a waste water is being evaluated.

3.2.2 Ground Water Sources

Ground water is water that occupies the voids within rocks and the soil. Subsurface material containing ground water may be divided into zones of saturation and aeration. Voids within the zone of saturation are completely filled with water, while the zone of aeration consists of voids occupied partially by water and partially by air. Because only the zone of saturation contains drainable water, ground water for irrigation comes from the zone of saturation (and not from the zone of aeration).

The zone of aeration is subdivided into the root, intermediate, and capillary zones as in Figure 3.1. The root zone is characterized by soil water content ranging from the permanent wilting point to saturation. The top of the intermediate zone begins where the influence of plants ends (bottom of the root zone) and extends to the capillary zone. The water content of the intermediate zone varies little with depth and is approximately equal to field capacity, since there is no evapotranspiration (ET) from this zone. The capillary zone, which is immediately above the saturated zone, results from upward movement of water from the zone of saturation due to capillarity. The water content of this zone varies with height. Just above the zone of saturation nearly all voids are filled, while at the top of the capillary zone the water content approaches field capacity.

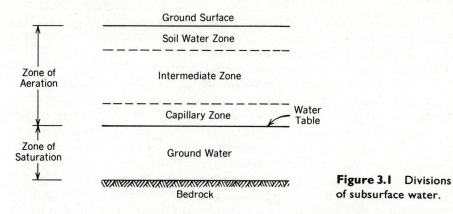

Figure 3.1 Divisions of subsurface water.

Portions of the zone of saturation that yield significant quantities of water are called *aquifers*. Aquifers can be unconsolidated rock (chiefly sand and gravel), fractured zones in dense plutonic rocks, porous sandstone beds, open caverns in limestone, and many other geologic formations.

Aquifers may be classified as *unconfined* or *confined*, depending upon the presence or absence of a water table. Examples of unconfined and confined aquifers are illustrated in Figure 3.2.

The water table of an unconfined aquifer separates the zones of aeration and saturation. Water at the water table is at atmospheric pressure, while water below the water table exists at pressures in excess of atmospheric pressure and water in the zone of aeration above the water table is subject to subatmospheric pressures.

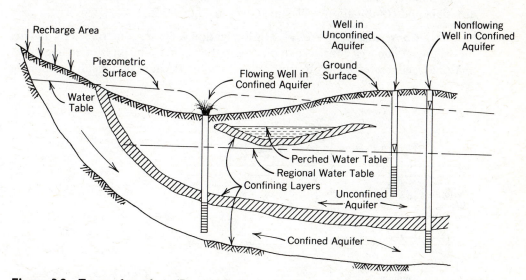

Figure 3.2 Types of aquifers. (From U.S. Geological Survey, Bureau of Reclamation, 1977.)

The water table undulates depending upon the amount of water withdrawal (i.e., pumping) and recharge (i.e., amount of precipitation, irrigation, and seepage from lakes, reservoirs, and streams percolating to the aquifer).

Confined aquifers occur where ground water is confined by relatively impermeable layers above and below the aquifer, as in Figure 3.2. Because water is under pressures greater than atmospheric, water will rise above the bottom of the confining material when a well penetrates the aquifer. If the pressure is high enough, water may flow from the well. The elevation to which water rises in the well equals the hydrostatic pressure level of water in the aquifer and is called the piezometric head. The piezometric surface of a confined aquifer is an imaginary surface that shows the areal variation of piezometric head within an aquifer.

Water enters a confined aquifer in an area where the confining bed rises to the ground surface or ends underground and the aquifer becomes unconfined. A region supplying water to a confined aquifer is called a *recharge area.*

A *perched aquifer* is a special type of unconfined aquifer that occurs whenever a body of ground water is separated from the main ground water by a relatively impermeable stratum of limited areal extent. A typical perched aquifer is diagrammed in Figure 3.2. Wells tapping perched aquifers often yield only temporary or small quantities of water.

3.2.2a Wells

An *irrigation well* is a conduit that conveys water from an aquifer to the ground surface (*drainage wells* convey water to an aquifer). Wells are either gravity, artesian, or a combination of gravity and artesian, depending on the type of aquifer supplying the water. *Gravity wells* penetrate unconfined aquifers, while *artesian wells* tap confined aquifers. *Combination wells* result when a well obtains water from confined and unconfined aquifers simultaneously. Examples of gravity, artesian, and combination wells are shown in Figure 3.3.

(i) Construction of Wells Most irrigation wells are constructed by digging or drilling. A *dug well* consists of a pit dug to the aquifer. The pit is often lined with masonary, concrete, or steel to support excavation. Because of difficulty in digging

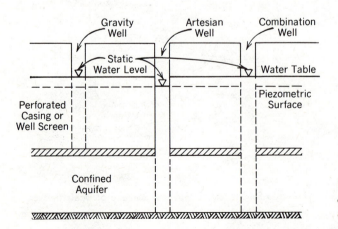

Figure 3.3 Examples of gravity, artesian, and combination wells.

below the water level, dug wells normally do not penetrate the zone of saturation to a depth sufficient to produce a high yield.

Drilled wells are constructed via the cable tool and rotary methods. With cable tools a bit sometimes weighing more than 500 kg (over 1000 lb) is repeatedly dropped onto material at the bottom of the hole. Periodically, a short section of pipe with a flap valve arrangement at its bottom (the entire assembly is called a *bailer*) is lowered into the hole to remove material that has been broken loose and crushed by the bit. Cable tool rigs are generally limited to drilling wells with maximum diameters of 600 to 650 mm (about 24 in) and depths of less than 600 m (about 2000 ft). The cable tool method is best suited to drilling through consolidated rock formations.

The rotary method is a faster and more expensive way of drilling wells (than the cable tool method). The direct circulation rotary method uses a hollow bit rotated by a string of pipe. A drilling fluid consisting of water or drilling mud (a mixture of water, clay, and a commercial organic thickener) is pumped down the pipe, through the bit and up the outside of the pipe. The drilling fluid lubricates and cools the bit, jets material from the bottom of the hole, and carries cuttings to the ground surface. The drilling fluid also forms a thin layer of mud on the wall of the hole that stabilizes it and reduces seepage into the hole.

The reverse rotary method is probably the most rapid drilling method available for unconsolidated formations. With this method, water is pumped up through the drill pipe rather than down the pipe as in the rotary method. Drilling mud is seldom used. Because high velocities down the hole can result in erosion, the minimum diameter of well that can be drilled with this method is 400 mm (about 16 in). The maximum well diameter and depth that can normally be obtained are 1800 mm (about 6 ft) and 150 m (about 500 ft), respectively.

Air rotary drilling was developed as a rapid drilling technique in hard rock in arid areas. The primary difference between this method and direct circulation rotary drilling is that air is circulated down the drill pipe to cool the bit and carry cuttings to the surface rather than water or drilling mud. Wells up to 900 mm (about 3 ft) in diameter have been drilled with air rotary rigs when foams and other air additives have been used. The depth of wells is limited by available air pressure (which must be greater than the pressure exerted by water in the well).

(ii) Components of a Well   The major components of a well include surface and pump chamber casings, screens, and a gravel pack (see Figure 3.4). All wells should have a pump chamber casing and screen(s). A surface casing and gravel pack are, however, not always needed.

A *surface casing* is a pipe installed to simplify and facilitate drilling a well in unstable, unconsolidated, or fractured materials. It extends from near the ground surface through unstable material into more stable material. When permanently installed, it can provide a sanitary seal that protects the ground water from undesirable surface waters.

A *pump chamber casing* is a pipe that extends from the ground surface to the source aquifer and contains the pump. It is an essential part of every well that prevents surface water from entering the well (when there is no surface casing),

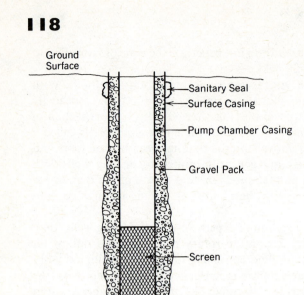

Ground
Surface

—Sanitary Seal
—Surface Casing

—Pump Chamber Casing

— Gravel Pack

—Screen

— Gravel or Concrete Base

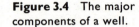

Figure 3.4 The major components of a well.

provides support to the sides of the hole, and isolates the source aquifer from other water-bearing strata. By isolating the source aquifer, a pump chamber casing prevents the loss of source aquifer water to "thief" zones with lower heads. In other situations, the entrance of water that cascades down the hole from shallower zones or the surface is prevented. Cascading water is a problem when it contains suspended materials and chemical contaminants that clog and/or diminish the quality of the source aquifer or when the chemical composition of the cascading and source aquifer waters are such that precipitates that clog the source aquifer are formed when the two waters are combined. In addition, air entrained by cascading water can adversely affect pump performance. A gravel or concrete base at the bottom of the well is often installed to support the pump chamber casing.

A screen that allows water from the source aquifer to enter the pump chamber casing without excessive head loss is needed especially in unconsolidated aquifers. Screens stabilize the hole (in the source aquifer), keep sand out of the well, and facilitate flow into and within the well. Screens range from perforated pump chamber casing to carefully fabricated cage-type wire-wound screens with accurately sized slot openings. A screen should be sized so the average velocity through the screen, neglecting blockage by aquifer or gravel pack material, is 0.03 m/s (about 0.1 ft/s) or less (to minimize head loss through the screen) and to retain 40 to 45 percent of the aquifer material. The screen must also be corrosion-resistant. Water and aquifer material samples are needed to select the proper well screen.

A *gravel pack* is a gravel envelope around the outside of the screen. It is used in situations where it is necessary to stabilize an aquifer and minimize sand pumping, to permit use of the largest possible screen slot (to minimize velocities through the screen), and to provide an annular zone of high permeability, thus improving the yield of the well by increasing its effective radius.

Table 3.2   Types and Causes of Gravity Springs

Type	Cause
Depression springs	Formed where the land surface intersects the water table
Contact springs	Created by permeable water-bearing formation overlaying a less permeable formation that intersects the ground surface
Artesian springs	Results from releases of water under pressure from confined aquifers either at an outcrop of the aquifer or through an opening in the confining bed
Tubular or fracture spring	Flow from rounded channels, such as lava tubes, solution channels, or fractures in impermeable rock connecting with a ground water supply

3.2.2b Springs

A spring is a concentrated discharge of ground water appearing at the ground surface as flowing water. Springs may be the result of nongravitational forces such as those associated with volcanic rocks and deep fractures in the earth's crust. Such springs often discharge water that is highly mineralized and has elevated temperatures (e.g., hot springs and warm springs). Other major springs are the result of gravitational forces. Table 3.2 lists several types of "gravity springs."

The discharge of springs often fluctuates in response to variations in recharge rates. *Perennial springs* drain extensive permeable aquifers and flow throughout the year, whereas *intermittent springs* discharge only when there is sufficient recharge. Areas of volcanic rock are noted for their perennial springs of nearly constant discharge.

3.3 Water Quality

The quality of an irrigation water is judged by the amount of suspended and dissolved materials it contains. Suspended materials include eroded soil particles, seeds, leaves, and other debris. The most common cations (positively charged ions) dissolved in irrigation water are calcium, magnesium, sodium, and potassium. Bicarbonate, sulfate, and chloride are the most common anions (negatively charged ions). Other solutes including nitrates, carbonates, and trace elements, such as boron, are occasionally present.

Suspended materials larger than 50 to 100 microns can usually be removed with filters (see Chapter 6, "Trickle Irrigation"). Table 3.3 summarizes the operational characteristics of the major filter types. Particles smaller than 50 to 100 microns that are not retained by filters often accumulate in pipes and canals. These materials are usually removed by flushing or dredging.

Dissolved materials in irrigation water are described by the total concentration of ions (without reference to the specific ions) and by the identity and concentration of the specific ions present. Crop yield can be reduced significantly

Table 3.3 **Operational Characteristics of the Major Filter Types Used to Remove Suspended Substances.**

Filter type	Characteristics
Sediment basins	Large quantities of sand-sized particles with specific gravities greater than 1 can be removed. Large size required to store sediment. Algae growth and windblown contaminents can accumulate. Trash racks and/or screens are required to remove floating debris
Sand media	Effectively removes suspended sands, organic materials, and most other suspended substances. Does not remove silt and clay-sized particles or bacteria. Are relatively inexpensive and easy to operate. Cleaned by manual or automatic backflushing
Screens	Effectively removes most suspended substances. Minimum size retained depends on mesh size. Fifty-micron particles are the smallest size that can be practically retained. Cleaned by manual or automatic through flushing
Cartridge	Retains organic materials and fine particles that cannot be removed with media filters. Are either disposable or washable
Centrifugal separator	Effectively removes large quantities of suspended substances with specific gravities in excess of 1.2. Cleaned manually or automatically

when the total concentration of ions dissolved in the irrigation water, usually called the salinity of the irrigation water, is high enough. High amounts of exchangeable sodium can cause soil particle dispersion that reduces soil structure and restricts air and water movement into and within the soil. Sodium, chloride, boron, and other ions are toxic to many plants when present in sufficient concentrations.

3.3.1 Salinity

The *salinity* of an irrigation water is the sum of all the ionized dissolved salts in the water without reference to the specific ions present. The electrical conductivity (EC), of an irrigation water is often used to characterize salinity, since the ability of a water to conduct electricity is directly related to the number of ions present. EC has units of decisiemens per meter (dS/m) or millimhos per centimeter (mmho/cm). Decisiemens per meter is the SI unit and is preferred even though mmho/cm has been the traditional unit of EC in the United States. One dS/m equals one mmho/cm.

Salinity is also expressed as a concentration in units of moles per cubic meter (mol/m^3), milliequivalents per liter (meq/l), milligrams per liter (mg/l) or parts per million (ppm). Moles/m^3 and meq/l are equivalent as are mg/l and ppm. To convert from mol/m^3 or meq/l to mg/l or ppm multiply mol/m^3 or meq/l by the atomic weight of the ion given in Table 3.4.

3.3.1a Salinity Effects

The primary effect of salinity is to restrict the availability of soil water to the plant. The presence of salt in soil water increases the energy needed to remove

**Table 3.4   Determinations Required to Evaluate Water
Quality for Irrigation**

Determination	Symbol	Unit of Measure	Atomic Weight
Total salt content			
(1) Electrical conductivity	EC	dS/m[a]	—
(2) Concentration	C	mg/L or mol/m^3	—
Sodium Hazard			
(1) Sodium adsorption ratio	SAR		
(2) Adjusted sodium adsorption ratio	adj SAR	—	
Constituents			
(1) Cations			
calcium	Ca	mol/m^3	40.1
magnesium	Mg	mol/m^3	24.3
sodium	Na	mol/m^3	23.0
potassium[b]	K	mol/m^3	39.1
(2) Anions			
bicarbonate	HCO$_3$	mol/m^3	61.0
sulphate	SO$_4$	mol/m^3	96.1
chloride	Cl	mol/m^3	35.5
carbonate[b]	CO$_3$	mol/m^3	60.0
nitrate[b,c]	NO$_3$	mg/l	62.0
Trace elements[b,c]			
boron	B	mg/l	10.8
Acidity or alkalinity	pH	—	—

Source: G. J. Hoffman, R. S. Ayers, E. J. Doering, and B. L. McNeal, "Salinity in Irrigated Agriculture." In *Design and Operation of Farm Irrigation Systems* (1980), M. E. Jensen (Ed.), ASAE Monograph 3, p. 145. Copyright © 1980 by ASAE pp. 169. Reprinted by permission of ASAE.
[a] 1 mmho/cm, referenced to 25°C.
[b] Constituent that may be important in special situations only.
[c] Because of its low concentration, this constituent is usually expressed on a mg/l basis.

water from the soil. Figure 3.5 shows typical relationships between the energy with which a soil holds water and soil water content for various salt concentrations in the soil water. Figure 3.5 indicates that for a given soil water content it takes more and more energy to extract water as the concentration of salt increases.

Salinity influences crop physiology and yield. Cell enlargement and division, the production of proteins and nucleic acids, and the rate of increase in plant mass are physiological processes that are retarded by high levels of salinity. Visible injury symptoms such as leaf burn normally occur only at extremely high salinity levels.

Figure 3.6 shows a typical relationship between crop yield and salinity. In this relationship, crop yield is independent of salt concentration when salinity is below a threshold salinity level. Thus, the presence of salt will not affect crop yield when salinity does not exceed the threshold value. Above the threshold level, yield

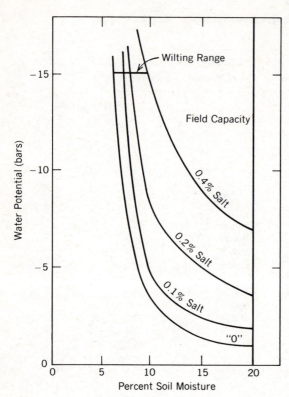

Figure 3.5 Soil characteristic curves for soil water with different salt concentrations. *Source*: From *Irrigation*, C. H. Pair, W. H. Hinz, K. R. Frost, R. E. Sneed, and T. J. Schiltz (Eds.), copyright © 1983 by The Irrigation Association, Silver Spring, MD, 82 pp. Reprinted by permission of The Irrigation Association.

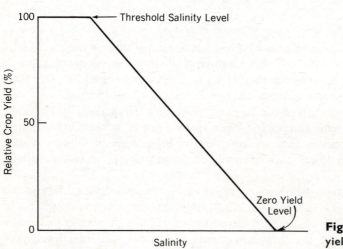

Figure 3.6 A crop yield–salinity relationship.

Table 3.5 Threshold and Zero Yield Salinity
Levels for the Four Salinity Rating Groups
in Table 3.6

Salinity Rating Group	Threshold Salinity (dS/m)	Zero Yield Salinity (dS/m)
Sensitive	1.4	8.0
Moderately sensitive	3.0	16.0
Moderately tolerant	6.0	24.0
Tolerant	10.0	32.0

Source: G. J. Hoffman, R. S. Ayers, E. J. Doering, and B. L. McNeal, "Salinity in Irrigated Agriculture." In *Design and Operation of Farm Irrigation Systems* (1981), M. E. Jensen (Ed.), ASAE Monograph 3, p. 145. Copyright © 1980 by ASAE, pp. 158–160. Reprinted by permission of ASAE.

decreases linearly as salinity increases. The salinity at zero yield is an estimate of the maximum salinity that a crop can tolerate. Crop production cannot normally be maintained when salinity exceeds the zero yield value.

3.3.1b Crop Tolerance of Salinity

The relationship between yield and salinity varies with crop. The yield of some salt-tolerant crops, like barley and sugarbeets, is not affected at salinity levels that severely reduce the yields of salt-sensitive crops, like apples and carrots. In fact, salt-tolerant crops usually have threshold salinity levels that exceed the zero yield values of salt-sensitive crops (see Table 3.5).

Relationships similar to the one in Figure 3.6 have been developed for most agricultural crops (see Hoffman et al., 1980). Four groups of crops with similar salt tolerances have been established using these relationships. Table 3.6 lists the crops belonging to each group. The maximum threshold salinity level and a zero yield salinity level for each group are given in Table 3.5. No crop within a group has a threshold salinity level that exceeds the maximum salinity level for the group. Most crops within a group have zero yield values less than the zero yield value for the group.

3.3.1c Salinity Control

The objective of salinity control is to maintain and/or improve soil water availability to the crop. This is usually accomplished by leaching salts from the profile, maintaining high soil water contents, selecting more salt-tolerant crops, and improving drainage.

(i) Leaching Leaching involves applying enough water to cause drainage from the root zone. Salts are carried out of the root zone in the drainage water.

Table 3.6 Salinity Tolerance Ratings for Selected Agricultural Crops

---

Sensitive

Almond	Lemon
Apple	Okra
Apricot	Onion
Avocado	Orange
Bean	Peach
Blackberry	Plum
Boysenberry	Raspberry
Carrot	Strawberry
Grapefruit	

Moderately Sensitive

Alfalfa	Pepper
Bentgrass	Potato
Broadbean	Radish
Broccoli	Rhodegrass
Cabbage	Rice, paddy
Clover	Sesbania
Corn (forage, grain, sweet)	Sorghum
Cowpea	Spinach
Cucumber	Sugarcane
Flax	Sweet potatoes
Grape	Timothy
Lettuce	Tomato
Lovegrass	Trefoil, big
Meadow Foxtail	Vetch
Millet, Foxtail	
Orchardgrass	
Peanut	

Moderately Tolerant

Barley (forage)	Safflower
Beet, garden	Soybean
Bromegrass	Sudongrass
Canarygrass, reed	Trefoil, Birdsfood, narrowleaf
Fescue, tall	Wheat
Hardinggrass	Wheatgrass, crested and slender
Olive	Wildrye, beardless
Ryegrass, perennial	

Tolerant

Barley, grain	Sugarbeet
Bermudagrass	Wheatgrass, fairway and tall
Cotton	Wildrye, Altai and Russian
Date palm	

Source: G. J. Hoffman, R. S. Ayers, E. J. Doering, and B. L. McNeal, "Salinity in Irrigated Agriculture." In *Design and Operation of Farm Irrigation Systems* (1981), M. E. Jensen (Ed.), ASAE Monograph 3, p. 145. Copyright © 1980 by ASAE, pp. 158–160. Reprinted by permission of ASAE.

Equation 3.1 is used to estimate the amount of leaching required to maintain salinity at a desired level (usually the threshold value). Equation 3.1 is

$$LF = \frac{D_D}{D_I} = \frac{EC_I}{EC_D} \tag{3.1}$$

where

  LF = leaching fraction (dimensionless);
  D_D = depth (volume per unit area) of water draining from the root zone (mm, in);
  D_I = depth (volume per unit area) of water entering the root zone (mm, in);
  EC_I = electrical conductivity of water entering the root zone (dS/m);
  EC_D = electrical conductivity of water draining from the root zone (dS/m).

    LF is the fraction of the total water applied that must be drained from the root zone to maintain salinity at the desired level. Equation 3.1 is based on the assumptions that there is adequate drainage and that salt added by mineral weathering and amendments (fertilizers, manure, chemicals, etc.) equals the salt removed by precipitation as insoluble minerals and with the harvested crop. It is also based on the assumption that the amount of salt carried into the root zone by capillarity from underlying ground water and by rainfall are negligible compared to that in the irrigation water.

    Since Eq. 3.1 is a steady-state equation, it is based on the assumption that the initial salinity of the root zone is at or below the desired value. Thus, an initial leaching to reduce the salinity of soils with high natural salt contents to the desired level will normally be necessary. The amount of water needed to accomplish this initial leaching depends largely on the initial soil salinity. Generally, about 80 percent of the soluble salts initially present in a soil profile will be removed by leaching with a depth of water equivalent to the soil depth to be treated. For crops sensitive to salinity, further leaching may be required, while somewhat less leaching may be required for salt-tolerant crops.

    The EC within the root zone can be determined from a solution extracted from a soil paste formed by adding water until an initially air-dry soil sample is saturated. This EC is called the saturation extract EC. The adequacy of the initial leaching can be accessed using the saturation extract EC. It should be noted that actual salinity levels during an irrigation cycle will exceed the saturation extract EC since the soil will not normally be saturated with water.

EXAMPLE 3.1   Use of Eq. 3.1 to Compute D_D and D_I

Given:
- crop water requirement = 30 in (762 mm)
- $EC_I = 0.6$ dS/m
- crop salinity rating = moderately tolerant

Required:

a. EC_D

b. D_D

c. D_I

Solution:

a. $EC_D = 6$ dS/m,   the threshold salinity for crops with moderate tolerance to salinity (from Table 3.5).

b. rearranging Eq. 3.1 yields

$$D_D = D_I\left(\frac{EC_I}{EC_D}\right)$$

$$D_I = 30 + D_D$$

$$D_D = (30 + D_D)\left(\frac{0.6}{6}\right) = 3.0 + 0.1D_D$$

$$D_D = \frac{3.0}{0.9} = 3.33 \text{ in } (84.6 \text{ mm})$$

c. $D_I = 30 + 3.33 = 33.33$ in (846.6 mm)

(ii) Maintaining High Soil Water Contents   Because the salt concentration of the root zone increases as soil water is depleted by evaporation and plant transpiration, lower salt concentrations result when the water content of the soil is maintained at high levels. This may make it possible to irrigate with waters that have relatively high salinity levels. Increasing the frequency of irrigation is an effective way of maintaining high soil water contents and making the use of more saline water possible. Caution must, however, be exercised when irrigation methods that wet plant surfaces, such as sprinkle irrigation, are used to apply saline water. Severe leaf burn may result. High-frequency trickle irrigation has been effectively and successfully used to apply extremely saline waters.

(iii) Salt-Tolerant Crops   Selecting crops or varieties (root stocks) of crops with high tolerances to salt is another widely used method of controlling the effects of salinity. Waters with much higher salinity levels can be used to irrigate crops with "tolerant" salt tolerance ratings (see Tables 3.5 and 3.6) than can be used with salt-sensitive crops. The development of more salt-tolerant crops and crop varieties is receiving considerable attention.

(iv) Drainage   Adequate drainage is essential for salinity control. The existence of a water table within or near the root zone may result in considerable salt accumulation within the root zone. Salts in the ground water, natural soil salts, or salts that have been leached from the root zone will be carried into the root zone as the water table rises. In addition, inadequate drainage may limit the amount of leaching that can be safely accomplished. The installation of a drainage system may be required to achieve salinity control.

3.3.2   Exchangeable Sodium

High concentrations of exchangeable sodium in irrigation waters and soils cause the eventual deterioration of soil structure and a resulting reduction in hydraulic conductivity. When calcium and/or magnesium are the predominant cations occupying soil exchange sites, soils tend to have a granular structure that is readily permeable to both air and water. As the concentration of exchangeable sodium in the soil increases, the ratio of sodium to calcium and magnesium ions rises and the number of exchange sites occupied by calcium or magnesium decreases. This causes soil mineral particles to disperse and hydraulic conductivity to decrease.

The exchangeable-sodium-percentage (ESP), the sodium-adsorption-ratio (SAR), and the adjusted SAR of soil extracts or irrigation waters are used to evaluate the exchangeable sodium status of soils and irrigation waters. Equation 3.2 can be used to compute ESP

$$\text{ESP} = \frac{\text{Exchangeable Na (meq/100 g of soil)}}{\text{cation-exchange-capacity (meq/100 g of soil)}} \, 100 \tag{3.2}$$

Generally, higher ESP levels can be tolerated in course-textured than in fine-textured soils.

The SAR is generally a good indicator of the exchangeable sodium status of the soil and it can be determined more readily than ESP. The SAR is defined by Eq. 3.3. Equation 3.3 is:

$$\text{SAR} = \frac{\text{NA}}{\sqrt{\dfrac{\text{Ca} + \text{Mg}}{2}}} \tag{3.3}$$

where all ion concentrations are in moles per cubic meter. For many soils, the SAR of the soil saturation extract is approximately equal to the ESP below ESP values of 25 or 30.

In some cases, the SAR of the irrigation water has not been a satisfactory guide to potential soil permeability problems because of the influence of carbonates and bicarbonates on the precipitation of calcium and magnesium. An adjusted SAR that includes the relative concentrations of carbonate and bicarbonate as well as sodium, calcium, and magnesium has been developed to estimate the permeability hazard of irrigation waters (see Hoffman et al., 1980).

Permeability effects are complicated by the interaction of SAR and salinity. Large concentrations of dissolved salt tend to neutralize the effect of sodium on soil dispersion. The interaction of EC and SAR is illustrated in Figure 3.7. Three ranges representing no problem, increasing problem, severe problem are shown. All water quality combinations of EC and SAR that lie above the curved band should have no permeability problem. Those that lie within the curved band have increasing problems. Those that lie below the band will present severe problems even though waters have low SAR.

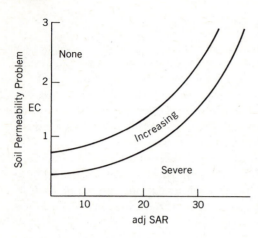

Figure 3.7 Guidelines for predicting possible permeability problems. *Source:* A. Marsh, "Guidelines for Evaluating Water Quality Related to Crop Growth." In: 1982 Technical Conference Proceedings, copyright © 1982 by The Irrigation Association, 72 pp. Reprinted by permission of The Irrigation Association.

3.3.3 Toxicity

Dissolved ions are absorbed into the roots with soil water, and are transmitted through the plant to the leaves where they accumulate. Toxicity occurs when the accumulation of an ion becomes large enough to cause leaf burn (i.e., drying of the leaf tissue). A chemical analysis of the leaf is often needed to identify the toxic ion.

Sodium, chloride, and boron are the most common phytotoxins (plant toxins) found in natural irrigation waters. Many other phytotoxins may be present in reclaimed sewage waters that are used for irrigation.

Table 3.7 lists sodium and chloride toxicity guidelines for sensitive crops. For sodium, there is no toxicity problem when the adjusted SAR is less than 3, an increasing problem in the range of 3 to 9, and a severe problem when the adjusted SAR exceeds 9. Woody perennial plants are sensitive to sodium toxicity.

Levels below 4 meq/l of chloride normally are not toxic to sensitive crops. There is an increasing toxicity problem between chloride levels of 4 and 10 meq/l and a severe toxicity problem when the chloride concentration exceeds 10 meq/l. Woody perennials such as tree fruits and grapes are sensitive to chloride toxicity.

Because of foliar absorption irrigation methods that wet plant leaves, such as overhead sprinkling, may cause toxicity problems at sodium and chloride concentrations lower than those that cause problems with surface irrigation methods. This occurs primarily during periods of high temperature and low humidity. Sprinkle

Table 3.7 Toxicity Guidelines for Toxicity to Sodium and Chloride

	Degree of Problem		
Problem	None	Increasing	Severe
Sodium (adjusted SAR)	3	9	
Chloride (meg/l)	4	10	

Table 3.8   **Relative Tolerance[a] of Crops to Boron**

Tolerant[b] 4.0 mg/l of boron	Semitolerant 2.0 mg/l of boron	Sensitive 1.0 mg/l of boron
Asparagus	Sunflower, native	Pecan
Date palm	Potato	Walnut, black and Persian or English
Sugarbeet	Cotton, Acala and Pima	Jerusalem artichoke
Garden beet	Tomato	Navy bean
Alfalfa	Radish	Plum
Broadbean	Field pea	Pear
Onion	Olive	Apple
Turnip	Barley	Grape
Cabbage	Wheat	Kadota fig
Lettuce	Corn	Persimmon
Carrot	Sorghum	Cherry
	Oat	Peach
	Pumpkin	Apricot
	Bell pepper	Thornless blackberry
	Sweetpotato	Orange
	Lima bean	Avocado
		Grapefruit
		Lemon
2.0 mg/l of boron	1.0 mg/l of boron	0.3 mg/l of boron

Source: G. J. Hoffman, R. S. Ayers, E. J. Doering, and B. L. McNeal, "Salinity in Irrigated Agriculture." In *Design and Operation of Farm Irrigation Systems* (1980), M. E. Jensen (Ed.), ASAE Monograph 3, p. 145. Copyright © 1980 by ASAE, pp. 165. Reprinted by permission of ASAE.
[a] Relative tolerance is based on the boron concentration in irrigation water at which boron toxicity symptoms were observed when plants were grown in sand culture. It does not necessarily indicate a reduction in crop yield.
[b] Tolerance decreases in descending order in each column between the stated limits.

irrigated citrus has been damaged with sodium and chloride concentrations as low as 3 meq/l even though these same conditions had no toxic effects when water was applied with furrow or flood irrigation.

Boron is an essential minor element that is toxic to many crops if present in excess. Boron sensitivity varies from crop to crop. The relative sensitivity of several crops to boron is listed in Table 3.8. Few surface waters contain toxic concentrations of boron.

3.4 Water Quantity

Both the total volume and the available flow rate (volume of water per unit of time) are as important as water quality in determining the suitability of surface and ground water sources for irrigation. The seasonal irrigation requirement of the farm (i.e., the total volume of water needed during an irrigation season to irrigate the farm) establishes the annual supply volume a source must provide, while the volumetric flowrate (i.e., volume of water per day) needed to achieve the desired level of irrigation on the farm during the growing season year after year (i.e., the design capacity of the irrigation system) determines the required source supply rate. Since irrigation needs must be provided year after year, water source evaluation must also consider year-to-year fluctuations in water supply and irrigation needs caused by variations in climate.

Hydrologically, the evaluation of a water source involves comparing the expected source supply volume and rate to the farm's seasonal irrigation requirement and system design capacity. In order for a source to be hydrologically suitable, its supply volume must exceed the seasonal irrigation requirement. Otherwise, the source, by itself, cannot be depended upon to supply the seasonal irrigation requirement year after year (such a source may, however, be adequate in some years and supply a major portion of the SIR in other years). Reservoirs may be used to store water during years of abundant water for use during periods when there is not enough water.

When the minimum expected source supply rate is less than the design capacity of the system during any part of the irrigation season (and the expected annual supply volume is sufficient), the source will be suitable if additional water can be obtained during periods of inadequate supply. Additional water is often obtained from reservoirs constructed to store water during periods of excess supply (i.e., when the source supply rate exceeds the system design capacity and/or when the irrigation system is not operating). Table 3.9 summarizes the conditions influencing the hydrologic suitability of surface and ground water sources.

Various legal factors can also determine the suitability of a water source. Stream flows as well as stream and lake levels mandated by law for fish, hydropower generation, sanitary purposes, navigation, domestic and industrial

Table 3.9 Hydrologic Conditions Affecting the Suitability of Surface and Ground Water Sources of Irrigation Water

Relationship Between		
SV[a] and SIR[b]	SR[c] and SDC[d]	**Hydrologic Suitability of Water Source**
—	SR > SDC	Hydrologically suitable for specified return period
SV > SIR	SR < SDC	Not suitable unless extra water is provided during periods of inadequate supply
SV < SIR	—	Cannot be depended upon to supply all the water needed for irrigation year after year. Reservoirs may provide a way of storing water during periods of oversupply for use in periods of undersupply

[a] SV = source supply volume.
[b] SIR = seasonal irrigation requirement.
[c] SR = minimum expected source supply rate.
[d] SDC = system design capacity.

uses, and other irrigators may limit the use of potential water sources that are otherwise suitable for irrigation by hydrologic and water quality standards. The use of ground water may also be similarly restricted.

In the sections that follow, procedures for determining the seasonal irrigation requirement and system design capacity of a farm and for estimating the supply rate and volume of surface and ground water sources are presented. These procedures include the influence of year-to-year fluctuations in climate.

3.4.1 Seasonal Irrigation Requirements

The total volume of water needed during an irrigation season (year) for a given crop grown at a specified location can be estimated using Eq. 3.4.

$$\text{SIR}_{j,k} = 100\left(\frac{(\text{ET}_s)_{j,k} - (\text{P}_e)_{j,k}}{\text{E}_i}\right) \qquad (3.4)$$

where

$\text{SIR}_{j,k}$ = seasonal irrigation requirement for crop j during year k (mm, in).
$(\text{ET}_s)_{j,k}$ = total seasonal evapotranspiration for crop j during year k (mm, in)
$(\text{P}_e)_k$ = annual effective precipitation for crop j during year k (mm, in)
E_i = irrigation efficiency (percent).

Equation 3.4 is a simplified form of Eq. 1.24 based on the assumption that $\theta_f = \theta_i$ (i.e., that the initial and final soil moisture levels for the irrigation season are identical) even though there is a continuous variation in soil water content during the season. Equation 3.4 further assumes that there are sufficient deep percolation losses to provide the leaching requirement. This is usually a reasonable assumption for most systems, except when there is intentional under-irrigation for an extended period of time.

ET_s for a crop can be computed using either the Penman, Jensen–Haise, Blaney–Criddle, or pan evaporation approaches. The Blaney–Criddle approach is often used for seasonal irrigation requirement calculations, since it provides adequate accuracy, is easy to use, uses data that are normally readily available, and requires less time to apply than the more sophisticated methods.

The total volume of water needed to irrigate the farm during a given year can be calculated using Eq. 3.5:

$$(\text{SI}\forall_f)_k = K \sum_{j=1}^{n} (\text{SIR}_{j,k})A_j \qquad (3.5)$$

where

$(\text{SI}\forall_f)_k$ = seasonal irrigation volume for farm during year k (m^3, acre in);
K = unit constant ($K = 10.0$ when $\text{SI}\forall_f$ is in m^3, SIR is in mm, and A is in hectares. $K = 1.0$ when $\text{SI}\forall_f$ is in acre in, SIR is in inches, and A is in acres);
A_j = area of crop j (hectares, acres).

A frequency analysis of several years of data is needed to account for year-to-year fluctuations in SIR and $\text{SI}\forall_f$. Procedures similar to those illustrated in

Example 2.2 can be used to determine the SIR (and hence SIV_f) for a specified probability of occurrence.

The following example illustrates the use of Eq. 3.4 and 3.5.

EXAMPLE 3.2 Computing Seasonal Irrigation Requirements for Several Crops and the Seasonal Irrigation Volume for a Farm

Given:
- 120-acre farm located at 46°N latitude
- farm has 20 acres of corn and 100 acres of alfalfa
- monthly temperature, P_e, a, and b data for growing season
- growing season for corn begins May 15
- growing season for alfalfa is May 1 to Oct 31
- Assume that leaching requirements are met by dormant season precipitation that does not run off.

Required:
- 1973 seasonal irrigation requirement for each crop
- 1973 seasonal irrigation volume for farm

Solution:

Month	$P_e{}^a$ ins	$\bar{T}$ °F	P	a	b	ET_0 in	Corn K_c	Corn ET	Alfalfa K_c	Alfalfa ET
1	2	3	4	5	6	7	8	9	10	11
May	0.31	62.1	0.34	−2.00	1.47	7.18	0.42	1.56^b	0.95	6.82
June	0.14	67.8	0.35	−2.02	1.48	8.15	0.55	4.48	0.95	7.74
July	0	74.5	0.34	−2.16	1.62	10.08	0.99	9.98	0.95	9.58
Aug.	0	71.9	0.32	−2.15	1.56	8.59	1.15	9.78	0.95	8.08
Sept.	0.46	64.8	0.31	−2.02	1.45	6.35	0.97	6.16	0.95	6.03
Oct.	0.71	54.1	0.31	−1.82	1.22	4.12	0.72	0.19^c	0.95	3.91
Totals	1.62 in							32.15 in		42.16 in

Column 2: effective precipitation (assumed equal to monthly precipitation)
  3: average daily temperature in Eq. 1.22
  4: P in Eq. 1.22 (Table 1.15)
  5: a in Eq. 1.22
  6: b in Eq. 1.22
  7: ET_0 computed with Eq. 1.22
  8: K_c values for corn (Table 1.4)
  9: (column 7) (column 8)
  10: K_c values for alfalfa (Table 1.6)
  11: (column 7) (column 10)

^{a} Only growing season precipitation is considered (dormant season precipitation is assumed to provide the leaching requirement)
^{b} May ET for corn = $(0.42)(7.18)(\frac{16}{31})$ = 1.56 in.
^{c} Oct. ET for corn = $(0.72)(4.12)(\frac{2}{31})$ = 0.19 in. (Corn is harvested Oct. 2, 140 days after May 15)

$$\text{SIR for corn} = 100\left(\frac{32.15 \text{ in} + 0 - 1.62 \text{ in}}{75}\right)$$

$$= 40.71 \text{ in} \sim 41 \text{ in}$$

$$\text{SIR for alfalfa} = 100\left(\frac{42.16 + 0 - 1.62}{75}\right)$$

$$= 54.05 \text{ in} \sim 54 \text{ in}$$

for 1973 at Richland

$$\text{SIV}_f = 1.0[(42 \text{ in})(20 \text{ acre}) + (54 \text{ in})(100 \text{ acre})]$$

$$= 6220 \text{ acre ins}[6.39(10)^5 \text{ m}^3]$$

3.4.2   Irrigation System Design Capacity

To be suitable for irrigation, an irrigation system (and hence water source) must provide water at the volumetric rate computed with Eq. 3.6. This rate is called the system design capacity (SDC). Equation 3.6 is:

$$\text{SDC} = \frac{(K)(\text{DDIR}_f)(A)}{(E_i)(\text{HPD})} \tag{3.6}$$

where

$\quad$ SDC = system design capacity (l/min, gpm);

DDIR$_f$ = design daily irrigation requirement for the farm (mm/day, in/day) (see Section 2.4.3);

$\quad\quad$ A = area irrigated (ha, acre);

$\quad\quad$ E$_i$ = irrigated efficiency (percent);

$\quad$ HPD = hours per day that the system operates

$\quad\quad$ K = unit constant (K = 16667 for SDC in l/min, DDIR in mm/day, and A in ha. K = 45254 for SDC in gpm, DDIR in in/day, and A in acres).

EXAMPLE 3.3   Determining the Design Capacity of a Farm Irrigation System

Given:
DDIR$_f$ = 9.3 mm/day
$\quad$ A = 72 ha
$\quad$ E$_i$ = 70%

Required:
design capacity for system operating
a. 24 hr/day
b. 20 hr/day

Solution:

a. $\text{SDC} = \dfrac{(16667)(9.3)(72)}{(70)(24)} = 6643 \ \text{l/min}$

b. $\text{SDC} = \dfrac{(16667)(9.3)(72)}{(70)(20)} = 7972 \ \text{l/min}$

3.4.3 Minimum Expected Source Supply Rate

3.4.3a Surface Streams and Ground Water Springs

A frequency analysis that uses a Type III Extreme Value (Weibull) probability distribution and several years of daily flow data for the irrigation season will yield the expected minimum supply rate for surface streams and ground water springs as a function of return period. The procedure is similar to that illustrated in Example 2.2 except that the log of daily flow values during the irrigation season for each year of record is arranged in descending order (rather than arranging DDIR values in ascending order). Daily flow data for larger streams can sometimes be obtained from the U.S. Geological Survey. The following example illustrates a procedure for computing the expected minimum supply rate for a surface stream (or ground water spring).

EXAMPLE 3.4 Estimating the Expected Minimum Supply Rate for a Surface Stream (or Ground Water Spring)

Given:
minimum daily stream flow during the irrigation season (April 1 to Oct 31) for each year of record arranged in descending order

Required:
expected minimum source supply rate for a 10-year return period

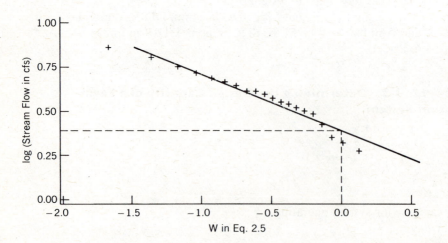

Solution:

Minimum daily flow (cfs)	Rank	P^a	Return years[a] period	W^a
2.33	1	95.24	1.05	−1.674
2.20	2	90.48	1.11	−1.362
2.09	3	85.71	1.17	−1.174
2.01	4	80.95	1.24	−1.037
1.96	5	76.19	1.31	−0.928
1.92	6	71.43	1.40	−0.835
1.88	7	66.67	1.50	−0.754
1.82	8	61.91	1.62	−0.681
1.82	9	57.14	1.75	−0.614
1.79	10	52.38	1.91	−0.552
1.75	11	47.62	2.10	−0.492
1.71	12	42.86	2.33	−0.434
1.70	13	38.10	3.00	−0.378
1.66	14	33.33	3.00	−0.321
1.63	15	28.57	3.50	−0.264
1.60	16	23.81	4.20	−0.205
1.51	17	19.05	5.25	−0.143
1.40	18	14.29	7.00	−0.073
1.36	19	9.52	10.50	0.009
1.30	20	4.76	21.00	0.121

[a] P, return period, and W were computed using Eqs. 2.4, 2.3, and 2.5, respectively.

$$P = \frac{100}{RP} = \frac{100}{10 \text{ yr}} = 10\%$$

$$W = \log\left(-\log\frac{10}{100}\right) = 0$$

from plot, flow = 1.44 cfs at W = 0.

Thus, the expected minimum supply rate for a 10-year return period for this stream is 1.44 cfs (on the average, the supply rate for this stream will be less than 1.44 cfs once each 10 years)

3.4.3b   Lakes

The supply rate at which lakes can provide water is normally unlimited when the supply volume is adequate. Thus, it is usually only necessary to establish the supply volume of lakes to evaluate the suitability of the source.

3.4.3c   Wells

The volumetric rate at which a well is able to supply water is determined by the physical characteristics of the aquifer and the maximum well drawdown allowed. *Drawdown* is the vertical distance the water table or piezometric surface

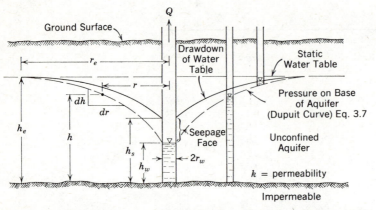

Figure 3.8 Radial flow in an unconfined aquifer to a fully penetrating well. *Source:* V. E. Hansen, O. W. Israelsen, and G. E. Stringham, *Irrigation Principles and Practices,* copyright © 1979 by John Wiley & Sons, Inc., New York, p. 266. Reprinted by permission of John Wiley & Sons, Inc.

drops when a well is pumped. Drawdown is highest in the well and decreases with increasing distance from the well as diagramed in Figures 3.8 and 3.9. The cone-shaped dewatered area is called the *cone of depression* and the radical distance from the center of the well to where drawdown becomes negligible is the radius of influence, r_e. Drawdown in the well equals the distance between the *static water level* (the level that exists 24 hours after pumping has ended) and the *pumping level*

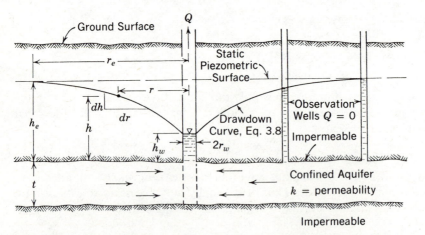

Figure 3.9 Radial flow in a confined aquifer to a fully penetrating well. *Source:* V. E. Hansen, O. W. Israelsen, and G. E. Stringham, *Irrigation Principles and Practices,* copyright © 1979 by John Wiley & Sons, Inc., New York, p. 265. Reprinted by permission of John Wiley & Sons, Inc.

(the level after at least 12 hours of pumping at a constant rate). The allowable drawdown is the maximum drop in water level within the well that can be tolerated.

Because static water levels fluctuate during the irrigation season and from year to year because of differences in climate and pumpage of neighboring wells, a frequency analysis similar to Examples 2.2 and 3.4 is needed to relate static water level to return periods. The maximum allowable drawdown for a specified return period is computed from the corresponding static water level. The American Society of Agricultural Engineers (ASAE) (Engineering Practice: ASAE EP400) recommends that the maximum allowable drawdown is the distance from the static water level to either

a. the top of the well inlet (the well inlet is the portion of the well that has openings through which water enters the pump chamber casing).
b. the water level needed to maintain the lower third of the aquifer saturated, or
c. the water level that results when the well is pumped at a rate equal to 125 percent of the desired rate for a minimum of 8 hours.

The smallest of these distances is the maximum allowable drawdown.

(i) Unconfined Aquifers The supply rate of a well that fully penetrates an unconfined aquifer is computed using Eq. 3.7:

$$SR = \frac{Kk(h_e^2 - h_w^2)}{\ln(r_e/r_w)} \qquad (3.7)$$

where

SR = supply rate (l/min, gpm);
 k = aquifer conductivity (m/day, ft/day);
 h_e = aquifer thickness (m, ft);
 h_w = distance of steady-state pumping level above bottom of aquifer (m, ft);
 r_e = radius of influence (m, ft);
 r_w = radius of the well (m, ft);
 K = unit constant ($K = 2.18$ for SR in l/min, k in m/day, and h_e, h_w, r_e, and r_w in m. $K = 1.63(10)^{-2}$ for SR in gpm, k in ft/day, and h_e, h_w, r_e, and r_w, in ft).

Figure 3.8 also defines the parameters in Eq. 3.7. The conductivity, k, for several aquifer materials is given in Figure 3.10. Well tests that involve pumping the well at different rates and measuring the resulting drawdowns can be used to determine aquifer conductivity. Recommended procedures for testing wells are presented in ASAE Engineering Practice: ASAE EP400.

Equation 3.7 assumes a constant radius of influence, r_e. Theoretically, however, r_e for an aquifer of infinite areal extent increases at a constantly diminishing rate as long as the well is pumped. Under field conditions, the rate of change typically becomes so slow after a sufficiently long period of pumping that it is negligible and the assumption that r_e is constant is valid. The value of r_e used in

Conductivity

Ft³/Ft²/Day (ft/day)

| 10^5 | 10^4 | 10^3 | 10^2 | 10^1 | 1 | 10^{-1} | 10^{-2} | 10^{-3} | 10^{-4} | 10^{-5} |

Ft³/Ft²/Min (ft/min)

| 10^1 | 1 | 10^{-1} | 10^{-2} | 10^{-3} | 10^{-4} | 10^{-5} | 10^{-6} | 10^{-7} | 10^{-8} |

Gal/Ft²/Day (gal/ft²/day)

| 10^5 | 10^4 | 10^3 | 10^2 | 10^1 | 1 | 10^{-1} | 10^{-2} | 10^{-3} | 10^{-4} |

Meters³/Meter²/Day (m/day)

| 10^4 | 10^3 | 10^2 | 10^1 | 1 | 10^{-1} | 10^{-2} | 10^{-3} | 10^{-4} | 10^{-5} |

Relative Conductivity

| Very High | High | Moderate | Low | Very Low |

Representative Materials

| Clean gravel | — | Clean sand and sand and gravel | — | Fine sand | — | Silt, clay and mixtures of sand, silt and clay | — | Massive clay |

| Vesicular and scoriaceous basalt and cavernous limestone and dolomite | — | Clean sandstone and fractured igneous and metamorphic rocks | | | — | Laminated sandstone shale, mudstone | — | Massive igneous and metamorphic rocks |

Figure 3.10 Comparison of the hydraulic conductivity (for water) of representative aquifer materials. (From U.S. Geological Survey, Bureau of Reclamation, 1977.)

Table 3.10 Radius of Influence, r_e, of Wells

Soil Formation and Texture	Radius of Influence (m)	Radius of Influence (ft)
Fine sand formations with some clay and silt	30–90	100–300
Fine to medium sand formations fairly clean and free from clay and slit	90–180	300–600
Coarse sand and fine gravel formations free from clay and slit	180–300	600–1000
Coarse sand and gravel, no clay or silt	300–600	1000–2000

Eq. 3.7 depends, in part, upon the aquifer material. Typical values of r_e for different aquifer materials are given in Table 3.10.

EXAMPLE 3.5 Computing SR for a Well That Fully Penetrates an Unconfined Aquifer

Given:
• clean fine sand aquifer
• 300 ft to the bottom of the aquifer
• static water level for a 10 year return period is 50 ft below the soil surface
• r_w is 8 inches
• the top of the well inlet is 200 ft below the soil surface

Required:
supply rate of the above well for 10-year return period

Solution:

$k = 30$ ft/day from Figure 3.10
$h_e = 300 - 50 = 250$ ft
$h_w = 250 -$ allowable drawdown   (allowable drawdown is the lesser of
$250 - 0.33(250) = 168$ ft
or $250 - 100$ ft $= 150$ ft)

$h_w = 250 - 150 = 100$ ft

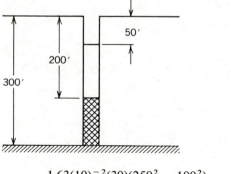

$$SR = \frac{1.63(10)^{-2}(30)(250^2 - 100^2)}{\ln(500/0.67)} = 3880 \text{ gpm}$$

(ii) _Confined Aquifers_   The supply rate of a well that fully penetrates a confined aquifer is computed using Eq. 3.8

$$SR = \frac{Kkt(h_e - h_w)}{\ln(r_e/r_w)}$$ (3.8)

where

t = thickness of aquifer (m, ft);
h_w = distance of steady-state pumping level above top of aquifer (see Figure 3.9) (m, ft);
h_e = distance of static piezometric surface above top of aquifer (see Figure 3.9) (m, ft);
k = conductivity (m/day, ft/day);
K = unit constant ($K = 4.36$ for SR in l/min, k in m/day, and t, h_e, h_w, r_e, and r_w in m. $K = 3.26(10)^{-2}$ for SR in gpm, k in ft/day, and t, h_e, h_w, r_e, and r_w in ft).

Values of r_e and k can be obtained from Table 3.10 and Figure 3.10, respectively. Well tests are also used to measure k (see ASAE Engineering Standard: ASAE EP400).

EXAMPLE 3.6   Computing SR for a Well That Fully Penetrates a Confined Aquifer

Given:
- clean fine sand aquifer
- upper confining layer in 200 ft below the soil surface
- aquifer is 75 ft thick
- r_w is 8 in
- static water level in well for a 10-year return period is 50 ft below the soil surface
- the top of the 75-ft-long well inlet coincides with the top of the aquifer

Required:
supply rate of the above well for a 10-year return period

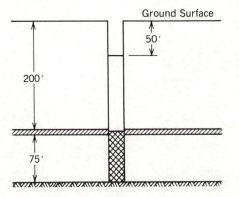

Solution:

$k = 30$ ft/day from Figure 3.10

$h_e = 200 + 75 - 50 = 225$ ft

$h_w = h_e -$ allowable drawdown   (allowable drawdown $= 200 - 50 = 150$ ft)

$h_w = 225 - 150 = 75$ ft

$$SR = \frac{3.26(10)^{-2}(30)(75)(225 - 75)}{\ln(2000/0.67)}$$

$$= 1375 \text{ gpm}$$

3.4.4 Source Supply Volume

3.4.4a Streams and Ground Water Springs

A frequency analysis on several years of annual flow data is needed to determine the volume of water that a stream or spring can supply each year. Procedures similar to those in Example 3.4 can be used to evaluate supply volume for specified return periods. Annual flow data for the analysis can normally be obtained from water supply data published by the U.S. Geological Survey.

3.4.4b Lakes

A storage volume versus stage (water level) curve, a minimum allowable stage, and several years of minimum daily lake stage data are needed to establish the volume of water a lake can supply. A typical storage–stage relationship for a lake is given in Figure 3.11. Minimum lake stages are often mandated by prior water and recreational rights. The minimum expected lake stage for a given return period is established from a frequency analysis of several years of minimum daily stage data. Assuming that all other users of water have been satisfied, the supply volume of a lake can be estimated using Eq. 3.9.

$$SV = \forall(S_L) - \forall(S_{\min}) - \text{Evap} \tag{3.9}$$

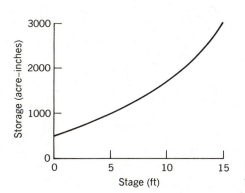

Figure 3.11 A storage-stage relationship for a lake.

where

$$SV = \text{supply volume}$$
$$S_L = \text{minimum expected lake stage for specified return period}$$
$$S_{min} = \text{minimum allowable lake stage}$$
$$\forall(S_L), \forall(S_{min}) = \text{storage volumes where lake is at stages } S_L \text{ and } S_{min},$$
$$\text{respectively}$$
$$\text{Evap} = \text{volume of lake evaporation.}$$

EXAMPLE 3.7 Computing Supply Volume for a Lake

Given:
- storage versus stage relationship in Figure 3.14
- minimum allowable stage is 10 ft
- minimum expected lake stage for 10 year return period is 11.4 ft
- neglect evaporation

Required:
Supply volume for 10-year return period (i.e., the supply volume that, on the average, will be exceeded only once each 10 years)

Solution:
from Figure 3.11

$$\forall(11.4) = 1910 \text{ acre-in}$$
$$\forall(10) = 1740 \text{ acre-in}$$
$$SR = 1910 - 1740 = 170 \text{ acre-in.}$$

3.4.4c Wells

The concept of safe yield will be used to determine the supply volume for an aquifer. Safe yield is the volume of water that can be withdrawn annually from an aquifer without producing an undesired result. Hydrologically, safe yield equals the long-time mean annual water supply to the aquifer. Withdrawals exceeding this supply must come from storage within the aquifer.

Economic, water quality, and legal considerations can also determine safe yield. Economic considerations may govern safe yield in an aquifer where, because of lowered water levels, the cost of pumping ground water becomes excessive and wells are abandoned in favor of more economical sources of water. Safe yield can also be exceeded if draft on an aquifer causes water of inferior quality to enter the aquifer. For example, pumping in a well in a coastal area could induce sea water intrusion of the aquifer. Legal considerations limit safe yield when pumpage interferes with the prior water rights of others within an aquifer or in adjacent aquifers.

There are several methods for estimating the hydrologic safe yield (hence, supply volume of an aquifer). One method, called the Hill (Todd, 1967) method (for R. A. Hill, who developed the method), is to plot annual changes in ground water levels against annual draft. Safe yield is the draft corresponding to zero change in water level elevation. For the safe yield to be representative, the supply during the period of record shall approximate the long-time mean supply.

3.5 Water Law

The availability of surface and ground water sources for irrigation is often established by laws governing the right to use water. Water source evaluations should therefore include a thorough study of the water right laws that apply to the source being considered. Major components of this study should be an identification of holders of vested water rights and an investigation of procedures for obtaining a water right for the proposed project.

In the United States, laws relating to the ownership and use of water are the responsibility of the courts and legislative bodies of the states. There are no federal statutes under which a water right can be acquired. In other countries, water right laws may exist in the form of rules and customs developed and accepted by the people in their everyday activities.

Two entirely different systems for acquiring water rights are followed in the contiguous 48 states. These are the doctrine of riparian rights, recognized in the 31 predominantly eastern states, and the doctrine of prior appropriation, recognized in the 17 western states.

3.5.1 Riparian Doctrine

The essence of the *riparian doctrine*, which is based on the common law of England, is that a landowner contiguous to a stream is entitled to have the water of the stream flow by his land undiminished in quantity and unpolluted in quality. For ground water, ownership of land overlying an aquifer is sufficient to establish a ground water right. A strict application of this rule would not permit consumptive use of water from a stream or well for any purpose (including irrigation). As a result, many states that recognize the riparian doctrine have modified or limited it.

3.5.2 Appropriation Doctrine

The *appropriation doctrine* declares that water belongs to the public, but a right to use it may be obtained by individuals or agencies provided they comply with certain procedures and principles. Beneficial use is the measure and limit of the right. Rights may be lost by failure to use the water. In times of water scarcity, prior rights of appropriators must be satisfied before subsequent appropriators are entitled to the water. The elements of an appropriation right include: quantity of flow, time of use, point of diversion, nature of use, place of use, and the priority of the right. In general, none of these elements may be changed by the water user without prior approval by the state authority, since changes may infringe on prior rights. Appropriation doctrine permits storage of water, transfer of water from one basin to another, commingling of transferred water with waters of natural streams, and water for human consumption having first preference. Theoretically, a higher preferential use may permit condemnation of water from a lesser preferential use by eminent domain proceedings.

Homework Problems

3.1 During an irrigation season 110 cm of irrigation and rain water was caught in a rain gauge located in a sprinkle irrigated field. The seasonal water requirement for the crop was estimated to be 88 cm. Determine the depth of leaching and the leaching fraction. Neglect any changes in soil water content that may have occurred during the irrigation season.

3.2 During an irrigation season 90 cm of irrigation water was applied to a field of corn. The electrical conductivity of the irrigation water was 0.5 dS/m. Determine the required leaching fraction and depth of leaching. Do not consider the effects of rainfall.

3.3 Determine the required leaching fraction and depth of leaching for the conditions in Problem 3.2 if the annual rainfall was 30 cm and the electrical conductivity of the rain water was:
a. 1.0 dS/m,
b. 0.3 dS/m,

Assume that there is no runoff.

3.4 Use data from Problem 1.31 to estimate the seasonal irrigation requirement for furrow irrigated corn. The electrical conductivities of the irrigation and rain waters were 1.0 and 0.5 dS/m, respectively. A total of 10 cm of rain fell during the months October through April. Assume that all rainfall is effective.

3.5 Repeat Problem 3.4 for grain barley. Compare the depths of leaching and the leaching fractions of the two crops. The growing season begins April 1. ET_0 for April is 94.3 mm.

3.6 Use information from Problems 3.4 and 3.5 to compute the seasonal irrigation volume for the following farm.

Crop	Area Irrigated (ha)
Corn	5
Grain barley	15

The overall efficiency of the irrigation system is 75 percent.

3.7 Determine the system design capacity of an irrigation system for the farm in Problem 3.6 if the system is operated.
a. 24 h per day,
b. 20 h per day.

The readily available water holding capacity for both crops is 11.0 cm.

3.8 Given the following information
- a farm with 75 ha of alfalfa and 25 ha of apples with cover crop
- sprinkle irrigation system that operates 24 h/day
- the soil consists of 55 cm of loam overlaying 25 cm of clay loam
- the electrical conductivity of the irrigation water and precipitation is 0.4 dS/m
- the farm is located in a humid area with light winds
- the overall efficiency of the irrigation system is 75 percent
- the growing season for apples with a cover crop usually begins April 1 and ends September 30
- the growing season for alfalfa usually begins April 1 and ends October 31
- The following data

Month	30-yr Mean ET_0 (Grass) (mm)	P_e (mm)
Apr.	95	25
May	130	21
June	208	12
July	220	6
Aug.	170	9
Sept.	91	18
Oct.	65	32

determine:
a. the seasonal volume of water required to irrigate the farm, and
b. the design capacity of the sprinkle irrigation system.

3.9 Use information from Problems 3.6 and 3.7 to evaluate the hydrologic suitability of the stream in Example 3.4 for a 10-year return period (1 cfs = 1699 l/min).

3.10 Use information from Problems 3.4 and 3.5 to evaluate the ability of the stream in Example 3.4 to supply the needs of a surface irrigated farm with 50 ha of corn and 100 ha of grain barley. The system operates 24 hours per day during peak use periods. What additional information is required to establish the suitability of the stream for irrigating this farm?

* Indicates that a computer program will facilitate the solution of the problems so marked.

*3.11 Given the following streamflow and climatic data

Year	Mean Annual Streamflow (m/s)	Minimum Streamflow (m/s)
1	3.1	1.5
2	4.2	1.3
3	6.1	1.2
4	5.3	1.9
5	3.2	1.0
6	2.9	1.5
7	3.4	1.0
8	1.9	1.0
9	2.8	1.2
10	1.9	1.0

- a flow of $1.0 \ \mathrm{m^3/s}$ must be maintained for fish and sanitary flow purposes
- the average electrical conductivity of the stream during the 10-year period was 0.9 dS/m

Year	Apr.	May	June	July	Aug.	Sept.	Oct.	Annual Precipitation (mm)
			mm of Monthly Pan Evaporation[a]					
1	109	182	233	248	215	124	89	25
2	128	209	257	288	201	110	85	26
3	134	204	226	231	213	105	85	25
4	106	193	232	239	217	113	77	24
5	128	173	253	270	239	107	80	28
6	114	173	229	243	207	114	81	22
7	106	171	235	269	227	113	78	24
8	121	173	224	261	211	110	72	25
9	122	181	259	265	218	125	72	24
10	106	166	228	260	225	123	73	24

[a] Use a pan coefficient of 0.8.

- the electrical conductivity of the rain was 0.4 dS/m

Determine how many hectares of alfalfa can be irrigated with water from the stream. The readily available water holding capacity of the soil for alfalfa is 20 cm.

3.12 A 50-cm-diameter well completely penetrates a 100-m-deep unconfined sandstone aquifer. The conductivity of the aquifer is 3.0 m/day. Determine the discharge of the well when the static and pumping water levels are 15 and 50 m below the ground surface, respectively. Use a radius of influence of 1000 m.

3.13 Repeat Problem 3.12 using a radius of influence of
a. 500 m,
b. 100 m.

How does the radius of influence affect the computation of well discharge? What is the minimum value of radius of influence that should be used with this aquifer?

3.14 Determine the maximum rate at which the well in Problem 3.12 can be pumped.

3.15 A 110-m-deep, 50-cm-diameter well completely penetrates a 10-m-thick confined aquifer. The conductivity of the aquifer is 3.0 m/day. Determine the well discharge when the static and pumping levels are, respectively, 15 and 50 m below the ground surface.

3.16 Determine the maximum rate at which the well in Problem 3.15 can be pumped assuming that the aquifer is able to "safely" supply the annual volume pumped from the well.

3.17 A well is being designed to supply irrigation water to a 50-ha farm. The design daily irrigation requirement for the farm is 9 mm/day. Data from a test well drilled in the area indicates that the unconfined aquifer is about 200 m deep and that the static water level is approximately 50 m below the ground surface. The conductivity of the aquifer is estimated to be 0.8 m/day. Determine the minimum diameter of the well required for sprinkle irrigation. Assume that the aquifer is able to "safely" supply the volume pumped from the well.

3.18 Use information from Problem 3.17 to determine the number of hectares of the farm that can be sprinkle irrigated with a 20-cm-diameter well. Assume that the aquifer is able to "safely" supply the annual volume pumped from well.

References

American Society of Agricultural Engineers (1985). *Designing and Constructing Irrigation Wells*. ASAE Standard EP400.

Davis, S. N., and R. J. M. De Wiest (1966). *Hydrogeology*. Wiley, New York, 463 pp.

Hansen, V. E., O. A. Isrealsen, and G. Stringham (1979). *Irrigation Principles and Practices*. Wiley, New York, 417 pp.

Hillel, D. (1971). *Soil and Water Physical Principles and Practices*. Academic Press, New York, 288 pp.

Hoffman, G. J., R. S. Ayres, E. J. Doering, and B. L. McNeal (1980). Salinity in irrigated agriculture. In *Design and Operation of Farm Irrigation Systems*. M. E. Jensen (Ed.), ASAE Monograph 3, St. Joseph, MI. pp. 145–188.

James, L. G., J. M. Erpenbeck, D. L. Bassett, and J. E. Middleton (1982). Irrigation requirements for Washington—Estimates and methodology. Research Bulletin XB 0925, Agricultural Research Center, Washington State University, Pullman, 37 pp.

James, L. G., J. M. Erpenbeck, and D. L. Bassett (1983). Estimating seasonal irrigation requirements for Washington. *Trans. ASAE*, **26**(5); pp. 1380–1385.

Marsh, A. W. (1982). Guidelines for evaluating water quality related to crop growth. In *Proc. 1982 Tech. Conf.* The Irrigation Association, Silver Spring, Md., 686 pp.

Pair, C. H., W. H. Hinz, K. R. Frost, R. E. Sneed, and T. J. Schiltz (1983). *Irrigation.* The Irrigation Asssociation, Silver Spring, Md., 686 pp.

Replogle, J. A., J. L. Merriam, L. R. Swarner, and J. T. Phelan (1980). Farm water delivery systems. In *Design and Operation of Farm Irrigation Systems*, M. E. Jensen (Ed.), ASAE Monograph 3, St. Joseph, MI. pp. 317–343.

Schwab, G. O., R. K. Frevert, T. W. Edminster, and K. K. Barnes (1981). *Soil and Water Conservation Engineering.* Wiley, New York. 525 pp.

Todd, D. K. (1967). *Ground Water Hydrology.* Wiley, New York, 336 pp.

U.S. Geological Survey, Bureau of Reclamation (1977). *Ground Water Manual.* 480 pp.

4

Pumps

4.1 Introduction

Pumps are mechanical devices that impart energy to a fluid. Irrigation pumps lift water from one elevation to a higher level, overcome friction losses during conveyance, provide pressure for sprinkler (trickle emission device) operation, and meter (inject) chemicals into irrigation systems. Irrigation pumps use mechanical energy, usually from electric, gasoline, diesel, liquid petroleum gas, or natural gas motors to increase the potential (pressure) and/or kinetic energy of the irrigation water. In some parts of the world, human and animal power is used to pump irrigation water.

Pumps may be classified as rotary, reciprocating, or centrifugal. Rotary pumps use gears, vanes, lobes, or screws to trap and convey fluid from the inlet to outlet sides of the pump. Pumps that use the back and forth motion of mechanical parts, such as pistons or diaphragms, to pressurize the fluid are known as reciprocating pumps. Centrifugal pumps use the centrifugal force imparted to the fluid by one or more rotating elements (called impellers) to increase the kinetic and pressure energy of the fluid.

Rotary and reciprocating pumps, often called positive displacement pumps, are not normally used to pump irrigation water, primarily because of their relatively low discharge capacity and susceptibility to sediment laden water. They are, however, used to meter (inject) chemicals into irrigation systems. Centrifugal pumps are used for both chemical injection and irrigation pumping. Since the remainder of this chapter deals with irrigation pumping, only centrifugal pumps are considered.

This chapter describes the operation, performance, and selection of centrifugal pumps. Sufficient information is presented to allow the reader to

1. become familiar with the common types of centrifugal pumps used with farm irrigation systems,

2. define the parameters typically used to characterize centrifugal pump performance,
3. describe the performance characteristics of centrifugal pumps,
4. predict the performance characteristics of two or more pumps connected in series or parallel,
5. compute system curves and use them with pump performance curves to determine the head and discharge at which a pump or system of pumps will operate,
6. evaluate the effect of changes in impeller speed and diameter on pump performance,
7. select the most suitable pump or combination of pumps for a farm irrigation system.

4.2   Centrifugal Pumps

A centrifugal pump consists of a set of rotating vanes, called impellers, enclosed within a stationary housing called a casing. Water is forced into the center (eye) of the impeller by atmospheric or other pressure and set into rotation by the impeller vanes. The resulting centrifugal force accelerates the fluid outward between the vanes until it is thrown from the periphery of the impeller into the casing. The casing collects the liquid, converts a portion of its velocity energy into pressure energy, and directs the fluid to the pump outlet.

Centrifugal pumps are either single- or multistage depending on the number of impellers. Single-stage pumps have only one impeller, while multistaged pumps have several impellers connected in series (i.e., the outflow of the first impeller is directed into the eye of the second, the outflow of the second impeller into the third, etc). Centrifugal pumps are classified as either horizontal or vertical according to the orientation of their axis of rotation. Horizontal pumps are sub-classified according to the location of the suction nozzle (inlet) as end-suction, side-suction, bottom-suction, or top-suction. In addition, pumps are also classified by casing and impeller type.

4.2.1   Casings

Centrifugal pump casings are either of the volute or diffuser type. Two typical volute casings are diagrammed in Figure 4.1. These casings have carefully designed volute (spiral) shaped passages for water that increase in cross-sectional area toward the outlet of the pump. The rate of area increase within a volute casing is normally sufficient to reduce the velocity of the fluid as it approaches the outlet. This increases pressure at the outlet. A majority of single-stage pumps have volute-type casings.

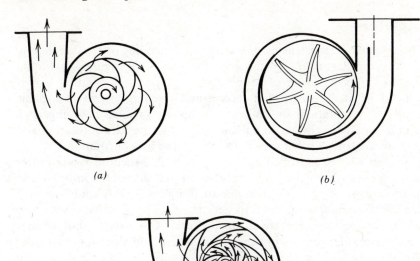

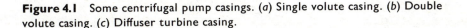

Figure 4.1 Some centrifugal pump casings. (*a*) Single volute casing. (*b*) Double volute casing. (*c*) Diffuser turbine casing.

Casings like the one in Figure 4.1*a*, called single-volute casings, are normally used when pumps operate continuously at or near design capacity (i.e., operated at or near peak efficiency). Double-volute casings (Figure 4.1*b*) should be used when pumps are expected to operate at part-capacity for extended periods of time, because radial forces are more balanced. The double-volute design is also used in situations where the rib that separates the first and second volutes is needed to strengthen the casing. In single volute casings, uniform or near uniform pressures act on the impeller when the pump is operated at design capacity.

Diffuser-type casings have stationary vanes surrounding the impeller that guide water from the impeller into volute or circular shaped casings. Fluid pressure usually rises as it passes through diffusers because of the progressive increase in cross-sectional area between vanes in the direction of flow. Diffuser vanes provide a more uniform distribution of pressure in the casing and are used in multistaged pumps.

Pump casings are either axially or radially split. Pump casings are made of two or more parts that can be "split" (by removing the bolts that hold them together) to provide access to impellers, bearings, seals, and other internal parts. Pumps with casings that can be split in a plane parallel to the axis of rotation are called axially split pumps. In radially split pumps, the casing split is in a plane perpendicular to the axis of rotation.

4.2.2 Impellers

Impellers are classified as radial, axial, or mixed flow depending on the direction of flow through the impeller relative to the axis of rotation (see Figure 4.2). Liquid enters radial flow impellers in a plane parallel to the axis of rotation and is discharged at 90° to it. Impellers that receive and discharge water in a plane parallel to the axis of rotation are known as propeller or axial flow impellers. Mixed flow impellers receive water in a plane parallel to the axis of rotation and discharge it at an angle that is between 0 and 90° to the axis of rotation.

Impellers can also be classified as single- or double-suction. Water enters single-suction impellers through an inlet located on one side of the impeller and flows symmetrically into both sides of double-suction impellers. A double-suction impeller is, in effect, two single-suction impellers arranged back-to-back in a single casing. Double-suction impellers are normally used with axially split casings because of their hydraulic balance in the axial direction and since they normally have low net positive suction requirements (see Section 4.3.2d). Single-suction impellers are preferred when waters with high concentrations of suspended materials are being pumped, for multistaged pumps, and for small pumps, because single-suction impellers have large waterways that are easier to manufacture. Because of low first cost and ease of maintenance, single-suction impellers are utilized in most end-suction radially-split casing pumps.

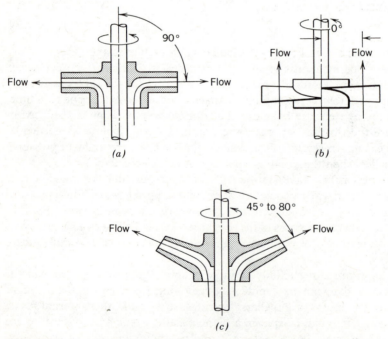

Figure 4.2 Impellers for centrifugal pumps. (*a*) Radial flow impeller. (*b*) Axial flow impeller. (*c*) Mixed flow impeller.

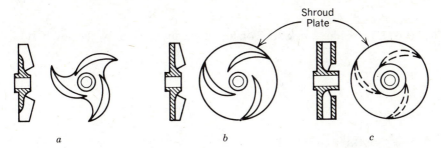

Figure 4.3 (a) Open, (b) semiopen, and (c) enclosed impellers for centrifugal pumps. *Source:* G. O. Schwab, R. K. Frevert, T. W. Edminster, and K. K. Barnes, *Soil and Water Conservation Engineering*, copyright © 1981 by John Wiley & Sons, Inc., New York, p. 370. Reprinted by permission of John Wiley & Sons, Inc.

Impellers are also classified according to their mechanical construction. Enclosed impellers, like the one in Figure 4.3, have shrouds (sidewalls) enclosing the waterways between vanes. Semienclosed (semiopen) impellers have one full shroud (backwall), while open impellers consist only of vanes attached to a hub (without a shroud or sidewall). The vanes of many open impellers are, however, strengthened by ribs or partial shrouds. Open and semienclosed (semiopen) impellers are most suitable for pumping suspended material or trashy water. Enclosed impellers are not normally suitable for use with such waters because the abrasiveness of the suspended materials greatly increases impeller wear and hence, deterioration of pump performance. Enclosed impellers are, however, more widely used with clear waters than open or semiopen impellers because less adjustment is required to maintain efficient operation.

4.3 Centrifugal Pump Performance

4.3.1 Performance Parameters

Capacity, head, power, efficiency, required net positive suction head, and specific speed are parameters that describe a pump's performance. These parameters are discussed in the following sections.

4.3.1a Pump Capacity

The capacity of a pump (Q) is the volume of water per unit time delivered by the pump. In SI units, Q is usually expressed in liters per minute (l/min) or, for larger pumps, cubic meters per second (m³/s). The corresponding units in the English system are gallons per minute (gpm) and cubic feet per second (cfs).

4.3.1b   Head

The head (H) is the net work done on a unit weight of water by the pump. It is given by

$$H = \left(\frac{P}{\gamma} + \frac{V^2}{2g} + Z\right)_d - \left(\frac{P}{\gamma} + \frac{V^2}{2g} + Z\right)_s \tag{4.1}$$

where

P = water pressure (kPa, psi);
γ = specific weight of fluid (kN/m^3, lb/ft^3);
V = water velocity (m/s, ft/s);
g = acceleration of gravity (9.81 m/s^2, 32.2 ft/s^2);
Z = elevation head in meters or feet above a datum. (For horizontal
      pumps, the datum is a horizontal plane through the centerline of the shaft.
      For vertical pumps, the datum is a horizontal plane through the entrance
      eye of the first-stage impeller.)

The subscripts d and s identify the discharge and suction sides of the pump, respectively.

4.3.1c   Power

The power imparted to the water by the pump is called water power. Equation 4.2 is used to compute water power (WP).

$$WP = \frac{QH}{K} \tag{4.2}$$

where

WP = water power (kW, hp);
Q = pump capacity (l/min, m^3/s, gpm, cfs);
H = head (m, ft);
K = unit constant (K = 6116 for WP in kW and Q in l/min, K = 0.102 for
      WP in kW and Q in m^3/s, K = 3960 for WP in hp and Q in gpm, and
      K = 8.81 for WP in hp and Q in cfs).

4.3.1d   Efficiency

Pump efficiency (E_p) is the percent of power input to the pump shaft (the brake power) that is transferred to the water. E_p can be computed using Eq. 4.3:

$$E_p = 100\left(\frac{WP}{BP}\right) \tag{4.3}$$

where

E_p = pump efficiency (percent);
WP = water power (kW, hp);
BP = brake power (kW, hp).

4.3.1e Required Net Positive Suction Head

The required net positive suction head ($NPSH_r$) is the amount of energy required to prevent the formation of vapor-filled cavities of fluid within the eye of single- and first-stage impellers. These cavities, which form when pressures within the eye drop below the vapor pressure of water, collapse within higher pressure areas of the pump. The formation and subsequent collapse of these vapor-filled cavities is called cavitation. When these collapses occur violently on interior surfaces of the pump they produce ring-shaped indentations in the surface called pits. Continued cavitation and pitting can severely damage pumps, and must be avoided.

The net positive suction head required to prevent cavitation is a function of pump design and is usually determined experimentally for each pump. Manufacturers conduct laboratory tests to determine $NPSH_r$ values for each pump model they manufacture. Cavitation is prevented when heads within the eye of single- and first-stage impellers exceeds the $NPSH_r$ values published by the manufacturers.

4.3.1f Specific Speed

Specific speed (N_s) is an index to pump performance derived using dimensional analysis. It consolidates a pump's speed, design capacity, and head into one term. N_s is computed from

$$N_s = \frac{NQ^{1/2}}{H^{3/4}} \tag{4.4}$$

where

N_s = specific speed (rpm);
N = pump speed (rpm);
Q = pump design capacity (gpm);
H = design head (ft).

Geometrically similar pumps have similar performance characteristics and identical specific speeds regardless of their size. For example, pumps with radial flow impellers generally deliver relatively small discharges at high heads and have low specific speeds ranging between 500 and 2000 rpm regardless of impeller diameter. Similarly, pumps with axial flow impellers normally deliver relatively large discharges against low heads and have large specific speeds ranging between 5000 and 15,000 rpm for all impeller diameters.

4.3.2 Performance (Characteristic) Curves

The operating properties of a pump are established by the geometry and dimensions of the pump's impeller and casing. Curves relating head, efficiency, power, and required net positive suction head to pump capacity are utilized to describe the operating properties (characteristics) of a pump. This set of four curves is known as the pump's *characteristic curves*. Characteristic curves for a typical single-stage

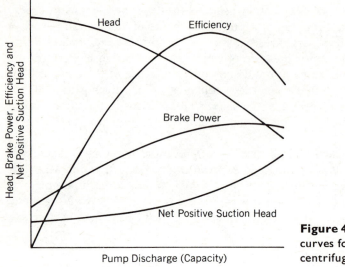

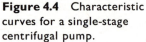

Figure 4.4 Characteristic curves for a single-stage centrifugal pump.

pump are illustrated in Figure 4.4. Characteristic curves for multi-stage pumps are considered in Section 4.3.4.

Pump manufacturers normally publish a set of characteristic curves for each pump model they make. Data for these curves are developed by testing several pumps of a specific model. Some manufacturers' curves represent the average performance of all pumps of a specific model tested, while other manufacturers prepare their curves for the pump having the poorest performance.

4.3.2a   Head Versus Pump Capacity

This curve relates the head produced by a pump to the volume per unit time of water being pumped. Generally, the head produced steadily decreases as the amount of water pumped increases. The shape of the $H-Q$ curve varies with specific speed.

Figure 4.5 shows typical $H-Q$ curves for various specific speeds and impeller designs. For radial flow impellers, head decreases only slightly and then drops rapidly as Q increases from zero. Slope changes along $H-Q$ curves for mixed and axial flow impellers are not as dramatic as those for radial flow impellers. Radial flow impellers operating on the flat portion of their $H-Q$ curves work well in situations where head must remain essentially constant as Q fluctuates (as in set-move systems where the number of operating laterals varies during the irrigation season). In situations where a relatively constant Q is desired and H is expected to fluctuate (such as in a well, small stream, or small reservoir), impellers with higher specific speeds will probably perform best.

The head generated when Q is zero (i.e., when the pump is operating against a closed valve) is the shutoff head (see Figure 4.5). For pumps with steadily declining

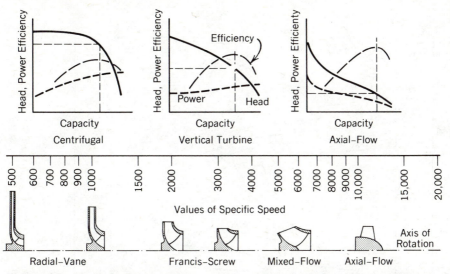

Figure 4.5 Head–discharge curves as a function of specific speed and impeller design. *Source:* R. Walker, *Pump Selection: A Consulting Engineer's Manual,* copyright © 1972 by Butterworth Publishers, Stoneham, MA, 58 pp. Reprinted by permission of Butterworth Publishers.

H–Q curves, the shutoff head is the maximum head and must be known to design piping on the dischargeside of the pump. In such situations, discharge-side piping must be able to withstand the shutoff head when the discharge-side valve is closed. Note that pump efficiency is zero at the shutoff head, since energy is being used to turn the pump.

4.3.2b Efficiency Versus Pump Capacity

An E_p–Q curve for a typical pump is illustrated in Figure 4.4. E_p for a pump steadily increases to a peak, and then declines as Q increases from zero. There is generally only one peak efficiency for a specified impeller. The E_p–Q relationship for a pump is sometimes drawn as a series of envelopes on the H–Q curves of different impeller diameters (see Figure 4.6).

Theoretical efficiencies as a function of specific speed, impeller design, and pump capacity are shown in Figure 4.7. Data in this figure indicate that larger capacity pumps with specific speeds of about 2500 rpm can be expected to have the highest efficiencies.

E_p is also related to the types of materials used in construction, the finish on castings, the quality of machining, and the type and quality of bearings used. For example, impellers with extremely smooth surfaces tend to be more efficient than rougher surfaced impellers.

An E_p–Q curve is usually for a specific number of stages. If a different number of stages is needed for a particular situation, efficiencies must be adjusted up- or downward depending on the number of stages. Manufacturers usually provide

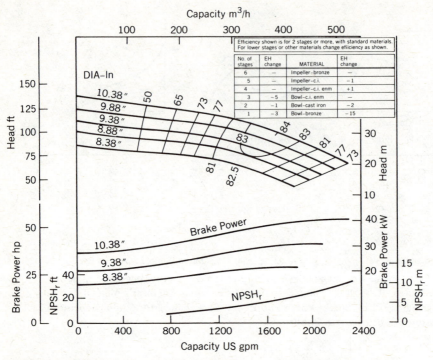

Figure 4.6 Example characteristic curves from a pump manufacturer's catalog.

information for making these adjustments. For example, information in Figure 4.6 indicates that graphed efficiencies must be lowered by three percentage points for a single-stage pump, lowered by one percentage point for a two-stage pump, and would remain unchanged for more than three stages.

4.3.2c Brake Power Versus Pump Capacity

The brake power (BP) versus capacity curve for a pump is derived from its H-Q and E_p-Q curves. An equation for computing BP from H, Q, and E_p is obtained by solving Eq. 4.3 for BP and substituting Eq. 4.2 for WP. The resulting equation is:

$$BP = \frac{100\,WP}{E_p} = \frac{(100)(Q)(H)}{(E_p)(K)}$$

(4.5)

where K is the unit constant in Eq. 4.2.

The shape of the BP-Q curve depends on the pump's specific speed and impeller design. Figure 4.5 shows that for radial flow impellers, BP generally increases from a nonzero value to a peak and then declines slightly as Q increases from zero. BP increases steadily from a nonzero value as Q increases for mixed flow impellers. For axial flow impellers, however, BP is maximum when Q is zero and

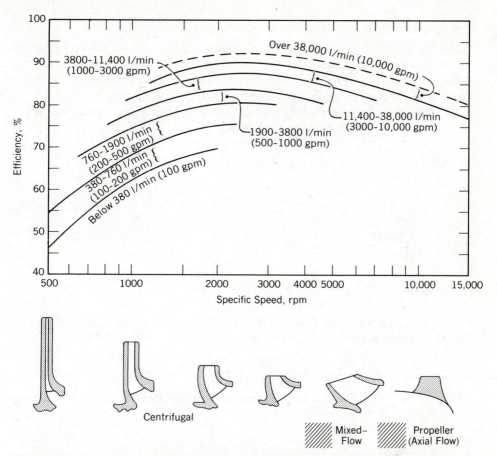

Figure 4.7 Theoretical pump efficiencies as a function of a specific speed, impeller design, and pump capacity. *Source:* G. O. Schwab, R. K. Frevert, T. W. Edminster, and K. K. Barnes, *Soil and Water Conservation Engineering*, copyright © 1981 by John Wiley & Sons, Inc., New York, 368 pp. Reprinted by permission of John Wiley & Sons, Inc.

steadily declines as Q increases from zero. Thus, the discharge side of the pump should be open to the atmosphere when axial flow impellers are started to minimize the start-up load. Similarly, a discharge-side valve should be closed when radial and mixed flow pumps are started.

4.3.2d Required Net Positive Suction Head Versus Pump Capacity

The fourth characteristic curve typically published by manufacturers is the $NPSH_r$ versus Q curve. Figure 4.4 shows that $NPSH_r$ steadily increases as Q increases for a typical radial flow pump. (The actual net positive suction head for a pump installation is discussed in Section 4.4.2.)

4.3.3 Affinity Laws

Changing the diameter and/or speed of an impeller alters its characteristic curves. This allows pump manufacturers to use a single impeller for a variety of head and discharge conditions and pump owners to alter pump performance to match changes in the configuration and/or operation of their irrigation systems.

Changes in impeller performance resulting from changes in pump speed can be estimated using the following equations.

$$Q_2 = Q_1\left(\frac{N_2}{N_1}\right) \tag{4.6}$$

$$H_2 = H_1\left(\frac{N_2}{N_1}\right)^2 \tag{4.7}$$

$$BP_2 = BP_1\left(\frac{N_2}{N_1}\right)^3 \tag{4.8}$$

$$(NPSH_r)_2 = (NPSH_r)_1\left(\frac{N_2}{N_1}\right)^2 \tag{4.9}$$

where the subscripts 1 and 2 refer to the original and new performance points. The following example demonstrates the use of Eqs. 4.6, 4.7, 4.8, and 4.9.

EXAMPLE 4.1 Estimating Changes in Pump Performance Resulting from a Change in Pump Speed

Given:
- original pump speed is 1750 rpm
- original discharge is 1000 gpm
- original head 300 ft
- original NPSH$_r$ is 12 ft
- original brake power is 100 hp

Required:
Q, H, BP, and NPSH$_r$ for a speed of 2000 rpm

Solution:

$$Q_2 = 1000 \text{ gpm} \left(\frac{2000}{1750}\right) = 1143 \text{ gpm}$$

$$H_2 = 300 \text{ ft} \left(\frac{2000}{1750}\right)^2 = 392 \text{ ft}$$

$$BP_2 = 100 \text{ hp} \left(\frac{2000}{1750}\right)^3 = 149 \text{ hp}$$

$$(NPSH_r)_2 = 12 \text{ ft}\left(\frac{2000}{1750}\right)^2 = 16 \text{ ft}$$

   Changes in pump performance due to changes in impeller diameter can be estimated using the following equations.

$$Q_2 = Q_1\left(\frac{D_2}{D_1}\right) \tag{4.10}$$

$$H_2 = H_1\left(\frac{D_2}{D_1}\right)^2 \tag{4.11}$$

$$BP_2 = BP_1\left(\frac{D_2}{D_1}\right)^3 \tag{4.12}$$

$$(NPSH_r)_2 = (NPSH_r)_1\left(\frac{D_2}{D_1}\right)^2 \tag{4.13}$$

Because trimming (reducing) the diameter of impellers may alter impeller geometry, Eqs. 4.10 through 4.13 approximate changes in pump performance. Eqs. 4.10 through 4.13 are most reliable for diameter changes of less than 20 percent. The following example illustrates the use of these equations.

EXAMPLE 4.2   Estimating Changes in Pump Performance Resulting from a Change in Impeller Diameter

Given:
• Information from Example 4.1
• Original diameter 8 in
Required:
Q, H, BP, and $NPSH_r$ for a diameter of 7.5 in
Solution:

$$Q_2 = 1000 \text{ gpm}\left(\frac{7.5}{8.0}\right) = 938 \text{ gpm}$$

$$H_2 = 300 \text{ ft}\left(\frac{7.5}{8.0}\right)^2 = 264 \text{ ft}$$

$$BP_2 = 100 \text{ hp}\left(\frac{7.5}{8.0}\right)^3 = 82 \text{ hp}$$

$$(NPSH_r)_2 = 12 \text{ ft}\left(\frac{7.5}{8.0}\right)^2 = 10.6 \text{ ft}$$

   Equations 4.6 through 4.13 are known as the affinity laws. These laws and the procedures demonstrated in Examples 4.1 and 4.2 can be used to generate shifts in entire head–pump capacity curves similar to those in Figure 4.8.

4.3.4   Performance Curves for Pumps Operating in Series

Two or more pumps operate in series when they are linked inlet-to-outlet as shown in Figure 4.9. Series hookups are used when the head required by the irrigation

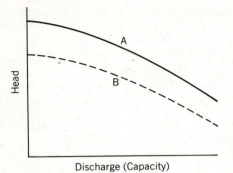

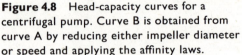

Figure 4.8 Head-capacity curves for a centrifugal pump. Curve B is obtained from curve A by reducing either impeller diameter or speed and applying the affinity laws.

system exceeds that which can be supplied by individual pumps. Figure 4.10 shows the individual head–discharge curves for pumps A and B and the combined head–discharge curve for pumps A and B operating in series. The combined curve is obtained by adding individual pump heads as shown in Figure 4.10.

The brake power and efficiency characteristic curves for pumps operating in series are obtained using Eqs. 4.14 and 4.15, respectively.

$$BP(Q)_c = \sum_{i=1}^{n} BP(Q)_i \qquad (4.14)$$

$$(E_p)_c = \frac{(100)(Q)}{(K)(BP(Q)_c)} \left(\sum_{i=1}^{n} H(Q)_i \right) \qquad (4.15)$$

where

$BP(Q)_c$ = combined brake power for discharge Q(kW, hp);
n = number of pumps operated in series;
$BP(Q)_i$ = brake power required by pump i to produce discharge Q(kW, hp);
$(E_p)_c$ = combined efficiency (percent);
Q = discharge (l/min, gpm);
$H(Q)_i$ = head corresponding to discharge Q for pump i(m, ft);
K = unit constant for Eq. 4.2.

When pumps are close together, that is, in the same pumping station, it is normally only necessary to provide the NPSH$_r$ needs of the first pump (i.e., the furthest pump upstream). When pumps are widely separated, however, the head available at the inlet of each pump should be determined and compared to the

Pump A   Pump B

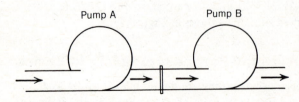

Figure 4.9 Two centrifugal pumps hooked in series. (The arrows indicate the direction of flow.)

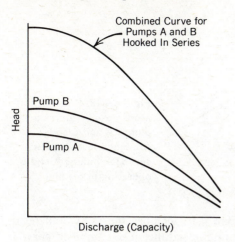

Head

Combined Curve for
Pumps A and B
Hooked In Series

Pump B

Pump A

Discharge (Capacity)

Figure 4.10 Individual head-capacity curves for pumps A and B and the combined head-capacity curve for pumps A and B operating in series.

NPSH$_r$ needs of the pump. It is also important that pressures not exceed those that can be withstood by pump seals and piping.

4.3.5 Performance Curves for Pumps Operating in Parallel

Pumps operate in parallel when they obtain water from a common source and discharge it into a single outlet as in Figure 4.11. Parallel operation of two or more pumps is a common method of meeting variable discharge requirements, since brake power and energy consumption can be minimized by operating only those

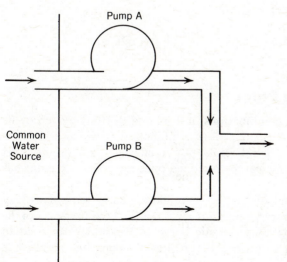

Pump A

Common
Water
Source

Pump B

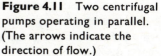

Figure 4.11 Two centrifugal pumps operating in parallel. (The arrows indicate the direction of flow.)

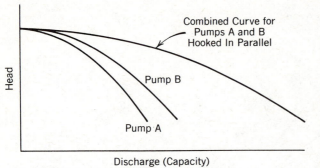

Figure 4.12 Individual head-capacity curves for pumps A and B and combined head-capacity curve for pumps A and B operating in parallel.

pumps needed to meet the demand. Figure 4.12 includes the individual head-discharge curve for pumps A and B operating in parallel. The combined curve is obtained by adding individual pump discharges as shown in Figure 4.12.

The brake power and efficiency characteristics curves for pumps operating in parallel are obtained using the following equations

$$\mathrm{BP}(H)_c = \sum_{i=1}^{n} \mathrm{BP}(H)_i \tag{4.16}$$

$$(E_p)_c = \frac{(100)(H)}{(K)(\mathrm{BP}(H)_c)} \sum_{i=1}^{n} Q(H)_i \tag{4.17}$$

where

$\mathrm{BP}(H)_c$ = combined brake power for head H(m, ft);

$\quad\quad n$ = number of pumps operated in parallel;

$\mathrm{BP}(H)_i$ = brake power required by pump i to produce head H(kW, hp);

$\quad(E_p)_c$ = combined efficiency (percent);

$\quad\quad H$ = head (m, ft);

$\quad Q(H)_i$ = discharge corresponding to head H for pump (l/min, gpm);

$\quad\quad K$ = unit constant for Eq. 4.2.

4.3.6   Pump Operating Point

A centrifugal pump operates at combinations of head and discharge given by its H–Q characteristic curve. The particular H–Q combination at which a pump is operating is the pump's operating point. Brake power, efficiency, and required net positive suction head for the pump can be obtained once the operating point has been determined.

The operating point depends on by the head and discharge requirements of the irrigation system. A system curve, which describes the H–Q requirements of the irrigation system, and the H–Q characteristic curve of the pump are used to determine the operating point. As shown in Figure 4.13, the operating point is the head and discharge at the point where the system and pump H–Q curves intersect

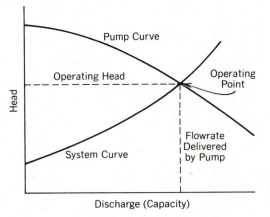

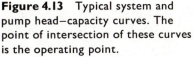

Figure 4.13 Typical system and pump head–capacity curves. The point of intersection of these curves is the operating point.

(i.e., where the H–Q requirements of irrigation system equal the H–Q produced by the pump).

    System curves are constructed by computing the heads required by the irrigation system to deliver different volumes of water per unit of time. System head is computed using the equation

$$H_s = \text{SL} + \text{DL} + \text{DD} + H_l + M_l + H_o + \text{VH} \tag{4.18}$$

where

  H_s = system head (m, ft);
  SL = suction-side lift (m, ft);
  DL = discharge-side lift (m, ft);
  DD = water source drawdown (m, ft);
  H_l = head loss due to pipe friction (m, ft);
  M_l = minor losses through fittings (m, ft);
  H_o = operating head (m, ft);
  VH = velocity head (m, ft).

The suction- and discharge-side static lifts are independent of system flow, while DD, H_l, M_l, H_o and VH all increase with increasing Q. H_s is independent of the pump with the exception of friction loss that occurs in the column pipe of vertical turbine (diffuser-type) pumps.

    The suction-side static lift, SL, is the vertical distance between the center line of horizontal pumps and the static water surface elevation of the water source (determined when the pump is not operating). For vertical pumps, SL is measured from the top of the discharge pipe rather than the centerline of the pump. SL is positive when the water source is below the pump and negative when the water source is above it. The water surface elevation for a well is the static water level in the well. The static water level of a water source is normally assumed to be constant even though there are fluctuations in static water levels as streams, lakes, reservoirs, and wells respond to changing hydrologic conditions. Considerable judgment is needed to choose the proper water level.

Lift on the discharge-side of the pump, DL, is the elevation difference between the point of delivery and the centerline and top of the discharge pipe of horizontal and vertical pumps, respectively. DL is computed using

$$DL = Elev_u - Elev_p \qquad (4.19)$$

where

$Elev_u$ = elevation of the point of delivery (m, ft);
$Elev_p$ = elevation of pump as defined for horizontal and vertical pumps (m, ft).

Drawdown, DD, is the decline in the water surface elevation of the water source due to pumping. For large surface bodies of water, drawdown is extremely small and is normally neglected. Drawdown versus Q relationships for wells and surface sources can be determined by test pumping the source at several pumping rates. Equations 3.7 and 3.8 can be used to estimate drawdown in unconfined and confined wells, respectively.

Head losses due to pipe friction (H_l) and minor losses (M_l) in fittings as a function of Q are determined using Eq. 5.12 through 5.15. The relationship between system operating head and discharge is normally established by the pressure-discharge relationships of individual sprinklers, the number of sprinklers, and pressure variations within the irrigation system. The systems H_0-Q relationship must be evaluated for the entire range of operating conditions. Velocity head is computed using

$$VH = \frac{Q^2}{KD^4} \qquad (4.20)$$

where

Q = system discharge (l/min, gpm);
D = diameter of discharge pipe at pump (cm, in);
K = unit constant (K = 435.7 for VH in m, Q in l/min, and D in cm.
$\quad K$ = 385.9 for VH in ft, Q in gpm, and D in in).

EXAMPLE 4.3 Determining the System Curve for a Sprinkle Irrigation System

Given:
• irrigation system with 100 sprinklers
• $Q_s = 1.41\ P^{0.5}$ (Q_s is sprinkler discharge in gpm; P is operating pressure in psi)

- 2000 ft long, 8 in diameter, PVC supply line ($C = 150$)
- minor losses are 10 percent of pipe friction
- water supply is a large reservoir
- water is lifted 200 ft to field

Required:
system curve

Solution:
quantifying the terms in Eq. 4.18

$$SL + DL = 200 \text{ ft}$$

$$H_l = \frac{(0.285\,C)^{-1.852}LQ^{1.85}}{D^{4.87}} \qquad (5.15 \text{ and } 5.16b)$$

$$= \frac{(0.285(150))^{-1.852}(2000)Q^{1.85}}{8^{4.87}}$$

$$= 7.63(10)^{-5}Q^{1.85}$$

$$M_l = (0.1)H_l$$

where

C = Hazen-Williams coefficient (from Table 5.7);

L = length of pipe (m, ft);

Q = flowrate (l/min, gpm);

D = diameter of pipe (mm, in).

Operating head (H_o):

$$Q = (1.41P^{0.5})100 = 141P^{0.5}$$

$$P = \left(\frac{Q}{141}\right)^2$$

$$H_o = (2.31)\left(\frac{Q}{141}\right)^2 = 1.16(10)^{-4}Q^2$$

$$VH = \frac{Q^2}{385.9(8)^4} = 6.33(10)^{-7}Q^2$$

$$H_s = 200 + 7.63(10)^{-5}Q^{1.85} + 0.763(10)^{-5}Q^{1.85}$$
$$+ 1.16(10)^{-4}Q^2 + 6.33(10)^{-7}Q^2$$

$$H_s = 200 + 8.39(10)^{-5}Q^{1.85} + 1.17(10)^{-4}Q^2$$

Tabular solution of this equation is as follows.

Q (gpm)	H_s (ft)
0	200
100	202
200	206
300	214
400	224
500	238
600	254
700	273
800	295
900	319
1000	347
1100	377
1200	410
1300	446
1400	485
1500	526

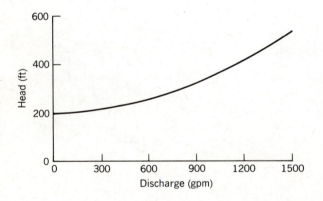

Operating points can be altered by changing either the H–Q curve for the pump or the irrigation system. Pump curves are altered by changing pump speed or impeller diameter as per the affinity laws. The following example illustrates a procedure, based on the affinity laws, for obtaining a new operating point.

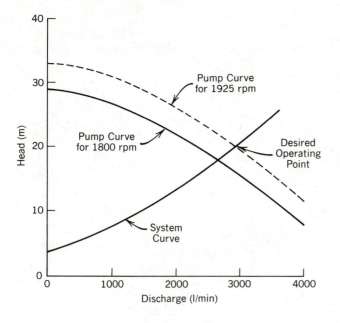

EXAMPLE 4.4   Obtaining a New Operating Point by Changing Pump Speed (or Impeller Diameter)

Given:
head–discharge curve for pump operating at 1800 rpm and system–head curve in the following diagram

Required:
pump speed to achieve an operating point of 2900 l/min and 20 m of head

Solution:

$$\frac{Q_1}{Q_2} = \frac{N_1}{N_2}$$

$$\left(\frac{H_1}{H_2}\right)^{1/2} = \frac{N_1}{N_2}$$

$$\frac{Q_1}{Q_2} = \left(\frac{H_1}{H_2}\right)^{1/2}$$

$$Q_1 = Q_2\left(\frac{H_1}{H_2}\right)^{1/2}$$

$$Q_1 = \frac{2900}{\sqrt{20}} H_1^{1/2} = 648.5\sqrt{H_1}$$

Since Q_1 and H_1 are unknowns in the previous equation, an additional equation is required to evaluate Q_1 and H_1. The pump curve provides the required relationship. The following table summarizes an iterative procedure for evaluating Q_1 and H_1:

Trial H_1 (m)	Computed Q_1 (l/min)	Q_1 from Pump Curve (l/lim)
21	2972	2150
15	2511	3100
19	2827	2500
17.5	2712	2712 ← solution

$$N_2 = N_1\left(\frac{Q_2}{Q_1}\right) = (1800)\left(\frac{2900}{2712}\right) = 1925 \text{ rpm}$$

Pump curve for $N_2 = 1925$ rpm.

1800 rpm		1925 rpm	
Q_1 (l/min)	H_1 (m)	Q_2 (l/min)	H_2 (m)
0	29.0	0	33.2
1000	26.5	1069	30.3
2000	22.0	2139	25.2
3000	15.8	3208	18.1
4000	8.0	4278	9.2

The operating point can also be altered by changing system head loss. This is accomplished with different pipe and fitting sizes in the design phase or by throttling (valve adjustment) after the system is in place. Increasing head loss results in steeper system head curves. Figure 4.14 shows how the operating point was shifted from point 1 to 2 by increasing head loss.

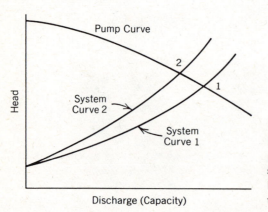

Figure 4.14 A shift in operating point resulting from a change in system head loss. There is more head loss associated with system curve 2 than system curve 1.

4.4 Pump Selection

Pump selection is the process of choosing the most suitable pump for a particular irrigation system. It involves specifying the performance requirements of the irrigation system, selecting the required pump type, and identifying alternate pumps that meet the requirements of the irrigation system. Normally, the most suitable pump is chosen from these pumps on the basis of economics.

4.4.1 Performance Requirements

The discharge and head requirements of the irrigation system must be known to select the most suitable pump(s). In systems where discharge and/or head requirements vary (depending on the number and identity of the fields being irrigated, for example), the range of discharges and/or heads required by the irrigation system must be determined. Discharge and head requirements are computed with Eqs. 3.6 and 4.18, respectively.

4.4.2 Pump Type

Horizontal volute and vertical diffuser (turbine) pumps are the main pump types used with farm irrigation systems. Because horizontal volute pumps are usually less expensive and cost less to install (than vertical pumps) they are normally used whenever possible. Vertical turbine pumps, which can be positioned below the water surface, are used in deep wells or with surface sources where it is not practical (economical) to position horizontal volute pumps so that their $NPSH_r$ needs are provided. Vertical turbine pumps are also used to eliminate the need for priming horizontal pumps.

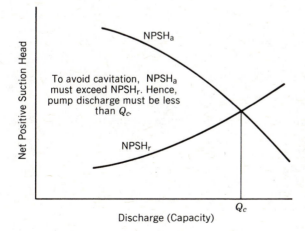

Figure 4.15 Relationship between pump discharge and required net positive suction head, $NPSH_r$, and available net positive suction head, $NPSH_a$. (The relationship between drawdown and discharge from the water source must be known to compute the $NPSH_a$–Q relationship.)

Available net positive suction head ($NPSH_a$) is often used to determine if a pump's $NPSH_r$ can be provided. A pump will operate without cavitation for all discharges where $NPSH_a \geq NPSH_r$ (see Figure 4.15). $NPSH_a$ is defined by the following equation

$$NPSH_a = BP - VP_w - (H_l)_s - (M_l)_s - VH_s - SL - DD \tag{4.21}$$

$$BP = K_1 - 1.17(10)^{-3}h + K_2 h^2 \tag{4.21a}$$

where

BP = barometric pressure (m or ft of water);

VP_w = vapor pressure of water (see Appendix F) (m or ft of water);

$(H_l)_s$ = head loss due to pipe friction in the suction line (m, ft);

$(M_l)_s$ = minor losses in suction line (m, ft);

VH_s = velocity head in suction line (m, ft);

SL = suction-side lift (m, ft);

DD = drawdown (m, ft);

h = elevation above sea level (m, ft);

K_1, K_2 = unit constants ($K_1 = 10.33$ and $K_2 = 5.55(10)^{-8}$ for BP in m of water and h in m. $K_1 = 33.89$ and $K_2 = 1.69(10)^{-8}$ for BP in ft of water and h in ft).

VP_w can be obtained from Appendix F. The following example illustrates the use of Eq. 4.21.

EXAMPLE 4.5 Determining $NPSH_a$

Given:

- $Q = 1000$ gpm

- 8-in-diameter, 25-ft-long aluminum suction line (Hazen-Williams $C = 135$)

- water temperature $= 65°F$

- water level remains constant during pumping

- elevation is 1000 ft above sea level

- the following diagram

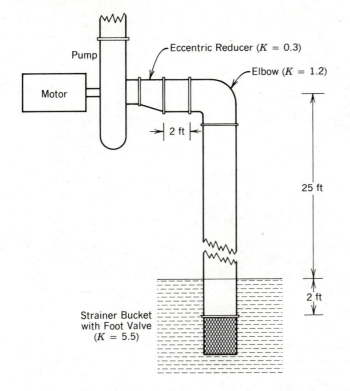

Required:
NPSH$_a$

Solution:
use Eq. 4.21

$$BP = 33.89 - 1.17(10)^3(1000) + 1.69(10)^8(1000)^2$$
$$= 32.74 \text{ ft}$$

$$VP_w @ 65°F = 0.31 \qquad (\text{From Appendix F})$$
$$= (0.31)(2.31 \text{ ft/psi}) = 0.72 \text{ ft}$$

$$(H_l)_s = \frac{((0.285)(135))^{-1.852}(1.0)(29)(1000)^{1.85}}{8^{4.87}} = 0.48 \text{ ft} \qquad (5.15)$$

$$(M_l)_s + VH_s = (0.3 + 1.2 + 5.5 + 1.0)\frac{Q^2}{385.9D^4} = (8.0)\frac{1000^2}{385.9(8)^4}$$

$$= 5.06 \text{ ft}$$

$$SL + DD = 25 + 0 = 25 \text{ ft}$$

$$NPSH_a = 32.74 - 0.72 - 0.48 - 5.06 - 25 = 1.49 \text{ ft}$$

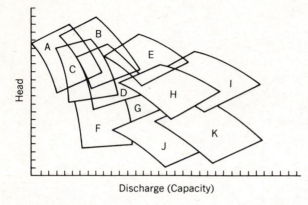

Head

Discharge (Capacity)

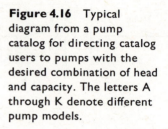

Figure 4.16 Typical diagram from a pump catalog for directing catalog users to pumps with the desired combination of head and capacity. The letters A through K denote different pump models.

Since NPSH$_r$ for most 1000-gpm-capacity horizontal volute pumps exceeds 1.49 ft, reducing the suction lift from 25 ft by repositioning the pump closer to the water surface should be considered. If this is not practical or too expensive, a vertical turbine pump should probably be considered.

4.4.3   Identifying Suitable Pumps

Manufacturer catalogs are consulted to identify pumps of the proper type that are capable of supplying the discharge and head requirements of the irrigation system. Most pump catalogs have tables or graphs similar to Figure 4.16 to direct catalog users to pumps with the desired capacity and head combinations. Characteristic curves for these pumps are examined to determine which of these pumps are suitable for the irrigation system.

4.4.4   Selecting the Most Suitable Pump or Combination of Pumps

Economics is often the primary criterion for selecting the most suitable pump or combination of pumps for a farm irrigation system. In such cases, a detailed analysis of the total annual ownership and operation costs is necessary. In other situations, minimizing energy use or first cost may be more important. Selection criteria must be specified before the most suitable pump or combination of pumps can be selected. The following example illustrates a selection process and the effect of different criteria on pump selection.

EXAMPLE 4.6   Selecting the Most Suitable Pump

Given:
- irrigation system in which the lift ranges between 100 and 130 ft
- system curves for each lift

- pump operates: $\frac{1}{4}$ of time at 100-ft lift
    - $\frac{1}{4}$ of time at 130-ft lift
    - $\frac{1}{2}$ of time at 115-ft lift
- crop requires 30 inches of water per year
- head–discharge and efficiency–discharge curves for pumps A and B
- pump A is a 3-stage, 7.5-in-diameter, 1760-rpm, vertical turbine pump that costs $12,200 (cost includes 50-hp electric motor)
- pump B is a single-stage, 6.25-in-diameter, 3500-rpm vertical turbine pump that costs $9,200 (cost includes 50-hp electric motor)

Required:
a. pump with lowest energy cost
b. pump with lowest annual fixed cost
c. pump with lowest total annual fixed and operating cost

Solution:
On the basis of the economic data summarized in the following table, pump A has the lowest energy costs and total annual fixed and operating costs. Pump B has the lowest total annual fixed costs.

Fixed Costs	Pump A	Pump B
Annual depreciation and interest	$1555	$1173
Annual taxes and insurance	244	184
Total Fixed Costs	$1799	$1357
Operating Costs		
Annual maintenance and repair	$ 366	$ 276
Energy	4176	4539
Total operating costs	$4542	$4539
Total annual fixed and operating costs	$6341	$5896

Annual Depreciation and Interest Costs (ADIC):

ADIC = CRF(PW of pump and motor) (Eq. 2.18)

CRF is computed with Eq. 2.17a using an interest rate of 12 percent and an analysis period of 25 yr

$$CRF = \frac{(0.12)(1 + 0.12)^{25}}{(1 + 0.12)^{25} - 1} = 0.13$$

Assume SV = 0 and a useful life of 25 yrs (from Table 2.4)
PW = IC when SV = 0 and AP = UL (Eq. 2.19)
ADIC = 0.13 (IC)
Annual taxes and insurances = 2 percent of initial cost (see Section 2.4.6a)
Maintenance and Repair = 3 percent of initial cost (see Section 2.4.6b)
Energy = ($0.03/kWh)(kWh)

Kilowatt-hour calculations are summarized in the following table.

Pump	H (ft)	Q^a (gpm)	E_p^a	BP^b (kW)	t^c (hrs)	kWh^d	Total kWh
A	166	820	79	32.5	497	16121	
	158	940	81	34.5	867	29930	
	150	1000	81	34.9	407	14207	60,258
B	165	810	77	32.7	503	16440	
	158	940	79	35.4	867	30688	
	150	1000	80	35.3	407	14385	61,513

[a] Q and E_p were obtained from the pump characteristic curves.

[b] To obtain parametric pressure

$$^b BP = \frac{(100)(Q)(H)}{(E_p)(3960)} (0.746 \text{ kW/hp})$$

[c] To obtain duration

$$t = \frac{(452.5)(120 \text{ acres})(\text{inches applied})}{Q \text{ in gpm}}$$

[d] To obtain Kilowatt-hours

$$kWh = (BP)(t)$$

Homework Problems

4.1 Water is being pumped from a reservoir into 25-cm-diameter pipe and conveyed to a sprinkle irrigation system. The system design capacity is 4000 l/min. Determine

a. the head that the pump supplies, and

b. the power that the pump imparts to the water (i.e., the water power)

when a pressure gauge located immediately downstream of the pump reads 600 kPa. The water surface of the reservoir is 5 m below the centerline of the pump.

4.2 Determine the specific speed of the pump in Problem 4.1 for a pump speed of 1750 rpm. Does the pump have a radial, mixed, or axial flow impeller?

4.3 Determine the brake power requirement of the pump in Problem 4.1 if it is 70 percent efficient.

4.4 Determine the demand power requirement of the pump in Problems 4.1 and 4.3 if the electric motor that powers it is 90 percent efficient. Demand power = brake power/motor efficiency.

4.5 Water is pumped from a reservoir into a 1000-m-long, 25-cm-diameter steel pipe and conveyed to an open canal. The water surface in the canal is 15 m above the water surface in the reservoir. Determine
a. the total head that the pump must supply,
b. the water power,
c. the brake power, and
d. the demand power

if 5000 l/min is being pumped, the pump speed is 3600 rpm, and the pump and motor efficiencies are 75 and 92 percent, respectively. There are three elbows ($k = 1.2$) in the pipeline and a strainer bucket with a foot valve ($k = 10$) on the pump inlet pipe (i.e., on the suction line).

4.6 A vertical turbine pump is being used to pump water from a 200-m-deep, 50-cm-diameter well that completely penetrates a 50-m-thick confined aquifer. The conductivity of the aquifer is 2.0 m/day and the static water level is 100 m below the ground surface. A pressure of 1000 kPa is required on the discharge side of the pump to operate a sprinkle system with a design capacity of 3000 l/min. Minor and pipe friction losses on the suction side of the pump total 20 m. Neglect velocity head. Determine
a. the total head that the pump must supply, and
b. the water power.

4.7 Use data from Figure 4.6 to develop
a. the head–discharge,
b. the brake power–discharge, and
c. the net positive suction head–discharge

curves for a 10.38-in-diameter impeller operating at 1200 rpm. The characteristic curves in Figure 4.6 are for a pump speed of 1800 rpm.

4.8 Use data from Figure 4.6 to develop
a. head–discharge and
b. brake power–discharge

curves for two identical pumps with 10.38-in-diameter impellers hooked in series.

4.9 Repeat Problem 4.8 for two identical pumps with 10.38-in-diameter impellers hooked in parallel.

*4.10 Develop a system curve for the conveyance system in Problem 4.5.

*4.11 Repeat Problem 4.10 for a 30-cm-diameter pipeline. How does a change in minor and/or pipe friction loss affect the system curve?

*4.12 Repeat Problem 4.10 using a difference in water surface elevations between the reservoir and the canal of 20 m. How does a change in lift affect the system curve?

* Indicates that a computer program will facilitate the solution of the problems so marked.

4.13 Use the system curve from Problem 4.10 to determine the operating point for the pump in Figure 4.6. The pump has a 10.38-in-diameter impeller and a speed of 1800 rpm.

4.14 Repeat Problem 4.13 using the system curves from
   a. Problem 4.11, and
   b. Problem 4.12.

4.15 Use data from Problem 4.13 to determine the speed at which the pump would have to be operated to obtain a discharge of 5000 l/min.

4.16 The pump in Figure 4.6 will be used to pump water from a very large reservoir into a sprinkle irrigation system with a design capacity of 5000 l/min. The suction side pipe is a 25-cm-diameter, 7-m-long aluminum pipe with a bucket strainer (without a foot valve), an elbow, and an eccentric reducer arranged as in Example 4.5. The temperature of the water in the reservoir is 20°C. The water surface in the reservoir is 500 m above mean sea level. Determine the maximum vertical distance that the pump can be located above the water surface of the reservoir.

*4.17 The pump in Problem 4.16 is being used to pump reservoir water through a 500-m-long, 25-cm-diameter aluminum pipe to a sprinkle irrigated field located 1 m above the pump. The minor loss coefficient, k, for the pipe network downstream of the pump is 15. The discharge of individual sprinklers is given by the following equation

$$Q = 2.31P^{0.5}$$

where

Q = sprinkler discharge in l/min
p = operating pressure in kPa.

Determine the operating point when
   a. 100 sprinklers,
   b. 150 sprinklers

are operating.

4.18 The sprinkle system in Problem 4.17 operates with 150 sprinklers approximately 65 percent of the time and 100 sprinklers the other 35 percent of the time. The pump operates 1000 hours per irrigation season. Which of the following pumps has
   a. the lowest operating costs,
   b. the lowest fixed costs, and
   c. the lowest total annual costs.

Q	Pump A		Pump B	
(l/min)	Head	Efficiency	Head	Efficiency
0	51.0		37.5	
1000	48.5		39.5	
2000	44.3		40.2	70
3000	40.5	67	39.8	78
4000	36.3	78	36.2	81
5000	32.0	84	32.1	78
6000	27.4	85	23.5	62

The cost of pump A and B, including a 40-kW electric motor, are $5000 and $4350, respectively. Use an energy cost of $0.04 per kWh.

References

Anon. (1959). Irrigation pumping plants. Section 15 of the *National Engineering Handbook*. U.S. Department of Agriculture, Soil Conservation Service, 70 pp.

Anon. (1977). *Ground Water Manual*. U.S. Department of the Interior, Bureau of Reclamation, 480 pp.

Dornaus, W. L. (1976). Intakes and suction piping. In *Pump Handbook*, I. J. Karassik, W. G. Krutzsch, W. H. Fraser, and J. P. Messina (Eds.), McGraw-Hill, New York, pp. 11-1–11-19.

Hansen, V. E., I. O. Isrealsen, and G. Stringham (1979). *Irrigation Principles and Practices*. Wiley, New York. 417 pp.

Hicks, T. G., and T. W. Edwards (1971). *Pump Application Engineering*. McGraw-Hill, New York, 435 pp.

Karissik, I. J. (1971). *Centrifugal Pump Clinic*. McGraw-Hill, New York, 479 pp.

Karissik, I. J. (1976). Centrifugal pump construction. In *Pump Handbook*, I. J. Karissik, W. C. Krutzsch, W. H. Fraser, and J. P. Messina (Eds.), McGraw-Hill, New York, pp. 2-31–2-124.

Longenbaugh, R. A., and H. R. Duke (1980). Farm Pumps. In *Design and Operation of Farm Irrigation Systems*, M. E. Jensen (Ed.), ASAE Monograph 3, St. Joseph, MI pp. 347–391.

Pair, C. H., W. H. Hinz, K. R. Frost, R. E. Sneed, and T. J. Schiltz (1983). *Irrigation*. The Irrigation Association, Silver Spring, Md., 686 pp.

Schwab, G. O., R. K. Frevert, T. W. Edminster, and K. K. Barnes (1981). *Soil and Water Conservation Engineering*. Wiley, New York, 525 pp.

Simon, A. L. (1976). *Practical Hydraulics*. Wiley, New York, 306 pp.

Walker, R. (1972). *Pump Selection: A Consulting Engineer's Manual*. Ann Arbor Science Publishers, Ann Arbor, Mich., 118 pp.

5

Sprinkle Irrigation Systems

5.1 Introduction

A sprinkle irrigation system uses pressure energy to form and distribute "rainlike" droplets over the land surface. Although they are normally designed to supply the irrigation requirements of the farm, sprinkle systems are also used for crop and soil cooling, frost protection, delaying fruit and bud development, controlling wind erosion, providing water for germinating seeds, application of agricultural chemicals, and land application of waste waters. (See Section 2.2 for a description of sprinkle systems uses.)

In a sprinkle system, water is conveyed from a pump or other source of water under pressure through a network of pipes, called mainlines and submains, to one or more pipes with sprinklers called laterals. The sprinklers distribute the water over the land surface. A typical sprinkle system is diagrammed in Figure 5.1.

This chapter describes sprinkle irrigation systems and their design. The operating characteristics of several types of sprinkle systems, procedures for selecting sprinklers and pipelines, and the basic steps in the design of set-move, solid-set, traveler, center pivot, and linear-move systems are presented in this chapter. After completing this chapter the student should be able to: describe the operating characteristics of sprinkle systems and sprinklers; choose sprinklers and sprinkler spacings that provide the required application amount and uniformity without runoff; design pipe networks that provide the discharge and pressure requirements of sprinklers; understand waterhammer and air entrapment in pipelines and how to control them; and design set-move, solid-set, traveler, center-pivot, and linear-move sprinkle systems.

180

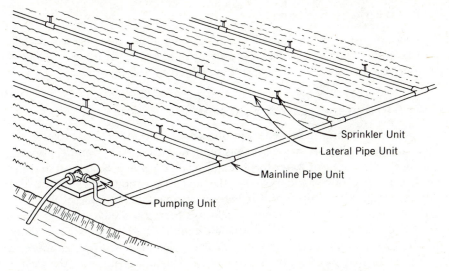

Figure 5.1 A typical sprinkle irrigation system consists of four basic units: a pumping unit, mainline pipes, lateral pipes, and one or more sprinklers.

5.2 Types of Sprinkle Systems

Sprinkle irrigation systems may be classified as portable, semiportable, semipermanent, or permanent. They are also classified as set-move, solid-set, or continuous move.

5.2.1 Portable System

A *portable system* has portable mainlines, submains, laterals, and a portable pumping plant. The entire system can be moved from field to field.

5.2.2 Semi-portable System

A *semiportable system* is similar to a fully portable system except that the location of the water source and pumping plant is fixed. Such a system may be used on more than one field where there is an extended mainline, but may not be used on more than one farm unless there are additional pumping plants.

5.2.3 Semipermanent System

A *semipermanent* system has portable lateral lines, permanent mainlines, and a stationary water source and pumping plant. The mainlines are usually buried, with risers (located at suitable intervals) for connecting laterals.

5.2.4 Permanent System

A fully *permanent system* has buried mainlines, submains, and laterals with a stationary pumping plant and/or water source. Sprinklers are permanently located on each riser. Such systems are costly and are suited to automation.

5.2.5 Set-Move Irrigation Systems

Set-move sprinkle systems are moved from one set (irrigation) position to another by hand or mechanically. Set-move systems remain stationary as water is applied. When the desired amount of water has been applied, the water is shut off and the sprinkler laterals are drained and moved to the next set position. When the move is complete the water is turned on and irrigation resumed at the new set position. This sequence is repeated until the entire field has been irrigated.

Set-move systems commonly have a single mainline laid through the center of the field with one or more laterals on each side of the mainline. Multiple laterals are equally spaced so that when a lateral reaches the starting position of the lateral ahead of it, the entire field has been irrigated once.

5.2.5a Hand-Move

Hand-move laterals are moved by uncoupling, picking-up, and carrying sections of lateral pipe by hand to the next set position where the lateral sections are reconnected. Most hand-move sprinkler laterals are aluminum 50 to 150 mm (2 to 6 in) in diameter, and either 6, 9, or 12 m (20, 30, or 40 ft) in length. Short lengths increase the walking during moves while longer lengths are difficult to handle and may not provide proper spacing for the common sprinkler sizes.

5.2.5b Tow-Move

Tow-move sprinkle systems are the least expensive type of mechanically moved set-move system. Each section of a *tow-move lateral* has skids or wheels so that the entire lateral can be pulled to the next set position. The conventional way of moving of tow-move lateral is diagrammed in Figure 5.2. Usually a tractor is hooked to the mainline end of the lateral and the lateral pulled around the capstands in a S-shaped curve across the mainline to the next lateral position. For the next setting, the cap stands are moved and the lateral is dragged in the other direction across the mainline in an opposite S-shaped curve. The moves are made easier by buried mainlines.

Tow-move systems are not used extensively because moving the laterals is tedious, requires careful attention, and also damages many crops. Tow-move systems have been used successfully in some forage crops and in row crops.

5.2.5c Side-Roll

A *side-roll* or *wheel-move system*, like the one in Figure 5.3, is an extremely popular type of mechanically moved set-move system. Each section of pipe in a side-roll lateral has a wheel, with the pipe serving as the axle of the wheel. A

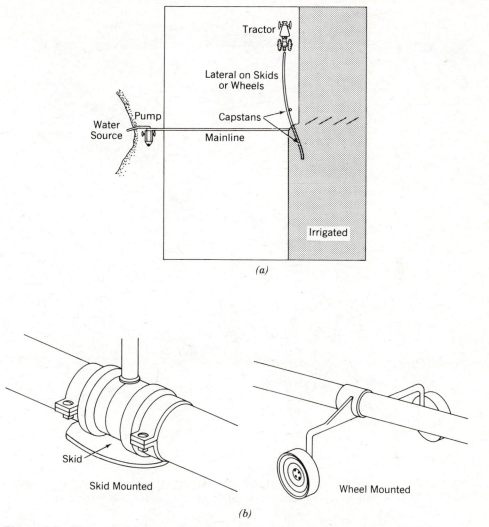

Figure 5.2 The essential parts of a tow-move sprinkle irrigation system. (*a*) Field layout for multi-sprinkler, tractor-moved system. (*b*) Skid-mounted and wheel-mounted tractor-moved system.

gasoline engine and transmission with a reverse gear at the center or the end of the lateral supplies the power needed to roll the lateral, which may be as long as 800 m (about one-half mile), from one set position to the next. The lateral is commonly 100 or 125 mm (4 or 5 in) diameter aluminum pipe with a wall thickness of at least 1.8 mm (0.072 in). Each lateral section is usually 12.2 m (40 ft) long with a wheel at its center and a sprinkler mounted on a short riser at one end. Often the sprinklers have self-levelers to "right" the sprinkler when the lateral is stopped so that the riser is "tilted" from its upright position. A drain valve that opens automatically

(a)

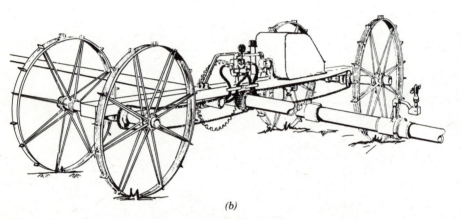

(b)

Figure 5.3   The essential parts of a side-roll (wheel-move) sprinkle irrigation system. (*a*) Wheels, pipe, self-plumbing sprinklers. (*b*) Power mover (usually located at center or end of lateral).

when there is a loss of pressure is usually located opposite each riser. This allows the lateral to be quickly drained and permits moving the lateral with a minimum time loss. The most common spacing along the mainline is 18.3 m (60 ft).

Common wheel diameters for side-roll laterals are 1.2, 1.5, 1.6, and 1.9 m (46, 58, 64, and 78 in). The wheel diameter must be large enough to allow the lateral to pass over the crop without damaging it and provide the desired spacing between sets.

Water can be supplied at the end or middle of side-roll laterals. The friction loss when water is admitted at the center of the lateral is one-fifth as much as when water enters at the lateral's end. Additional mainline may be required to admit water at the center of the lateral, however.

In one of the common operating schemes for side-roll systems, the lateral is moved along the mainline using each outlet valve. When the lateral reaches the last outlet in its zone, the lateral is drained and moved to the opposite side of the mainline. When side-roll laterals can operate only on one side of the mainline, they are sometimes rolled (walked) back empty to the first outlet in its zone. In another scheme, the lateral is connected to every even-numbered outlet value on the mainline as it is moved across its zone, and then connected to the odd-numbered outlet valves on its way back across the zone.

5.2.5d Gun-type

Another type of set-move system consists of a *large-volume* (*big-gun*) *sprinkler* mounted on a wheeled cart or trailer that is moved from set to set with a tractor or by hand. Sprinklers with capacities as large as 4700 l/m (about 1250 gpm), wetted diameters of as much as 180 m (about 600 ft), and a recommended operating pressure range of 480 to 896 kPa (70 to 130 psi) are commonly used. These systems are sometimes used for waste water disposal.

5.2.6 Solid-Set Systems

A *solid-set system* has enough laterals and sprinklers to irrigate the entire field simultaneously (although simultaneous operation of all sprinklers usually occurs only during frost protection). These systems can be portable, semiportable, semipermanent or permanent.

Portable, semiportable, and semipermanent solid-set systems usually have aboveground aluminum laterals that are placed in the field at the start of the irrigation season and left until harvest. Permanent systems have underground mainlines and laterals with only the sprinklers and a portion of the risers above ground.

Although it is possible to irrigate the entire field simultaneously, the field is usually divided into blocks of adjacent laterals. During irrigation, water is switched sequentially from block to block until the field is irrigated. Individual "prescription irrigations" for each block are possible.

An example of a block-type solid-set system is diagrammed in Figure 5.4. A mainline brings water from the source to an on/off valve (either manual or automatic) that controls flow into each of the 10 blocks. Each block contains a submain that distributes water to eight laterals. During irrigation, water is cycled from block-to-block until the entire field has been irrigated. Only one block receives water at a time.

Another scheme of solid-set irrigation is illustrated in Figure 5.5. Again there are enough sprinklers to simultaneously irrigate the entire field. In this scheme, however, individual laterals are controlled by valves so that each lateral can be operated individually (rather than as a block). Normally, several laterals located at widely separated positions in the field are operated simultaneously. The advantage of the individual lateral control solid-set system is that the size of some mainline pipe can be reduced (compared to the block system). The disadvantages are the

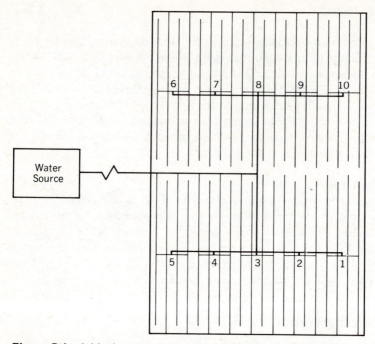

Figure 5.4 A block-type solid-set sprinkle irrigation system. This arrangement is also used with trickle irrigation systems.

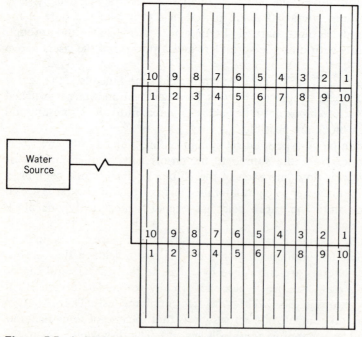

Figure 5.5 Individual lateral control solid-set irrigation system.

186

number of valves required; and the time to open and close valves when a manual valve system is used.

The cost of solid-set systems can be reduced by purchasing only enough sprinklers to irrigate a portion of the field. Sprinklers are moved manually from one location to another across the field. Laterals are buried as in the previously described systems. Such systems cannot be used for frost protection nor can they be fully automated.

5.2.7 Continuous-Move Systems

Continuous-move systems have laterals and sprinklers that remain connected to the mainline and move continuously as water is supplied. The popularity of these systems has steadily increased as labor costs have risen and shortages of labor for moving portable laterals and sprinklers have continued. Three major types of continuously moving sprinkle systems are discussed here. These are center-pivot, traveler, and linear-move systems.

5.2.7a Center-Pivot Sprinkle System

A *center-pivot irrigation* system consists of a sprinkler lateral that rotates in a circle around a fixed pivot structure. Water is supplied to the lateral at the pivot point. The lateral is supported by towers spaced each 24 to 76 m (80 to 250 ft) along the lateral and trusses or cables. Towers are usually mounted on wheels and are driven by 0.4, 0.75, 1.1 kW ($\frac{1}{2}$, 1, or $1\frac{1}{2}$ hp) electric motors. They can also be driven by hydraulic oil or water drive motors. Laterals are generally about 365 to 400 m (about 1200 to 1300 ft) long, but can be as short as 60 m (about 200 ft) or as long as 790 m (about 2600 ft). Sprinklers operate at pressures ranging from 140 kPa (about 20 psi) to more than 790 kPa (about 100 psi).

The pivoting lateral is kept in a straight line as it moves around the pivot point by an alignment system that starts and stops movement of the towers as required to maintain alignment. The speed at which the system rotates is determined by the amount of time that the outermost tower operates. The remaining towers operate as needed to maintain alignment with the outer tower.

Center-pivot (CP) systems are best suited to coarse textured soils with high infiltration capacities and are especially useful where light, frequent irrigations are required. When properly designed, CP systems can be successfully used on steep and undulating terrain, and on some finer textured soils.

The primary disadvantage of center-pivot irrigation systems is the circular shape of the irrigated area (unless special equipment, described later, is used to irrigate the corners of square fields). In a square-shaped 65 ha (160 acre) field, for example, only 51 ha (126 acres), about 79 percent of the field, can be irrigated with a conventional center-pivot system. Some center-pivot laterals have a "corner-catcher" lateral added to the main lateral that folds out as the corner is approached and folds back toward the main lateral as the machine moves out of the corner toward the near side of the field. Wires buried around the edge of the field control the position of the corner-catcher unit. Some systems use telescoping laterals rather than folding sections. Large-gun sprinklers mounted at the end of the main lateral or corner lateral are also used to increase the area irrigated. These sprinklers are turned on in the corners and off along the near sides of the field. This cycling of sprinklers on and off can reduce pump efficiency.

5.2.7b   Linear-Move Systems

Linear-move systems have been developed in response to the problems of covering the corners and runoff associated with center-pivot irrigated fields (the average application rate is half that at the end of the pivot). *Linear-move systems* have towers with electric motors and alignment systems, just as do center-pivot systems. The lateral moves continuously in a linear fashion across the field rather than rotating about a central pivot point. Water is supplied to the laterals through a flexible hose hooked to a mainline or by a traveling pumping plant that pumps water from an open ditch. The pump is often powered by a diesel engine. The engine also drives an electric generator that powers the electric motors of each tower. Recently, a linear-move system that moves continuously and obtains water from risers connected to a buried mainline by automatically connecting and disconnecting itself from riser valves has been developed. Because this system eliminates the need for a graded or level open ditch, it can operate on more extreme topography.

Linear-move systems are not as adaptable to daily irrigation as are center-pivot systems, because a center-pivot system is at its starting position when an irrigation is completed, while a linear-move system is at the opposite end of the field when an irrigation is completed.

Several management options for linear-move systems are diagrammed in Figure 5.6. In Option 1, the left half of the field is irrigated as the machine moves from the field's left edge to its approximate center. At the center of the field, the water is shut off and the system is moved as rapidly as possible across the right half of the field (the right half of the field is not irrigated). The water is then restarted and the direction of travel reversed. When the system returns to the center of the field the water is again shut off and the system quickly moved to the field's left edge. Option 1 minimizes deep wheel-track rutting caused by running the machine across recently irrigated portions of the field.

In Option 2, the entire field is irrigated as the machine moves across the field. The machine is turned off when it reaches the opposite side of the field and not restarted until the soil has dried enough to minimize wheel-track rutting. To begin the next irrigation the machine is quickly moved to the original side of the field "dry" (i.e., without applying water). Option 2 increases system cost by increasing the design discharge, but minimizes operator attention.

Option 3 involves moving the machine back and forth across the field, irrigating in both directions. This option is best suited to frequent shallow applications on soils with relatively high infiltration capacities that do not "rut."

Option 4 is similar to Option 1, except that the machine does not run dry during any portion of an irrigation. The machine is started at one end and run "wet," applying a relatively "deep" application to the middle of the field. At the

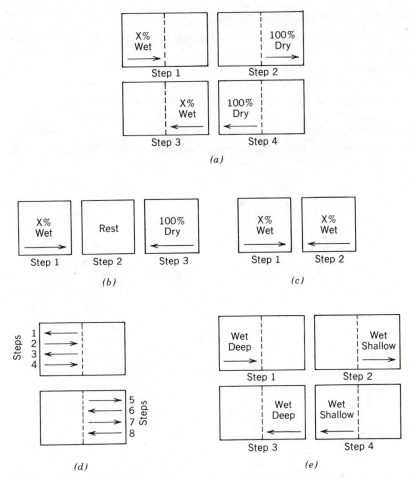

Figure 5.6 Management options for linear move sprinkle systems.
(a) Option 1. (b) Option 2. (c) Option 3. (d) Option 4. (e) Option 5.

middle of the field, the speed is increased to reduce the depth applied. Upon reaching the end of the field, the system is reversed and slowed down to increase the application depth. The speed is increased at the middle of the field so that less water is applied as the machine moves to the original starting point. Option 4 is used in situations where it is not possible or desirable to turn the water "off" while the system is being repositioned.

Option 5 can be used when the supply ditch or mainline bisect the field and the lateral can be moved from one side of the supply ditch or mainline to the other. The machine starts irrigating at one end of the field, and when it reaches the opposite end of the field the system is stopped and drained. Its wheels are rotated 90 degrees and the machine moved to the opposite side of the supply ditch or mainline.

5.2.7c Traveler Sprinkle Systems

Travelers are continuous-move sprinkle systems with a high-capacity gun-type sprinkler mounted on a cart and a hose that conveys water from a buried or portable mainline to the sprinkler. The cart is pulled across the field by either a cable and winch or the supply hose itself. Travelers are classified as either soft- or hard-hose systems, depending on the type of hose utilized and the method used to pull the cart across the field.

A typical soft-hose system, also called a hose-drag or cable-tow system, is diagrammed in Figure 5.7*a*. Soft-hose travelers have a cart-mounted cable winch

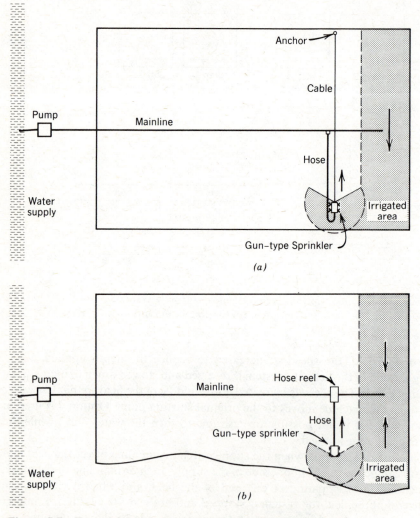

Figure 5.7 Typical field layouts for (*a*) soft-hose (cable-tow) and (*b*) hard-hose (hose-pull) traveler irrigation systems.

that pulls the cart and collapsible hose across the field (as shown in Figure 5.7*a*). To operate a soft hose system, the sprinkler cart is placed on one side of the field and the cable unreeled from the cart and anchored firmly at the field's opposite side. The hose is connected to the mainline. During irrigation, the winch pulls the cart and hose across the field as the cable is wrapped around the reel. The winch is powered by a cart-mounted water turbine, water piston, or internal combustion engine.

A hard-hose traveler, like the one in Figure 5.7*b*, has a hose reel that pulls the sprinkler cart across the field. To operate the system, a semirigid polyethylene hose is unreeled from the reel as the sprinkler cart is deployed to the opposite side of the field from the hose reel. During irrigation, the cart is pulled across the field by the hose as it is wound around the reel. The hose reel is powered by a water piston, water turbine, or internal-combustion engine. Sufficient pressure to provide 550 kPa (about 80 psi) at the sprinkler is required to operate most hard-hose travelers.

The advantages of soft-hose machines include their light weight and the compactness of the hose when rolled up. Disadvantages include the requirement that all hose must be unrolled before water can be pumped through it. Because of the hose curl behind the sprinkler cart, at least a 3 m (10 ft) wide travel lane must be left through the crop.

Hard-hose travelers require smaller travel lanes because the hose pulls in a straight line. This straight-line operation presents less opportunity for hose damage than the hose curling associated with soft-hose systems. Another important attribute of hard-hose systems is that better drive life can be expected, since the sprinkler cart is separate from the drive mechanism. This is especially important when manures and other slurries are being sprayed. A disadvantage associated with hard hoses is that polyethylene becomes increasingly brittle as its temperature decreases. Operation of hard-hose systems at temperatures less than 4°C (40°F) should be avoided.

Traveler systems are adaptable to many field sizes, shapes, and terrains. A traveler system can irrigate a rectangular strip as long as 800 m (one-half mile) and can be transported from field to field at highway speed. Traveler systems usually require a crop-free towpath.

5.3   Sprinkle System Components

Sprinklers, laterals, submains, and mainlines are the primary components of a sprinkle irrigation system. Sprinklers spread water as "rainlike" droplets over the land surface. Laterals convey water from the mainlines and submains to the sprinklers. Mainlines convey water from the water source to the submains and laterals. Sprinklers, laterals, mainlines, and submains are discussed in the following sections.

5.3.1 Sprinklers

The function of *sprinklers* is to distribute water uniformly over the land without runoff or excessive drainage (i.e., deep percolation) from the root zone. Many different types of sprinklers have been developed to accomplish this.

5.3.1a Types of Sprinklers

Most sprinklers are either rotating or fixed-head sprinklers. *Rotating sprinklers* include impact, gear-driven, and reaction-type sprinklers. *Fixed-head sprinklers* include most of the spray-type sprinklers currently available.

(i) *Impact Sprinklers* Impact sprinklers have one or more nozzles that discharge jets of water into the air. These jets are rotated in a start and stop manner as a spring-loaded arm strikes (impacts) and then is bounced out of one of the jets. The spring returns the arm to strike the jet again and the process is repeated.

Several different nozzle types have been developed for impact sprinklers. These include constant-diameter, constant-discharge, and diffuse-jet nozzles. Constant-diameter nozzles, made of brass or plastic, are the most common type of nozzles used with impact sprinklers. The discharge from these nozzles is proportional to the square root of the operating pressure.

Constant-discharge nozzles are also used with impact sprinklers. These nozzles are constructed so that as long as the operating pressure exceeds a threshold value, changes in pressure do not affect sprinkler discharge significantly. Constant-discharge nozzles can be used to minimize the variation in sprinkler discharge along laterals with fluctuating pressure caused by undulating terrain, for example.

Diffuse-jet nozzles are designed so that droplets are formed at a lower pressure than with other impact nozzles. This is accomplished by using noncircular-shaped nozzle openings or turbulence inducer at the orifice to spread (diffuse) the jet as it leaves the nozzle. A fan-shaped stream of water like the one in Figure 5.8 results. Diffuse-jet nozzles do not wet as large an area as do constant-diameter and constant-discharge nozzles.

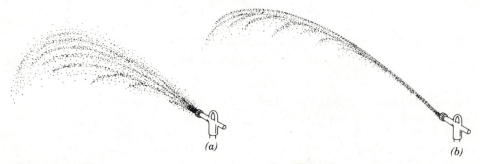

(a)   *(b)*

Figure 5.8 Typical "jets" from impact sprinklers with a (*a*) diffuse-jet nozzle and (*b*) a constant-diameter nozzle.

(ii) Gear-driven Sprinklers   Some rotating sprinklers are driven by a small water turbine located in the base of the sprinkler. These sprinklers are called gear-driven sprinklers because the high rotational speed of the turbine is reduced through a series of gears. Like impact sprinklers, gear-driven sprinklers have one or more jets that rotate around the vertical axis of the sprinkler. Unlike the start and stop rotation of impact sprinklers, gear-driven sprinklers rotate smoothly without the splash that occurs each time the arm of an impact sprinkler strikes the jet.

(iii) Reaction Sprinklers   Small reaction-type rotating sprinklers are normally rotated by the torque produced by the reaction of the water leaving the sprinkler. These sprinklers usually do not wet as large an area as do impact or gear-driven sprinklers and usually operate at much lower pressures (70 to 210 kPa or 10 to 30 psi).

(iv) Fixed-head Sprinklers   Fixed-head sprinklers depend on smooth and grooved cones, deflector plates, and slots to produce full- or nearly full-circle sprays or several small streamlets that are discharged around the circumference of the sprinkler. An example of a fixed-head spray-type sprinkler is illustrated in Figure 5.9, as is a multistreamlet-type fixed-head sprinkler. Many fixed-nozzle sprinkers that produce small droplets and that operate at low pressures (less than 210 kPa or 30 psi) are currently available for center-pivot and linear-move irrigation systems.

5.3.1b   Sprinkler Performance

Operating pressure and nozzle geometry (i.e., nozzle opening size, shape, and angle) are the primary factors that control the operation of sprinklers. The performance of a sprinkler is described by its discharge, distance of throw, distribution pattern, application rate, and droplet size. After each of these performance parameters is defined, the effect of operating pressure and nozzle geometry on each parameter is discussed below.

(i) Sprinkler Discharge   Sprinkler discharge is the volume per unit time passing out of the sprinkler. Common units for sprinkler discharge are liters per minute (l/min) and gallons per minute (gpm) in the SI and English systems,

(a)   (b)

Figure 5.9   Distribution patterns from (a) fixed-head spray sprinkler and (b) a multistreamlet fixed-head sprinkler.

respectively. Equation 5.1 can be used to relate sprinkler discharge to operating pressure and nozzle geometry.

$$Q = \sum_{i=1}^{n} KC_i A_i P_i^{x_i} \tag{5.1}$$

where

Q = sprinkler discharge;
n = number of nozzles;
K = constant that depends on units used;
C = coefficient that depends on shape and roughness of opening in nozzle i;
A = cross-sectional area of the opening in nozzle i;
P = operating pressure in nozzle i;
x = exponent for nozzle i.

Thus, discharge of a multinozzled sprinkler is the sum of the nozzle discharges.

Values of C and x for each nozzle are normally determined empirically. Since x is about 0.5 for most sprinklers, higher pressures and/or larger nozzle openings will increase sprinkler discharge. Sprinkler manufacturers commonly publish tables of pressure and discharge data for various nozzle diameters. Sprinkler discharge is not related to nozzle angle.

(ii) Distance of Throw   The spacing between adjacent sprinklers depends, in part, on the distance sprinklers throw water. Spacing usually increases as the distance of throw rises.

The operating pressure, and the size, shape, and angle of the nozzle opening determine the distance a sprinkler throws water. Distance of throw rises as pressure is increased within the range of recommended operating pressures (for a constant nozzle opening size, shape, and angle). Distance of throw also tends to increase as nozzle size increases (other things remaining constant). Nozzle opening shapes that create smaller diameter droplets tend to wet a smaller area than nozzles that emit larger droplets. Distance of throw usually increases and then declines as nozzle angle rises above horizontal. Sprinkler manufacturers commonly publish wetted diameter or other measures of distance of throw for different operating pressures, and nozzle sizes, shapes, and angles.

(iii) Distribution Pattern   The volume and rate of water application beneath a sprinkler normally varies with distance from the sprinkler. The pattern of this variation, called the *distribution pattern*, is normally consistent for a given operating pressure, nozzle geometry, and wind. Typical distribution patterns beneath a conventional impact sprinkler with a fixed-nozzle geometry for various operating pressures are diagrammed in Figure 5.10. Nozzles operating at low pressures that emit an essentially uniform droplet size often have "donut-shaped" distribution patterns (i.e., most water is deposited in a ring near the outer edge of the pattern). A wider range of droplet sizes caused by higher nozzle operating pressures will normally result in triangular shaped distribution patterns (i.e.,

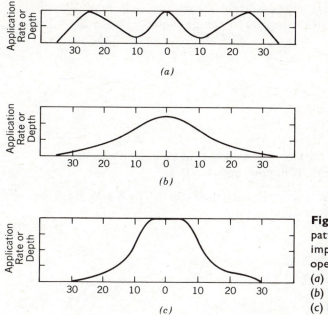

(a)

(b)

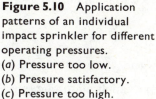

Figure 5.10 Application patterns of an individual impact sprinkler for different operating pressures.
(*a*) Pressure too low.
(*b*) Pressure satisfactory.
(*c*) Pressure too high.

(c)

patterns in which the depth of application increases linearly from the outer edge of the pattern toward the sprinkler). Extremely high pressures increase the percentage of small droplets (relative to the number of large droplets). A pattern similar to the one in Figure 5.10*c* results, since smaller droplets usually fall closer to the sprinkler than do larger ones.

Nozzle shape and opening size usually do not affect distribution patterns as much as operating pressure. Plates and pins that deflect water as it leaves the nozzle can be used to create various distribution pattern shapes. Proper selection of such devices can produce an essentially uniform pattern like the one in Figure 5.11. Varying the nozzle angle from horizontal will normally affect pattern size, that is, wetted area, rather than pattern shape.

(iv) *Application Rate* Application rate is an extremely important parameter that is used to properly match sprinklers to the soil, crop, and terrain on which they operate. When sprinkler application rates are too high, runoff and erosion can occur (this is discussed in Section 5.3.1d(ii)).

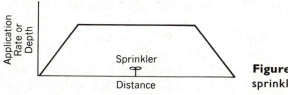

Figure 5.11 An essentially uniform sprinkler application pattern.

Application rate has dimensions of length per unit time. The average application rate of an individual sprinkler can be computed using Eq. 5.2.

$$A = K \frac{Q}{a}$$ (5.2)

where

A = application rate (mm/h, in/h);
Q = sprinkler discharge (l/min, gpm);
a = wetted area of sprinkler (m^2, ft^2);
K = unit constant (K = 60.0 for A in mm/h, Q in l/min, and a in m^2.
$\qquad$ K = 96.3 for A in in/h, Q in gpm, and a in ft^2).

When several identical sprinklers are spaced in a L by S grid, Eq. 5.3 can be used to compute the average application rate.

$$A = \frac{KQ}{LS}$$ (5.3)

where

A = application rate (mm/h, in/h);
Q = discharge of individual sprinklers (l/min, gpm);
L = distance between sprinklers along the lateral (m, ft);
S = spacing between adjacent sprinkler lines or lateral set positions (m, ft);
K = K in Eq. 5.2.

The average application rate beneath a lateral of sprinklers can be computed using Eq. 5.4.

$$A = \frac{KQ_l}{L_l S}$$ (5.4)

where, A and S are as previously defined

Q_l = total flowrate into upstream end of lateral (l/min, gpm);
L_l = length of lateral (m, ft);
K = K in Eq. 5.2.

For most sprinklers, variation in operating pressure has little, if any, effect on the average application rate of an individual sprinkler. When operating pressure increases, for example, the increase in Q tends to be offset by the increase in wetted area. The average application rate of several identical overlapping sprinklers, however, tends to be directly related to operating pressure, since L and S remain constant as Q increases. This is also true for the average application rate beneath a sprinkler lateral.

The average application rate for an individual sprinkler varies widely depending upon nozzle geometry. Deflector plate sprinklers, for example, have relatively high average application rates, since they wet a much smaller area than do other types of sprinklers. Conversely, conventional impact sprinklers are normally designed to achieve the maximum wetted area, and thus lowest possible average application rate. Nozzle opening shapes that create smaller droplets and wet a smaller area tend to have the highest average application rates. Average application rate will usually decrease and then increase as nozzle angle increases above horizontal. Increasing nozzle diameter usually increases the average application rate, since Q normally increases more rapidly than wetted area.

There is also considerable variation in the *instantaneous application rate* (i.e., the rate at which water is applied to a given point on the soil surface during an instant in time) from sprinkler type to sprinkler type (James and Stillmunkes, 1980). Conventional impact sprinklers, for example, apply water as concentrated streams that wet a point on the soil surface once or twice each rotation of the sprinkler head, while deflector-plate sprinklers usually apply water as a spray that covers all or nearly all of the wetted area continuously. Thus, the rate at which water is applied to a given point on the soil surface during an instant of time, the instantaneous application rate, is less for deflector-plate sprinklers than for conventional impact sprinklers. On some soils, lower instantaneous rates of application can decrease the potential for runoff and erosion by reducing soil splash and water ponding on the soil surface.

(v) Droplet Size Droplet size is an important factor affecting the formation of "seals" on bare soil surfaces that restrict water movement into the soil. Because small droplets possess less power when they impact the soil surface, "seals" that limit infiltration form more slowly than with larger droplets. For these reasons, it is sometimes possible to reduce runoff and erosion by converting from sprinklers that emit large droplets to ones with smaller droplets.

Droplet size is especially important when sprinklers must operate in winds. Distribution patterns from sprinklers that emit smaller droplets are more subject to wind distortion and lower application uniformity. In addition, increased losses due to wind drift usually occur with small droplet sprinklers.

Higher operating pressures normally increases the volume of water applied as smaller droplets while decreasing the volume of larger droplets. A similar, but a significantly smaller effect occurs on the larger droplets (not on the volume of water) as nozzle opening size is decreased. Nozzle opening shape can have an important effect on droplet size while nozzle angle has little, if any effect.

5.3.1c   Performance Characteristics of Sprinkler Types
The performance characteristics of several sprinkler types are compared in Table 5.1. In general, impact sprinklers tend to operate at higher pressures, have larger wetted diameters and droplet sizes, and have lower application rates than deflector-plate-type sprinklers. When operated within the recommended pressure range, impact sprinklers generally have triangular shaped distribution patterns, while most deflector-plate sprinklers have donut shaped patterns (see Figure 5.10).

5.3.1d   Sprinkler Selection
Sprinkler selection is the process of choosing sprinklers for the farm irrigation system. Sprinklers are normally selected on the basis of cost, operating pressure

Table 5.1 Sprinkler Types and Performance Characteristics

Sprinkler Type	Pressure Range (kPa)	(psi)	Discharge Range (l/min)	(gpm)	Distance of Throw (m)	(ft)	Relative Application Rate	Relative Droplet Size
Impact								
Low pressure								
Single nozzle	103–207	15–30	119–19	0.5–5	18–24	60–80	Low	Large
Double nozzle	103–207	15–30	11–38	3–10	21–25	70–100	Medium	Large
Medium pressure								
Single nozzle	207–414	30–60	15–76	4–20	21–43	70–140	Low–Medium	Medium
Double nozzle	207–414	30–60	15–360	4–80	21–61	70–200	Medium	Medium
High pressure								
Single nozzle	345–690	50–100	15–416	4–110	27–73	90–240	Medium	Small
Double nozzle	345–690	50–100	15–530	4–140	27–73	90–240	Medium–High	Small
Constant-discharge nozzle	276–552	40–80	8–38	2–10	27–37	90–120	Low–Medium	Medium
Diffused-jet nozzles	172–345	25–50	8–195	2–25	20–40	65–130	Medium	Small
Gun-type	276–896	40–130	197–4542	25–1200	61–183	200–600	Medium–High	Small
Spray Sprinklers								
180° spray nozzles	35–276	5–40	1–95	0.3–25	2–11	8–35[a]	Very High	Fine
					3–12	10–40[b]	Very High	Fine
360° spray nozzles with smooth, flat deflector plate	35–276	5–40	1–95	0.3–25	3–12	10–40[a]	High–Very High	Fine
					6–17	20–55[b]	High–Very High	Fine
360° spray nozzles with serrated, flat deflector plate	35–276	5–40	1–95	0.3–25	4–15	12–50[a]	High	Small
					8–21	25–70[b]	Medium–High	Small

[a] Sprinklers are mounted 2 m (6.6 ft) above ground surface.
[b] Sprinklers are mounted 3.6 m (12 ft) above ground surface.

requirements, and ability to provide the design daily irrigation requirement (DDIR) with acceptable uniformity and without runoff. In addition, sprinklers must have the proper nozzle (trajectory) angle, droplet size, distance of throw, and application pattern characteristics for the crop, soil, and wind conditions in which they are to operate.

   (i) Sprinkler Discharge Capacity The sprinkler must have sufficient capacity to supply the DDIR plus wind drift and evaporative losses that occur after water leaves the sprinkler and before it reaches plant and soil surfaces. Equation 5.5 can be used to estimate required sprinkler discharge (capacity).

$$Q_s = \frac{(K)(D_a)(L)(S)}{(H - T_m)(E_a)} \tag{5.5}$$

where

Q_s = sprinkler capacity (l/min, gpm);
D_a = depth to be applied (mm, in);
 L = spacing (distance) between laterals (m, ft);
 S = spacing (distance) between sprinklers on a lateral (m, ft);
 H = time interval between the beginnings of successive irrigations of a given set (hr);
T_m = downtime for moving set-move systems and/or maintenance (hr);
E_a = application efficiency (percent);
 K = unit constant ($K = 1.67$ for Q_s in l/min, D in mm, and L and S in m.
       $K = 1.04$ for Q_s in gpm, D in in, and L and S in ft).

The interval H in Eq. 5.5 can be determined using Eq. 5.6.

$$H \leq \frac{(0.24)(P_f)(D)}{\text{DDIR}} \tag{5.6}$$

where

P_f = percent of total field irrigated when the system is operating;
 D = desired depth of irrigation (mm, ins);
DDIR = design daily irrigation requirement for D mm or ins of soil storage (mm/day when D is in mm or in/day when D is in ins).

   DDIR is determined using either Eq. 2.6 or the procedures used in Example 2.1. The value of D used in Eq. 5.6 must be identical to the one used to evaluate DDIR. Thus, D normally equals readily available water (RAW).
   The depth applied, D_a in Eq. 5.5 is computed with Eq. 5.7 once a value of H has been selected (this value of H must, of course, be less than or equal to the value computed with Eq. 5.6). Eq. 5.7 is.

$$D_a = \frac{(H)(\text{DDIR})}{(0.24)(P_f)} \tag{5.7}$$

The term P_f in Eqs. 5.6 and 5.7 equals 100 percent for continuous-move sprinkle systems or when an entire solid system is operating (i.e., all sprinklers are operating simultaneously). Equation 5.8 can be used to determine P_f for solid-set and set-move systems.

$$P_f = \frac{(L_l)(L)(N_l)}{(K)(A_f)} \tag{5.8}$$

where

L_l = length of lateral (m, ft);
N_l = number of laterals operating simultaneously;
A_f = total field area (ha, ac);
K = unit constant ($K = 100.0$ when L_l and L are in m, and A_f is in ha.
$\quad\quad K = 435.6$ when L_l and L are in ft, and A_f is in ac).

The following example illustrates the use of Eqs. 5.5, 5.6, 5.7 and 5.8.

EXAMPLE 5.1   Determining Sprinkler Capacity for a Set-Move System

Given:
set-move sprinkle system that is 75 percent efficient

DDIR = 0.30 in/day
$\quad S = 40$ ft
$\quad L = 60$ ft
$\quad L_1 = 1320$ ft
$\quad N_1 = 5$ laterals
$\quad D = 2$ inches
$\quad T_m = 0.5$ hours
$\quad A_f = 100$ acres

Required:
sprinkler capacity

Solution:

   Solution Steps
    1. Compute P_f using Eq. 5.8
    2. Choose H using Eq. 5.6
    3. Compute D_a using Eq. 5.7
    4. Compute sprinkler capacity, Q_s, using Eq. 5.5.

$$P_f = \frac{(1320)(60)(5)}{(435.6)(100)} = 9.09\%$$

$$H \leq \frac{(0.24)(9.09)(2)}{0.30}$$

$$H \leq 14.54 \text{ hours}$$

Use $H = 12$ hours, since this allows the irrigator to move the system twice per day

$$D_a = \frac{(12)(0.30)}{0.24(9.09)} = 1.65 \text{ in}$$

$$Q_s = \frac{(1.04)(1.65)(60)(40)}{(12 - 0.5)(75)} = 4.77 \text{ gpm}$$

L and S for set-move, solid-set, and traveler sprinkle systems can be determined using Eqs. 5.9 and 5.10.

$$L \le K_l D \tag{5.9}$$

$$S \le K_s D \tag{5.10}$$

where

$L = $ distance between laterals (m, ft);
$S = $ distance between sprinklers on lateral (m, ft);
$K_l, K_s = $ constants that depend on sprinkler spacing pattern and wind (see Table 5.2);
$D = $ diameter of wetted area (m, ft).

Use of the K_l and K_s values in Table 5.2 assures acceptable application uniformity for different winds and with various spacing patterns.

Triangular, square, and rectangular are three basic types of sprinkler spacing patterns for set-move and solid-set systems. These patterns are illustrated in Figures 5.12, 5.13, and 5.14, respectively.

In the triangular pattern in Figure 5.12, the three sprinklers that form the pattern are a distance S apart. The spacing between adjacent laterals, L, is $0.86S$. Because sprinklers along adjacent laterals are offset from one another, sprinklers

Table 5.2 Values of K_l and K_s for Triangular, Square, and Rectangular Sprinkler Spacing Patterns

Wind Velocity Range		Triangular		Square		Rectangular[b]	
(m/s)	(mph)	K_l	K_s	K_l	K_s	K_l	K_s
0–1.3	0–3	[a]	0.60	0.55	0.55	0.60	0.50
1.8–3.1	4–7	[a]	0.55	0.50	0.50	0.60	0.45
3.6–5.4	8–12	[a]	0.50	0.45	0.45	0.60	0.40

Source: D. D. Davis, *Irrigation Systems Design Handbook*, Copyright © (1976), Rain Bird Sprinkler Manufacturing Corporation, Glendora, Calif., pp. 29–30. Reprinted by permission of Rain Bird Sprinkler Manufacturing Corporation.
[a] Constant $K_l = 0.86 K_s$
[b] Assumes laterals are perpendicular to prevailing wind direction.

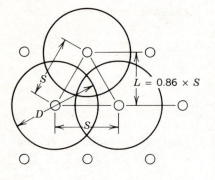

Figure 5.12 An equilateral triangle spacing pattern for irrigation sprinklers. *Source:* D. D. Davis, *Irrigation Systems Design Handbook,* copyright © 1976 by Rain Bird Sprinkler Manufacturing Corporation, Glendora, CA, p. 29. Reprinted by permission of Rain Bird Sprinkler Manufacturing Corporation.

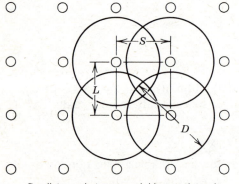

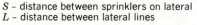

Figure 5.13 A square spacing pattern for irrigation sprinklers. *Source:* D. D. Davis, *Irrigation Systems Design Handbook,* copyright © 1976 by Rain Bird Sprinkler Manufacturing Corporation, Glendora, CA, p. 29. Reprinted by permission of Rain Bird Sprinkler Manufacturing Corporation.

S – distance between sprinklers on lateral
L – distance between lateral lines

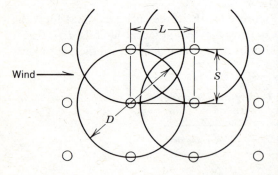

Figure 5.14 A rectangular spacing pattern for irrigation sprinklers. *Source:* D. D. Davis, *Irrigation Systems Design Handbook,* copyright © 1976 by Rain Bird Sprinkler Manufacturing Corporation, Glendora, CA, p. 30. Reprinted by permission of Rain Bird Sprinkler Manufacturing Corporation.

can be spaced further apart (relative to square and rectangular patterns). This reduces the number of set-move lateral positions, and solid-set laterals as well as the number of sprinklers required to irrigate a field. Triangular sprinkler spacing patterns are not commonly used with set-move systems because of the inconvenience of shifting the entire lateral to or from the mainline by a distance equal to $S/2$ each time the lateral is moved. Triangular spacing patterns are best suited to solid-set systems.

Set-move systems usually have square or rectangular sprinkler spacing patterns. In a square pattern L equals S, and the distance across the diagonal of the pattern is approximately 1.4 times L or S. The distances L and S must therefore be reduced (relative to triangular patterns) to avoid "dry" spots in the center of the pattern.

In a rectangular sprinkler spacing pattern L and S are not equal. Rectangular patterns are used in areas with elevated wind velocities. Usually laterals are placed perpendicular to the prevailing wind direction and L exceeds S.

Values of K_l and K_s for different sprinkler spacing patterns and wind velocity ranges are given in Table 5.2. In general, both K_l and K_s decrease as wind velocity increases.

An application efficiency term is often used to account for wind drift and evaporate losses that occur after water leaves the sprinkler and before it reaches plant and soil surfaces during application. The efficiency term assumes that recommended values of S and L have not been exceeded and thus, that the uniformity of application is sufficiently high. The usual value of E_a for set-move and solid-set systems is 75 percent. For continuous-move systems with closely spaced sprinklers, E_a is increased to 85 percent.

The spacing between sprinklers on center-pivot laterals is often much less than the maximum recommended spacing values obtained using Eq. 5.10 and K_s values from Table 5.2. Maximum spacing between sprinklers, especially at the downstream end of center pivot laterals, results in the use of large-capacity sprinklers that are expensive and operate at extremely high pressures. Because of the desirability of using several closely spaced, smaller capacity sprinklers that operate at lower pressures (rather than larger sprinklers spaced further apart) in such situations and the need for maximum flexibility in sprinkler selection, center-pivot-system manufacturers usually provide closely spaced sprinkler outlets along center pivot laterals.

The term L in Eq. 5.5 equals the distance that a linear-move system travels per irrigation. This is usually the length of the field. For center-pivot systems L is computed using Eq. 5.11.

$$L = 2\pi r \tag{5.11}$$

where, r is the distance along the lateral from the pivot to the sprinkler. Thus, L is different for each sprinkler along a center-pivot lateral.

(ii) Allowable Application Rate Normally, sprinkle irrigation systems are designed so that no runoff occurs. Thus, the rate at which a sprinkle system is designed to apply water is less than the infiltration capacity of the soil (i.e., the maximum rate at which water can enter the soil at a given time) or the application is ended before all surface depressions are filled with water and a sufficient depth of water to cause runoff over the soil surface has accumulated. Figure 5.15 illustrates these concepts.

Curve A in Figure 5.15 shows that the infiltration capacity of a typical soil is highest immediately after infiltration begins and then decreases steadily with time

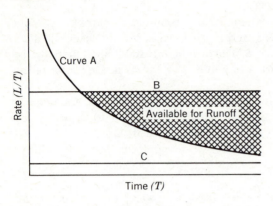

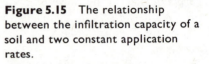

Figure 5.15 The relationship between the infiltration capacity of a soil and two constant application rates.

toward an asymptote that is often called the basic infiltration rate of the soil. In a very deep, homogeneous soil the basic infiltration rate equals the saturated hydraulic conductivity of the soil.

Consider the rate of application shown as the horizontal line B in Figure 5.15. Initially, all water that is being applied by the sprinkler system enters the soil, since the application rate is less than the infiltration capacity of the soil. There is no possibility of runoff until line B crosses line A and the application rate exceeds the infiltration capacity of the soil. Runoff does not begin until depressions in the soil surface (caused by various tillage operations, for example) have been filled with water and a sufficient depth of water to cause flow has accumulated on the soil surface. The amount of water that can accumulate in depressions and on the soil surface depends on the condition (amount of vegetation and/or residue and the depth and extent of surface depressions) as well as the slope of the soil surface. The presence of vegetation and/or residue, and tillage operations that have an irregular surface tend to delay runoff by increasing the amount of water that can be stored. The crosshatched area below line B and above line A in Figure 5.15 is water that is available for runoff.

Line C in Figure 5.15 shows a system application rate that never exceeds the infiltration capacity of the soil. Because such an application rate can be continued indefinitely without runoff, set-move systems with long duration sets are commonly designed to apply water at rates less than the basic infiltration rate of the soil.

Table 5.3 lists basic infiltration rates of five soil textures for bare soil surfaces (without vegetative or residue cover) with and without a surface "seal". These seals, which impede infiltration, form as a result of sprinkler droplet impact and soil wetting during irrigation. Since most bare soils seal when sprinkler irrigated, the values for soils with seals should be used in most situations. The values for soils without seals should be used only when there is vegetative or residue cover over the field for the entire irrigation season.

Application rates beneath center-pivot laterals increase to a peak and then decline to zero as the lateral approaches and then moves past a particular point in the field. Many times the peak rate of application exceeds the recommended values

Table 5.3 Basic Infiltration Rates for Two Bare Soil Conditions

	Basic Infiltration Rate for			
	Condition A		Condition B	
Soil	(mm/h)	(in/h)	(mm/h)	(in/h)
Coarse sand	19–25	0.75–1.00	8.9	0.35
Fine sands	13–19	0.50–0.75	6.4	0.25
Fine sandy loams	8.9–13	0.35–0.50	5.1	0.20
Silt loams	6.4–10.2	0.25–0.40	3.8	0.15
Clay loams	2.5–7.6	0.10–0.30	2.5	0.10

Source: C. H. Pair, W. H. Hinz, K. R. Frost, R. E. Sneed, and T. J. Schiltz (Eds.), *Irrigation*, copyright © 1983, The Irrigation Association, Silver Spring, Md., p. 74. Reprinted by permission of the Irrigation Association.

Note: Condition A is for well aggregated soils with high organic matter, open granular structure, and no evidence of surface sealing. Condition B is for poorly aggregated soils with low organic matter contents and a thin sealed layer at the surface

in Table 5.3, especially at the downstream end of the lateral. The variation of application rate with time as a center-pivot lateral passes over a point in the field is diagrammed in Figure 5.16 for two different locations along the center pivot lateral. Curves A and B are for positions near the up- and downstream ends of the lateral, respectively. The depth of application (i.e., the area under the curves in Figure 5.16) is the same at each location.

The peak application rate for curve A exceeds the basic infiltration rate but never exceeds the infiltration capacity of the soil. There is no possibility for runoff in this case. The peak application rate for curve B, however, exceeds the infiltration capacity. In this case, runoff will occur if the volume applied exceeds the volume of water that can be stored in surface depressions plus that required to cause flow over the ground surface (represented by the crosshatched area in Figure 5.16). The

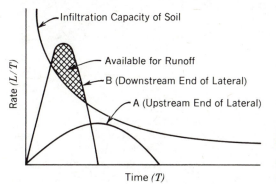

Figure 5.16 Typical relationship between soil infiltration capacity and application rates beneath center pivot laterals.

Table 5.4 Allowable Surface Storage Value for Various Slopes

Slope (%)	Allowable Surface Storage (mm)	(in)
0–1	12.7	0.5
1–3	7.6	0.3
3–5	2.5	0.1

Source: R. C. Dillion, E. A. Hiler, and G. Vittetoe, "Center Pivot Sprinkler Design Based on Intake Characteristics," Trans. ASAE (1972), Vol. 15, No. 5, pp. 996–1000. Copyright © 1972. Reprinted with permission of ASAE.

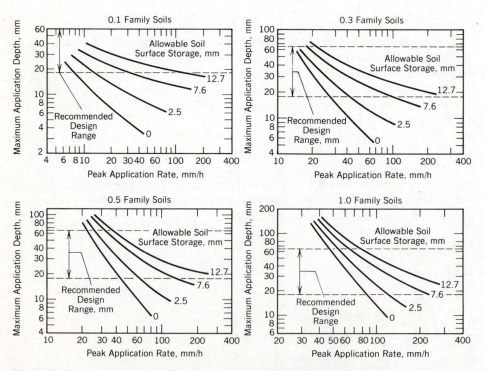

Figure 5.17 Maximum depth of water that can be applied with center-pivot and linear-move sprinkle systems per irrigation for soil conservation service (SCS) intake families 0.1, 0.3, 0.5, and 1.0. *Source:* J. R. Gilley, "Suitability of Reduced Pressure Center-Pivots," copyright © 1984 by American Society of Civil Engineers. Reprinted by permission of ASCE from the Journal *IR/ASCE*, January 1984, pp. 30–31.

Table 5.5 Values of *a* and *b* in Equation 5.12 for SCS Intake Families 0.1, 0.3, 0.5, 1.0, and 1.5

SCS Intake Family	*a* for *f* in mm/h	*a* for *f* in in/h	*b*
0.1	6.83	0.269	−0.485
0.3	15.16	0.597	−0.381
0.5	21.77	0.857	−0.340
1.0	36.59	1.441	−0.305
1.5	47.90	1.886	−0.290

amount that can be stored without runoff for various slopes and an unspecified soil condition is listed in Table 5.4.

Gilley (1984) developed a series of relationships (Figure 5.17) between the depth of water that can be applied per irrigation without runoff for surface storage amounts of 0, 2.5, 7.6, and 12.7 mm (0, 0.1, 0.3, and 0.5 in), peak application rates ranging between 4 and 400 mm/h (0.16 to 15.7 in/h), and four soil types. The four soil types were the 0.1, 0.3, 0.5, and 1.0 Soil Conservation Service (SCS) intake families. These relationships assume an elliptical relationship between application rate and time (similar to Curve B in Figure 5.16) and that Eq. 5.12 describes the variation of infiltration rate with time. Equation 5.12 is

$$f = at^b \tag{5.12}$$

where

f = infiltration rate of the soil (mm/h, in/h);
t = time since infiltration began (h);
a, b = constants.

Values of *a* and *b* for SCS intake families 0.1, 0.3, 0.5, and 1.0 are listed in Table 5.5.

EXAMPLE 5.2   Determining the Depth of Water that can be Applied per Center Pivot Irrigation Without Runoff

Given:
• SCS intake family 0.3
• 2.5 mm of surface storage can accumulate before runoff begins
• 30 mm/h peak application rate

Required:
Maximum amount of water that can be applied without runoff

Solution:
From Figure 5.17 for intake family 0.3, 24 mm can be applied without runoff.

(iii) Operating Pressure   In order to reduce energy consumption and lower operating costs, sprinkle systems should operate at the lowest pressure at which acceptable application uniformity and efficiency can be achieved. Sprinkler manufacturers' catalogs usually identify a recommended range of operating pressure that results in acceptable performance for each sprinkler. The design operating pressure should be as low as possible and within the recommended range.

(iv) Other Performance Parameters The nozzle angle, droplet size, distance of throw, and application pattern must also be known to select the proper sprinkler. These parameters are qualitatively evaluated on the basis of such factors as wind, crop, and system type.

The nozzle angle used depends on wind speed and the height of the sprinkler relative to crop height. Low nozzle angles are used in windy areas and in orchards with undertree sprinkle systems. High nozzle angles are used where wind speeds are normally low and/or when sprinklers are mounted approximately at crop height.

Sprinklers that emit fine droplets should be considered when it is necessary to irrigate bare soil surfaces that tend to "seal" when sprinkled. Sprinklers with relatively large droplets are recommended for windy conditions since their distribution patterns are less affected by wind than those of finer droplet sprinklers.

Sprinklers with large distances of throw are used when low application rates and/or wider sprinkler spacings are desired. Large sprinklers with high operating pressures generally have large distances of throw, while low-pressure spray-type sprinklers have relatively small distances of throw.

Sprinklers with triangular shaped distribution patterns are used with most types of sprinkle systems. The use of sprinklers with donut and other irregular shaped patterns is, however, normally limited to continuous-move systems. Sprinklers with donut patterns are also recommended for under-tree systems in orchards where tree interference tends to increase water application near the sprinklers.

EXAMPLE 5.3   Sprinkler Selection

Given:
- desired sprinkler discharge is 4.77 gpm (from Example 5.1)
- information from Example 5.1
- alfalfa is being irrigated with hand lines
- prevailing wind speed is 5 mph
- 50 psi is available
- fine sandy loam soil is being irrigated
- sprinkler manufacturer catalog information in the following tables

Required:
sprinkler

Solution:

  Solution Steps

  1. estimate desired wetted diameter
  2. select nozzle angle, droplet size, application pattern

Sprinkler A

Nozzle psi	Nozzle 9/64″		Nozzle 5/32″		Nozzle 11/64″		Nozzle 3/16″		Nozzle 13/64″		Nozzle 7/32″	
	gpm	Diam.	gpm	Diam.	gpm	Diam.	gpm	Diam.	gpm	Diam.	gpm	Diam.
25	2.88	80	3.52	82	4.24	83	5.00	85	5.90	86	6.85	88
30	3.15	81	3.85	85	4.64	88	5.50	91	6.50	94	7.55	96
35	3.40	82	4.16	87	5.02	90	5.96	94	7.05	97	8.20	100
40	3.64	83	4.45	88	5.37	92	6.38	96	7.55	99	8.80	102
45	3.86	84	4.72	89	5.70	94	6.78	98	8.00	101	9.35	104
50	4.07	85	4.98	90	6.01	95	7.16	100	8.45	103	9.90	106
55	4.27	86	5.22	91	6.30	96	7.52	101	8.85	104	10.40	107
60	4.46	87	5.45	92	6.57	97	7.85	102	9.25	105	10.75	108
65	4.65	88	5.68	93	6.83	98	8.18	103	9.60	106	11.10	109
70	4.83	89	5.90	94	7.09	99	8.50	104	9.95	107	11.40	110
75	5.00	90	6.11	95	7.34	100	8.80	105	10.25	108	11.70	111
80	5.17	91	6.30	96	7.58	101	9.09	106	10.50	109	12.00	112

Sprinkler A is a single-nozzle impact sprinkler with a 27° nozzle angle.

3. select sprinkler
4. compare sprinkler application rate to maximum allowable application rate

Solution Step 1
Rearranging Eqs. 5.9 and 5.10 yields

$$D \geq \frac{L}{K_1} \quad \text{and} \quad D \geq \frac{S}{K}$$

$$\geq \frac{60}{0.6} = 100 \text{ ft}$$

$$\geq \frac{40}{0.45} = 89 \text{ ft}$$

wetted diameter ≥ 89 ft

Solution Step 2
• because the sprinkler mounting height on a handline is approximately at crop height, a large nozzle angle should be used
• because the crop canopy protects the soil surface and wind speeds are low, all except possibly very fine droplet sizes are acceptable
• a triangular-shaped application pattern is preferred for a set-move system.

Solution Step 3
• sprinkler A is preferred since it has the highest nozzle angle (27°).
• sprinkler A with 5/32-in nozzle operating at 46 to 47 psi is the best choice

Sprinkler B

Nozzle psi	Nozzle 9/64″ × 3/32″ 7° gpm	Diam.	Nozzle 5/32″ × 3/32″ 7° gpm	Diam.	Nozzle 11/64″ × 3/32″ 7° gpm	Diam.	Nozzle 3/16″ × 3/32″ 7° gpm	Diam.	Nozzle 3/16″ × 1/8″ 20° gpm	Diam.	Nozzle 13/64″ × 1/8″ 20° gpm	Diam.	Nozzle 7/32″ × 1/8″ 20° gpm	Diam.
25	4.18	80	4.83	82	5.55	83	6.31	85	7.40	85	8.30	86	9.25	88
30	4.58	81	5.28	85	6.07	88	6.93	91	8.15	91	9.15	94	10.20	96
35	4.95	82	5.71	87	6.57	90	7.51	94	8.83	94	9.93	97	11.10	100
40	5.29	83	6.11	88	7.03	92	8.03	96	9.46	96	10.65	99	11.90	102
45	5.61	84	6.48	89	7.46	94	8.53	98	10.05	98	11.30	101	12.65	104
50	5.91	85	6.82	90	7.87	95	9.00	100	10.60	100	11.90	103	13.35	106
55	6.20	86	7.15	91	8.25	96	9.45	101	11.10	101	12.50	104	14.00	107
60	6.48	87	7.47	92	8.60	97	9.86	102	11.60	102	13.00	105	14.50	108
65	6.75	88	7.78	93	8.93	98	10.27	103	12.10	103	13.50	106	15.00	109
70	7.00	89	8.08	94	9.25	99	10.66	104	12.50	104	14.00	107	15.45	110
75	7.25	90	8.37	95	9.56	100	11.02	105	12.90	105	14.40	108	15.90	111
80	7.49	91	8.65	96	9.87	101	11.36	106	13.25	106	14.80	109	16.30	112

Sprinkler B is a double-nozzle impact sprinkler with a range nozzle angle of 27°.

Solution Step 4

$$A = \frac{(96.3)(4.77)}{(60)(40)} = 0.19 \text{ in/h}$$

From Table 5.3, maximum allowable application rate = 0.50 in/h. Sprinkler A with a 5/32 in nozzle is acceptable, since 0.19 in/h is less than the basic infiltration rate of 0.50 in/h.

Sprinkler C

Nozzle psi	Nozzle 9/64″ gpm	Diam.	Nozzle 5/32″ gpm	Diam.	Nozzle 11/64″ gpm	Diam.	Nozzle 3/16″ gpm	Diam.	Nozzle 13/64″ gpm	Diam.	Nozzle 7/32″ gpm	Diam.
25	2.88	72	3.52	75	4.24	78	5.00	80	5.90	82	6.85	84
30	3.15	75	3.85	78	4.64	81	5.50	84	6.50	86	7.55	88
35	3.40	77	4.16	81	5.02	84	5.96	87	7.05	89	8.20	91
40	3.64	79	4.45	83	5.37	86	6.38	89	7.55	91	8.80	93
45	3.86	81	4.72	85	5.70	88	6.78	91	8.00	93	9.35	95
50	4.07	82	4.98	86	6.01	90	7.16	93	8.45	95	9.90	97
55	4.27	83	5.22	87	6.30	91	7.52	94	8.85	96	10.40	98
60	4.46	84	5.45	88	6.57	92	7.85	95	9.25	97	10.75	99
65	4.65	85	5.68	89	6.83	93	8.18	96	9.60	98	11.10	100
70	4.83	86	5.90	90	7.09	94	8.50	97	9.95	99	11.40	101
75	5.00	87	6.11	91	7.34	95	8.80	98	10.25	100	11.70	102
80	5.17	88	6.30	92	7.58	96	9.09	99	10.50	101	12.00	103

Sprinkler C is a single-nozzle impact sprinkler with a 15° nozzle angle.

Sprinkler D

Nozzle psi	Nozzle 9/64" × 3/32" 7°		Nozzle 5/32" × 3/32" 7°		Nozzle 11/64" × 3/32" 7°		Nozzle 3/16" × 3/32" 7°		Nozzle 3/16" × 1/8" 20°		Nozzle 13/64" × 1/8" 20°		Nozzle 7/32" × 1/8" 20°	
	gpm	Diam.	gpm	Diam.	gpm	Diam.	gpm	Diam.	gpm	Diam.	gpm	Diam.	gpm	Diam.
25	4.18	72	4.83	75	5.55	78	6.31	80	7.40	80	8.30	82	9.25	84
30	4.58	75	5.28	78	6.07	81	6.93	84	8.15	84	9.15	86	10.20	88
35	4.95	77	5.71	81	6.57	84	7.51	87	8.83	87	9.93	89	11.10	91
40	5.29	79	6.11	83	7.03	86	8.03	89	9.46	89	10.65	91	11.90	93
45	5.61	81	6.48	85	7.46	88	8.53	91	10.05	91	11.30	93	12.65	95
50	5.91	82	6.82	86	7.87	90	9.00	93	10.60	93	11.90	95	13.35	97
55	6.20	83	7.15	87	8.25	91	9.45	94	11.10	94	12.50	96	14.00	98
60	6.48	84	7.47	88	8.60	92	9.86	95	11.60	95	13.00	97	14.50	99
65	6.75	85	7.78	89	8.93	93	10.27	96	12.10	96	13.50	98	15.00	100
70	7.00	86	8.08	90	9.25	94	10.66	97	12.50	97	14.00	99	15.45	101
75	7.25	87	8.37	91	9.56	95	11.02	98	12.90	98	14.40	100	15.90	102
80	7.49	88	8.65	92	9.87	96	11.36	99	13.25	99	14.80	101	16.30	103

Sprinkler D is a double-nozzle impact sprinkler with a range nozzle angle of 15°.

5.3.2 Pipelines for Sprinkle Systems

Pipelines for sprinkle systems are normally pressurized (rather than low head pipelines). Sprinkle system pipelines are classified as either mainlines, submains, or laterals. Mainlines convey water from the source and distribute it to the submains. The submains provide water to the laterals that supply water to the sprinklers. Some systems do not have submains. In these systems, laterals are connected directly to the mainline.

Pipelines must supply water at the desired pressure to each sprinkler and lateral, be strong enough to withstand expected operating and surge pressures, and have a life expectancy that equals or exceeds the life of other system components. Buried pipes must resist overburden and dynamic surface loads, while portable laterals and mainlines must be light and durable (i.e., resist denting, etc.). Pipe materials and loading of buried pipes are important factors that affect the design and operation of piplines for sprinkle systems.

5.3.2a Material

Although the personal preference of the system owner can be important, pipe material selection usually involves identifying the least expensive pipe for the conditions in which the pipe is to operate. Conditions that can affect pipe material selection include the chemical composition of the soil, the amount of rocks in the soil, and the type of irrigation system. Aluminum is generally used for portable systems, while steel pipe is usually used for center pivot laterals. Asbestos-cement, PVC, and wrapped steel are typical choices for buried laterals and mainlines. Wrapped aluminum has also been used for buried pipelines.

Asbestos-cement (AC) pipe, with asbestos fiber to increase the tensile strength of the concrete, is commonly used for buried pipelines. It combines strength with light weight (as compared to steel pipe) and is immune to rust and corrosion. AC pipe may be cut, drilled, and tapped in the field.

Three different types of AC pipe are manufactured. Type I pipe is for use where moderately aggressive water and soil of moderate sulfate content are expected to come into contact with the pipe. Type II pipe is for use where highly aggressive water or water and soil of high sulfate content or both are expected to come into contact with the pipe. Type III pipe is for use where contact with aggressive waters and sulfates is not expected. See ASTM Specification C-29G-Asbestos-Cement Pressure Pipe for details.

Because it is so brittle, care must be exercised in the installation of AC pipe. It must be laid on a stable bed that is free of rocks and supports the pipe along its full length. Backfill around the pipe cannot contains rocks or other debris that could damage the pipe. Concrete thrust blocks are required at all turns, dead ends, and changes in grade so that internal water pressures and/or expansion and contraction will not separate pipe joints or fittings. Manuals that detail installation techniques and requirements for AC pipe are available from manufacturers.

Aluminum is used for most portable laterals and mainlines because it is light and durable. Minimum standards for pressure rating, deflection, and denting resistance of aluminum irrigation tubing are published by ASAE (ASAE Standard: ASAE S263.2). Exposure to saline or acid conditions can corrode aluminum. Aluminum pipe can be protected from some corrosive factors by cladding. Cladding involves metallurgically bonding an aluminum alloy coating to the inside and/or outside surface of the pipe. Cathodic protection, which is an induced flow of direct electrical current from buried electrodes to the pipe, has been used to control corrosion of buried aluminum (or steel) pipes. Various wrappings can also be used to protect buried aluminum pipes.

Steel pipes are strong but subject to corrosion on both the inside and outside of the pipe. "Flakes" of corrosion that break loose from the inside of the pipe can adversely affect valve and sprinkler performance. Some center-pivot manufacturers line steel center-pivot laterals with epoxy to reduce corrosion and friction loss. Steel pipes buried in saline or acid soils should be wrapped or coated to protect the pipe from corrosion. Cathodic protection as described in the previous paragraph can also be used for buried steel pipe.

Polyvinyl chloride plastic (PVC) pipe resists corrosion for most water conditions, has smooth walls with relatively low friction loss characteristics, and has a long life expectancy when protected from surge pressures. Trenches in which PVC is buried should have a relatively smooth, firm bottom free of rocks. Where rough rock edges cannot be avoided, the trench should be overexcavated and filled to the bedding depth with sand or finely graded soils. Minimum standards for the design, installation, and performance of underground thermoplastic irrigation pipelines are published by ASAE (ASAE Standard: ASAE S376.1). PVC pipe manufacturers also provide installation manuals.

5.3.2b Pressure Variation

In most situations, the irrigation system must apply water uniformly over the entire field. Since the performance of most sprinklers is related to operating pressure, high uniformities of application require that the pressure needed for the desired sprinkler performance be provided. Friction loss in pipes and fittings, and differences in elevation cause pressures to vary in a field. Friction loss causes the pressure to decrease in the downstream direction, while changes in elevation can cause either an increase or decrease in pressure (depending on whether the elevation change is downhill or uphill). Equation 5.13 can be used to estimate the difference in pressure between locations along a pipeline.

$$P_d = P_u - K(h_l \pm \Delta Z) \tag{5.13}$$

where

P_d, P_u = pressure at down- and upstream positions, respectively (kPa, psi);
h_l = energy loss in pipe between the up- and downstream positions (m, ft);
ΔZ = difference in elevation between up- and downstream positions (m, ft);
K = unit constant ($K = 9.81$ for P_d, and P_u in kPa, and h_l and ΔZ in m.
$K = 0.43$ for P_d and P_u in psi, and h_l and ΔZ in ft).

When the change in elevation between the up- and downstream positions is uphill, the sign in front of ΔZ is plus ($+$). Conversely, this sign is negative ($-$) when the elevation at the upstream location exceeds the elevation at the downstream location.

Equation 5.14 can be used to estimate the energy loss term, h_l.

$$h_l = FH_l + M_l \tag{5.14}$$

where

F = constant that depends on the number of outlets (sprinklers or laterals) removing water from the pipe between the up- and downstream location. F also depends on the method used to estimate H_l;
H_l = friction loss in pipe between up- and downstream locations (m, ft);
M_l = minor losses through fittings (m, ft) (see Appendix E).

Appendix E can be used to estimate M_l. Minor losses caused by sprinkler risers are extremely small and are usually neglected. Either the Darcy-Weisbach, Hazen-Williams, or Scobey equation can be used to compute H_l. These equations can be written in the form of Eq. 5.15.

$$H_l = \frac{(K)(c)(L)(Q^m)}{D^{2m+n}} \tag{5.15}$$

Table 5.6   Information Needed to Use Equation 5.15

Method of Computing H_l	c (SI Units)	(English Units)	m	n
Darcy–Weisbach	277778	1.235	2.0	1.0
Hazen–Williams	591722	1.000	1.85	1.17
Scobey	610042	1.000	1.90	1.10

where

K = friction factor that depends on pipe material;
L = length of pipe (m, ft);
Q = flowrate (l/min, gpm);
D = diameter of pipe (mm, in);
c, m, n = constants from Table 5.6.

For the Darcy–Weisbach equation, K in Eq. 5.15 is given by Eq. 5.16a.

$$K = 0.811\left(\frac{f}{g}\right) \tag{5.16a}$$

where

f = friction factor from the Moody diagram;
g = acceleration of gravity (9.81 m/s^2 or 32.2 ft/s^2).

A Moody diagram and other information for evaluating f can be found in most fluid mechanics textbooks.

Equation 5.16b can be used to compute K when the Hazen-Williams equation is used to determine H_1.

$$K = (0.285\ C)^{-1.852} \tag{5.16b}$$

Values of C for several pipe materials are listed in Table 5.7.

When H_1 is computed using the Scobey equation, K is determined using Eq. 5.16c.

$$K = \frac{Ks}{348} \tag{5.16c}$$

Values of Ks for several pipe materials are listed in Table 5.8.

Table 5.5 contains values of c, m, and n for the Darcy–Weisbach, Hazen-Williams, and Scobey equations. H_1 in Equation 5.15 can also be obtained from friction loss tables published by pipe manufacturers.

There is less friction loss along a pipe with several equally spaced discharging outlets such as submains and sprinkler laterals than along a pipe of equal diameter, length, and material with constant discharge (constant discharge means that inflow to the pipe section equals the outflow from the section). This occurs because the

Table 5.7 Values of C (in Eq. 5.16*b*) For the Hazen–Williams Formula

Type of Pipe	C
Asbestos-cement	140
Brass	130–140
Brick sewer	100
Cast iron	
New, unlined	130
Old, unlined	40–120
Cement lined	130–150
Bitumastic enamel lined	140–150
Tar-coated	115–135
Concrete or concrete lined	
Steel forms	140
Wooden forms	120
Centrifually spun	135
Copper	130–140
Fire hose (rubber lined)	135
Galvanized iron	120
Glass	140
Lead	130–140
Plastic	140–150
Steel	
Coat-tar enamel lined	145–150
New unlined	140–150
Riveted	110
Tin	130
Vitrified clay	100–140

Table 5.8 Values of *Ks* (in Eq. 5.16*c*) for Scobey Equation

Pipe	Ks
2- and 2½-in O.D. welded steel	0.34
3-in O.D. welded steel	0.33
4-, 5- and 6-in O.D. welded steel	0.32
Aluminum tubing without couplers	0.33
Aluminum tubing with couplers each 20 ft	0.43
Aluminum tubing with couplers each 30 ft	0.40
Aluminum tubing with couplers each 40 ft	0.39

Table 5.9 Values of *F* (in Eq. 5.17a) Used When the Distance to the First Sprinkler Equals the Sprinkler Head Spacing

Number of Outlets	*m* = 1.85	*m* = 1.90	*m* = 2.00
1	1.0	1.0	1.0
2	0.639	0.634	0.625
3	0.535	0.528	0.518
4	0.486	0.480	0.469
5	0.457	0.451	0.440
6	0.435	0.433	0.421
7	0.425	0.419	0.408
8	0.415	0.410	0.398
9	0.409	0.402	0.391
10	0.402	0.396	0.385
11	0.397	0.392	0.380
12	0.394	0.388	0.376
13	0.391	0.381	0.373
14	0.387	0.381	0.370
15	0.384	0.379	0.376
16	0.382	0.377	0.365
17	0.380	0.375	0.363
18	0.379	0.373	0.361
19	0.377	0.372	0.360
20	0.376	0.370	0.359
22	0.374	0.368	0.357
24	0.372	0.366	0.355
26	0.370	0.364	0.353
28	0.369	0.363	0.351
30	0.368	0.362	0.350
35	0.365	0.359	0.347
40	0.364	0.357	0.345
50	0.361	0.355	0.343
100	0.356	0.350	0.338
More than 100	0.351	0.345	0.333

quantity of water in the submain or lateral diminishes in the downstream direction because of outlet discharge.

The term *F* in Eq. 5.14 equals 1 when there are no outlets between the up- and downstream locations along a pipe (i.e., discharge along the pipe is constant). Equation 5.17a and 5.17b and Tables 5.9 and 5.10 can be used to determine *F* when there are more than one equally spaced outlet; each removing approximately the same amount of water from the pipe. (When the discharge varies widely from outlet-to-outlet Eq. 5.13 should be applied between successive outlets working from the known pressure to the unknown pressure.) Equations 5.17a and 5.17b are

$$F = \frac{1}{m+1} + \frac{1}{2N} + \frac{\sqrt{m-1}}{6N^2} \tag{5.17a}$$

Table 5.10 Values of F (in Eq. 5.17b) Used When the Distance to the First Sprinkler Equals One-half of the Sprinkler Head Spacing

Number of Sprinkler Heads on Lateral N	F		
	For $m = 1.85$	For $m = 1.9$	For $m = 2.0$
1	1.000	1.000	1.000
2	0.518	0.512	0.500
3	0.441	0.434	0.422
4	0.412	0.405	0.393
5	0.397	0.390	0.378
6	0.387	0.381	0.369
7	0.381	0.375	0.363
8	0.377	0.370	0.358
9	0.374	0.367	0.355
10	0.371	0.365	0.353
11	0.369	0.363	0.351
12	0.367	0.361	0.349
13	0.366	0.360	0.348
14	0.365	0.358	0.347
15	0.364	0.357	0.346
16	0.363	0.357	0.345
17	0.362	0.356	0.344
18	0.361	0.355	0.343
19	0.361	0.355	0.343
20	0.360	0.354	0.342
22	0.359	0.353	0.341
24	0.359	0.352	0.341
26	0.358	0.351	0.340
28	0.357	0.351	0.340
30	0.357	0.350	0.339
35	0.356	0.350	0.338
40	0.355	0.349	0.338
50	0.354	0.348	0.337
100	0.353	0.347	0.335

$$F = \frac{1}{2N - 1} + \frac{2}{(2N - 1)N^m} \left(\sum_{i=1}^{N-1} (N - i)^m \right) \tag{5.17b}$$

where

m = appropriate m value from Table 5.6;
N = number of sprinklers.

Either Eq. 5.17a or Table 5.9 is used when the distance from the pipeline to the first outlet equals the outlet spacing. When the distance to the first outlet is half of the outlet spacing, either Eq. 5.17b or Table 5.10 should be used.

EXAMPLE 5.4   Determining the Pressure at the Downstream End of a Sprinkler Lateral with a Constant Upward 2 percent Slope

Given:

$L = 1040$ ft.
$D = 4$ in
$Q = 150$ gpm
$N = 26$ sprinklers
   40 ft. sections of aluminum pipe
   first sprinkler is 40 ft downstream of main line
   P at mainline $= 50$ psi

Solution:

Solution Steps

1. Determine H_1 (using Eq. 5.15)
2. Determine h_1 (using Eq. 5.14)
3. Determine P_d (using Eq. 5.13)

Determining H_1

$m = 1.90$
$n = 1.10$ from Table 5.5

$$K = \frac{Ks}{348} = \frac{0.39}{348} = 1.12(10)^{-3}$$

$$H_1 = \frac{1.12(10)^{-3}(1.0)(1040)(150)^{1.90}}{(4)^{2(1.90)+1.10}} = 17.81 \text{ ft}$$

Determining h_1

$h_1 = (F)(H_1) + (M_1)$ Eq. 5.14
$M_1 = 0$ since there are no minor losses
$F = 0.37$ from Table 5.8
$h_1 = (0.37)(17.81) + 0 = 6.59$ ft

Determining P_d

$P_d = P_u - 0.43(h_1 + \Delta Z)$
$\quad = 50 - 0.43(6.59 + 0.02(1040)) = 38.1$ psi

5.3.2c   Surge Pressures

Cyclic oscillations of pressure above and below the normal operating pressure, caused by changes in fluid velocity, can significantly affect pipeline design. Thicker walled pipes, larger pipe diameters, pressure relief valves and/or surge tanks may be required to protect the pipeline. Typical causes of velocity changes in pipelines include (but are not limited to) accidental or planned changes in valve settings, starting and stopping of pumps, and unstable pump operation.

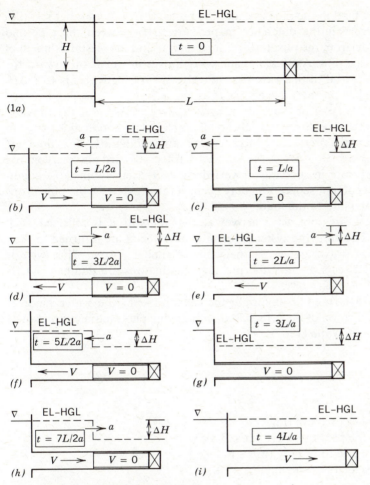

Figure 5.18 Graphical portrayal of water hammer phenomenon.
(From The Johns-Manville Corporation, 1978.)

The system in Figure 5.18 will be used to illustrate the water hammer phenomenon. The system is a single pipe fed by a reservoir. The elevation head, H, causes water to flow in the pipeline with a velocity, V, through a valve located a distance L downstream of the reservoir.

When at some time $t = 0$ the valve is closed suddenly, the layer of fluid nearest the valve is compressed as it slams into the valve. This compression causes the pipe to expand under an increased pressure, ΔH. After the first layer of fluid is compressed, the process is repeated with the next layer. A pressure wave propagates upstream toward the reservoir as successive layers of fluid are stopped and compressed. The fluid upstream of the wave flows with velocity V. When the wave reaches the reservoir the time that has elapsed (since the valve was closed) equals L divided by the speed, a, with which the water hammer wave travels. Thus, $t = L/a$.

At this time all the fluid in the pipe is under the pressure $H + \Delta H$ and at rest (i.e., $V = 0$). At the reservoir the difference in pressure, ΔH, between the pipe and reservoir causes flow from the pipe toward the reservoir and the reestablishment of the pressure H in the pipe. As this pressure wave moves downstream toward the valve, the pressure ahead of the wave (on the valve side of the wave) is $H + \Delta H$, while the pressure behind the wave equals H.

At time $t = 2L/a$ the pressure wave arrives at the valve and the pressure equals H all along the pipe. Since the valve remains closed and no fluid is added to the pipe, flow toward the reservoir reduces the pressure within the layer of fluid just upstream of the valve to $H - \Delta H$. A pressure wave travels toward the reservoir at a speed, a, as the pressure in successive layers declines. Extremely unstable conditions, such as fluid column separation, may occur if the pressure in the pipe drops below the vapor pressure of the fluid.

When the low pressure wave arrives at the reservoir it is reflected back toward the valve. This time the system pressure and velocity are returned to their original values. The wave arrives at the valve at time $t = 4L/a$. This cycle is repeated every $4L/a$ seconds until the pressure fluctuations are damped out by friction losses and pipe elasticity.

Equation 5.18 is used to compute the magnitude of the water hammer pressure, ΔH. A derivation of Eq. 5.18, based on the principle of momentum, can be found in Streeter and Wylie (1967).

$$\Delta H = \frac{a}{g} \Delta V \tag{5.18}$$

where

$\Delta H =$ water hammer pressure (m, ft);
$\quad a =$ velocity of the pressure wave (m/s, ft/s);
$\quad g =$ acceleration due to gravity (m/s^2, ft/s^2);
$\Delta V =$ change in velocity of the fluid (m/s, ft/s).

$$a = \frac{K\sqrt{B/\rho}}{\sqrt{1 + \left(\dfrac{B}{E}\right)\left(\dfrac{D}{t}\right)C_1}} \tag{5.19}$$

where

$\quad B =$ bulk modulus of water (kN/m^2, psi) (see Appendix F);
$\quad \rho =$ density of water (kg/m^3, slugs/ft^3) (see Appendix F);
$\quad K =$ unit constant ($K = 1.0$ for B in kN/m^2 and ρ in kg/m^3;
$\qquad\quad K = 12$ for B in psi and ρ in slug/ft^3);
$\quad D =$ internal diameter of pipe (mm, in);
$\quad t =$ pipe wall thickness (mm, in);
$\quad E =$ modulus of elasticity of pipe material (kN/m^2, psi);
$\quad C_1 =$ constant that depends on how the pipe is constrained.

Streeter and Wylie (1967) use the principle of continuity to derive Eq. 5.19.

Table 5.11 Modulus of Elasticity and Poisson's
Ratio of Some Common Piping Materials

Material	Modulus of Elasticity (kN/m^2)	(psi)	Poisson's Ratio (μ)
Asbestos-cement	$20.7(10)^6$	3×10^6	0.20
Cast iron	$10.3(10)^7$	15×10^6	0.29
Ductile Iron	$16.5(10)^7$	24×10^6	0.29
PVC	$27.6(10)^5$	4×10^5	0.46
Polyethylene	$69.0(10)^4$	1×10^5	0.40
Steel	$20.7(10)^7$	30×10^6	0.30

Equation 5.20a can be used to compute C_1 when the pipe is anchored at each end so that there can be no movement in the axial direction.

$$C_1 = 1 - \mu^2 \qquad (5.20a)$$

where, μ is the Poisson ratio for the pipe material. Values of μ for common pipe materials are listed in Table 5.11.

When a pipe is anchored at both ends but has expansion joints, $C_1 = 1$. Equation 5.20b is used when the pipe is anchored at one end only.

$$C_1 = \tfrac{5}{4} - \mu \qquad (5.20b)$$

The magnitude of water hammer pressure can be reduced by decreasing ΔV in Eq. 5.18. When water hammer pressures are the result of changes in valve settings, ΔV can be reduced by slowing the rate of valve adjustment. In order to reduce the magnitude of water hammer pressure computed with Eq. 5.18, the time of valve adjustment must exceed $2 L/a$ (in order for the maximum pressure to be reduced by reflected waves). The following example illustrates this concept.

EXAMPLE 5.5 Computing the Minimum Time of Valve Closure

Given:

$L = 300$ ft
$a = 1500$ ft/s

Solution:

$$T_{min} = \frac{2(3000 \text{ ft})}{1500 \text{ ft/s}} = 4 \text{ s}$$

Thus, for the conditions of the example problem, the time to close the valve should be at least 4 seconds to reduce the magnitude of the water hammer pressure.

For most valve designs (including gate, cone, globe and butterfly valves), flow through the valve is not linearly related to valve stem travel. This is shown in

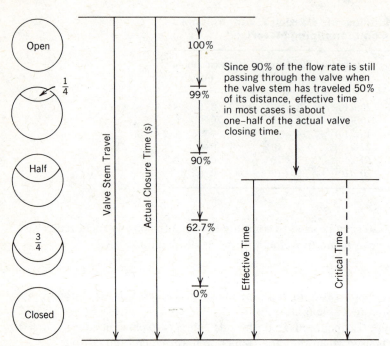

Since 90% of the flow rate is still passing through the valve when the valve stem has traveled 50% of its distance, effective time in most cases is about one-half of the actual valve closing time.

Figure 5.19 Valve stem travel versus flow stoppage for a gate valve. (From The Johns-Manville Corporation, 1978.)

Figure 5.19 for a gate valve. Since a majority of the flow is cut off, the final portion of stem travel is the most effective portion of the closure. Therefore, it is extremely important that the timing of the valve closing be based on the "effective closing time" of the valve rather than the actual closing time. Although there is variation between valve types, the "effective time" is usually assumed to be one-half the actual valve closure time.

Figure 5.20 can be used to estimate water hammer pressures when the effective time of valve closure exceeds $2 L/a$. Either the effective time of closure needed to limit the water hammer pressure to a certain level or the water hammer pressure that results from a certain effective time of closure can be determined using Figure 5.20. The following example illustrates the use of Figure 5.20.

EXAMPLE 5.6 Computing the Actual Closure Time Needed to Limit Water Hammer Pressure to a Desired Valve

Given:

$a = 3500$ ft/s

$V = 4$ ft/s

$$C = 0.517 \quad \text{for Reduced Area Globe Valve}$$
$$C = 0.486 \quad \text{for Full Area Cone Valve}$$
$$C = 0.392 \quad \text{for Full Area Gate Valve}$$

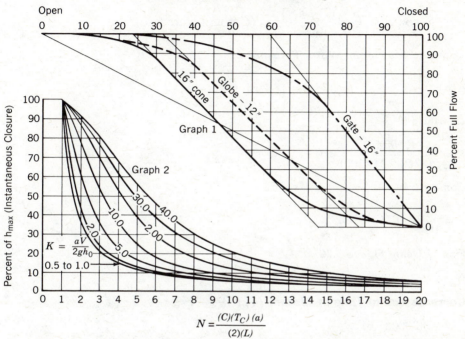

$$N = \frac{(C)(T_C)(a)}{(2)(L)}$$

Figure 5.20   Calculation of valve closing time, T_c, for limiting surge pressures. *Source:* S. L. Kerr, "Water Hammer . . . A Problem in Engineering Design," copyright © 1958 by Technical Publishing, a Division of Dun-Donnelley Publishing Corp., a company of the Dun & Brandstreet Corp. Reprinted by permission from *Consulting Engineer* (May 1958) Vol. 10, No. 5, pp. 88–92.

Normal operating pressure of pipe = 100 psi

$L = 3000$ ft

Water hammer pressure limit = 50 psi

Required:

T_c for a full area gate valve

Solution:

Solution Steps

1. compute maximum water hammer pressure

2. compute $K = \dfrac{aV}{2gh_0}$ (where $h_0 = 2.31 \times$ normal operating pressure of pipe)

3. compute $P = \dfrac{\text{water hammer pressure limit}}{\text{maximum water hammer pressure}}$

4. find N from Figure 5.20

5. compute $T_C = \dfrac{(2)(L)(N)}{(c)(a)}$ (C from Figure 5.20)

Compute Maximum Water Pressure

$$\Delta H = \frac{aV}{g} = \frac{3500(4)}{32.2} = 435 \text{ ft} = 188 \text{ psi}$$

Compute K

$$K = \frac{(3500)(4)}{2(32.2)(2.31)(100)} = 0.94$$

Compute P

$$P = \frac{50}{188} = 0.27$$

Find N Using Figure 5.20

$N \sim 2.6$

Compute Actual Time of Closure (T_c)

$$T_c = \frac{(2)(3000)(2.6)}{(0.392)(3500)}$$

$$= 11.4 \text{ s}$$

Thicker walled pipes, larger pipe diameters, pressure relief valves, and/or surge tanks can be used to protect pipelines from water hammer pressures. Thicker walled pipes are able to withstand higher pressures than thinner walled pipes of the same nominal diameter. Changing the thickness of the pipe wall can, however, change the pressure wave velocity and hence, expected water hammer pressure (as indicated by Eqs. 5.18 and 5.19). Thus, the expected water hammer pressure should be recomputed each time the wall thickness is changed.

Changing the pipe diameter also affects pressure wave velocity and the expected water hammer pressure. Larger pipe diameters will reduce the expected water hammer pressure by reducing the velocity change that occurs in a pipe. In addition, lower pressure wave velocities and hence water hammer pressures will result whenever larger pipe diameters increase the ratio of pipe diameter to wall thickness in Eq. 5.19. Thus, increasing pipe diameter is an extremely effective way of controlling water hammer pressure.

In situations where rapid changes in flow velocity are necessary, pressure relief valves, like the one in Figure 5.21, are often the most economical way of protecting pipelines from water hammer pressures. Pressure relief valves are

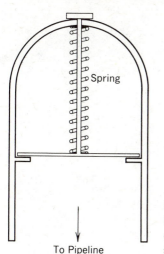

Spring

To Pipeline

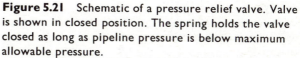

Figure 5.21   Schematic of a pressure relief valve. Valve is shown in closed position. The spring holds the valve closed as long as pipeline pressure is below maximum allowable pressure.

designed to open at a certain preset pressure and discharge fluid to relieve the surge. They may close immediately when pressure drops below the setting, or the closure may be damped to extend the closure time.

A surge suppressor is a relief valve that opens automatically in response to a signal. The valve, for example, may open when the power to a pumping station is lost. The response of the pipeline that the surge suppressor is protecting to transient flow should be carefully studied, since surge suppressors that open too soon may cause column separation.

For protection against low pressures, air inlet valves (also called vacuum relief valves) are often provided at critical points in a pipeline (see Figure 5.22). The air that enters the pipeline can be troublesome to remove, however. Open tanks connected to the pipeline, called surge tanks, are also used to protect pipelines from surge pressures. Water flows in and out of surge tanks as water hammer pressure waves move back and forth through the pipeline. A simple surge tank is generally sized so that water levels within the tank do not fluctuate in resonance with valve regulation (due to governor action, for example) and so that the tank will not be drained or overtopped by flows to and from the pipeline. Some surge tanks have orifices, that restrict both tank inflow and outflow to promote energy dissipation and dampening of pressure surges. Frequently, inflow is restricted more than outflow to reduce the danger of column separation. A differential surge tank is a combination of an orifice surge tank and a simple surge tank of small cross-sectional area. One-way surge tanks that allow only outflow are used primarily with pumping plants to prevent column separation.

5.3.2d   Air Entrapment

The presence of air in a pipeline affects its operation by reducing the cross-sectional area through which flow occurs. This can increase pumping costs

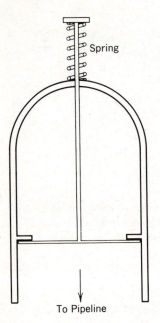

Spring

To Pipeline

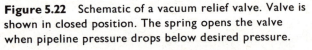

Figure 5.22 Schematic of a vacuum relief valve. Valve is shown in closed position. The spring opens the valve when pipeline pressure drops below desired pressure.

(because of increased head loss) and/or reduce the carrying capacity of the pipe. In addition, the movement of air pockets within the pipe and their sudden entrapment at high points along the pipeline can cause fluctuation in flow and water hammer.

Air may enter a pipeline during filling, where water enters the line (at a pump or gravity inlet, for example), or through air release valves, vacuum breakers, and leaky joints and fittings (during a negative surge that drops the pressure within the pipe below atmospheric). Changes in temperature or pressure can also cause dissolved air from the fluid to accumulate in the pipeline.

Air should be prevented from entering pipelines. This can be accomplished by careful design of the pump or gravity inlet, maintaining the average water velocity of between 0.3 to 0.6 m/s (1 to 2 ft/s) during pipeline filling, and laying pipe to a grade that results in a minimum number of high points. In order to remove any air that does enter the pipeline, properly sized air and vacuum release valves should be installed at all high points or other locations where air would be expected to accumulate.

Air release valves are designed to exhaust air under various pressure conditions during normal pipeline operation while restricting the outflow of liquid. Exhaust ports for such valves are on the order of 1.6 to 6.4 mm (1/16 to 1/4 in) in diameter. Vacuum release valves, which have orifice sizes on the order of 25 to 200 mm (1 to 8 in) in diameter, are designed to exhaust large volumes of air during pipeline filling and to close when the filling process is complete. Both air release and vacuum release valves are closed either by impact of water against the valve closure element or by a float.

5.3.2e   Valving

Valves are integral parts of the pressurized pipelines utilized in farm irrigation systems. They provide on–off service, throttling (i.e., flow regulation), pressure regulation, surge control, pressure relief, air release, vacuum relief, and backflow prevention. Pressure relief, air release, and vacuum relief valves are discussed in Sections 5.3.2c and 5.3.2d.

(i) On-Off Valves   On–off valves may be manual or automatic. Figure 5.23 shows examples of an automatic and several manual on–off valves.

On-off valves are used for a variety of purposes. When located at the upstream end of mainlines, submains, and laterals, they provide on–off service to pipes downstream. This allows water to be cycled to different parts of the system to meet irrigation requirements of each portion of the farm. It also allows portions of the system to be isolated for maintenance and repair (i.e., water can be shut off to one part of the system to enable maintenance and repair while other parts of the system continue to operate). On–off valves also allow set-move systems to be moved without stopping pumps.

Automatic on–off valves at the upstream end of submains and laterals allow an irrigation system to be automated. An electromechanical or electronic controller located in the farm office, for example, can be programmed to control the operation of several automatic on–off valves. Once programmed by the irrigator, the controller opens and closes each valve according to the irrigator-supplied program until reprogrammed. Communication between the controller and valves is via electric wires, hydraulic or pneumatic conduit, or radio telemetry.

On–off valves are also located at the downstream end of permanent pipes to allow flushing of sediment and debris. Provision for draining permanent pipes is provided by installing on–off valves at low points within the pipe network. Portable pipes are drained and flushed by disconnecting pipe sections and removing end plugs.

(ii) Throttling   Throttling can be accomplished by partially closing the manual valves in Figure 5.23. The flow will remain constant as long as the pressure and valve setting remain unchanged. Automatic valves which supply a constant flow regardless of changing pressures are also sometimes used with pressurized pipe networks. As with manual valves, the flow rate can be adjusted by changing the opening (i.e., the valve setting). Throttling valves are located at the upstream end of mainlines, submains, and laterals.

(iii) Pressure Regulating Valves   Pressure regulating valves are automatic valves that hold a constant downstream pressure regardless of changing flow and/ or upstream pressure. They are used where system pressure fluctuations make it difficult to apply water uniformly as in fields (or solid-set blocks) where available pressure depends upon which other fields (or blocks) are being irrigated. In such situations, a pressure regulating valve supplies a constant pressure to the field regardless of which other fields are being irrigated (assuming the system has adequate capacity to supply the requirements of all fields or blocks). Frequently,

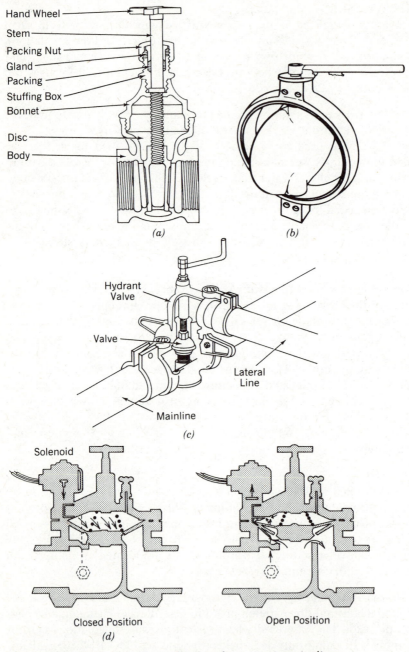

Figure 5.23 Examples of on–off valves for irrigation pipelines.
(*a*) Gate valve. (*b*) Butterfly valve. (*c*) Hydrant valve. (*d*) Solenoid
activated, automatic on–off valve.

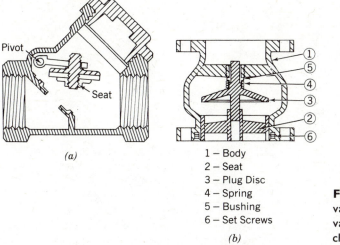

1 — Body
2 — Seat
3 — Plug Disc
4 — Spring
5 — Bushing
6 — Set Screws

(a)

(b)

Figure 5.24 Check valves. (*a*) Swing check valve. (*b*) Spring-loaded check valve.

pressure regulating valves provide surge control by protecting downstream pipes from upstream originating surges.

(iv) Check Valves Check valves, like the ones in Figure 5.24, control reverse flow (backflow) in pipelines. Backflow protection is essential when fertilizers and/or agricultural chemicals are injected into irrigation systems to prevent contamination of the water source during pump failure (see ASAE Engineering Practice EP409).

(v) Surge Control Valves Surges are controlled during pipeline filling and emptying with automatic valves installed downstream of pumps. For centrifugal pumps, a normally closed automatic valve opens slowly during pipe filling. This controls surges by allowing full system pressure to develop gradually. When the system is being shut down, the automatic valve closes slowly, gradually reducing flow as the pump continues to run. For turbine pumps, a normally open automatic valve is installed on a short length of pipe that tees from the main pipeline. A check valve is installed downstream of the tee. When the pump is started, the normally open valve begins to slowly close. Initially, all air and water from the pump column is discharged into the atmosphere through the normally open valve. More and more pump outflow is diverted through the check valve and into the irrigation system as the automatic valve closes. This provides surge control by allowing system pressures to develop slowly. During shutdown the automatic valve slowly opens diverting an increasing amount of pump output to the atmosphere. This controls surge pressures by allowing system pressure to gradually decrease as the pipeline empties.

5.3.2f Economics

Economics is an extremely important consideration in pipeline design. Piplines are normally designed to deliver water at the required pressure and

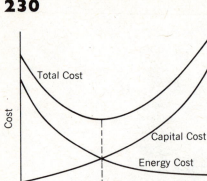

Figure 5.25   Typical relationship between pipe diameter and annual capital, operating, and total costs of pipelines.

flowrate throughout the irrigation system with minimum total (capital plus operational) costs. Pipe diameter and wall thickness, and the amount of energy loss (due to pipe friction and minor losses), and the size and type of pump are major factors affecting total cost.

Figure 5.25 shows a typical relationship between pipe diameter and annual capital, operational, and total costs. Annual capital costs steadily increase as pipe diameter increases, while operational costs steadily decline as pipe diameters become larger. Total cost, which equals the sum of the capital and operational costs, decreases to a minimum and then steadily increases. The recommended pipe diameter is the one corresponding to the lowest total cost.

Annual capital cost is computed using the methods described in Chapter 2. The following example problem illustrates a procedure for identifying the most economical pipe.

EXAMPLE 5.7   Determining the Most Economical Pipe for a Pipeline (Constant Annual Energy Costs)

Given:
- 10,000-ft-long mainline
- design flowrate is 1000 gpm
- there is no change of elevation along the pipe
- assume 20-year life and no salvage value
- 15 percent interest rate
- energy cost = 2.3 cents/kWh
- the pipe costs in column 2 of the following table
- 50 psi (116 ft of head) must be available at downstream end of the mainline
- irrigation system operates 1500 hr/year
- assume cost of pump is $60/bhp

Required:
most economical pipe size (assuming a pumping plant efficiency of 80 percent and that minor losses are negligible)

Solution:

Diameter (in)	Cost/ft	H_l (ft)[a]	Cost of Pump	Annual Capital Cost	Annual Energy Cost	Total Annual Cost[b]
1	2	3	4	5	6	7
	($)		($)	($)	($)	($)
4	2.00	4497	85,182	16,829	37,461	54,290
6	2.80	624	11,824	6372	6011	12,383
8	4.75	154	2913	8066	2191	10,257[c]
10	7.40	52	983	11,997	1363	13,360
12	10.35	21	404	16,785	1115	17,900

[a] It is assumed that one pumping plant operating at 80 percent efficiency can provide these total heads.

[b] Does not include labor, water, maintenance, and repair costs.

[c] Most economical pipe (8 in diameter).

Column 3

$$H_1 = \frac{(0.285C)^{-1.852}LQ^{1.85}}{D^{4.87}}$$

$$= \frac{(0.285(140))^{-1.852}(10000)(1000)^{1.85}}{D^{4.87}}$$

$$= \frac{3.85(10)^6}{D^{4.866}}$$

Column 4

$$= \frac{(Q)(H_l)(\$60/bhp)}{3960E_p} = \frac{(1000)(H_l)(60)}{(3960)(0.80)}$$

$$= 18.94(H)$$

Column 5

Total annual capital cost $= ((\text{cost/ft})(10000) + \text{cost of pump})CRF$

$$= (\text{cost/ft})(10000)\left(\frac{0.15(1 + 0.15)^{20}}{(1 + 0.15)^{20} - 1}\right)$$

$$= 0.16((\text{cost/ft})(10000) + \text{cost of pump})$$

Column 6

$$\text{Annual energy cost} = \left(\frac{QH}{3960E_p}\right)(hr)\left(0.746\,\frac{kW}{Hp}\right)(\$0.023/kWh)$$

$$= \frac{1000(H + 116)}{3960(0.80)}\,(1500)(0.746)(0.023)$$

$$= 8.12(H + 116)$$

In situations where energy costs rise at a constant rate over the life of the pipeline the equivalent annual cost factor (Eq. 2.22) is applied to the annual energy cost for the first year.

The following example illustrates a procedure of accounting for the effect of escalating energy costs in the pipe selection process.

EXAMPLE 5.8 Determining the Most Economical Pipe for a Pipeline when Energy Costs are Increasing at a Constant Rate Over the Life of the Pipeline

Given:
• information from Example 5.7
• columns 1, 2, 3, 4, 5, 6 of the solution table for Example 5.7
• energy costs are expected to rise at a constant 10 percent per year

Required:
most economical pipe diameter (assuming a pumping plant efficiency of 80 percent and that minor losses are negligible).

Solution:

Diameter (in)	Annual Capital Costs	Annual Energy Cost	Annual Energy Cost with Escalating Energy Costs	Total Annual Cost
1	2	3	4	5
4	16829	37461	70,427	87,256
6	6372	6011	11,301	17,673
8	8066	2191	4119	12,185
10	11,997	1363	2563	14,560
12	16,785	1115	2097	18,721

Note: Columns 1, 2, and 3 are from solution table for Example 5.7.

Column 4

Annual Energy Cost with Escalating Energy Cost = (Annual Energy Cost in Column 3) (Equivalent Annual Cost Factor Computed with Eq. 2.2)

$$\text{Equivalent Annual Cost Factor} = \left(\frac{(1+0.10)^{20}-(1+0.15)^{20}}{(1+0.10)-(1+0.15)}\right)\frac{0.15}{(1+0.15)^{20}-1}$$

Equivalent Annual Cost Factor = 1.88

Equivalent Annual Cost Factor = $(AEC_O)(1.88)$

Annual Energy Cost with Escalating Energy Cost = (Annual Energy Cost in Column 3) (1.88)

Eight inch pipe remains the most economical pipe diameter.

5.4   Sprinkle System Design

Sprinkle system design involves identifying alternative layouts of laterals, submains, and mainlines for the farm and then developing design specifications for the most feasible layouts. As discussed in Section 2.4.4, this process begins once all the necesssary data have been assembled, a suitable water source identified, and the DDIR determined. The total annual cost of owning and operating each alternate system is developed when design specifications for sprinklers, pipes, valves, and pumping plants are available. The design of set-move, solid-set, traveler, center-pivot, and linear-move systems are discussed.

5.4.1   Design of Set-Move and Solid-Set Systems

The basic strategy outlined in this section applies to the design of the set-move and solid-set systems described in Section 5.2.5. The major steps in this process are: laying out mainlines, submains, and laterals; determining the number of laterals to be operated per irrigation; selecting sprinklers; and developing pipeline, valve, and pumping plant specifications for the most feasible layouts.

5.4.1a   System Layout

Key factors affecting system layout are topography, field shape, and the location of the water source. Identifying the best layout often requires consideration of several alternate layouts and careful pipe-size analysis. This is especially true for large odd-shaped fields. The layout selected is usually the one with the lowest total ownership and operation costs.

(i) Water Source and Pumping Plant Location   When there is a choice, the water source and pumping plant should be located to minimize pipe and pumping costs. Wells should therefore be located as close to the center of the irrigated area as possible. When water is obtained from lakes and streams pumping plants should be located at a central point for delivery to all points of the design area. On flat or gently sloping lands where water is to be pumped from canals, ditches, or streams, mainline costs are normally minimized by delivering water to the center of the field via canal or ditch.

Booster pumps should be considered when, because of high elevations, a small fraction of the total system discharge must be delivered at a higher pressure than the remainder of the flow. The use of booster pumps in such situations lowers pumping costs by reducing the pressure that must be supplied by the main-pumping plant.

(ii) Laterals   Laterals should be laid across prominent land slopes to minimize the variation of pressure along the lateral. The American Society of Agricultural Engineers (ASAE) recommends that along-the-lateral pressure variation in set-move and solid-set systems not exceed ± 10 percent of the design lateral pressure. Thus, if the design lateral pressure was 400 kPa, the pressure at any sprinkler along a set-move or solid-set lateral could not be less than 360 kPa or greater than 440 kPa.

When it is necessary to run laterals up and down prominent slopes, it is preferrable for water to flow downslope rather than upslope. The elevation head gained as water flows downslope allows longer laterals for a given pipe size or smaller pipe for a given length of lateral. Running laterals uphill should be avoided wherever possible. Where water must flow uphill, however, laterals need to be shortened unless pressure or flow regulators are used. Whenever possible, laterals should be located at right angles to the prevailing wind direction.

(iii) Mainlines and Submains   Mainline and submain layout is keyed to lateral layout (see Figure 5.26). When laterals run across prominent land slopes, mainlines or submains will normally run up and down the slopes (as in Figures 5.26a and b). When it is necessary to run laterals up and down hill, mainlines or submains should be located on ridges (as in Figures 5.26c, d, e, and f) to avoid laterals that run uphill.

It is also desirable to locate mainlines or submains so that set-move laterals may operate on either side of them (see Figures 5.26a, b, and e). Such layouts, called split lateral layouts, minimize friction loss because of shorter laterals. Split layouts also allow set-move laterals to be rotated around mainlines (see Figures 5.26a and b). This reduces labor requirements by eliminating the need for moving lateral pipes back to the starting point (as is necessary in Figures 5.26c and d).

5.4.1b   Number of Laterals Operated Per Irrigation Set

With set-move and solid-set systems, the irrigated area is subdivided into sets and each set irrigated separately. An *irrigation* consists of cycling water from set to set until all sets have been irrigated. Several days are normally needed to complete

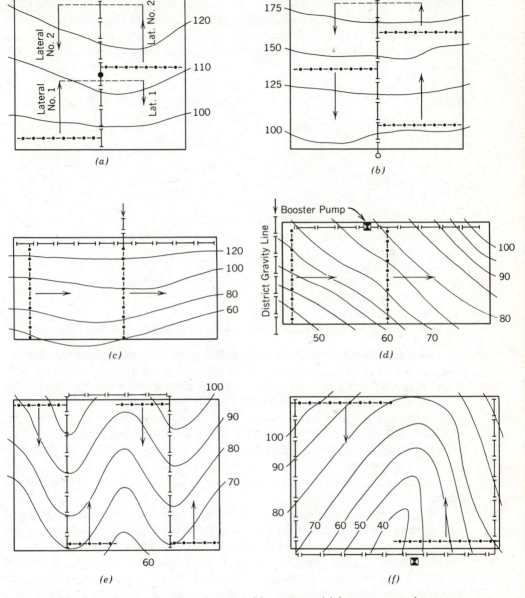

Figure 5.26 Some layouts for set-move sprinkle systems. (a) Layout on moderate, uniform slopes with water supply at center. (b) Layout illustrating use of odd number of laterals to provide required number of operating sprinklers. (c) Layout with gravity pressure where pressure gain approximates friction loss and allows running lateral downhill. (d) Layout illustrating area where laterals have to be laid downslope to avoid wide pressure variation caused by running laterals upslope. (c) Layout with two main lines on ridges to avoid running laterals uphill. (f) Layout with two main lines on the sides of the area to avoid running the laterals uphill.

an irrigation with set-move systems, while solid-set irrigations can be completed in times ranging from less than a day to several days. *Set length* is the time required to apply the desired amount of water to a set.

The number of laterals operated on a set can be computed using the following equation.

$$N_{1/s} = \frac{N_1}{N_s}$$

$$N_s = \text{integer value of } \frac{(24)(\text{II})}{H} \tag{5.22}$$

$$H = T_{s_1} + T_m$$

where

$N_{1/s}$ = number of laterals operating on a set;
N_1 = total number of lateral positions in the irrigated area;
N_s = number of sets;
II = irrigation interval (days);
H = hours between successive irrigations of a set position ($H \leq$ the value computed with Eq. 5.6);
T_{s1} = set length (h);
T_m = downtime for moving set-move laterals and/or maintenance (h).

N_1 is determined for each alternate layout once the spacing between adjacent lateral positions (L in Eqs. 5.5 and 5.8) has been determined. The length, L, usually ranges between 9.1 and 24.4 m (30 and 80 ft), depending on the wetted radius of the sprinkler and the prevailing wind (see Sections 5.3.1c and d). Because the length of portable pipe is standardized, L for portable mainlines or submains is either 9.1, 12.2, 15.2, 18.3, or 24.4 m (30, 40, 50, 60, or 80 ft). Tables 5.1 and 5.2 and Eq. 5.9 can be used to estimate L. It may, however, be necessary to adjust L once sprinkler selection is complete.

The irrigation interval (II), is computed using Eq. 2.1, with D equal to the desired depth of application, D_a.

$N_{1/s}$ is determined using a value of H that is less than or equal to the upper limit computed with Eq. 5.6. For set-move systems, this value of H (and hence T_{s1}) is chosen on the basis of labor cost and/or availability, and capital cost. Long set lengths increase $N_{1/s}$ and therefore the cost of the irrigation system. Conversely, the use of short set lengths reduces system cost but increases labor requirements by increasing the number of times set-move laterals must be moved. In addition, H for set-move systems is chosen so that the number of sets irrigated per day is an integer. This simplifies system management by allowing set-move laterals to be moved at the same time each day.

5.4.1c Sprinkler Selection

Sprinklers can be selected using the procedures presented in Section 5.3.1 when the value of S, the spacing between sprinklers along a lateral, is available (since values of D, H, T_m, and L were evaluated and used to determine $N_{1/s}$ in the previous section). Application efficiency E_a for set-move and solid-set systems is about 75 to 80 percent.

Because the length of portable pipe is standardized, S for portable set-move and solid-set systems is either 9.1, 12.2, or 18.3 m (30, 40, or 60 ft). Tables 5.1 and 5.2 and Eq. 5.10 are used to estimate S.

5.4.1d Design of Pipelines

Laterals, submains, and mainlines are designed using the procedures presented in Section 5.3.2. Pipeline design begins with the laterals, proceeds to the submains, if present, and ends with mainlines.

(i) Laterals The design of laterals involves using Eq. 5.13 to estimate along-the-lateral pressure variations for several possible pipes. Pressure variations should be determined for laterals operating at locations with large changes in elevation, steep uphill slopes (in the direction of flow), and/or numerous sprinklers. Ownership and operating costs are estimated for those pipes in which the pressure variation is acceptable (i.e., those in which the difference between the minimum and maximum pressure is less than 20 percent of the design pressure). Total ownership and operation costs of these pipes are compared and the one with lowest total cost is selected for the laterals. When only one pipe material is considered, the smallest diameter pipe with acceptable pressure variation is usually selected.

A single pipe diameter and material is normally used for all laterals of portable set-move and solid-set systems. Two or more pipe diameters may be used for buried solid-set laterals when acceptable pressure variation is possible and it is less expensive than using a single diameter.

In situations where it is not possible or convenient to keep the pressure variation within 20 percent of the design pressure, pressure regulators or constant-discharge (flow control) nozzles can be used to minimize the variation in sprinkler discharge along a lateral. Pressure regulators located at the base of each sprinkler control discharge by dissipating all pressure in excess of the desired operating pressure (which equals the sprinkler design pressure). The pressure at the upstream end of each lateral must be sufficient to supply a pressure to the inlet of each regulator equal to the desired operating pressure plus the pressure loss through the regulator.

When constant discharge nozzles are used, a certain minimum pressure must be supplied to each nozzle to assure proper nozzle operation. This threshold value is usually about 275 kPa (approximately 40 psi). Pressures in excess of 550 kPa (about 80 psi) should be avoided, since they result in poor performance and may damage nozzles. When at least the threshold pressure is available to each sprinkler, discharge should not vary significantly along the lateral.

Along-the-lateral variations in pressure can also be controlled by land smoothing (to reduce changes in elevation) or by relocating submains and mainlines. The choice between land smoothing, relocating submains and mainlines, pressure regulators, and constant discharge nozzles is usually made on the basis of economics.

The discharge and upstream pressure requirements of each lateral are needed to design submains and mainlines. In addition, the discharge of a lateral is needed to compute pressure variations along it using Eq. 5.13 (for determining lateral diameters).

The discharge of a lateral can be determined by summing the discharges of all sprinklers along the lateral. This involves solving for the pressure and discharge at each sprinkler simultaneously using an iterative procedure that is most easily performed on a computer. A simpler, but less accurate method of estimating lateral discharge is based on Eq. 5.23.

$$(Q_1)_i = (Q_s)(N_{s/1}) \tag{5.23}$$

where

$(Q_1)_i$ = estimated discharge of lateral i (l/min, gpm);
   Q_s = design sprinkler discharge (l/min, gpm);
$(N_{s/1})$ = number of sprinklers along lateral i.

(ii) *Mainlines and Submains*   Laterals are connected to submains or a mainline. When laterals receive water from a single pipe, as in Figs. 5.26*a*, *b*, *c*, and *d*, the pipe feeding the laterals is called a mainline. Such systems do not have submains. A *submain* is a pipe that supplies only a portion of the laterals in a system, as in Figs. 5.26*e* and *f*. Although these systems each have two submains, there is no theoretical limit on the number of submains in a system. In systems with submains, mainlines supply water to the submains.

A submain (or mainline that feeds laterals) is designed to supply the discharge and pressure requirements of each lateral that it serves. Similarly, mainlines for submains must provide the flow and pressure requirements of each submain. Equations 5.13 and 5.15 are used to estimate the energy used by various combinations of submain and/or mainline pipes. An economic analysis similar to the one illustrated in Examples 5.7 and 5.8 and includes the cost of the pumping plant is used to identify the pipeline–pumping plant combination with the lowest total cost of ownership and operation.

A single pipe diameter and material is normally used for portable submains and mainlines. When buried, submains and mainlines (which feed either laterals or submain) are often subdivided into reaches, each with a different constant pipe diameter. The diameter of each successive reach increases in a stepwise manner from the downstream to upstream end of the submain or mainline. The same pipe material is generally used for all submains and/or mainlines within a set-move or solid-set system.

5.4.2   Design of Traveler Sprinkle Systems

The design process for traveler systems is similar to the one used for set-move and solid-set systems. The initial step of traveler system design is to lay out towpaths (rather than set-move or solid-set laterals), submains, and mainlines. Next, sprinklers are selected and specifications for travelers, hoses, submains, mainlines, and pumping plant are developed.

5.4.2a System Layout
Figure 5.7 shows a typical traveler system layout. As with other systems, traveler systems should be laid out to be convenient to operate and to minimize total ownership and operation costs. Odd-shaped fields and broken terrain complicate this process.

(i) Water Source and Pumping Plant Location   When there is a choice, the water source and pumping plant for a traveler system should be located as described in Section 5.4.1a(i).

(ii) Towpaths   A towpath is a crop-free lane on which the traveler unit and sprinkler travel as they either pull themselves or are pulled across the field (see Figure 2.5). One or more travelers are moved from towpath to towpath to irrigate the field.

Towpaths should be laid across land slopes or along field contours and, if possible, in the same direction as crop rows. When traveler systems are to operate for extended periods of time in 2.2 m/s (about 5 mph) or more winds, towpaths should be perpendicular to the prevailing wind direction. Towpaths are normally 400 m (about one-quarter mile) long, but can be as long as 800 m (about one-half mile).

(iii) Mainlines and Submains   Submains or a mainline (as defined in Section 5.4.1a(ii)) should bisect towpaths to minimize hose length whenever possible. Submains and mainlines that feed traveler hoses should be laid up and down prominent land slopes to allow towpaths to be across the slope.

5.4.2b Sprinkler Selection
The gun-type sprinklers that are typically used with traveler systems operate at pressures in excess of 500 kPa (about 70 psi) and have capacities between 380 and 3800 l/min (100 and 1000 gpm). They have wetted diameters ranging from 61 to 183 m (200 to 600 ft) and trajectory angles of 18 to 32 degrees.

Equations 5.5, 5.6, 5.7, and 5.8 are used to select the sprinkler. In Eq. 5.8, L_1 is the length of the longest towpath, L is the spacing between adjacent towpaths, and N is the number of travelers. Spacing L is computed with Eq. 5.9 with approximate values of WD from Table 5.1 and K_1 values from Table 5.12. Spacing L is adjusted so that there are an integer number of towpaths. When more than one traveler is used, the number of towpaths must be an integer multiple of the number of traveler units.

Traveler systems are normally designed to irrigate one or at most two sets per day. The maximum set-length should be 23 h for one set per day systems and 11 h

Table 5.12 Recommended Values of K (in Eq. 5.9) for Big Gun-Type Sprinklers Used with Traveler Systems

Wind Speed		
(m/s)	(mph)	K_t
0	0	0.80
<2.2	<5	0.70–0.75
2.2 to 4.5	5 to 10	0.60–0.65
<4.5	710	0.50–0.55

Source: Nelson Irrigation Corporation. Nelson Big Guns® for Traveling Systems. Undated Sales Brochure.

for two set per day systems. The peak application rate of a traveler sprinkler is computed with the following equation.

$$A_p = \frac{4}{\pi} A = 1.27A \tag{5.24}$$

where

$$A = \frac{(K)(Q_s)}{(WD)^2(\theta)}$$

A_p = peak application rate (mm/h, in/h);
A = application rate (mm/h, in/h);
Q_s = sprinkler discharge (l/min, gpm);
WD = wetted diameter discharge (l/min, gpm);
θ = arc setting (i.e., portion of circle receiving water) (degrees);
K = unit constant. (K = 33,953 when A is in mm/h, WD is in m, and Q_s is in l/min. K = 54,470 when A is in in/h, WD is in ft, and Q_s is in gpm).

The equation for computing A_p is based on the assumption that the distribution pattern for a moving sprinkler is elliptical. Because the average application rate computed with Eq. 5.24 is based on 45 percent of the wetting diameter, A is an approximate application rate for a major portion of the pattern rather than the average application rate over the whole wetted area.

Part circle sprinklers are used to improve application uniformity and/or to provide a dry path for the traveler unit. Part-circle sprinklers do, however, have higher application rates (which increase the potential for runoff and erosion) than do full-circle sprinklers with identical discharges (see Eq. 5.24).

5.4.2c Traveler Speed

The traveler unit must move along a towpath at constant speed for uniform water distribution over the irrigated area. Traveler speed can, however, vary as much as 200 to 300 percent because of differences in hose pull, water pressure, flowrate, and the amount of cable build-up on the cable reel. Hose pull varies because of differences in soil type, terrain, and towpath condition.

The traveler speed needed to apply a given depth of water is computed using Eq. 5.25.

$$S_T = \frac{(K)(Q_s)}{(L)(D_a)}$$ (5.25)

$$D_a \leq D_m$$

where

S_T = required traveler speed (m/min, ft/min);
Q_s = sprinkler discharge (l/min, gpm);
L = towpath spacing (m, ft);
D_a = depth of application (mm, in);
D_m = amount that can be applied without runoff from Figure 5.17;
K = unit constant ($K = 1.0$ for S_T in m/min, Q_s in l/min, L in m, and D_a in
   mm. $K = 1.60$ for S_T in ft/min, Q_s in gpm, L in ft, and D_a in in).

Traveler speed is usually adjusted to about 0.3 m/min (1 ft/min) by varying L, H, N_l, and DDIR.

When it is not permissible, or practical, to irrigate over the boundaries of the field, the traveler and sprinkler should be stopped a distance equal to the sprinkler wetter radius from each edge of the field and remain stationary until the desired depth water has been applied. In such cases, the length of the towpath, L_1, should be reduced by a distance equal to the wetted diameter of the sprinkler.

5.4.2d   Hose Selection

Flexible, tough-skinned hoses that are capable to withstanding high pressures are utilized to convey water to the traveler unit and sprinkler. The standard hose length is 200 m (about 660 ft). Hoses are extremely expensive and often have a short life due to being punctured by sharp objects along towpaths, or being run over by the traveler and farm equipment. Other causes of reduced hose life include oil and grease coming into contact with the hose, winding the hose onto the reel before all water has been expelled, and rodent damage during the off-season.

The selection of hose diameter is made on the basis of tolerable pressure loss. Pressure loss is estimated with Eq. 5.13.

5.4.2e   Design of Mainlines and Submains

Traveler systems have either portable or buried mainlines and submains. Portable pipes are made of aluminum, while buried mainlines and submains are usually PVC. The procedures described in Section 5.4.1d apply to the design of traveler system mainlines and submains.

5.4.3 Design of Center-Pivot Systems

A center-pivot system includes one or more center pivot units. System design involves locating individual center-pivot units and laying out mainlines and submains. It also includes specifying lateral diameters, selecting sprinklers, and recommending speeds of rotation for each unit. The final step in the design process is the sizing of submains and mainlines.

5.4.3a System Layout

A typical field layout for a center-pivot unit is diagrammed in Figure 5.27. The lateral rotates continuously in either direction around the field as described in Section 5.2.7a. Water is normally conveyed from the water source to the pivot point in an underground pipe. Surface pipes can be used when pipe bridges such as the one in Figure 5.28 are utilized to carry towers across the pipeline or when the water source is located so that towers do not cross the pipeline.

Figure 5.29 shows alternate layouts for farms with several center pivot units. Nesting center pivots (Figure 5.29*b*), mixing field sizes (Figure 5.29*c*), and using part circle fields (Figure 5.29*b*) maximize the area irrigated.

Alternate pipeline layouts for a center-pivot irrigated farm are shown schematically in Figure 5.30. Connecting submains from several center-pivot units to the mainlines at a cluster point (Figure 5.30*b*) reduces labor by allowing valving, electric control panels, chemical injection equipment, etc., for each center pivot unit served by the cluster point to be located at one location.

5.4.3b Diameter of Lateral

Pipe diameters for center-pivot laterals usually range between 100 and 250 mm (4 and 10 in), with 168 mm ($6\frac{5}{8}$ in) being the most common size. The size of

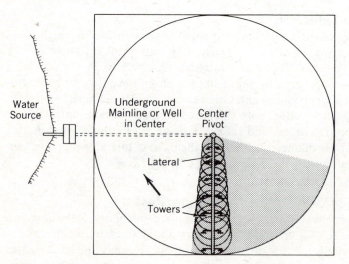

Figure 5.27 A layout for a center-pivot irrigated field.

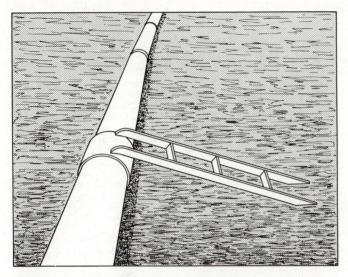

Figure 5.28 A bridge for assisting the wheels or tracks of center-pivot laterals to cross over a mainline.

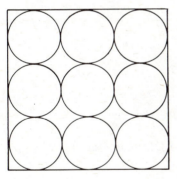

(a)

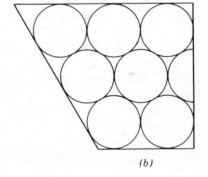

(b)

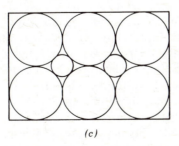

(c)

Figure 5.29 Alternate layouts for farms with several center-pivot units. (a) Non-nested. (b) Nested and part circle fields. (c) Mixing field sizes.

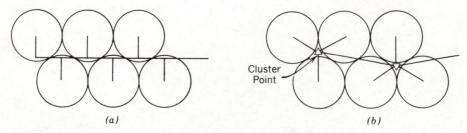

Figure 5.30 Alternate mainline/submain layouts for farms with several center-pivot units. (*a*) Without cluster points. (*b*) With cluster points.

a particular center-pivot lateral is most often chosen on the basis of economics using procedures similar to those utilized in Examples 5.7 and 5.8. In addition to the increased costs due to larger diameter, thicker walled pipes, this analysis should include the cost of larger structural members and tower motors required to support and propel larger pipes and heavier water loads.

The pressure required to operate a center pivot is computed with Eqs. 5.13 and 5.14 (with $F = 0.543$, as per Chu and Moe, 1972) and 5.15. System discharge is estimated with Eq. 5.26.

$$Q_{cp} = \frac{(K)(A)(\text{DDIR})}{E_a} \qquad (5.26)$$

where

Q_{cp} = system discharge (l/min, gpm);
A = area irrigated (ha, acres);
DDIR = design daily irrigation requirement (mm/day, in/day);
E_a = application efficiency (percent);
K = unit constant ($K = 694.4$ for Q in l/min and DDIR in mm/day. $K =$ 1886 for Q in gpm and DDIR in in/day).

5.4.3c   Sprinkler Selection

Most center-pivot machines have one of the water application packages diagrammed in Table 5.13. This table provides guidance for selecting a package for the expected crop, soil, topographic, and climatic conditions. In general, the package with the lowest total ownership and operating costs that provides the desired capacity and an acceptable application rate and uniformity is chosen.

The selection of individual sprinklers begins after a water application package and lateral diameter have been chosen. Many sprinkler manufacturers routinely use computer programs to select sprinklers for center pivots since the discharge requirement and availability of pressure at each sprinkler outlet is unique and because of the large number of sprinklers along center-pivot laterals. The basic strategy of these programs is to

1. determine the discharge required from each sprinkler,
2. then, starting with a design pressure at either end of the lateral, determine the pressure available at each sprinkler outlet, and
3. from the required discharge and available pressure, select the appropriate sprinkler for each outlet.

Equation 5.5 can be used to determine the discharge at each sprinkler. Terms H and L are computed using Eq. 5.6, with P_f equal to 100 percent, and Eq. 5.11, respectively. Term S equals half the distance to the next upstream sprinkler plus half the distance to the next downstream sprinkler. For the up- and downstream-most sprinklers, the sprinkler's effective radius of coverage is used in lieu of half the distance to the upstream or downstream sprinkler, respectively. A value of T_m must also be estimated to use Eq. 5.5.

Although pressure calculations can begin at either end of the lateral, it is normally easier to begin at the downstream end. Table 5.13 provides guidance in choosing an operating pressure for the downstream-most sprinkler. This pressure and the required discharge are substituted into Eq. 5.13 and the pressure available at the next sprinkler upstream computed. The required discharge of this sprinkler is then added to the previous sprinkler's discharge to compute the pressure at the next sprinkler upstream. This process progresses upstream until the pressure at all outlets has been computed.

Because it is not always possible to identify a sprinkler with the exact combination of pressure and discharge required at a location, sprinkler selection is normally done simultaneously with pressure calculations. This improves the selection process by allowing differences due to the substitution of sprinklers with slightly different discharge and pressure relationships to be considered.

Large gun-type sprinklers are often located at the downstream end of center-pivot laterals to extend the area irrigated by the center pivot unit. These are normally part-circle sprinklers (see Figure 5.31) that operate only in the corners of the field. They may require a booster pump when their required pressure exceeds the pressure needed by the sprinkler upstream of the gun sprinkler.

The design discharge of an end gun can be computed using Eq. 5.5. With the effective radius of coverage of the end gun, which is normally estimated to be 40 percent of the end gun's wetted diameter, substituted for S. Term L is computed with the following equation.

$$L = 2\pi\left(R + \frac{R_e}{2} \right) \tag{5.27}$$

where

$R =$ distance from the pivot point to the end gun (m, ft);
$R_e =$ effective radius of end-gun coverage (m, ft).

The respective values of H and T used in end-gun calculations must be identical to those used to select the other sprinklers.

Table 5.13 Water Application Packages for Center-Pivot and Linear-Move Sprinkle Systems

Water Application Package	Sprinkler Type	Outlet Spacing	Normal Min Pressure	Relative Droplet Size	Application Rate	Uniforming (no-wind)
	Single and Double Nozzle Impact	Wide Constant 12 m (40 ft)	310–450 kPa (45–65 psi)	Large	Lowest	Very good
	Diffuse-jet Impact	Close Variable	205–345 kPa (30–50 psi)	Medium	Medium–low	Excellent
		Variable 1.5–6.0 m (5–20 ft)	205–345 kPa (30–50 psi)	Medium	Medium	Very good
	Spray Nozzle on Top of Lateral	1.5–3.0 m (5–10 ft)	105–210 kPa (15–30 psi)	Fine	High (subject to wind drift)	Good
				Small–medium	High	Good

	Spray Nozzle on Drop tube	1.5–3.0 m (5–10 ft)	105–210 kPa (15–30 psi)	Fine	Very high[a]	Good
				Small-medium	Very high[a]	Good
	Spray Nozzles on Drop Booms[a]	Boom every 140–210 m (20–30 ft) with 3–7 spray nozzles per boom	85–175 kPa (12–25 psi)	Fine	Medium-high	Very good

[a] Application rate can be reduced by extending drop tubes several meters alternately fore and aft of the lateral.

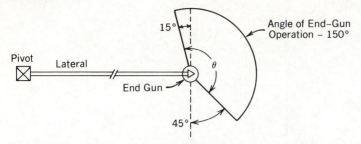

Figure 5.31 Top view of an end-gun sprinkler wetting pattern for a center-pivot sprinkle system.

5.4.3d   Speed of Center Pivot Rotation

The speed at which a center-pivot lateral rotates around the field determines the depth of water application. High speeds of rotation increase machine wear, while slow speeds may result in runoff. The number of hours per rotation must not exceed the value of H computed with Eq. 5.6. Many times, however, the number of hours will have to be less than H to prevent runoff. The maximum hours per revolution is the lesser of these two values. The following equation provides an estimate of the maximum hours per revolution to prevent runoff.

$$S_r \leq \frac{(K)(D_m)}{\text{DDIR}} \tag{5.28}$$

where

$\qquad S_r$ = speed of rotation (h/revolution);
$\qquad D_m$ = amount that can be applied per irrigation without runoff from Figure
$\qquad\qquad$ 5.17 (mm/revolution);
DDIR = design daily irrigation requirement (mm/day, in/day);
$\qquad K$ = unit constant (K = 24 for DDIR in mm/day. K =
$\qquad\qquad$ 0.95 for DDIR in in/day).

The following equation was developed by Dillion et al. (1972) to compute the peak application rate:

$$A_p = \frac{(K)(Q_{cp})}{(R)(R_e)} \tag{5.29}$$

where

$\qquad A_p$ = peak application rate (mm/h, in/h);
$\qquad Q_{cp}$ = system discharge (l/min, gpm);
$\qquad R$ = distance from pivot to the downstream-most sprinkler (m, ft);
$\qquad R_e$ = effective radius of coverage of downstream-most sprinkler (m, ft);
$\qquad K$ = unit constant (K = 60 for Q_{cp} in l/min, and R and R_e in m. K =
$\qquad\qquad$ 96.3 for Q_{cp} in gpm, and R and R_e in ft).

Equation 5.29 is based on the assumptions that the relationship between application rate and time is elliptical and the peak application rate beneath a center-pivot lateral occurs at the downstream-most sprinkler.

The application rate of a part-circle end sprinkler can be adjusted by varying its angle of coverage. The following equation can be used to make such adjustments.

$$A_{eg} = \frac{(K)(Q_{eg})}{\theta (R_e)_{eg}^2} \tag{5.30}$$

where

A_{eg} = application of end gun (mm/h, in/h);
Q_{eg} = discharge of end gun (l/min, gpm);
θ = arc setting (i.e., angle of coverage in degrees) (see Figure 5.3);
$(R_e)_{eg}$ = effective radius of coverage of end gun (m, ft);
K = unit constant [$K = 6876$ for Q_{eg} in l/min and $(R_e)_{eg}$ in m. $K = 11031$ for Q_{eg} in gpm and $(R_e)_{eg}$ in ft].

EXAMPLE 5.9   Determining the Speed of Center-Pivot Rotation

Given:
- 400-m-long center-pivot machine that irrigates 54 ha
- impact sprinklers are used
- wetted diameter of sprinklers at the downstream end of lateral is about 35 m
- RAW = 8 cm
- soil belongs to the 0.5 *SCS* intake family
- assume no surface storage
- DDIR = 10 mm/day

Required:
maximum hours per center pivot revolution

Solution:
   Solution Steps

1. Compute H using Eq. 5.6
2. Compute Q_{cp} using Eq. 5.26
3. Compute peak application rate using Eq. 5.29
4. Determine D_m using Figure 5.17
5. Compute S_r using Eq. 5.28

Solution Step 1

$$H \le \frac{0.24(100)(80)}{10} = 192 \text{ h}$$

Solution Step 2

$$Q = \frac{(694.4)(54)(10)}{85} = 4411 \text{ l/min}$$

Solution Step 3

$$A_p = \frac{(60)(4411)}{(400)(0.40)(35)} = 47.3 \text{ mm/h}$$

where $R_e = (0.40)(35)$.

Solution Step 4 (from Figure 5.17)

$D_m = 18$ mm

Solution Step 5

$$S_r = \frac{(24)(18)}{10} = 43.2 \text{ h}$$

    The recommended speed of rotation is 43.2 hours per revolution or less (since H is 192 h). When S_r is extremely small it may be necessary to switch to a water application package with a lower application rate. A 24-h per revolution speed of rotation is not recommended, since the same portion of the field is irrigated at the same time each irrigation. When the system is started at the same time. This may reduce the uniformity of application because of diurnal variations in evaporation and/or wind.

5.4.3e   Mainlines and Submains

    Procedures for designing mainlines and submains for other sprinkle systems also apply to center-pivot systems. On large farms with several center-pivot units, each submain which feeds a center pivot unit should have an on–off valve where it joins the mainline so that the center pivot can be isolated from the remainder of the system if the center-pivot unit malfunctions. Such valves, called *isolation valves*, allow the remainder of the system to operate while the malfunctioning unit is repaired. Large isolation valves should also be included at the upstream end of mainline branches.

5.4.4   Design of Linear-Move Systems

Linear-move systems are described in Section 5.2.7b. The design of these systems include field layout, lateral pipe diameter selection; sprinkler selection; determining the speed of travel, and developing mainline or supply canal specifications.

5.4.4a   Layout

    Linear-move systems obtain water from pipelines or open channels constructed through the center or along the edge of the field. Center-fed systems normally require less energy per unit of water applied, since flow is split to each side of the system. Center-fed systems are, therefore, generally used for larger fields. End-fed units, however, leave the field unobstructed by a supply canal, surface pipe, or risers from an underground pipeline.

5.4.4b   Lateral Diameter
The hydraulics of flow through linear-move laterals is similar to that in set-move and solid-set laterals, since discharge and outlet spacing is uniform. Equations 5.14 and 5.15 are therefore used for pipe friction computations. Lateral diameters are selected on the basis of economics.

5.4.4c   Sprinkler Selection
The water application packages in Table 5.13 can also be used with linear-move systems. Some systems have large part-circle sprinklers at the downstream end of laterals similar to those used with center-pivot units.

Since outlets are evenly spaced along most linear-move systems, the discharge requirements of all sprinklers (except part-circle end sprinklers) are the same. Equations 5.5, 5.6, and 5.7 with P_f equal to 100 percent can be used to compute the required sprinkler discharge. Spacing L in Eq. 5.5 is the distance the moving lateral system covers "wet" per irrigation.

Because pressure varies from outlet-to-outlet along linear-move laterals different sprinkler nozzle diameters are required at each outlet. The procedures used to compute the available pressure and to select sprinkler nozzle diameters for outlets along center-pivot laterals are also used for linear-move systems. In practice, however, it is more common to use a single nozzle diameter that provides the system discharge for the computed pressure distribution at all outlets along the lateral. When terrain is such that pressure regulators are used, a sprinkler nozzle diameter that provides the desired discharge and operates at the outlet pressure of the regulator is used at all outlets.

5.4.4d   Travel Speed
The application depth is determined by the speed at which the system moves across the field. The depth of application is increased by reducing the travel speed.

The travel speed equals the length of the field divided by the time required for the system to complete an irrigation. This time cannot exceed the value of S_r computed with Eq. 5.28 or the maximum value of H computed with Eq. 5.6 When S_r is extremely small (and less than the value of H computed with Eq. 5.6), it may be necessary to switch to a water application package with a lower application rate to increase S_r.

The following equation gives the peak application rate needed to evaluate D_m.

$$A_p = \frac{(K)\,(Q_s)}{(S)\,(W_e)} \tag{5.31}$$

where

A_p = peak application rate (mm/h, in/h);
Q_s = design discharge of sprinklers (l/min, gpm);
 S = outlet spacing (m, ft);
W_e = effective width of water pattern (m, ft);
 K = unit constant ($K = 60$ for A_p in mm/h, Q_s in l/min, and S and W_e in m.
     $K = 96.3$ for A_p in in/h, Q_s in gpm, and S and W_e in ft).

Equation 5.31 is based on the assumption that the application pattern beneath a linear-move system is elliptical.

5.4.4e Water Delivery System

Water is conveyed to linear-move systems in either ditches or pipelines. Ditches are earthen or lined with concrete, asphalt, or plastic membrane. The most commonly used pipeline materials are PVC, aluminum, and steel.

In general, soil type and terrain determine whether water is delivered in an earth ditch, lined canal, or pipeline. Pipelines are normally required in rolling terrains, while either pipelines or sloping canals are utilized in "slightly" sloping terrains. The use of earth ditches is limited to flat terrains and soils in which seepage losses will be small.

(i) Earth Ditches The earth ditches used with linear-move systems are essentially long, narrow reservoirs located along one edge or through the middle of the field. They have trapezoidal cross sections and are level along their length. Water is pumped from then by a pump that travels with the linear-move lateral as it moves across the field. Because earth ditches are level, water can be diverted into them anywhere along their length.

Earth ditches must be deep enough to provide freeboard for overflow protection and a water depth of at least 90 cm (about 36 in) to properly submerge the pump intake pipe. The depth of freeboard should be at least 20 percent of the total depth and may be as large as 30 to 35 percent of the total depth of larger ditches. Recommended side slopes for earth ditches are listed in Table 5.14.

Maintenance of earth ditches is essential. Ditches must be periodically sprayed, burned, or clearned out by hand to control vegetation that restricts flow and plugs screens, sprinklers, etc. Sediment that accumulates in ditches must be removed regularly.

(ii) Lined Canals Canals are lined to prevent seepage losses, reduce maintenance, and decrease the size of the canal. Although concrete is the most commonly used lining material, asphalt, rock masonry, brick, soil cement, rubber, and plastic are also used. Lined canals may or may not have a longitudinal slope.

In sloping canals, a weir that travels in the canal just ahead of the pump intake pipe causes water to pond in the canal. This allows a water depth to be established that is large enough to properly submerge the pump intake pipe and minimize canal outflow. Provisions for canal outflow must be included, however, to allow leakage around the weir to leave the canal.

Table 5.14 Recommended Side Slopes for Earth Ditch Water Delivery Systems for Linear-Move Sprinkle Systems

Soil	Depth of Channel			
	Up to 4 ft (122 cm) Horizontal : Vertical		Over 4 ft (122 cm) Horizontal : Vertical	
Stiff (heavy) clay	$\frac{1}{2}$: 1		1 : 1	
Silt loam or clay	1 : 1		$1\frac{1}{2}$: 1	
Sandy loam	$1\frac{1}{2}$: 1		2 : 1	
Loose sand	2 : 1		3 : 1	

Source: From *Zimmatic Design Manual* (1982), Lindsay Manufacturing Company, Lindsey, Nab.

Table 5.15 Recommended Dimensions for Lined Canal Water Delivery Systems for Linear-Move Sprinkle Systems

Maximum Flow		Depth		Top Width		Water Depth	
(gpm)	(l/min)	(in)	(cm)	(in)	(cm)	(in)	(cm)
1500	5678	24	61	60	152	20	51
2000	7570	26	66	64	163	20	51
2500	9468	28	71	68	173	20	51
3000	11355	30	76	72	183	20	51

Source: From *Zimmatic Design Manual* (1982), Lindsay Manufacturing Company, Lindsay, Neb.

Note: These recommendations are valid for canals with a 30 cm (12 in) bottom width and slopes greater than 0.01 percent.

Table 5.16 Approximate Values of Hazen–Williams C (in Eq. 5.16b) for Hoses Commonly Used with Linear-Move Systems

Hose Type	C
4.5 inch (114 mm) diameter soft hose	180
4.75 inch (121 mm) diameter hard hose	150
5.9 inch (149 mm) diameter hard hose	160

Source: From *Zimmatic Design Manual* (1982), Lindsay Manufacturing Company, Lindsay, Neb.

Lined canals have trapezoidal cross sections with 1 to 1 side slopes and a bottom width of 30 cm (12 in). Recommended dimensions for lined canals with slopes of 0.01 percent or greater are listed in Table 5.15.

(iii) Pipelines The procedures used to design delivery pipelines for linear-move systems are the same as those used for other pressurized mainlines. Friction loss in the connection between the linear move lateral and delivery pipe is an important consideration. This connection may be either a hose or a device that automatically connects and disconnects the lateral from delivery pipe risers.

Approximate values of Hazen-Williams C for hoses commonly used with linear move systems are given in Table 5.16.

Homework Problems

5.1 Sprinklers that discharge 25 l/min and have a wetted diameter of 30 m are spaced 15 m apart along a lateral. The spacing between laterals is 18 m. Determine
a. the application rate of the sprinkler,
b. the average application rate along the lateral.

5.2 A 300-m-long sprinkler lateral discharges 500 l/min. The spacing between laterals is 15 m. Determine the average application rate of the sprinkler lateral.

5.3 Ten 300-m-long laterals with sprinklers in a 15-m square spacing pattern are operated simultaneously to irrigate a 25-ha field. The system is designed to deliver a design daily irrigation requirement of 7 mm/day and a desired depth of irrigation of 15 mm. Determine the maximum time between successive irrigations.

5.4 Use the information from Problem 5.3 to determine the sprinkler capacity needed for a set length of 8 hours. Assume that 0.5 hours per set is required to move each lateral. What is the depth applied per 8 hour set?

5.5 Use information from Problems 5.3 and 5.4 to determine the number of laterals that must be operated simultaneously to obtain a set length of 12 hours. What is the sprinkler capacity and the depth applied per 12 hour set?

5.6 What else (besides changing the number of laterals) can be done to the system in Problems 5.3 and 5.4 to obtain a set length of 12 hours?

5.7 Determine the required wetted diameter for the sprinkler in Problem 5.4. The average wind speed is 3 m/s.

5.8 Determine the average application rate for the sprinkler in Problems 5.3 and 5.4. What soils can be safely irrigated (i.e., irrigated without runoff) with this sprinkler?

5.9 Determine the average application rate for the sprinkler in Problem 5.4 for a triangular sprinkler spacing pattern in which sprinklers are spaced 15 m apart along sprinkler laterals. Determine the minimum wetted diameter for the sprinkler when the average wind speed is 3 m/s.

5.10 A hand-move sprinkle system is being designed to irrigate a 500 by 500 m (25 ha) field of alfalfa. The soil is a 150-cm-deep loam. Use pan evaporation data (with pan coefficient = 0.8) and wind speed data from Appendix D to determine
  a. the number of laterals needed for a 12 hour set,
  b. the sprinkler capacity for a 12 hour set,
  c. the best sprinkler from the catalog information provided in Example 5.3, and
  d. the maximum allowable applicable rate.

Can this sprinkler be used to irrigate this alfalfa field?

5.11 A 190-m, 15-mm diameter steel pipe supplies 2000 l/min of water to an irrigation system. The pressure at the upstream end of the pipe is 500 kPa. Determine the pressure at the downstream end of the pipe for a difference in elevation between the up- and downstream ends of the pipe of
  a. 0 m,
  b. 5 m, and
  c. −5 m (i.e., water is flowing downhill).

5.12 A 120-mm-diameter aluminum sprinkler lateral has 20 equally spaced 40 l/min sprinklers. The spacing between sprinklers is 12 m. The first sprinkler is 6 m from the submain. Determine the difference in pressure between the up- and downstream ends of the lateral for a difference in elevation between the up- and downstream ends of the lateral of
  a. 0 m,
  b. 5 m, and
  c. −5 m (i.e., water is flowing downhill).

5.13 Determine the pressure at the downstream end of the pipe in Problem
  a. 5.12a,
  b. 5.12b, and
  c 5.12c

if the pressure at the upstream end of the pipe is 280 kPa. Which of these variations in pressure are acceptable (using the ASAE criterion)? What could be done to reduce these pressure variations?

5.14 Determine the minimum pipe diameter needed to limit the pressure variation along the pipe in Problem 5.13b to 20 percent of the desired operating pressure. Assume that the desired operating pressure equals the pressure at the upstream end of the pipe.

5.15 Determine the pressure at the tenth and last sprinklers along the lateral in Problem 5.12a.

5.16 The elevations at the upstream end, the sixth, and downstream end of the sprinkler lateral in Problem 5.12 are 100 m, 95 m, and 98 m, respectively. Determine the pressures at the sixth and last sprinklers when the pressure at the upstream end of the lateral is 450 kPa.

*5.17 Write a computer program to determine the pressure at each sprinkler along a lateral given the pressure at the downstream end of the lateral, the elevation of each sprinkler, the diameter and roughness coefficient of the lateral, and the discharge–pressure relationship of each sprinkler. The discharge–pressure relationship for the sprinklers is given by

$$Q = 0.0648 \; d^2 \; P^{0.5}$$

where

Q = sprinkler discharge in l/min;
d = nozzle diameter in mm;
P = operating pressure in kPa.

*5.18 Use the computer program from Problem 5.17 to determine the pressure at each sprinkler along the lateral in Problem
a. 5.15, and
b. 5.16.

Use a sprinkler nozzle diameter of 6 mm.

5.19 Use the program in Appendix G to solve Problems 5.15 and 5.16.

5.20 Irrigation water is delivered to three equally spaced sprinkler laterals in a 15-mm-diameter, 600-m-long PVC mainline. Each lateral supplies twenty 35-l/min sprinklers and requires a pressure of 350 kPa at its upstream end. The laterals are spaced 200 m apart and are moved in unison across the field (i.e., they are always 200 m apart). There is a 3.33 percent uphill slope from the up- to downstream end of the mainline. Determine the pressure that a pump located at the upstream end of the mainline must provide to meet the needs of the three sprinkler laterals. Neglect all minor losses.

5.21 Repeat Problem 5.20 for a 3.33 percent downhill slope from the up- to downstream end of the mainline.

*5.22 Use the computer program in Appendix G to determine the pressure that a pump located at the upstream end of the mainline in Problem 5.20 must deliver if the diameter of the first 200 m of mainline is 30 mm, the second 200 m is 25 mm, and the third 200 m is 20 mm. Use the program to determine the combination of mainline pipe diameters (for each 200-m section of mainline) that minimizes the pressure that the pump must provide.

* Indicates that a computer program will facilitate the solution of the problems so marked.

*5.23 Water is pumped from a reservoir through a 30-mm-diameter steel pipe to two gun-type sprinklers. The pipeline tees 250 m downstream from the pump. One of the legs from the tee conveys water to one of the guns, while the other leg supplies the second sprinkler. Both tee legs are 100-m-long, 30-mm diameter steel pipe. The elevation at the pump, tee, gun 1, and gun 2 are 100, 102, 107, and 97 m, respectively. The pressure–discharge relationship for the gun sprinklers is given by the equation in Problem 5.20 with $d = 18$ mm. The head–discharge relationship for the pump is

$$H = 100 - 6.67 \,(10)^{-3}Q - 1.73 \,(10)^{-6}Q^2$$

where

H = head in m;
Q = pump discharge in l/min.

Determine the pressure and discharge at which each sprinkler operates.

5.24 Five thousand l/mins of water is flowing from a reservoir through a 250-mm-diameter steel pipe into an open channel. The pipeline is 500 m long and anchored at one end. The temperature of the water is 20°C. Determine the water hammer pressure when a full area gate valve at the downstream end of the pipeline is closed instantaneously. The wall thickness of the pipe is 1 mm.

5.25 Repeat Problem 5.24 for a valve closure time of 5 s. Assume that the normal operating pressure is 300 kPa.

5.26 Repeat Problem 5.24 for
a. a pipe diameter of 500 mm,
b. a 250-mm-diameter PVC pipe with a wall thickness of 1 mm,
c. a 250-mm-diameter steel pipe (with a 1-mm-thick pipe wall) anchored at both ends without expansion joints, and
d. a 250-mm-diameter steel pipe (with a 1-mm-thick pipe wall) anchored at both ends with expansion joints.

Which of these methods is the most effective way of controlling water hammer pressure.

5.27 Determine the valve closing time required to limit the water hammer pressure in the pipeline in Problem 5.23 to 350 k-Pa (50 psi).

5.28 An irrigator plans to grow corn in a 100-cm-deep loam soil. A hand-move sprinkle system with a design daily irrigation requirement of 8 mm/day is being designed. Determine the number of sets required to irrigate the field if a 12-hour set length is desired.

5.29 Determine the sprinkler capacity for a single-unit traveler system to irrigate a 50-ha field that is 600 m wide. The readily available water holding capacity of the soil is 100 mm and the design daily irrigation requirement is

7 mm/day. The prevailing wind speed and direction are, respectively, 2 m/s and normal to the 600-m-long side of the field. Use a travel path length of 600 m.

5.30 Use information from Problem 5.29 to determine the average and peak application rates for arc settings of
a. 360,
b. 330, and
c. 270 degrees.

What soil textures can be safely irrigated (i.e., irrigated without runoff) with this sprinkler?

5.31 Determine the travel speed for the traveler system in Problems 5.29 and 5.30. Determine the travel speed for a travel path length of 300 m. Assume that $D_a < D_m$. List several other ways of adjusting travel speed to 0.3 m/s and discuss the advantages and disadvantages of each alternative.

5.32 Use information from Problem 5.29 and a travel path length of 300 m to select a hose for a hard-hose traveler system. Use a Hazen–Williams C of 150.

5.33 Determine the design capacity of a center-pivot irrigation system for a 53-ha field. The design daily irrigation requirement is 7 mm/day.

5.34 Determine the discharge and operating pressure of sprinklers along a center-pivot lateral. The 5-cm-diameter steel lateral is 95 m long and has 10 sprinklers spaced 10 m apart. The first sprinkler is located 5 m downstream of the pivot. The design daily irrigation requirement is 7 mm/day and the readily available water holding capacity of the soil is 75 mm. The desired operating pressure at the last sprinkler is 350 kPa.

5.35 Determine the maximum number of hours per revolution at which the center-pivot system in Problem 5.34 can be operated on a 0.5 SCS intake family soil with no surface storage. The effective wetted radius of the downstream most sprinkler is 5 m.

5.36 A gun sprinkler is mounted at the downstream end of the center pivot system in Problems 5.34 and 5.35. Determine the sprinkler capacity required to irrigate an additional 40 m of field. Determine the minimum arc setting for the sprinkler.

*5.37 Write a computer program to determine the discharge and operating pressure of sprinklers along center-pivot laterals when the operating pressure of the downstream-most sprinkler is known. Use this program to solve Problem 5.34.

5.38 Determine the required capacity for full-circle low-pressure spray sprinklers mounted on the top of a 400-m-long linear-move system. The readily available water holding capacity of the soil is 90 mm and the design daily irrigation requirement is 8 mm/day. The length of the field is 1250 m.

5.39 Determine the maximum travel time for the linear-move system in Problem 5.38 for a 0.5 SCS intake family soil. The wetted diameter of the sprinklers is 12 m.

5.40 Repeat Problem 5.38 for management Option 2 in Figure 5.6.

References

Addink, J. W., J. Keller, C. H. Pair, R. E. Sneed, and J. W. Wolf (1980). Design and operation of sprinkler systems. In *Design and Operation of Farm Irrigation Systems*, M. E. Jensen (Ed), ASAE Monograph 3, St. Joseph, MI pp. 621–660.

American Society of Agricultural Engineers (1985). R264.2, Minimum requirements of the design, installation, and performance of sprinkler irrigation equipment. In *1985 Agricultural Engineers Yearbook*.

American Society of Agricultural Engineers (1985). Safety devices for applying liquid chemicals through irrigation systems. ASAE Engineering Practice EP409.

Chu, S. T., and D. L. Moe (1972). Hydraulics of a center pivot system. *Trans. ASAE*, **15(5)**, pp. 894–896.

Chu, S. T. 1980. Center pivot irrigation design. Technical Bulletin 56, Agricultural Experiment Station, South Dakota State University, Brookings, S.D., 55 pp.

Davis, D. D. (1976). Irrigation systems design handbook. Rain Bird Sprinkler Manufacturing Corporation, Glendora, Calif.

Dillion, R. C., E. A. Hiler, and G. Vittetoe. (1972). Center pivot sprinkler design based on intake characteristics. *Trans. ASAE*, **15(5)**, pp. 996–1000.

Gilley, J. R. (1984). Suitability of reduced pressure center-pivots. *J. Irrigation and Drainage Division of the American Society of Civil Engineers*, **110** (1), pp. 22–34.

Heermann, D. F., and R. A. Kohl (1980). Fluid dynamics of sprinkler systems. In *Design and Operation of Farm Irrigation Systems*, M. E. Jensen (Ed.), ASAE Monograph 3, St. Joseph, MI pp. 583–618.

James, L. G., and R. T. Stillmunkes (1980). Instantaneous Application Rates Beneath Center Pivot Irrigation Systems. ASAE Paper PNW 80–206, Great Falls, Mont.

Lindsay Manufacturing Co. (1982). *Zimmatic Design Manual*, Lindsay, Neb.

Nelson Irrigation Corporation. Nelson Big Guns® for Traveling Systems. Undated Sales Brochure 4 pp.

Pair, C. H., W. H. Hinz, K. R. Frost, R. E. Sneed, and T. J. Schiltz (1983). *Irrigation*. The Irrigation Association, Silver Spring, Md., 686 pp.

Schwab, G. O., R. K. Frevert, T. W. Edminster, and K. K. Barnes (1981). *Soil and Water Conservation Engineering*. Wiley, New York, 525 pp.

Simon, A. L. (1976). *Practical Hydraulics*. Wiley, New York, 306 pp.

Soil Conservation Service (1983). Sprinkler irrigation, Chapter 11, Section 15, *National Engineering Handbook*, U.S. Department of Agriculture, Soil Conservation Service, 121 pp.

Streeter, V. L., and E. B. Wylie. (1967). *Hydraulic Transients*. McGraw-Hill, New York, 329 pp.

Thompson, A. L. and L. G. James. (1985). Water droplet impact and its effect on infiltration. *Trans. ASAE*, **28**(5); 1506–1510 and 1520.

Vennard, J. K., and R. L. Street (1975). *Elementary Fluid Mechanics*. Wiley, New York, 740 pp.

Withers, B., and S. Vipond (1980). *Irrigation Design and Practice*. Cornell University Press, Ithaca, N.Y., 306 pp.

Trickle Irrigation

6.1 Introduction

Trickle irrigation is the frequent, slow application of water either directly onto the land surface or into the root zone of the crop. Trickle irrigation is based on the fundamental concepts of irrigating only the root zone of the crop (rather than the entire land surface) and maintaining the water content of the root zone at near optimum levels. Irrigating only a portion of the land surface limits evaporation (by limiting the evaporative surface), reduces weed growth (few weeds grow in unirrigated areas), and minimizes interruption of cultural operations (i.e., tillage, harvesting, etc., and irrigation can be carried out simultaneously). Maintaining a near-optimum water content in the root zone usually involves frequent applications of small amounts of water. Trickle irrigation systems are often designed to operate daily for nearly the entire day and to supply water to only the root zone of the crop.

The design and operation of drip, subsurface, bubbler, and spray-type trickle irrigation systems are described in this chapter. Sufficient information is presented to familiarize the reader with:

1. the benefits and problems of trickle irrigation,
2. the major components of trickle systems,
3. various types of emission devices,
4. emission device and pipeline hydraulics and selection,
5. the use of emission uniformity to evaluate trickle system design and performance.
6. procedures and equipment for controlling trickle component clogging, and
7. trickle system design.

6.1.1 Benefits of Trickle Irrigation

Trickle irrigation has many desirable features. Higher yields, improved crop quality, and reduced water and energy use have all been attributed to trickle

260

irrigation. Crop yield experiments have shown wide differences varying from little or no difference to as much as 50 percent increases compared to other methods of irrigation. In addition, there is evidence that the quality of some crops is improved by trickle irrigation. Because there is usually less deep percolation, runoff, and evaporation (i.e., higher application efficiencies), and since only a portion of the potential root zone is irrigated, trickle irrigation systems usually use less water than other types of irrigation systems. Trickle systems generally have lower energy requirements than do sprinkle systems because of reduced water use and lower operating pressure requirements.

It is possible to use more saline water with trickle systems, since the soil is kept at a higher water content and water contact with the plant is minimized compared to other types of irrigation systems. Because aboveground portions of the plant are normally completely dry, bacteria, fungi, and other pests and diseases that depend on a moist environment are reduced when trickle irrigation is used.

6.1.2 Problems Associated with Trickle Irrigation

There are several problems associated with trickle irrigation. The most severe problem is the clogging of system components by particulate, chemical, and biological materials. Clogging can cause poor uniformity of application and if it continues long enough, can severely damage the crop. The special equipment needed to control clogging as well as the amount of pipe, emitters, valves, etc., typically used in trickle systems often makes the per acre cost of these systems high. Costs are, however, generally comparable to solid-set sprinkle systems, but are higher than those of surface irrigation systems except when extensive land leveling is needed.

Because a trickle irrigation system normally wets only part of the potential root zone, crop root development is normally limited to the wetted portion of the root zone. The resulting concentrated distribution of roots may reduce the plants' ability to withstand winds and not provide sufficient water when, because of unexpected problems such as system malfunctions, water in the wetted zone is depleted. A salt accumulation problem can occur in trickle irrigation when only a portion of the root zone is wet and saline waters are being used for irrigation. In such cases, salts may accumulate along the edge of the wetted zone during irrigation and be moved into the root zone to cause severe crop damage during a rain (when the irrigation system is not operating). Such problems can normally be avoided by operating the trickle system during the rainy period.

6.2 Trickle Irrigation Methods

Trickle irrigation encompasses several methods of irrigation, including drip, subsurface, bubbler, and spray irrigation.

6.2.1 Drip Irrigation

Drip irrigation is the slow, nearly continuous application of water as descrete drops. Water can be applied at a single point (small wetted area) on the land surface through devices called emitters or as a line source from either closely spaced emitters or tubes with continuous or equally spaced openings that discharge water a drop at a time. Discharge rates for point source emitters are generally less than 12 l/h (3 gph) and less than 12 l/h per meter (1 gph per foot) of lateral for line source emitters.

6.2.2 Subsurface Irrigation

Subsurface irrigation involves the use of point and line source emitters to apply water below the soil surface. Subsurface irrigation is different than *subirrigation*, which is the method wherein the root zone is irrigated through or by water table control. Discharge rates for subsurface irrigation systems are generally in the same range as are drip irrigation rates.

6.2.3 Bubbler Irrigation

In *bubbler irrigation*, water is applied to the land surface as a small stream. An example of a bubbler system is diagrammed in Fig. 6.1. Water is delivered to the point of application in tubes that are attached to buried laterals. The tubes may be as large as 10 mm (about 3/8 in) in diameter or more. The rate of discharge from each tube is controlled by varying the tube diameter and/or length. Because of the large-diameter tubes, bubbler systems are not as prone to clog and normally have higher discharge rates than drip and subsurface systems. Discharge rates are, however, generally less than 225 l/h (1 gpm). Because these higher discharge rates result in larger application rates, basins or furrows are sometimes needed with bubbler systems to control runoff and erosion.

6.2.4 Spray Irrigation

Spray-type trickle systems are also less likely to clog than are drip and subsurface systems. In *spray irrigation*, small sprinkler like devices (often called micro-sprinklers) spray water as a mist over the land surface. Micro-sprinklers can be

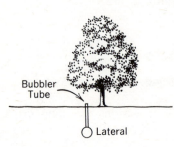

Bubbler
Tube

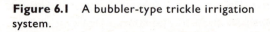

Lateral

Figure 6.1 A bubbler-type trickle irrigation system.

spaced to cover the entire land surface as with conventional sprinkler systems or a portion of the land surface like other trickle systems. Discharge rates are usually less than 115 l/h (0.5 gpm). The primary advantage of spray over bubbler irrigation is lower application rates which decrease the potential for runoff and erosion. Losses due to wind drift and evaporation are, however, greater with spray irrigation than with other trickle systems.

6.3 Trickle System Components

Several components of typical trickle irrigation systems are shown in Figure 6.2. Water is pumped into most systems and flows through valves, filters, mainlines, submains or manifold lines, and laterals before it is discharged into the field through point-source emitters, bubblers, or microsprinklers. Sometimes line sources of water are obtained with either porous tubes that discharge water continuously along their length or tubes that discharge water through closely spaced openings instead of laterals with point-source emitters or microsprinklers. Filters are sometimes omitted in bubbler and spray sprinkler systems. Trickle systems may or may not include chemical injection equipment.

In Figure 6.2, the check valve just downstream of the pump is open when flow is from the pump and closed when, because of pump failure or shutdown, flow is in the opposite direction toward the pump. This prevents water containing suspended materials and/or dissolved substances (such as fertilizer and pesticides) from flowing back through the pump to contaminate the water source. It should be

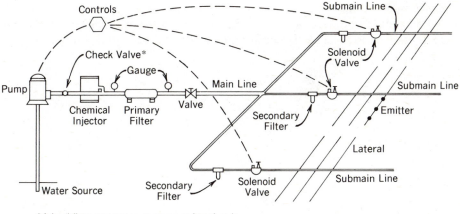

*A backflow preventer or vacuum breaker is
 required in some areas.

Figure 6.2 The components of a trickle irrigation system.

noted that when water is obtained from a domestic water supply a backflow protection device is often required by law.

Fertilizers and other chemicals are usually injected into trickle systems upstream of filters. Valves, flowmeters, secondary filters, pressure gauges and regulators, and flow control valves are also typically used in trickle systems. The primary components of trickle systems are discussed in the following sections of this chapter.

6.3.1 Emission Devices

Emission devices include point- and line-source emitters that operate either above or below the ground surface, bubblers that discharge small continuous streams of water, and microsprinklers that spray water over the land surface. Point- and line-source emitters generally have smaller passages for discharging water and are more prone to physical, chemical, and biologically induced clogging than are bubblers or microsprinklers.

6.3.1a Point-Source Emitters

There are many types and designs of point-source emitters available commercially. Most point-source emitters are either on-line or in-line emitters. Figure 6.3 shows a popular on-line emitter, while the emitter in Figure 6.4 is a widely used in-line emitter. The primary difference between on-line and in-line emitters is that the entire flow required downstream of the emitter passes through an in-line emitter. There is usually more head loss along a lateral with on-line emitters than one with in-line emitters, since the "barbs" of on-line emitters (Figure 6.3) obstruct flow and create additional head loss in the lateral. On the other hand, on-line emitters can normally be replaced more easily when they fail or become permanently clogged. It is usually necessary to shut off flow to the lateral and cut the pipe to replace a malfunctioning in-line emitter.

Point-source emitters can also be classified as long path, orifice, or pressure-compensating emitters. The classification depends on the exponent in Eq. 6.1.

$$Q = kP^x \qquad (6.1)$$

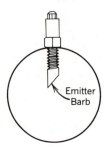

Figure 6.3 An on-line-type trickle emitter.

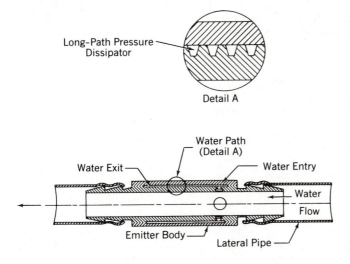

Figure 6.4   An in-line-type trickle emitter.

where

Q = emitter discharge (volume/time);
P = operating pressure (force/area);
k, x = constants for a specified emitter.

When x approaches 1, the emitter is considered a long-path or laminar-flow-type emitter. An orifice-type point-source emitter has a x of about 0.5, while x for a pressure-compensating emitter is positive and nearly zero.

The exponent in Eq. 6.1 provides a great deal of insight into the performance characteristics of the emitter that it describes. Table 6.1 lists theoretical values for k and x in Eq. 6.1 assuming laminar, turbulent, and fully turbulent flow.

The data in Table 6.1 indicate that laminar flow can be expected through emitters when x in Eq. 6.1 approaches 1. Because discharge and operating pressure are linearly related, the discharge of these emitters is sensitive to fluctuations in operating pressure (relative to emitters with exponents much less than 1). In addition, discharge can be very sensitive to fluid temperature since the theoretical value of k (in Eq. 6.1) depends on fluid viscosity.

When flow through an emitter is a turbulent or fully turbulent and the exponent in Eq. 6.1 is about 0.5, emitter discharge is not as sensitive to operating pressure and viscosity (hence fluid temperature) as are emitters with larger exponents.

Emitters with Eq. 6.1 exponents less than 0.5 are called *pressure-compensating emitters*, since the influence of pressure on discharge is reduced with smaller x values. The degree of pressure compensation (i.e., the insensitivity of discharge to pressure) increases as x (in Eq. 6.1) approaches 0. Pressure-compensating emitters are especially useful in minimizing emitter discharge variation when large pressure variations due to undulating terrain or system operation are expected.

Table 6.1. **Theoretical Values for k and x in Equation 6.1 for Laminar, Turbulent, and Fully Turbulent flow**

Flow Regime	k	x
laminar ($N_R < 2000$)	$\dfrac{\rho D^2 g A}{32 \mu L}$	1.00
turbulent $3000 < N_R < 10^5$	$2.87 A \left(\dfrac{g}{L}\right)^{0.57} D^{0.14} \left(\dfrac{\rho}{\mu}\right)^{0.14}$	0.57
fully turbulent $N_R > 10^6$	$A \left(\dfrac{2gD}{fL}\right)^{0.50}$	0.50

where

μ = dynamic viscosity of fluid;
L = length of emitter flow path;
A = cross-sectional area of emitter flowpath;
ρ = fluid density (M/L^3);
D = characteristic diameter of emitter flow path (L);
g = gravitational constant (L/T^2);
f = friction factor in Darcy–Weisbach equation;
N_R = Reynolds number.

The theoretical values of k for laminar and turbulent flow were obtained by substituting Eq. 6.10 into Equations 6.11a and 6.11b, respectively, and substituting the result into the Darcy–Weisbach form of Equation 5.15.

Several types of point-source emitters are shown in Figure 6.5. Values of x (in Eq. 6.1) and flowrate indices for 20°, 45°, and 65°C for these emitters are given in Table 6.2. The flowrate indices can be used to determine the percent change in flowrate expected when fluid temperature is increased from 20°C to 45°C and 20°C to 65°C.

6.3.1b   Line-Source Emitters

Porous pipes or tape, perforated pipes that discharge water along their entire length, laterals with closely spaced point-source emitters or microsprinklers, and bubblers discharging into furrows provide a "line source" of water. Although line sources are used primarily to irrigate row crops, they have also been used with other crops. Porous tubes are generally made of polymer compounds with small pores through which water seeps out of the pipe one drop-at-a-time. Complete filtration of water is required for proper operation of porous tubes since they are especially sensitive to clogging

Mono-walled and bi-walled perforated polyethylene pipe are commonly used line-source emitters. Water seeps from mono-walled pipe through perforations (orifices) in the pipe wall. Bi-walled pipe, emitter type l in Figure 6.5, has a secondary chamber "piggy-backed" onto a primary chamber. There are several exit orifices in the wall of the secondary chamber for each single orifice in the

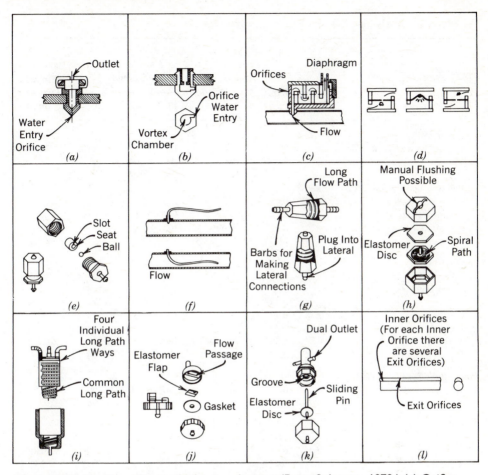

Figure 6.5 Sketches of several emission devices. (From Solomon, 1979.) (*a*) Orifice emitter. (*b*) Orifice–vortex emitter. (*c*) Emitter using flexible orifices in series. (*d*) Continuous flow principle for multiple flexible orifices. (*e*) Ball and slotted seat. (*f*) Long-path emitter small tube. (*g*) Long-path emitter. (*h*) Compensating long-path emitter. (*i*) Long-path multiple-outlet emitter. (*j*) Groove and flop short-path emitter. (*k*) Groove and disc short-path emitter. (*l*) Twin-wall emitter lateral. *Source*: K. Solomon, *Performance Comparison of Different Emitter Types*, copyright © 1977 by Rain Bird Sprinker Manufacturing Corporation, pp. 4–5. Reprinted by permission of Rain Bird Sprinkler Manufacturing Corporation.

primary chamber. The secondary chamber reduces the effect of clogging on application uniformity since water is able to exit from secondary chamber orifices even when several primary chamber orifices are clogged.

Equation 6.1 can be used to estimate the discharge of line-source emitters for a specified pressure. (Values of x for selected line-source emitters are given in Table 6.2.) Data for evaluating pressure loss along the length of line-source emitters can

Table 6.2 Emission Device Performance Data

Type of Emission Device[d]	Figure 6.5	Exp. x	Coefficient of Manufacturer Variation v	Flowrate Index[c] At Temperature $-°C$ 20°	45°	65°	Flow Passageway Minimum Diameter (mm)
Orifice Type							
Orifice/Vortex/Orifice	a	0.42	0.07	100	92	88	0.6
Multiple Flexible } 1[a]	c,d	0.7	0.05	100	104	107	—
Orifices } 11[a]	c,d	0.7	0.07	100	104	107	—
Ball and Slotted Seat 1	e	0.50	0.27	100	115	121	$(0.3)^b$
11	e	0.49	$(0.25)^b$	100	83	79	$(0.3)^b$
Compensating Ball } 1	e	0.15	0.35	100	85	81	0.3
and Slotted Seat } 11	e	0.25	0.09	100	90	89	$(0.3)^b$
Long-Path Types							
Small } 1	f	0.70	0.05	100	108	113	1.0
Tube } 11	f	0.80	0.05	100	116	122	1.0
Spiral Long } 1	g	0.75	0.06	100	119	118	0.8
Path } 11	g	0.65	0.02	100	$(110)^b$	$(115)^b$	0.7
Compensating } 1	h	0.40	0.05	100	119	133	$(0.75)^b$
Long Path } 11	h	0.20	0.06	100	111	124	$(0.75)^b$
Tortuous } 1	—	0.50	$(0.08)^b$	100	140	170	0.8
Long Path } 11	i	0.65	0.02	100	108	114	$(1.0)^b$
Short-Path Types							
Groove and Flap	j	0.33	0.02	100	100	100	0.3
Slot and Disc	k	0.11	0.10	100	106	108	0.3
Line-Source Devices							
Porous Pipe	—	1.0	0.40	100	270	380	(?)
Twin Wall } 1	l	0.61	0.17	100	$(105)^b$	$(110)^b$	$(0.4)^b$
Lateral } 11	l	0.47	$(0.10)^b$	100	$(104)^b$	$(108)^b$	$(0.4)^b$

[a] The 1 and 11 indicate different devices of the same general type.

[b] Numbers in parentheses are not based on measurements but reflect Solomon's estimate of probable values.

[c] Flowrate index is the % of flowrate at 20°C.

[d] Some of the devices are no longer available.

Source: K. Solomon, *Performance Comparison of Different Emitter Types*, copyright © 1977, Rain Bird Sprinkler Manufacturing Corp., p. 27. Reprinted by permission of Rain Bird Sprinkler Manufacturing Corp.

often be obtained from manufacturers literature. If such information is not available, Eq. 5.12 in Chapter 5, with $F = 0.33$ may be used to estimate pressure loss. Flowrate indices for the line-source emitters in Table 6.2 indicate that the discharge of emitters with flow exponents much less than 1 is relatively insensitive to fluid temperature, at least for these line-source devices.

6.3.1c Bubblers

Low-head bubbler systems use polyethylene tubes ranging in diameter from less than a millimeter to more than 10 mm (0.40 in) to apply small streams of water to the soil surface. The bubbler tubes discharge water on the soil surface and are connected to a buried lateral. Bubbler tubes have larger flowrates and are not as prone to clog as are drip emitters. These larger flowrates, however, increase the potential for runoff. Basins or furrows may be required to maintain control of the water.

The flowrate from a bubbler tube is determined by the tube's diameter and length, and the operating pressure available in the buried pipe. Equation 6.2 can be used to compute bubbler tube discharge:

$$Q = K\left(\frac{P}{L}\right)^{0.57} D^{2.74} \tag{6.2}$$

where

Q = bubbler tube discharge (l/min, gpm);
P = operating pressure (kPa, psi);
L = length of bubbler tube (cm, ft);
D = diameter of bubbler tube (mm, in);
K = unit constant ($K = 9.02(10)^{-2}$ for Q in l/min, P in kPa, L in cm, and D in mm. $K = 71.5$ for Q in gpm, P in psi, L in feet and D in in).

6.3.1d   Microsprinklers

Sprays, spitters, foggers, and microsprinklers are being used to overcome some of the disadvantages of drip systems. These devices have larger orifices than drip emitters, which reduces the need for filtration to control clogging, allows the crop to be irrigated in a shorter time, and provides the capacity of protecting the crop from frost. Spray systems also typically wet a larger soil volume than do drip systems and decrease the labor needed to check emitter operation. Furthermore, some microsprinklers apply less water per unit area than drip emitters because they spread water over a larger area. Such microsprinklers actually reduce the potential for runoff and erosion. The primary disadvantages of microsprinklers are the larger lateral diameter requirements and increased losses due to evaporation and wind drift.

Microsprinklers have flowrates that are greater than drip emitters and less than conventional sprinklers. Flowrates typically range from 15 to 200 l/h (4 to 50 gph). Microsprinklers normally require 35 to 300 kPa (5 to 40 psi) of pressure for proper operation and have wetted diameters from 2 to 9 m (6 to 30 ft).

Figure 6.6 shows several area application patterns that can be obtained with microsprinklers. Microsprinkler manufacturers normally provide pressure, flow-rate, and wetted diameter data similar to those provided by manufacturers of conventional sprinklers.

6.3.1e   Emission Device Selection

Emission device selection for trickle systems involves choosing the type of device to be used (i.e., choosing between point-source emitters, line-source emitters, bubblers, and microsprinklers) and then determining the capacity of the device.

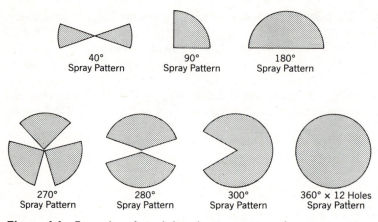

Figure 6.6 Examples of areal distribution patterns from microsprinklers.

The type of emission device depends on such factors as the crop to be irrigated, filtration requirements, the need for a cover crop and/or frost protection, cost, and grower preference. Microsprinklers should be strongly considered when a cover crop is needed for erosion, pest, or disease control or when frost protection is desired. Line-source emitters are especially well suited for row crops, although closely spaced point-source emitters, bubblers, and microsprinklers can also be used. In situations where filtration requirements are high, bubblers and micro-sprinklers may be the most viable alternatives.

The capacity of a trickle irrigation emission device may be computed using Eq. 6.3. Equation 6.3 is

$$C = \frac{(K)(D_a)(A_i)}{(H - T_m)(E_a)}$$
(6.3)

where

C = emission device capacity (l/h, gph);
D_a = depth of water applied (mm, in);
A_i = area irrigated by the emission device (m^2, ft^2);
H = hours of irrigation (i.e., the time used to apply D_a);
T_m = off time for maintenance, soil aeration, etc.;
E_a = application efficiency (percent);
K = unit constant (K = 100 for C in l/h, D in mm, A in m^2, H and T_m in hours. K = 62.33 for C in gph, D in in, A in ft^2, H and T_m in hours).

The depth of water applied per irrigation, D_a, is computed using the following equation.

$$D_a = \frac{(H)(\text{DDIR})}{(0.24)(P_f)}$$
(6.4)

The H term in Eqs. 6.3 and 6.4 is normally chosen on the basis of operator preference and/or convenience, subject to the conditions imposed by the following equation.

$$H \leq \frac{(0.24)(P_f)(D)}{\text{DDIR}} \tag{6.5}$$

where

DDIR = design daily irrigation requirement (corresponding to D) in mm/day or in/day;

D = desired depth of irrigation (mm, in);

$P_f = (100) \dfrac{\text{number of emission devices operating per irrigation}}{\text{total number of emission devices}}$.

Equation 2.6 or the procedures similar to those in Examples 2.1 and 2.2 can be used to determine the design daily irrigation requirement (DDIR). Water should not be applied continuously to any portion of the field, since continuous irrigation will normally result in a saturated (or nearly saturated) zone extending into the profile from the soil surface that restricts the exchange of air and other gasses between the atmosphere and root zone.

The area irrigated by an emission device, A_i, is computed with the following equation.

$$A_i = \frac{(L)(S)(P)}{(100)(N_e)} \tag{6.6}$$

where

A_i = area irrigated (m^2, ft^2);

L = spacing between adjacent plant rows (m, ft);

S = spacing between emission points (m, ft);

P = percent of cropped area being irrigated;

N_e = number of emission devices at each emission point.

An *emission point* is a location where a concentration of water is delivered by one or more emission devices. Configurations similar to those in Figure 6.7 are used to create either line sources or islands of water. Closely spaced emission points result in a line source of water, while islands of water are created by placing emission points further apart. Double lateral, zigzag, pigtail, multiexit, and other clustering configurations are used to enlarge the irrigated area in soils with poor lateral transmission properties (e.g., coarse textured soils) and where crops with widely spread areal root distributions are being trickle irrigated.

For mono-, bi-, and porous-walled tubing, S in Eq. 6.6 is one unit of tube length (i.e., 1 m or 1 ft) and N_e is 1. The resulting C computed with Eq. 6.3 equals the flowrate per unit length and has dimensions of volume per unit time per unit length.

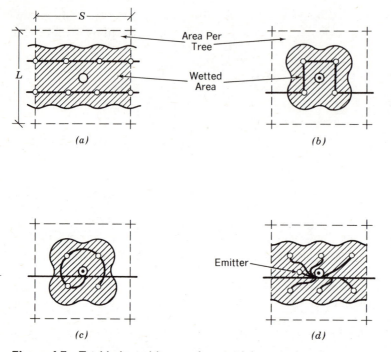

Figure 6.7   Trickle lateral layouts for a widely spaced permanent crop. (*a*) Double laterals for each tree row. (*b*) Zigzag lateral for each tree row. (*c*) Pigtail with four emitters per tree. (*d*) Multitext 6-outlet emitter with distribution tubing.

The P term in Eq. 6.6 is 100 percent when the entire surface of the field is irrigated; P normally varies between 30 and 100 percent, depending on the crop and its age. It will be larger for mature crops and crops with relatively close row spacings. For widely spaced vine, bush, and tree crops, 30 to 60 percent of the horizontal cross section of the root system is irrigated when it is necessary to keep the surface area between rows relatively dry. In areas that receive significant rainfall during the irrigation season, designs that irrigate less than one-third of the horizontal cross section of the root system may be adequate for medium to heavy textured soils. The P term often approaches 100 percent for crops spaced less than 1.8 m (6 ft) apart. Plant scientists should be consulted for the best value of P for a given crop and location.

Determining P and the number of emission devices per emission point, N_e, needed for the desired wetting pattern requires information describing the horizontal and vertical movement of water through the soil (from the emission device to be used). Figure 6.8 shows such information for a point-source emitter operating on the surface of a uniform sandy soil. These data indicate that the horizontal movement approaches a maximum value as vertical penetration steadily increases. Tests should be conducted at several representative sites around the field to obtain

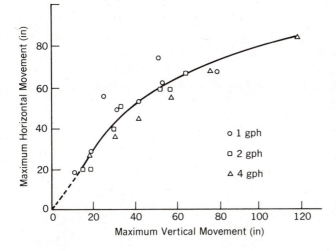

Figure 6.8 Relationship between vertical and horizontal water movement in a dry sandy soil for various amounts of water and various application rates. (From Section 15, Chapter 7 of the *SCS National Engineering Handbook*, 1984.)

these data. When such measurements are not feasible, data from Table 6.3 can be utilized to estimate the maximum horizontal distance of wetting from a single point-source emission device for 75 and 150 cm (30 and 60 in) root zone depths.

The N_e term is estimated by dividing the total area to be wet by the area wet per emission device. For single laterals with equally spaced emission points, the following equation relates the number of emission devices per plant to P and D_w.

$$N_e = \frac{(K)(P)(S)(L)}{(D_w)(S_e)}$$ (6.7a)

$$S_e = \leq 0.8\, D_w$$

Table 6.3 Expected Maximum Diameter of the Wetted Circle (D_w) Formed by a Single Emission Device Discharging Approximately 4 l/h (1 gph) on Various Soils

Soil or Root Depth and Soil Texture	Homogeneous		Varying Layers, Generally Low Density		Varying Layers, Generally Medium Density	
	(cm)	(in)	(cm)	(in)	(cm)	(in)
Depth 75 cm (30 in)						
Coarse	45	18	75	30	110	42
Medium	90	36	120	48	150	60
Fine	107	42	150	60	180	72
Depth 150 cm (60 in)						
Coarse	75	30	140	54	180	72
Medium	120	48	215	84	275	108
Fine	150	60	200	78	245	96

Source: Soil Conservation Service, "Trickle Irrigation," *National Engineering Handbook* (1984), Sec. 15, Chap. 7. U.S. Dept. of Agriculture, 129 pp.

where

N_e = number of emission devices per emission point;

D_w = maximum diameter of wetted circle formed by a single point source emission device (cm, in);

S_e = spacing between the emission devices of an emission point (cm, in);

P = percent of S times L irrigated;

K = unit constant. ($K = 100$ for S and L in m and D_w and S_e in cm. $K = 1.44$ for S and L in ft and D_w and S_e in in).

A line source of water is achieved when the spacing between emission points is less than or equal to $0.8D_w$, after which point, islands of water result.

For trickle systems with double laterals or zigzag, pigtail, or multiexit layouts (like those in Figure 6.7), N_e is computed with Eq. 6.7b.

$$N_e = \frac{(2)(K)(P)(S)(L)}{S_e(S_e + D_w)} \tag{6.7b}$$

where all parameters are as defined for Eq. 6.7a.

The spacing between double laterals should equal D_w. This spacing gives the largest A_i and leaves no extensive dry areas between the double lateral lines. For the greatest A_i with zigzag, pigtail, and multiexit layouts, the emission devices should be spaced a distance equal to D_w in each direction.

If the layout is not designed for maximum wetting and $S_e < D_w$, then D_w in Eq. 6.7b should be replaced by S_e.

For emission points with one or more microsprinklers, the number of micro-sprinklers per emission point is computed with the following equation.

$$N_e = \frac{(P)(S)(L)}{(100)\left(A_s + \dfrac{(D_w)(P_s)}{2K}\right)} \tag{6.7c}$$

$$S_e = D_T + \frac{D_w}{(2)(K)}$$

where

A_s = area wet by a microsprinkler (m^2, ft^2);

P_s = perimeter of area wet by microsprinkler (m, ft);

D_T = distance of throw (m, ft);

K = unit constant ($K = 100$ for S_e and D_T in m and D_w in cm. $K = 12$ for S_e and D_T in ft and D_w in in).

The following example illustrates the use of Eqs. 6.3 through 6.7.

EXAMPLE 6.1   Estimating the Capacity of a Trickle System Emission Device

Given:

$$RAW = 5.0 \text{ in}$$
$$DDIR = 0.30 \text{ in/day}$$
$$D_w = 60 \text{ in}$$
$$P = 30 \text{ percent}$$
$$20 \text{ ft by } 20 \text{ ft tree spacing}$$
$$P_f = 20 \text{ percent}$$
$$E_i = 85 \text{ percent}$$

Grower prefers a drip-type system

Required:
a. number of emission devices per plant
b. emission device capacity

Solution:

   Solution Steps
1. Determine N_e
2. Determine A_i
3. Choose H
4. Determine D_a
5. Determine C

   Computing N_e
   A double lateral, zigzag, pigtail, or multiexit layout is required to obtain $P = 30$ percent (since a line source 60 in wide would result in $P = 25$ percent). N_e for these layouts is estimated with Eq. 6.7a. Let

$$S_e = 0.8D_w$$
$$= (0.8)(60) = 48 \text{ in}$$

$$N_e = \frac{(2)(1.44)(30)(20)(20)}{48(48 + 60)} = 6.67 \sim 7$$

   Computing A_i   (with Eq. 6.6)

$$A_i = \frac{(20)(20)(30)}{(100)(7)} = 17.1 \text{ ft}^2$$

   Choosing H

$$H \leq \frac{(0.24)(20)(5.00)}{(0.30)} = 80.0 \text{ hr}$$

complete in 3 days = 72 hrs for the convenience of the operator

Computing D$_a$ (with Eq. 6.4)

$$D_a = \frac{(72)(0.30)}{(0.24)(20)} = 4.5 \text{ in}$$

Computing C   (with Eq. 6.3). Let

$$T_m = 0.5 \text{ h/day} \times 3 \text{ days} = 1.5 \text{ h}$$

$$C = \frac{(62.33)(4.5)(17.1)}{(72 - 1.5)(85)} = 0.80 \text{ gph}$$

This system would probably be operated 23.5 h/day so that each set (20 percent of field) was irrigated 4.7 h/day (23.5 × 0.20).

6.3.2 Trickle Irrigation Laterals

Laterals deliver water from main lines and submains to the emission devices. Laterals for point-source emitters are made of polyethylene, butylene, or PVC materials, and usually range in diameter from 10 to 20 mm (3/8 to 3/4 in). Laterals for microsprinklers and bubblers are also made of polyethylene, butylene, and PVC, but are often larger than 20 mm (0.8 in). PVC and corrugated, polyethylene drainage tubing (without perforations) have been used for bubbler system laterals.

Trickle laterals are designed to maintain an acceptable variation of emission device discharge along their length. As with sprinkler laterals pressure differences due to pipe friction, minor losses, and changes of ground surface elevation are important causes of along-the-lateral variation in emission device discharge. Other significant factors affecting emission device discharge include water temperature and the quality with which the emission device is manufactured. The effect of water temperature on the flowrate of several emission devices is summarized by the flowrate indices for 20°, 45°, and 65°C in Table 6.2.

The coefficients of manufacturer's variation in Table 6.2 describe the quality of the processes used to manufacture those emission devices. The manufacturer's coefficient of variation is determined from flowrate measurements for several identical emission devices and is computed with the following equation.

$$C_v = \frac{(q_1^2 + q_2^2 + \cdots + q_n^2 - n\bar{q}^2)^{1/2}}{\bar{q}(n - 1)^{1/2}} \qquad (6.8)$$

where

$$C_v = \text{manufacturer's coefficient of variation;}$$
$$q_1, q_2, \ldots, q_n = \text{discharge of emission devices (l/h, gph);}$$
$$\bar{q} = \text{average discharge of emission devices tested (l/h, gph);}$$
$$n = \text{number of emission devices tested.}$$

Table 6.4 lists ASAE recommendations for classifying the manufacturer's coefficient of variation.

Table 6.4   Recommended Classification of Manufacturer's Coefficient of Variation

Emitter type	C_v range	Classification
Point source	<0.05	Good
	0.05 to 0.10	Average
	0.10 to 0.15	Marginal
	>0.15	Unacceptable
Line source	<0.10	Good
	0.10 to 0.120	Average
	>0.20	Marginal to unacceptable

Source: American Society of Agricultural Engineers, "Design, Installation, and Performance of Trickle Irrigation Systems," *ASAE Engineering Practice*, 1985, ASAE EP 405, St. Joseph, MI. Copyright © 1985 by ASAE, p. 508. Reprinted by permission of ASAE.

Because water temperature, the quality of manufacturing, and the high sensitivity of laminar flow emitters to pressure variations, the ASAE pressure variation criteria for sprinkler laterals is not generally applicable to trickle laterals. Only in cases where emission device flow is fully turbulent, not sensitive to water temperature, and the manufacturer's coefficient of variation is either small or not important should the ASAE criterion for sprinkler laterals be used. In cases where the manufacturer's coefficient of variation and the effect of water temperature are not known or not important, a pressure variation of ± 5 percent of the design operating pressure has been used as a criterion for trickle lateral design. A more general criterion, called the *emission uniformity*, which depends on water temperature and the manufacturer's coefficient of variation, has been developed for evaluating trickle lateral design and emission device selection. The emission uniformity is defined for point- and line-source emitters by Eq. 6.9. Equation 6.9 is

$$EU = 100\left(1.0 - \frac{1.27}{\sqrt{N_e}}\,C_v\right)\frac{Q_{min}}{Q_{ave}} \tag{6.9}$$

where

EU = the design emission uniformity in percent;

N_e = number of point source emitters per emission point; the spacing between plants divided by the unit length of lateral line used to calculate C_v or 1, whichever is greater, for a line-source emitter;

C_v = the manufacturer's coefficient of variation for point-or-line source emitters;

Q_{min} = the minimum emitter discharge rate in the system (l/h, gph);

Q_{ave} = the average or design emitter discharge rate (l/h, gph).

Table 6.5 shows range for EU values recommended by ASAE for use in Eq. 6.9. Economic considerations may dictate a higher or lower uniformity than those given in Table 6.5.

Table 6.5 **Recommended Ranges of Design Emission Uniformity (EU)**

Emitter Type	Soil Topography	EU Range for Arid Areas[e]
Point source on permanent crops[a]	Uniform[c]	90 to 95
	Steep or undulating[d]	85 to 90
Point source on permanent or semipermenant crops[b]	Uniform	85 to 90
	Steep or undulating	80 to 90
Line source on annual row crops	Uniform	80 to 90
	Steep or undulating	70 to 85

Source: American Society of Agricultural Engineers, "Design, Installation, and Performance of Trickle Irrigation Systems," *ASAE Engineering Practice*, 1985, ASAE EP 405, St. Joseph, MI. Copyright © 1985 by ASAE, p. 509. Reprinted by permission of ASAE.
[a] Spaced >4 m (13.1 ft) apart.
[b] Spaced <2 m (6.6 ft) apart.
[c] Slope <2 percent.
[d] Slope >2 percent.
[e] For humid areas values may be lowered up to 10 percent.

The following example illustrates the use of the emission uniformity concept for evaluating trickle lateral design and emission device selection.

EXAMPLE 6.2   Computing the Emission Uniformity for a Trickle Lateral

Given:

- arid region
- point-source emitters on a permanent crop
- $C_v = 0.07$
- $Q_{min} = 10$ gph
- $Q_{ave} = 11$ gph
- $N_e = 1$
- Uniform terrain with slopes less than 2 percent.

Required:
determine if the design is acceptable

Solution:

$$EU = 100\left(1.0 - \frac{1.27}{\sqrt{1}} (0.07) \right)\left(\frac{10}{11} \right)$$

$$= 82.8 \text{ percent}$$

Conclusion:
Based on ASAE criteria in Table 6.5 this design is not acceptable, since *EU* should exceed 90 percent.

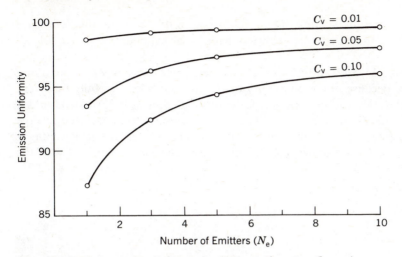

Figure 6.9   Relationship between emission uniformity, C_v, and number of emission devices for $Q_{min}/Q_{ave} = 1.0$.

The EU in the example could be improved by reducing the difference between Q_{min} and Q_{ave} (by using larger diameter and/or shorter laterals or by using pressure-compensating emitters), using an emitter with a lower C_v, or increasing N_e. Figures 6.9 and 6.10 illustrate how emission uniformity is affected by C_v, N_e, and the exponent in Eq. 6.1. Data in Figure 6.9 indicate that when emitters with higher C_v values are used, more emitters per emission point are required to achieve

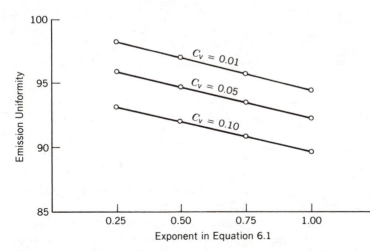

Figure 6.10   Relationship between emission uniformity, C_v, and emission exponent (x in Eq. 6.1). The graph is based on the assumptions of a 5-percent pressure variation and five emission devices per emission point.

acceptable emission uniformity. Data in Figure 6.10 illustrate that emission uniformity can be increased when emitters with lower exponents x in Eq. 6.1 are used.

Equations 5.11, 5.12, and 5.13 are used to determine pressure variations along trickle laterals. Because of the possibility of laminar, turbulent, or fully turbulent flow in trickle laterals (rather than only fully turbulent flow as in sprinkle laterals), the Darcy–Weisbach equation should be used to compute head loss due to pipe friction. The Darcy–Weisbach friction factor, f, for small-diameter trickle tubing is related to the Reynolds number, N_R. Reynolds number N_R is computed with the following equation.

$$N_R = \frac{(\rho)(D)(V)}{(K)(\mu)} \tag{6.10}$$

where

N_R = Reynolds number (dimensionless);
ρ = density of water (g/cm^3, slugs/ft^3);
D = diameter of pipe (cm, in);
V = average velocity (cm/s, ft/s);
μ = viscosity of fluid (N-s/m^2, lb-s/ft^2);
K = unit constant ($K = 10$ for ρ in g/cm^3, D in cm, V in cm/s, and μ in N-s/m^2. $K = 12$ for ρ in slug/ft^3, D in in, V in ft/s, and μ in lb-s/ft^2)

The equation used to compute f depends on the magnitude of N_R. For N_R less than 2000 (laminar flow)

$$f = \frac{64}{N_R} \tag{6.11a}$$

For N_R between 2000 and 100,000 (turbulent flow)

$$f = 0.32 N_R^{-0.25} \tag{6.11b}$$

For N_R greater than 100,000 (fully turbulent flow)

$$f = 0.80 + 2.0 \log\left(\frac{N_R}{\sqrt{f}}\right) \tag{6.11c}$$

The Hazen–Williams equation with $C = 150$ can also be used to estimate head loss due to pipe friction when $N_R > 100,000$.

Losses due to barbs from on-line emitters or supply tubes for bubblers or microsprinklers that extend through the pipe wall to obstruct flow must also be included. These losses are given in Figure 6.11 as equivalent pipe lengths for various sizes of barbs and inside diameters of laterals. The appropriate equivalent length value from Figure 6.11 is multiplied by the number of barbs and added to L in Eq. 5.13.

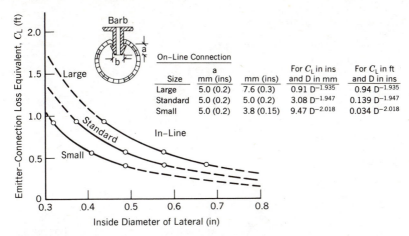

Figure 6.11 Emitter-connection loss (C_L) values for various sizes of barbs and inside diameters of laterals. (From Section 15, Chapter 7 of the *SCS National Engineering Handbook*, 1984.)

EXAMPLE 6.3 Computing Head Loss Due to Pipe Friction in a Trickle Lateral with In-Line Emitters

Given:
- 16-mm internal diameter trickle lateral
- the lateral is 200 m long with standard on-line emitters spaced each 1 m
- the design discharge of each emitter is 1 l/h
- water temperature is 20°C

Required:
Head loss due to pipe friction.

Solution:

$$Q = (1 \text{ l/h})(200)(1 \text{ h/60 min}) = 3.33 \text{ l/min}$$

$$V = \frac{Q}{A} = \frac{(3.33 \text{ l/min})\left(\dfrac{1 \text{ min}}{60 \text{ s}}\right)\left(\dfrac{1000 \text{ cm}^3}{1}\right)}{(1.6 \text{ cm})^2(\pi/4)} = 27.6 \text{ cm/s}$$

$$N_R = \frac{(0.998)(1.6)(27.6)}{(10)(1.002(10)^{-3})} = 4406 \qquad \text{(Eq. 6.10)}$$

Since N_R is between 2000 and 100,000, use Eq. 6.11b to compute f

$$f = (0.32)(4406)^{-0.25} = 3.93(10)^{-2} \qquad \text{Eq. 6.11b)}$$

$$K = \frac{(0.811)(3.93(10)^{-2})}{9.81} = 3.25(10)^{-3} \qquad \text{(Eq. 5.16a)}$$

Correcting L in Eq. 5.15 for barb losses

$L = 200$ m + (number of emission devices) C_L
$C_L = 0.36$ ft from Figure 6.11

Converting C_L to meters:

$C_L = (0.36$ ft$)(1$ m$/3.28$ ft$) = 0.11$ m

$L = 200 + 200(0.11) = 222$ m

$$H_L = \frac{(3.25(10)^{-3})(277778)(222)(3.33)^2}{(16)^5} = 2.12 \text{ m}$$ (Eq. 5.15)

$F = 0.33$ (from Table 5.8)

$h_L = (0.33)(2.12) = 0.70$ m   (Eq. 5.14)

6.3.3   Mainlines and Submains (Manifolds)

Laterals and line-source tubing are connected to a submain or in some cases directly to a mainline. Mainlines and submains are normally PVC, although asbestos-cement pipe is occasionally used for mainlines. Each mainline and submain should have a manual or solenoid activated valve at its upstream end to provide on–off service and for isolation purposes. An on–off valve at the downstream end of mainlines and submains for periodic flushing is recommended. Submains may also have pressure-regulating valves, flow control valves, secondary filters, flowmeters, and pressure gauges.

Pressure losses in trickle mainlines and submains are estimated using procedures similar to those used for trickle laterals. The pipe selection process used for sprinkle systems also applies to trickle systems.

6.3.4   Control Head

The *control head* of trickle system contains the pumping plant, primary filters, chemical injection equipment, backflow prevention devices, flowmeters, pressure gauges, valving, and automatic controllers.

6.3.4a   Valving

The control head may include pressure relief, on–off, pressure regulating, vacuum relief, flow regulation, and air relief valves. The operation and function of these valves is described in Section 5.3.2e.

6.3.4b   Filters

Filters are essential for controlling clogging in many trickle systems. Media, screen, and cartridge filters, and centrifugal separators are the main types of filters used with trickle systems. These filters as well as prefiltration devices including trash racks or screens and settling basins are described in Section 6.4.1.

6.3.4c Flowmeters and Pressure Gauges

Flow and pressure measurements are invaluable aids in system management and assessing the need for trickle system maintenance. Several types of devices can be used to measure the volume and/or volumetric flow rate (volume/time). Differential pressure flowmeters including venturi tubes, orifice plates, and elbow meters as well as various types of rotating mechanical and ultrasonic flowmeters can be used (see Chapter 8).

Pressure gauges located on the up- and downstream sides of filters are used to determine the need for filter maintenance (the need for filter maintenance is indicated by large pressure drops through the filter as described in Sections 6.4.1b and 6.4.1c). Table 6.6 lists other problems that may be associated with changes in flow.

6.3.4d Chemical Injection Equipment

Trickle systems often include equipment for metering into the irrigation system fertilizers, chemical treatments for controlling clogging, as well as other agricultural chemicals. Chemical injection requires that the pressure acting on the chemical be greater than the operating pressure within the trickle system. Three different methods for accomplishing this are shown in Figures 6.12, 6.13, and 6.14. Injection system components directly exposed to the chemicals should be corrosion resistant, since many of the chemicals that are injected into trickle systems are corrosive. Chemicals should be injected upstream of filters so that any precipitates or other clogging agents that may form do not enter laterals and emission devices. Check valves and vacuum release valves should be installed upstream of the injector and on injector lines to prevent backflow of chemicals into the water source. The American Society of Agricultural Engineers has specified required safety devices for applying chemicals through irrigation systems (ASAE, 1985).

Table 6.6 Possible Causes of Changes in Trickle System Flow

Increased Flow
    improperly adjusted/open valves
    pipeline leaks/breaks
    pressure downstream of pressure regulators is too high
    worn/oversized emission devices
    system on too long (as indicated by higher than expected volumes of flow)

Decreased Flow
    improperly adjusted valves
    clogged emission devices, filters, and other components
    pump wear
    pressure downstream of pressure regulators too low
    existence of entrapped air in system
    system not on long enough (as indicated by lower than expected volumes of flow)

Source: L. G. James and W. M. Shannon, "Flow Measurement and System Maintenance." In *Trickle Irrigation for Crop Production*, F. S. Nakayama and D. A. Bucks (Eds.), copyright © 1986, Elsevier Science Publishing Co., Inc., p. 305. Reprinted by permission of Elsevier Science Publishing Co., Inc. and authors.

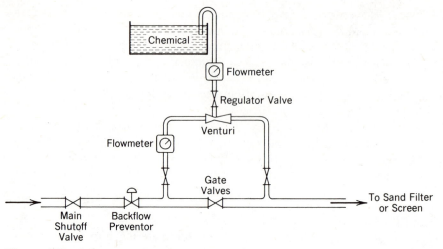

Figure 6.12 Schematic of a venturi injection system.

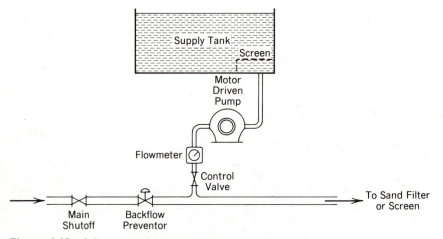

Figure 6.13 Schematic of a metering pump injection system.

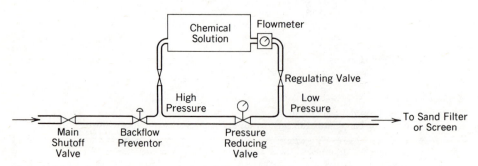

Figure 6.14 Schematic of a pressure-differential injection system.

(i) Eductors   The injection system in Figure 6.12 uses a venturi or other eductor to create an area of low pressure within the trickle system where chemicals may be added. The regulator valve in the injection line is adjusted to meter in the desired amount of chemical. The venturi principle is not very effective for low-pressure systems.

(ii) Injector Pumps   An injector (or metering) pump (Figure 6.13) is an effective way of providing sufficient pressure to inject chemicals. A properly selected injector pump can supply chemicals at a relatively constant concentration and rate for the required duration. Packed-plunger and mechanically or hydraulically activated pumps similar to those in Figure 6.15 are the most common types of

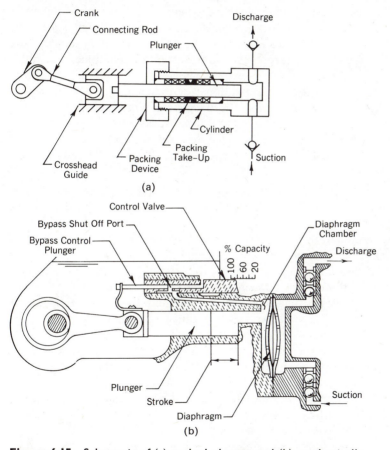

Figure 6.15   Schematic of (a) packed-plunger and (b) mechanically activated diaphragm pumps typically used for chemical injection. *Source*: L. G. James and W. M. Shannon, "Flow Measurement and System Maintenance." In *Trickle Irrigation for Crop Production*, F. S. Nakayama and D. A. Bucks (Eds.), copyright © 1986, Elsevier Science Publishing Co., Inc., p. 305. Reprinted by permnission of Elsevier Science Publishing Co., Inc. and others.

metering pumps used. Electric and water motors, gasoline engines, or belts driven by the pumping plant power unit power the metering pump.

(iii) *Pressure Differential Systems*   A pressure drop between the inlet and outlet of the chemical supply tank of the injection system in Figure 6.14 is created by the pressure reducing valve. This pressure difference causes flow through the tank and chemicals from the tank to be carried into the trickle system. A disadvantage of differential injection is that the concentration of chemical in the tank is diluted as injection continues.

6.3.4e   Backflow Prevention Equipment

Flow in the reverse direction through an irrigation system toward the water source (rather than away from it) is called *backflow*. Backflow is usually caused by pump, pipeline, or valve failure and can result in irrigation system damage and contamination of the water source. Backflow prevention equipment installed immediately downstream of the water source is often required by law to minimize the possibility of water source contamination when chemicals are injected into irrigation systems. Backflow prevention equipment can consist of a single check valve or two or more check valves hooked in series. Pressure relief, manual on–off, and vacuum relief valves are often included to provide surge protection and positive protection from backflow when the system is off.

6.3.4f   Automatic Controllers

The control head of automated trickle systems contain controllers. These controllers range from mechanical clocks that open/close a single valve on a preset time schedule to microcomputers, which are programmed to interrogate a series of soil and/or climatic sensors, decide when to begin and end irrigation, start/stop pumps and open/close valves to accomplish the irrigation, and remember how much water and fertilizer were applied to each block within the field. Many controllers are also able to diagnose system malfunctions and take corrective action. Some even turn the system off during rain storms and then restart the system when the storm ends.

A time-type of controller uses a clock (either solid state or motor driven electric) as the means for programming the starting and sequence of irrigation. The controller supplies electrical or hydraulic power to activate remote solenoid valves located on individual laterals or submains (manifolds). As many as 30 or more valves may be controlled by a controller. Communication between the controller and valves is via wires, hydraulic or pneumatic conduit, or radio telemetry.

Microprocessor/microcomputer-based controllers can be programmed to control pumps, injection equipment, filters, etc., as well as open/close solenoid activated on–off valves using data from tensiometers, pyranometers, evaporation pans, thermocouples, humidity meters, anemometers, flowmeters, pressure transducers, and other sensors. These controllers poll soil and/or climatic sensors according to a schedule specified by the irrigator. The controller is programmed to use these data to determine the need for irrigation in each field and block. It then operates the pumps, filters, injection equipment, and valves needed to accomplish the irrigation. Data from flowmeters and pressure sensors are used to determine the need for such things as flushing and to detect system malfunctions.

6.3.4g   Pumping Plants

Many trickle systems have pumping plants to lift water from ponds, reservoirs, lakes, streams, canals, and wells, and to provide pressure for trickle system operation. Pumping plants normally have either horizontal or vertical centrifugal pumps powered by either electric motors or internal-combustion engines (see Chapter 4).

6.4   Control of Trickle System Clogging

Trickle irrigation systems have very low flowrates and extremely small passages for water. These passages are easily clogged by mineral particles and organic debris carried in the irrigation water and by chemical precipitates and biological growths that develop within the system. The result of clogging is either the complete or partial stoppage of flow through clogged components.

Major physical contributors to clogging are summarized in Table 6.7. They include mineral particles of sand, silt and clay, and debris that are too large to pass through the small openings of filters and emission devices. Silt and clay particles that are usually much smaller than the smallest passages are often deposited in the

Table 6.7   Physical, Chemical, and Biological Contributors to Clogging of Trickle Systems

A. Physical (Suspended Particles)	B. Chemical (Precipitation)	C. Biological (Bacteria and Algae)
a. Organic	a. Calcium carbonate	a. Filaments
(1) Moss, aquatic plants, and algae	b. Calcium sulfate	b. Slime
(2) fish, snails, etc.	c. Heavy metal, hydroxides, oxides, carbonates, silicates, and sulfides	c. Microbial deposition
b. Inorganic		(1) Iron
(1) Sand	d. Fertilizers	(2) Sulfur
(2) Silt	(1) Phosphate	(3) Manganese
(3) Clay	(2) Aqueous ammonia	
	(3) Iron, zinc, copper, manganese	

Source: D. A. Bucks, F. S. Nakayama, and R. G. Gilbert, "Trickle Irrigation Water Quality and Preventive Maintenance,"
Agricultural Water Management copyright © 1979 by Elsevier Science Publishing Co., Inc., p. 151. Reprinted by permission of Elsevier Science Publishing Co., Inc. and authors.

low-velocity areas of laterals where they coagulate to form masses large enough to clog emission devices. Coatings of clay particles in filters and emission devices can also reduce water flow.

Precipitates that form within trickle systems and on the outside surface of emission devices are major chemical causes of clogging (see Table 6.7). When irrigation water contains soluble salts, crusts of salt often form on emission devices as water evaporates between irrigations. If the salt does not dissolve during the subsequent irrigation, crust accumulation will continue and clogging of the emission device will usually result. Carbonates may accumulate within trickle systems and cause clogging when high levels of calcium, magnesium, and bicarbonate are present in the irrigation water. Clogging can also occur when well waters with dissolved iron and manganese are exposed to the atmosphere and iron and manganese oxide precipitates are formed. Waters high in sulfides will also produce insoluble compounds that can cause clogging. In addition to naturally occurring compounds, precipitates may be formed when some types of liquid fertilizers or other chemicals are injected into the system.

Favorable environmental conditions within trickle systems can cause rapid growth of several species of algae and bacteria (see Table 6.7). These microorganisms often become large enough to cause complete clogging. Biological oxidation processes involving certain species of bacteria and waters with even very low concentrations of ferrous and manganous ions can produce deposits of iron and manganese oxides that promote clogging. Macroorganisms such as ants, spiders, fleas, and freshwater crustaceans can also contribute to clogging problems.

The corrective treatment for controlling clogging, depends on the type of clogging agent. Many physical agents can be removed using settling ponds, water filtration and/or periodic flushing of filters, mainlines, laterals, and emission devices. Injections of acids, oxidants, algaecides and bacteriacides are common treatments used to control chemically and biologically caused clogging.

6.4.1 Filtration

Removal of suspended particles is usually required for optimum trickle system performance, since irrigation water is rarely free of suspended material. Settling basins, sand or media filters, screens, cartridge filters, and centrifugal separators are the primary devices used to remove suspended material. Table 6.8 summarizes the minimum-sized particles that can be removed by several of these devices.

6.4.1a Settling Basins

Settling basins or reservoirs can remove large volumes of sand and silt. The minimum size of particle that can be removed depends on the time that sediment-laden water is detained in the basin. Longer detention times are needed to remove smaller particles. Removal of clay-sized particles requires several days and is not practical unless flocculating agents such as alum and/or polyelectrolites are used. Settling basins may need to be cleaned several times a year when large quantities of water with high concentrations of sediment are being passed through the basin.

Table 6.8 Filter Effectiveness

Filter Type	Size Range (microns)
Sediment basins	> 40
Slotted cartridge	>152
Sand media	5–100
Sand media	> 20
Screen (100–200 mesh)	75–150
Screen (200 mesh)	>100
Screen	> 75
Separator[a]	> 74
Separator[a] (two stage)	> 44

Source: W. M. Shannon, "Sedimentation in Trickle Irrigation Laterals," unpublished M. S. Thesis (1980), Washington State University, Pullman, 133 pp.
[a] Separators remove 98 percent of particles larger than size indicated.

Algae growth and windblown contaminants can be severe problems in settling basins. Chemical treatments with chlorine or copper sulfate may be required to control algae. Because of these problems, settling basins are recommended for use with trickle systems only in extreme circumstances.

6.4.1b   Media Filters

Sand or media filters like the one in Figure 6.16 consist of layered beds of graduated sand and gravel placed inside one or more pressurized tanks. They effectively remove suspended sands, organic materials, and most other suspended substances from surface and ground waters. They will not remove very fine substances (i.e., silt and clay sized particles) or bacteria. Media filters are also relatively inexpensive and easy to operate.

Figure 6.17 shows that the minimum particle size that passes through a typical media filter depends on the size and type of filter material and the rate at which water is flowing through the filter. In general, the use of finer materials and lower flowrates per unit surface area of filter decreases the maximum particle size that can pass through a filter. However, the capacity of a filter (i.e., the quantity of water filtered per unit time) is increased by using coarser textured filter materials. Thus, it is generally recommended that the filter material be as course textured as possible but fine enough to retain all particles larger than one-sixth the size of the smallest passageway in the trickle system (British Columbia Ministry of Agriculture, 1982). Filter materials should be large enough not to be removed during filter cleaning processes.

Filter cleaning involves removing filtered substances that have accumulated in the pores of the filter material. Cleaning is usually accomplished by reversing the direction of water flow through the filter and bypassing the effluent. The need for this cleaning process, called backflushing, can be detected by monitoring the

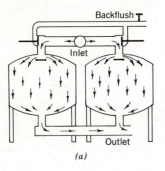

(a)

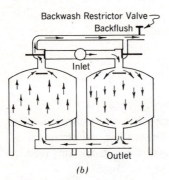

(b)

Figure 6.16 A sand medial filter. (a) Filtering process. (b) Backwash process.

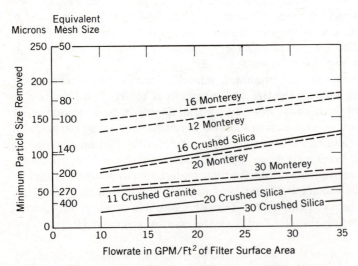

Figure 6.17 Filtration efficiency of different grades and types of sand at various flowrates. (From Yardney Electric Corporation, Water Management Divison.)

pressure drop across the filter. When the pressure drop has increased to a predetermined level (because of accumulating filtered material that partially clogs the filter), the filter should be backflushed. ASAE recommends (in ASAE Engineering Practice: ASAE EP405) that this pressure drop not exceed 70 kPa (10 psi).

Cleaning of media filters can be initiated either automatically or manually. Automatic systems similar to the one in Figure 6.16 usually have at least two tanks, one that is operating and one that is being backflushed. Flow is switched from one tank to another according to a time cycle or switching is automatically triggered when the pressure drop across the operating tank exceeds a predetermined level.

6.4.1c Screen Filters

Cylinder screens made of stainless steel or nylon similar to the one in Figure 6.18 are the most common type of screen filters used in trickle systems. Wye and basket screens (see Figure 6.19) are sometimes used as secondary filters in submains and laterals, while simple hand-cleaned screens placed in open delivery channels (usually located just upstream of the systems inlet pump) are also used in trickle systems. The size of screen openings and hence the minimum particle size retained by the screen is determined by the number of wires per inch (i.e., the screen's mesh number). Table 6.9 lists the minimum particle size retained for several screen mesh numbers. The screen mesh should be selected so that the screen retains all particles larger than one-sixth the size of the smallest passage (openings) in the trickle system (British Columbia Ministry of Agriculture, 1982).

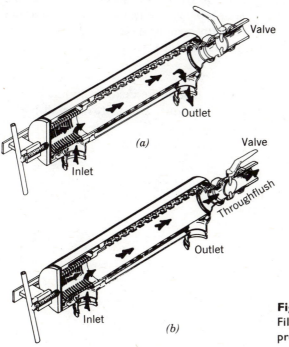

Figure 6.18 A screen filter. (*a*) Filtration process. (*b*) Throughflush process.

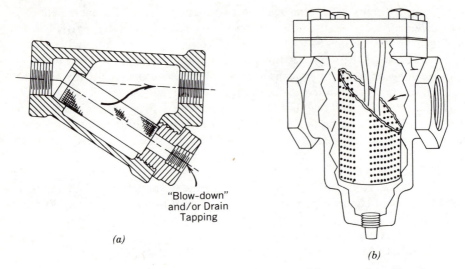

"Blow-down"
and/or Drain
Tapping

(a)

(b)

Figure 6.19   Typical (*a*) "Y" and (*b*) basket filters used for secondary filtration in trickle irrigation systems. They are normally installed at the upstream end of submains and laterals.

Like media filters, screen filters can be cleaned automatically or manually. One automatic system has dual cylinder filters, one of which operates while the other is being cleaned. The flow of water is switched from cylinder to cylinder automatically according to a time cycle or a predetermined pressure drop level as with media filters. Cleaning is accomplished by opening a valve on the outlet end of the screen (see Figure 6.18) and throughflushing the screen.

6.4.1d   Cartridge Filters
Cartridge filters can remove organic materials and extremely fine particles that cannot be removed with media filters. Cartridges, which are either disposable

Table 6.9   Filtration Capabilities of Different Screen Mesh Sizes

Mesh	Filtration to Micron Size
80	175
100	147
150	104
200	74

Note: For example, a 100-mesh screen will filter everything larger than 147 microns in size out of the irrigation water.

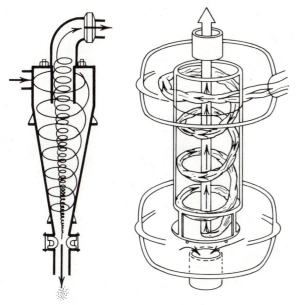

Figure 6.20   Centrifugal filters.

or washable, are available in a variety of materials including nylon, cotton, and fiberglass. Cartridges and screens are often the final filtering component before water enters the system.

6.4.1e   Centrifugal Separators

Centrifugal filters, also called sand or cyclonic separators, depend on vortex motion and centrifugal force to remove suspended substances with specific gravities in excess of 1.2. Organic materials with specific gravities of about 1 cannot, therefore, be removed with centrifugal filters. The filtration capacity of centrifugal filters, such as those in Figure 6.20 is greater (i.e., the size of particle that can be removed is smaller) for high-density particles than for lower density particles. Because centrifugal filters effectively remove large quantities of sand particles, they are often placed upstream of media, screen, and cartridge filters that remove finer sized particles and organic materials.

Table 6.10 lists general water filtration guidelines published by the British Columbia Ministry of Agriculture for irrigation waters containing suspended inorganic and organic solids. The British Columbia Ministry of Agriculture has also published, in Table 6.11, minimum filtration requirements for selected emission devices.

6.4.2   Flushing

Data in Table 6.8 indicate that silt and clay-sized materials (particles less than 50 microns) are usually not removed by settling basins, media filters, screens, cartridges, or centrifugal separators. These extremely fine materials are carried into

Table 6.10 **Filtration Requirements for Selected Physical Clogging Agents**

Water Quality	Suggested Treatment
Inorganic Solids	
< 10 mg/l	
Particles greater than 100 microns in diameter	Remove with stainless-steel screen. (Particles larger than one-sixth of emitter orifice diameter should be screened out)
Particles less than 100 microns in diameter	These particles may pass through the irrigation system if the Fe and S concentrations are not too high. If slug loadings are frequent, then automatic screen cleaning may be required
< 10 mg/l	
Particles over or under 100 microns	Automatic cleaning of screens suggested
Organic Solids	
< 10 mg/l	
Particles over 100 microns	Sand filters are required. Recommended flowrate through sand filter is 20 gpm/ft² of bed area. Manual backflushing should be satisfactory.
< 10 mg/l	
Particles over 100 microns	Sand filters with automatic backflush are required. Recommended flowrate through sand filter is 20 gpm/ft² of bed area
Slug loadings of organic solids with particles under 100 microns	High suspended solids loadings of this type of material may not pass through the trickle irrigation system. In this situation, sand filters will remove a large volume of material. Automatic backflushing of the sand filters will most likely be required
	Treatment with chlorine may be required periodically during the season to prevent accumulation of particles under 100 microns

Source: British Columbia Ministry of Agriculture, "Water Treatment Guidelines for Trickle Irrigation," Engineering Reference Information R512.000 (1982), 2 pp.

the system where they often come out of suspension and are deposited in low-velocity areas of submains and laterals. Clogging of emission devices can become a serious problem when individual particles are agglomerated by biological by-products. Submains and laterals must be periodically flushed to remove these deposits. The frequency of flushing depends on the rate at which these deposits accumulate and the size of pipe. It is usually recommended that the appearance of the flushed water be used to determine the frequency of flushing. In extreme situations daily flushing of laterals may be necessary. Normally, weekly or monthly flushing is sufficient.

Table 6.11 Minimum Filtration Requirements for Different Emission Devices

Emission Device	Orifice Size Inches	Orifice Size Microns	Screen Mesh	Sand Media Selection
Microjet	0.03	760	150	# 16 silica sand
	0.04	1120	80	# 8 crushed granite[a]
	0.05	1270	80	# 8 crushed granite[a]
	0.06	1525	80	# 8 crushed granite[a]
	0.07	1780	80	# 8 crushed granite[a]
Bi-Wall	0.019	480	200	# 20 silica sand
Microtube	0.020	500	200	# 20 silica sand
	0.025	635	150	# 16 silica sand
	0.035	890	100	# 11 crushed granite
	0.045	1140	80	# 8 crushed granite
Vortex			100	11 Crushed Granite
			200	50% # 30 silica sand
Viaflow			200	50% # 20 silica sand
Submatic	0.02	510	200	# 20 silica sand
	0.03	760	150	# 16 silica sand
Dripeze			200	# 20 silica sand
All long flow path emitters			200	# 20 silica sand

Source: British Columbia Ministry of Agriculture, "Water Treatment Guidelines for Trickle Irrigation," Engineering Reference Information R512.000 (1982), 2 pp.
[a] Emission devices with orifices larger than 1000 microns may not require a sand filtration system (except in extreme circumstances), but should always have at least a #80 or #100 mesh installed.

6.4.3 Chemical Treatment

The control of clogging caused by chemical and biological agents requires the periodic injection of chemicals into trickle systems. Chlorine and acids are the most commonly injected chemicals. It is important that American Society of Agricultural Engineer's specifications for required safety devices for applying chemicals through irrigation systems be observed (ASAE, 1985).

Chlorine is used to control algae and bacterial slimes that form when bacterial by-products agglomerate fine organic and inorganic particles. Injections of chlorine also prevent dissolved iron from being introduced into trickle systems by causing the resulting iron oxide precipitates to develop outside of the trickle system (rather than within the system where they can clog system components). Media filters located just downstream of the injector prevent these precipitates from entering the system.

Liquid sodium hypochlorite (NaOCl), granular calcium hypochlorite (CaOCl), and gaseous chlorine (Cl$_2$) are the primary sources of chlorine that are injected into trickle systems. Addition of sodium hypochlorite or calcium hypochlorite solutions raises the pH of a water, since both components quickly ionize into the hypochlorite (OCl$^-$) ion. If the pH becomes greater than 8.0 and concentrations of Ca in excess of 20 ppm are present, a media filter should be provided to remove the resulting precipitates before they can enter the trickle system. Sometimes acids are added simultaneously with the hypochlorite to stabilize the pH and control the formation of precipitates.

Calcium carbonate precipitates can result when calcium from calcium hypochlorite reacts with carbon dioxide dissolved in the supply water. The potential for this problem is higher when calcium hypochlorite is added to colder supply waters. This is because colder water can contain more dissolved carbon dioxide than warmer water. Sodium hypochlorite (rather than calcium hypochlorite) is therefore recommended for cold waters ($<5°C$) with high levels of dissolved carbon dioxide (British Columbia Ministry of Agriculture, 1982).

The rate at which chlorine needs to be added to a water depends on the chlorine concentration of the chemical solution to be injected, the water supply flowrate, and the desired chlorine concentration in the supply water. Equation 6.12 can be used to compute the required rate.

$$R_i = \frac{6(10)^{-3}QC_d}{C_i} \tag{6.12}$$

where,

R_i = rate at which the chemical solution is to be added to the supply water (l/h, gph);

Q = supply flow rate (l/min, gpm);

C_d = desired concentration of chemical in water supply (parts per million, ppm);

C_i = concentration of chemical in the solution to be injected (percent).

The following example illustrates the use of Eq. 6.12.

EXAMPLE 6.4   Estimating the Chemical Injection Rate

Given:
• 20 ppm of chlorine required in supply water
• supply flowrate = 50 gpm
• 12 percent solution of sodium hypochlorite is injected

Required:
the solution injection rate

Solution:

$$R_i = \frac{6(10)^{-3}(50)(20)}{12} = 0.50 \text{ gph}$$

Table 6.12 **Recommended Chemical Treatments for Selected Conditions**

Water Quality	Suggested Treatment
Ca > 50 ppm Mg > 50 ppm	Hard water, caused by high concentrations of Ca or Mg, can reduce flowrates by the buildup of scales on pipe walls and emitter orifices. Periodic injection of an HCl solution may be required throughout the season. Lower concentrations of Ca and Mg may require HCl treatment every few years
Fe > 0.5 ppm S > 0.5 ppm	Iron and sulphur, as well as other metal contaminants, provide an environment in water that is conducive to bacterial activity. The by-products of the bacteria in combination with the fine, less than 100-micron suspended solids can cause system plugging. Bacterial activity can be controlled by chlorine injection and line flushing on a regular basis throughout the irrigation season. Bacterial activity is prevalent in concentrations of Fe and S over 0.5 ppm, but also occurs at lower concentrations

Source: British Columbia Ministry of Agriculture, "Water Treatment Guidelines for Trickle Irrigation," Engineering Reference Information R512.000 (1982), 2 pp.

Sulfuric and hypochloric acids are injected to lower the pH of supply water and to reduce the amount of chemical precipitates. Regular acid treatments are necessary to clean emission device passages when concentrations of calcium or magnesium exceed 50 ppm or when supply water pH is greater than 8.0.

Table 6.12 lists general water treatment guidelines published by the British Columbia Ministry of Agriculture for selected irrigation waters. There are many water quality related problems not included in Table 6.12.

6.5 Fertilizer Injection

Water-soluble fertilizers can be effectively and efficiently applied through trickle irrigation systems. Reduced labor, energy, and equipment costs (as compared to conventional fertilizer application methods) are the primary benefits of fertilizer injection. Because fertilizers can be applied when and where they are required, plant nutrient levels can be maintained at an ideal level throughout the growing season. Many herbicides, insecticides, fungicides, bactericides, algicides, and other chemicals have been applied via trickle systems. Materials such as inorganic forms of phosphorous that form chemical precipitates and cause clogging should not be injected into trickle systems.

6.6 Trickle System Design

Most trickle systems can be classified as solid-set systems. Some are set-move systems. The procedures and information outlined for set-move and solid-set sprinkle systems in Chapter 5 apply to the design of most trickle systems.

Trickle system design begins with system layout and the determination of the number of laterals to be operated simultaneously during an irrigation. Next, emission device specifications are developed and laterals, submains, (manifolds), and mainlines designed. Filtration, chemical injection, automatic controller, valve, and pumping plant specifications are then developed.

Homework Problems

6.1 Determine the pressure at which a 3-m-long, 3-mm-diameter bubbler tube must be operated to deliver 1.3 l/min.

6.2 Determine the length of the 2.5-mm-diameter bubbler tube required to deliver 1 l/min from a lateral if the operating pressure is 150 kPa.

6.3 Determine the length of the 3-mm-diameter bubbler tube required to deliver 1.4 l/min at the up- and downstream ends of a lateral. The pressures at the up- and downstream ends of the lateral are 150 and 123 kPa, respectively.

6.4 Determine the capacity of emission devices for an apple orchard in which the trees are spaced in a 5 m by 5 m grid. The readily available water holding capacity of the soil is 20 cm and the design daily irrigation requirement is 8 mm/day. The irrigator desires to use a single emission device per tree and to irrigate 50 percent of the cropped area daily. (The irrigator plans to divide the orchard into 12 zones and irrigate each zone once per day for 2 hours.)

6.5 A line-source of water is required to trickle irrigate a row crop. Determine the maximum spacing between point source emission devices for homogeneous
a. sand,
b. loam, and
c. clay soils.

6.6 Repeat Problem 6.5 for microsprinklers that throw water 7 m.

6.7 Determine the number of point-source emission devices per emission point required to irrigate an orchard in which trees are spaced in a 5 m by 5 m grid for homogeneous
a. sand,
b. loam, and
c. clay soils.

6.8 Repeat Problem 6.7 for microsprinklers that throw water 4.2 m.

6.9 Determine the emission uniformity along a 100-m-long, 13-mm-diameter polyethylene lateral that has in-line point-source emission devices spaced each 1 m along its length. There is no difference in elevation along the lateral. The pressure at the upstream end of the lateral is 150 kPa. The discharge equation for the emitter is

$$Q = 0.21P^{0.6}$$

where

Q = emitter discharge (l/h);
P = operating pressure (kPa).

The manufacturers coefficient of variation is 0.05. The water temperature is 30°C. Is this an acceptable emission uniformity?

6.10 Determine the emission uniformity of the lateral in Problem 6.9 for a 1.5 percent uphill slope (in the direction of flow). How could the emission uniformity be improved?

6.11 Solve Problem 6.9 for an on-line point-source emission device with standard-size barbs. The discharge of the emission device is given by the equation in Problem 6.9.

6.12 Repeat Problem 6.10 using a pressure-compensating point-source emission device with the following discharge equation.

$$Q = 1.56P^{0.2}$$

where

Q = emission device discharge (l/h);
P = operating pressure (kPa).

6.13 Repeat Problem 6.10 using a 19 mm diameter polyethylene lateral.

*6.14 Solve Problem 6.9 using the computer program in Appendix G.

*6.15 Solve Problem 6.10 using the computer program in Appendix G.

*6.16 Solve Problem 6.11 using the computer program in Appendix G.

*6.17 Solve Problem 6.12 using the computer program in Appendix G.

*6.18 Solve Problem 6.13 using the computer program in Appendix G.

* Indicates that a computer program will facilitate the solution of the problems so marked.

References

American Society of Agricultural Engineers (1985). ASAE Engineering Practice, ASAE EP405.

British Columbia Ministry of Agriculture (1982). Water treatment guidelines for trickle irrigation. Engineering Reference Information R512.000, 2pp.

Bucks, D. A., F. S. Nakayama, and R. G. Gilbert (1979). Trickle irrigation water quality and preventative maintenance. *Agricultural Water Management*, **2**(2), pp. 149–162.

Bucks, D. A., F. S. Nakayama, and A. W. Warrick (1983). Principles, practices, and potentialities of trickle (drip) irrigation. In *Advances in Irrigation*, D. Hillel (Ed.), Vol. 1, Academic Press, New York, pp. 219–299.

Howell, T. A., D. S. Stevenson, F. K. Aljibury, H. M. Gitlin, I. P. Wu, A. W. Warrick, and P. A. C. Roots (1980). Design and operation of trickle (drip) systems. In *Design and Operation of Farm Irrigation Systems*, M. E. Jensen (Ed.), ASAE Monograph 3, St. Joseph, MI, pp. 663–717.

Keller, J., and D. Karmeli (1975). Trickle irrigation design. Rain Bird Sprinkler Manufacturing Corporation, Glendora, Calif.

Northeast Regional Agricultural Engineering Service (1980). Trickle irrigation in the eastern United States, NRAES-4, 23 pp.

Pair, C. H., W. H. Hinz, K. R. Frost, R. E. Sneed, and T. J. Schiltz (1983). *Irrigation*. The Irrigation Association. Silver Spring, Md., 686 pp.

Rawlins, S. L. (1977). Uniform irrigation with a low-head bubble system. *Agricultural Water Management*, **1**(2), pp. 167–168.

Schwab, G. O., R. K. Frevert, T. W. Edminster, and K. K. Barnes (1981). *Soil and Water Conservation Engineering*. Wiley, New York, 525 pp.

Shannon, W. M. (1980). Sedimentation in trickle irrigation laterals. Unpublished M.S. Thesis, Washington State University, Pullman, 133 pp.

Soil Conservation Service (1984). Trickle irrigation. U.S. Department of Agriculture, Soil Conservation Service, *National Engineering Handbook*, Chapter 15, Section 15, 129 pp.

Solomon, K. (1977). Performance comparison of different emitter types. Rain Bird Sprinkler Mfg. Corp., Glendora, Calif., 8 pp.

Vander Gulik, T. (1981). Reducing emitter plugging by chemical treatment of trickle irrigation systems. Engineering Notes, British Columbia Ministry of Agriculture, Abbotsford, B.C., Canada, 7 pp.

Vander Gulik, T. (1982). Trickle irrigation emitter selection. Engineering Notes, British Columbia Ministry of Agriculture, Abbotsford, B.C., Canada. 5 pp.

7

Surface Irrigation

7.1 Introduction

Surface irrigation is the oldest and most used method of irrigation. Farmers in Egypt, China, India, and countries of the Middle East are known to have irrigated lands at least 4000 years ago, most likely using surface methods. About two-thirds of the 25 million hectares (61 million acres) of irrigated land in the United States is surface irrigated. This percentage is higher in other parts of the world.

Surface irrigation systems convey water from the source to fields in lined or unlined open channels and/or low-head pipelines. Basins, borders, and furrows are the primary methods of applying water. A smaller investment is normally required for surface systems than for sprinkle and trickle systems, except possibly when extensive land smoothing is needed. Although traditionally regarded as labor intensive (relative to sprinkle and trickle systems), concepts and equipment for automating surface irrigation systems and reducing labor requirements are now available. Surface irrigation systems are best suited to soils with low to moderate infiltration capacities and lands with relatively uniform terrain and slopes less than 2 to 3 percent (Booher, 1974).

This chapter discusses surface irrigation processes, systems, and design. Sufficient information is presented for the reader to

1. become familiar with basin, border, and furrow irrigation,
2. describe surface irrigation processes,
3. become familiar with open channel and low-head pipeline delivery systems,
4. design lined and unlined open channels,
5. evaluate the effectiveness of surface irrigations,
6. perform land smoothing calculations,
7. be introduced to cablegation and surge flow irrigation,
8. be introduced to equipment and concepts for automating surface irrigation systems,

9. perform field tests to obtain information for designing basin, border, and furrow systems,
10. be introduced to mathematical modeling of surface irrigation,
11. design basin, border, and furrow systems with constant streamsize,

7.2 Surface Irrigation Methods

Basin, border, and furrow irrigation are the primary methods of surface irrigation. Other surface irrigation methods include water spreading and contour ditch irrigation.

7.2.1 Basin Irrigation

In *basin irrigation*, the field to be irrigated is divided into units surrounded by small levees or dikes. Gated outlets, siphon tubes, spiles, and hydrants (see Figure 7.1) conduct water from delivery channels or pipelines into each basin.

Basins may be either level or graded. In level basins, water is introduced into the basin as rapidly as possible and then held until it infiltrates or is drained away. High application efficiencies are possible primarily because runoff losses are minimized.

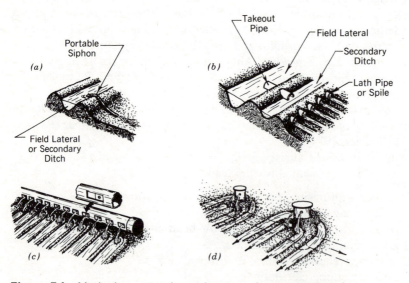

Figure 7.1 Methods commonly used to introduce water into furrows.
(*a*) Portable siphon tube. (*b*) Forbay with spiles. (*c*) Gated pipe.
(*d*) Multiple outlet risers.

Graded basins are constructed with two levees parallel and two perpendicular to field contours. Water enters graded basins along the upper contour and flows to the lower contour until the irrigation is complete. Water is then removed with surface drains located along the low contour levee. Graded basins are sometimes tiered so that water drained from upper basins is used to irrigate lower basins. For paddy rice, water is usually circulated through basins throughout most of the irrigation season. Graded basin irrigation is sometimes called contour levee irrigation.

Basin irrigation is best suited to soils of moderate to low infiltration capacities and land with smooth, gentle slopes. Basin size must be reduced to successfully irrigate soils with moderately high to high infiltration capacities. Undulating terrain or steep slopes can be prepared for basin irrigation provided soils are deep enough to permit land forming (i.e., land smoothing).

Basins vary in size from 1 square meter for intensive crops such as vegetables to as much as 16 hectares (40 acres) for the production of rice and other crops. Many different crops including cotton, grains, maize, orchards, and pastures are suited to this system of irrigation. Basin irrigation is not recommended for crops that are sensitive to wet soil conditions around the stems, soils that crust badly when flooded.

The principal disadvantage of basin irrigation is that the levees interfere with the movement of farm equipment. The presence of levees and ditches can also reduce the area available for crop production.

7.2.2   Border Irrigation

Border irrigation makes use of parallel earth ridges, called borders, to guide a sheet of flowing water across a field. The area between two borders is the border strip. These strips may vary from 3 to 30 meters (10 to 100 feet) in width and from 100 to 800 meters (300 to 2600 feet) in length. These strips have no cross slope and are either level or graded from inflow to outflow end. Level borders are normally diked at the downstream end to prevent runoff. As with basin irrigation, border irrigation is not recommended for crops that are sensitive to wet soil conditions around the stems. The operation of machinery can be a problem in borders as in basins (even though border levees run in one direction).

7.2.3   Furrow Irrigation

Furrow irrigation is accomplished by running water in small channels (furrows) that are constructed with or across the slope of a field. Water infiltrates from the bottom and sides of furrows moving laterally and downward to wet the soil and to move soluble salts, fertilizers and herbicides carried with the water. Land smoothing to provide uniform slopes can greatly improve the effectiveness of this method.

Water is diverted into furrows from open ditches or pipes. Two of the most common methods of introducing water into furrows from open ditches are siphon tubes and forbays with spiles. Portable gated pipes and single or multiple outlet risers are two popular ways of distributing water from low pressure underground pipe. These are diagrammed in Figure 7.1.

Furrows may be classified as level, contour, or graded furrows. Furrows that are level lengthwise are called *level furrows*. A dike is usually constructed at the downstream end of level furrows to pond water in the furrow and minimize runoff from the furrow. Large stream sizes are desired to achieve high uniformity of application. Stream sizes must not, however, exceed the capacity of the furrow or cause excessive erosion. This method is suited to fine-textured, very slowly permeable soils on relatively flat land.

Contour furrows are curved to fit the topography of the field. Contour furrows have a gentle slope along their length. Fields with slopes of up to 15 percent can be irrigated with contour furrows (Bishop et al., 1967).

Graded furrows are usually straight channels constructed down the prevailing land slope (grade). Land smoothing (i.e., filling low areas with material removed from high spots) is usually required for acceptable water application efficiency and uniformity. Excessive erosion, low efficiency, and/or poor uniformity can result when slopes exceed 2 percent and/or in coarse-textured soils.

Furrows are particularly suitable for irrigating crops that are subject to injury if water covers the crown or stems of the plants. Row crops such as vegetables, cotton, corn, sugar beets, maize, potatoes, and seed crops planted on raised beds are irrigated by furrows placed between the plant rows. Orchards and vineyards can be irrigated by placing one or more furrows between the tree or vine rows in order to wet a major area of the root zone. A variation of the furrow method is the use of small rills, or corrugations, for irrigating close-spaced crops such as grains, alfalfa, or pasture.

The labor required is generally greater for furrows than for other surface irrigation methods, except, possibly irrigation with small basins. Considerable experience is needed to divide water in the supply ditch into a number of furrow streams and to maintain correct rates of flow until irrigation is complete.

7.2.4 Water Spreading

Water spreading involves turning a stream of water onto a relatively flat field and allowing the water to spread naturally. This is normally a very inefficient method of irrigation providing little or no control of water distribution over the field. Water spreading is sometimes employed in low-lying areas with a stream or small river where water can be diverted during periods of high water, such as associated with spring runoffs, and used for preirrigations or initial irrigations. This practice allows use of short-term, excess supplies of water that would otherwise be lost.

7.2.5 Contour Ditch Irrigation

In this system of irrigation, water is released onto the field to be irrigated from a series of slightly sloping ditches spaced 25 to 200 m (75 to 300 ft) across the field contours. Water is released onto the land between ditches using large siphon tubes, through openings in the downslope ditch bank or over downslope ditch banks constructed lower than upslope banks. In theory, water moves over the land as a sheet. In practice, however, application uniformity is reduced as water is channeled by depressions in the land surface. Contour ditch irrigation is more thoroughly discussed by Hart et al. (1980).

7.3   Delivery Systems for Surface Irrigated Farms

Delivery systems for surface irrigated farms convey water from the farm water source to field(s) in open canals and/or pipelines. Delivery systems may include structures for measuring and regulating flow, controlling head (pressure) and erosion, and diverting water into basins, borders, and furrows. The capacity of a delivery system must be sufficient to deliver the required amount of water to any point in the field whenever it is needed. Delivery systems should be convenient to operate and maintain and be economically justified by the returns expected from the crops to be grown.

7.3.1   Open Channel Delivery Systems

Lined canals and unlined ditches are widely utilized for on-farm conveyance of irrigation water. Where land has been graded to a uniform plane, canals and ditches are normally placed in a straight line across the upper edge of the field being irrigated. On undulating terrain, they follow the general contour of the field to obtain a uniform channel slope.

7.3.1a   Unlined Ditches

Unlined ditches are popular because of their low capital cost and ease of construction. They are best suited to relatively flat lands with cohesive soils with low infiltration capacities. On steep lands or unstable soils, ditch erosion can be a serious problem. In porous, high infiltration capacity soils, large amounts of water may seep through the bottom and sides of unlined ditches to create or intensify drainage problems and to reduce the conveyance efficiency. Costs for maintenance of unlined ditches are normally more than for lined canals or pipelines.

Weed control is essential for efficient unlined ditch operation. Weeds growing along ditch banks reduce the carrying capacity of the ditches and are a source of seeds which may infest irrigated fields. Even with lined canals it is desirable that a soil sterilant be applied to the subgrade before placing the lining to inhibit possible weed growth through cracks or small openings which might develop.

Unlined ditches may or may not be permanent. Often ditches supplying water to annual crops are installed at the beginning of the irrigation season and removed prior to harvest. Because of the relative ease with which ditches can be relocated, irrigation systems with unlined delivery ditches are more easily modified than other systems.

7.3.1b Lined Canals

Linings reduce seepage through the bed and walls of canals and, thus reduce the potential for drainage problems. Smooth surface linings reduce friction losses and increase the carrying capacity of the canal. Canals can have steeper slopes when they are lined with material such as concrete or asphalt because they are more resistant to erosion. Because lined canals can have steeper banks and smaller cross-sectional areas than unlined ditches, they occupy less land. Linings reduce water losses caused by burrowing animals, particularly when the canals are placed on elevated fills. Finally, maintenance costs are generally less for lined canals.

Canal linings may be classified as hard-surface linings, covered membranes, and soil sealant linings. The selection of a lining material depends largely on availability of materials, soil conditions, cross-section and length of the canal, and comparative annual costs. Decisions on whether to line a channel are normally made on the basis of average annual cost, including maintenance costs and the value of the water saved. Lining materials are thoroughly discussed by Kraatz (1977).

7.3.1c Open Channel Design

Open canal design involves determining the amount of water the canal must convey (i.e., its capacity), the flow velocity, and the canal slope, shape, and cross-section dimensions.

(i) Channel Capacity Equation 7.1 can be used to estimate canal capacity. Equation 7.1 is

$$Q = \frac{(K)(\text{DDIR})(A)}{(\text{HPD})(E_i)} \tag{7.1}$$

where

$\quad\quad Q$ = canal capacity (l/min, gpm);
DDIR = design daily irrigation requirement (mm/day, in/day);
$\quad\quad A$ = irrigated area supplied by canal or ditch (ha, acres);
HPD = hours per day that water is delivered
$\quad\quad E_i$ = irrigation efficiency including conveyance efficiency of canal or ditch (see Eqs. 2.7, 2.8, and 2.9 and Table 7.1) in percent
$\quad\quad K$ = unit constant. ($K = 16667$ for Q in l/min, DDIR in mm/day, and A in ha. $K = 45254$ for Q in gpm, DDIR in in/day, and A in acres.

(ii) Velocity of Flow The velocity of flow in a canal or ditch should be low enough to prevent erosion of the canal bed and sides, but high enough to prevent the deposition of suspended substances. A flow velocity in excess of 0.6 m/s (2 ft/s)

Table 7.1 Typical Delivery System Conveyance Efficiencies

Waterway	Efficiency (%)
Unlined canals	70–80
Lined canals	80–85
Unlined large laterals	80–85
Lined large laterals and unlined small laterals	85–90
Small lined laterals	90
Pipelines	100

Source: U.S. Department of the Interior, Report on the Water Conservation Opportunities Study (1978), U.S. Bureau of Reclamation and Bureau of Indian Affairs.

will normally minimize deposition. The maximum velocity that does not cause excessive erosion depends on the erodibility of the soil or lining material. Local experience is often the most reliable way of determining maximum allowable velocities for particular soils. The maximum allowable velocities for lined canals and unlined ditches listed in Table 7.2 can be used when local experience is not available.

Table 7.2 Limiting Velocities for Essentially Straight Canals After Aging

	Velocity			
	Clear	Water	Water Transporting Colloidal	Silts
Material	m/s	ft/s	m/s	ft/s
Fine sand, colloidal	0.46	1.50	0.76	2.50
Sandy loam, noncolloidal	0.53	1.75	0.76	2.50
Silt loam, noncolloidal	0.61	2.00	0.91	3.00
Alluvial silts, noncolloidal	0.61	2.00	1.07	3.50
Ordinary firm loam	0.76	2.50	1.07	3.50
Volcanic ash	0.76	2.50	1.07	3.50
Stiff clay, very colloidal	1.14	3.75	1.52	5.00
Alluvial silts, colloidal	1.14	3.75	1.52	5.00
Shales and hardpans	1.83	6.00	1.83	6.00
Fine gravel	0.76	2.50	1.52	5.00
Graded loam to cobbles when noncolloidal	1.14	3.75	1.52	5.00
Graded silts to cobbles when colloidal	1.22	4.00	1.68	5.50
Coarse gravel, noncolloidal	1.22	4.00	1.83	6.00
Cobbles and shingles	1.52	5.00	1.68	5.50

Source: G. O. Schwab, R. K. Frevert, T. W. Edminster, and K. K. Barnes, *Soil and Water Conservation Engineering*, copyright © 1981, John Wiley & Sons, Inc., p. 295. Reprinted by permission of John Wiley & Sons, Inc.

Table 7.3 Side Slopes Ratios (Horizontal:Vertical) for Unlined Open Ditches

Soil	Shallow Ditches (with depths up to 1.2 m or 4 ft)	Deep Ditches (with depths over 1.2 m or 4 ft)
Peat and muck	vertical	0.25:1
Stiff (heavy) clay	0.5:1	1:1
Clay or silt loam	1:1	1.5:1
Sandy loam	1.5:1	2:1
Loose sandy	2:1	3:1

Source: G. O. Schwab, R. K. Frevert, T. W. Edminster, and K. K. Barnes, *Soil and Water Conservation Engineering,* copyright © 1981, John Wiley & Sons, Inc., p. 293. Reprinted by permission of John Wiley & Sons, Inc.

(iii) *Channel Shape* The most hydraulically efficient cross-sectional shape for an open channel is a semicircle. Since the sides of a semicircle are nearly vertical, this shape is not practical for unlined ditches and complicates the construction of lined canals. Thus, in general, the use of semicircular cross sections are limited to precast concrete and preshaped metal or plastic flumes. The most efficient trapezoidal section is the half-hexagon, but the side slopes of 0.58 (horizontal) to 1.0 (vertical) are too steep for most channels. Recommended side slopes for several types of lined canals and unlined ditches are given in Table 7.3. It has been recommended that the bottom width of lined trapezoidal canals be from half to equal the depth of flow.

(iv) *Channel Dimensions* Once the capacity, lining material, slope and the shape of the canal or ditch are known, channel dimensions can be determined. Equation 7.2, the Manning equation, and Eq. 7.3, the continuity equation, are used to relate flow velocity, capacity, and channel slope, shape, and cross section dimensions.

$$V = \frac{K}{n} R^{2/3} S^{1/2} \tag{7.2}$$

$$Q = AV \tag{7.3}$$

where

V = flow velocity (m/s, ft/s);
n = Mannings roughness coefficient (from Table 7.4);
R = hydraulic radius from Table 7.5 (m, ft);
S = slope (m/m, ft/ft);
A = cross-sectional area of canal perpendicular to flow from Table 7.5 (m², ft²);
Q = capacity of the channel (l/min, gpm);
K = unit constant. (K = 1.00, V is in m/s, and R is in m.
K = 1.49, V is in ft/s, and R is in ft).

Table 7.4 Typical Values of *n* in Eqs. 7.2 and 7.4 for Various Surfaces

Surface	Best	Good	Fair	Bad
Uncoated cast-iron pipe	0.012	0.013	0.014	0.015
Coated cast-iron pipe	0.011	0.012[a]	0.013[a]	
Commercial wrought-iron pipe, black	0.012	0.013	0.014[a]	0.015
Commercial wrought-iron pipe, galvanized	0.013	0.014	0.015	0.017
Smooth brass and glass pipe	0.009	0.010	0.011	0.013
Smooth lockbar and welded "OD" pipe	0.010	0.011[a]	0.013[a]	
Riveted and spiral steel pipe	0.013	0.015[a]	0.017[a]	
Vitrified sewer pipe	$\begin{cases} 0.010 \\ 0.011 \end{cases}$	0.013[a]	0.015	0.017
Common clay drainage tile	0.011	0.012[a]	0.014[a]	0.017
Glazed brickwork	0.011	0.012	0.013[a]	0.015
Brick in cement mortar; brick sewers	0.012	0.013	0.015[a]	0.017
Neat cement surfaces	0.010	0.011	0.012	0.013
Cement Mortar surfaces	0.011	0.012	0.013[a]	0.015
Concrete pipe	0.012	0.013	0.015[a]	0.016
Wood stave pipe	0.010	0.011	0.012	0.013
Plank Flumes:				
Planed	0.010	0.012[a]	0.013	0.014
Unplaned	0.011	0.013[a]	0.014	0.015
With battens	0.012	0.015[a]	0.016	
Concrete-lined channels	0.012	0.014[a]	0.016[a]	0.018
Cement-rubble surface	0.017	0.020	0.025	0.030
Dry-rubble surface	0.025	0.030	0.033	0.035
Dressed-ashlar surface	0.013	0.014	0.015	0.017
Semicircular metal flumes, smooth	0.011	0.012	0.013	0.015
Semicircular metal flumes, corrugated	0.0225	0.025	0.0275	0.030
Canals and Ditches:				
Earth, straight and uniform	0.017	0.020	0.0225[a]	0.025
Rock cuts, smooth and uniform	0.025	0.030	0.033[a]	0.035
Rock cuts, jagged and irregular	0.035	0.040	0.045	
Winding sluggish canals	0.0225	0.025[a]	0.0275	0.030
Dredged earth channels	0.025	0.0275[a]	0.030	0.033
Canals with rough stony beds, weeds on earth banks	0.025	0.030	0.035[a]	0.040
Earth bottom, rubble sides	0.028	0.0275	0.033[a]	0.035
Natural Stream Channels:				
(1) Clean, straight bank, full stage, no rifts or deep pools	0.025	0.0275	0.030	0.033
(2) Same as (1), but some weeds and stones	0.030	0.033	0.035	0.040
(3) Winding, some pools and shoals, clean	0.033	0.035	0.040	0.045
(4) Same as (3), lower stages, more ineffective slope and sections	0.040	0.045	0.050	0.055
(5) Same as (3), some weeds and stones	0.035	0.040	0.045	0.050
(6) Same as (4), stoney sections	0.045	0.050	0.055	0.060
(7) Sluggish river reaches, rather weedy or with very deep pools	0.050	0.060	0.070	0.080
(8) Very weedy reaches	0.075	0.100	0.125	0.150

[a] Values commonly used for design.

Source: V. E. Hansen, I. O. Israelsen, and G. Stringham, *Irrigation Principles and Practices*, copyright © 1979, John Wiley & Sons, Inc., pp. 238–239. Reprinted by permission of John Wiley & Sons, Inc.

Table 7.5 Shape Parameters and Equations for Computing Cross-Sectional Area and Hydraulic Radius for Rectangular, Trapezoidal, and Triangular Canals

Shape	Parameters to be measured	Area, A (l^2)	Hydraulic Radius R (l)
Rectangular (y, b)	b, y	by	$\dfrac{by}{b + 2y}$
Trapezoidal (y, b, z)	b, y, z	$(b + zy)y$	$\dfrac{(b + zy)y}{b + 2y\sqrt{1 + z^2}}$
Triangular (y, z)	y, z	zy^2	$\dfrac{zy}{2\sqrt{1 + z^2}}$

Equation 7.2 is substituted into Eq. 7.3 to obtain Eq. 7.4. Equation 7.4 is

$$Q = \frac{KK_1}{n} AR^{2/3}S^{1/2} \tag{7.4}$$

where

K_1 = unit constant. (K_1 = 60,000 when Q is in l/min, A is in m^2, and R is in m. K_1 = 449 when Q is in gpm, A is in ft^2, and R is in ft).

The following example demonstrates the use of Eq. 7.4.

EXAMPLE 7.1 Determining Open Canal Dimensions

Given:
• $Q = 1500$ gpm
• triangular-shaped canal constructed in a sandy loam soil
• $S = 0.50$ percent

Required:
Canal dimensions

Solution:
$V = 2.5$ ft/s   (from Table 7.2)
$N = 0.03$ (from Table 7.4)
$Z = 1.5$ (from Table 7.3)

From Table 7.5 for triangular-shaped ditches

$$A = Zy^2 = 1.5y^2$$

$$R = \frac{Zy}{2\sqrt{1 + z^2}} = \frac{1.5y}{2\sqrt{1 + 1.5^2}} = 0.42y$$

$$Q = \frac{667}{n} AR^{2/3}S^{1/2}$$

$$1500 = \frac{667}{0.030}(1.5y^2)(0.42y)^{2/3}(0.005)^{1/2}$$

$$Y = (1.14)^{3/8} = 1.05 \text{ ft}$$

Checking velocity

$$A = 1.5(1.05)^2 = 1.66 \text{ ft}^2$$

$$V = \frac{Q}{A} = \frac{(1500 \text{ gpm})(1 \text{ cfs}/448.8 \text{ gpm})}{1.58 \text{ ft}^2} = 2.02 \text{ ft/s}$$

V is okay since $2 \leq V \leq 2.5$ ft/s.

7.3.1d Inlet Structures

Canal/ditch inlets are structures that direct the flow of supply pipes from turnouts and pumps into an open channel (division boxes, which are described in the next section, direct flow into two or more channels). Flow regulation and debris removal (see Section 7.3.2b) are normally done upstream of the inlet structure. Inlet structures often include weirs, orifices, or flowmeters for flow measurement (flow measuring devices for open channels are discussed in Chapter 8).

7.3.1e Conveyance Structures

Conveyance structures are needed where flows must be carried under or across draws (topographic depressions), roads, railways, other water courses or other permanent obstructions. Conveyance structures, in addition to the canal itself, include inverted siphons, culverts, flumes, and elevated ditches.

(i) Inverted Siphons Inverted siphons are used to convey water under permanent structures and topographic depressions (see Figure 7.2). They consist of relatively short reinforced concrete inlet and outlet sections, connected by comparatively long barrels. The inlet and outlet sections are transitions between the canal and the circular barrel of the siphon. The barrels are usually constructed of either precast concrete pipe, cast in place monolithic reinforced concrete, or steel pipe. Wood-stave barrels were sometimes used in early construction. Siphon design is described in Aisenbrey et al., (1978).

(ii) Culverts Culverts are installed to carry canal flows under roads, railways, or other canals. Cross sections may be rectangular, square, or circular. Barrels may be single or multiple units. Transition sections of relatively simple design are usually adequate at the inlets and outlets (see Schwab et al., 1981).

(iii) Flumes Flumes are artificial troughs built on or above ground level to carry water along routes where other types of conveyance would be more expensive. Flumes usually have open rectangular or semicircular cross sections and

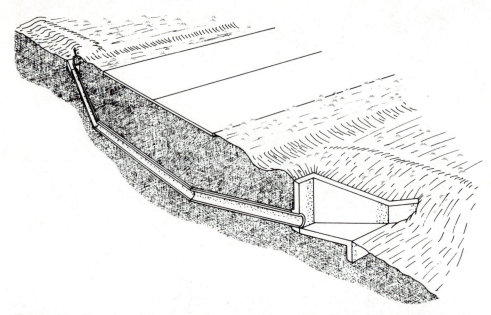

Figure 7.2 A section of a concrete inverted siphon. (From Section 15, Chapter 3 of the *SCS National Engineering Handbook*, 1967.)

are commonly built with timber, wood staves, sheet metal, or concrete. Flumes are designed using the procedures in Section 7.3.1c.

(iv) Elevated Canals and Ditches Elevated canals/ditches are open channels built on compacted earth fill to convey water across shallow depressions. They are sometimes less costly than flumes, siphons, or pipelines.

7.3.1f Control Structures

Control structures are included in open canal/ditch delivery systems to regulate velocity, head, and the quantity of water released into distribution laterals, basins, borders and furrows. The number and types of structures required depends on the type of canal or ditch, the slope, and the layout of the irrigation system. Detailed designs for the control structures described in the following sections can be found in Aisenbrey et al., (1978).

(i) Division Boxes Division boxes direct or divide flow from a supply pipe or channel between two or more distribution laterals. A fixed proportional flow divider-type division box is shown in Figure 7.3*a*. Other division boxes have movable splitters to change the flow proportions. Divisors that give accurate proportions often divide the flow at a control section where super critical flow exists, such as at the nappe of a free overfall. Flow can be divided accurately without creating super critical flow if (a) the approach channel is long and straight for at least 5 to 10 m (15 to 30 ft) upstream, so that flow approaches the divisor in

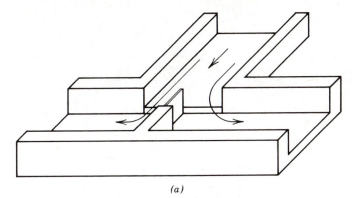

(a)

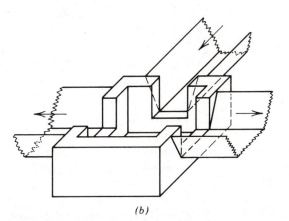

(b)

Figure 7.3   Division boxes with (*a*) a fixed proportional flow divider and (*b*) weir-type overflow outlets.

parallel paths without cross currents; (b) there is no backwater effect that would favor one side or the other; and (c) the flow section of the structure is of uniform roughness (Kruse et al., 1980). Detailed divisor designs are described by Kraatz and Mahajan (1975).

After water enters the division box shown in Figure 7.3*b*, it flows through the outlet gates and into the distribution laterals. Flashboards and weirs are also used as outlets. The width of outlet openings are chosen to supply the desired flow to each lateral. Once division boxes are constructed, flow is regulated by adjusting gate settings (openings), removing and adding flashboards, and varying the height of the weir crest.

Flashboards and weirs do not control flow as closely as gates. Flow over flashboards and weirs is proportional to the depth of water above the flashboards/ weir crest to the 1.5 power and is proportional to the depth of water above the centerline of slide gate openings to the 0.5 power. Thus, a change in water depth causes a larger change in flow when flashboards or weirs are used than when slide

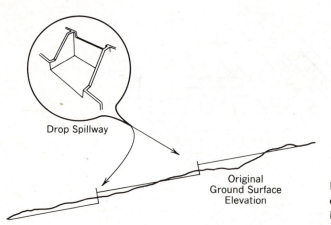

Drop Spillway

Original
Ground Surface
Elevation

Figure 7.4 A series of
drop structures along an
irrigation canal or ditch.

gates are used. Floating trash that passes over flashboards and weirs can, however, be a problem in a gated structure, as jetting of water through gates can pull debris below the water surface to obstruct the gate opening.

(ii) Drops Drop structures are needed in canals and ditches to convey water down steep slopes without erosive velocities. This is accomplished by subdividing the slope into several reaches with relatively flat slopes and constructing a drop structure at the end of each reach to lower water abruptly into the next reach. Water is conveyed down the slope in the stepwise manner shown in Figure 7.4. Although many different types of drop structures are used, most include an inlet section, a vertical or inclined drop, a stilling pool or other means of dissipating energy, and an outlet section for discharging water into the next reach. Several different drop structures are illustrated in Figure 7.5. See Ainsenbrey et al., (1978) for design details.

Kruse et al., (1980) recommend that drop heights in conveyance canals and ditches be limited to a maximum of 0.6 to 1 m (2 to 3 ft) and that drop height in distribution laterals be less than 15 to 30 cm (6 to 12 in). They also recommend that where the distance between drops is 100 m (300 ft) or more, the crest elevation should not be lower than the bottom of a stable ditch 100 m (300 ft) upstream. For reach lengths less than 100 m (300 ft), Kruse et al., (1980) recommended that the crest elevation not be lower than the sill of the next upstream structure nor less than 10 cm (4 in) above the apron of the next upstream structure.

(iii) Chutes Chutes are lined, high-velocity, open channels. They have an inlet, a steep-sloped section of lined canal where the elevation change occurs, a stilling pool or other energy dissipation device, and an outlet section. Chutes can be used on short, steep channel reaches and where drop structures would be so close together that a lined canal section is more practical. Chutes are expensive and are designed individually. Detailed procedures for designing various types of chutes are presented in Aisenbrey et al., (1978).

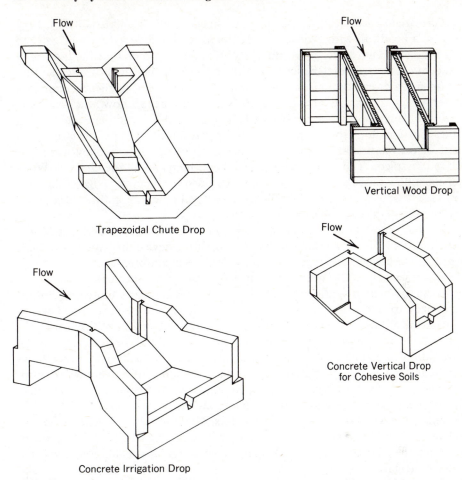

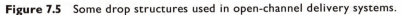

Figure 7.5 Some drop structures used in open-channel delivery systems.

(iv) **Relifts** Relifts are used when the water supply is not high enough to provide water to all areas of the farm. Relifts include a pump for lifting water from a canal or ditch to a higher elevation channel. A pump sump is located in the lower channel and a stilling basin in the upper canal or ditch.

(v) **Checks** Checks are permanent or portable structures placed in canals or ditches to control the upstream water level. They are used when the water level in canals and ditches must be raised above the normal depth of flow to provide head for operating outlets. There may or may not be flow past a check. When there is no flow past them, checks act as dams that confine water release to the area along the canal or ditch being irrigated. One or more sections along a canal or ditch can be irrigated when water is able to flow past the checks.

Permanent checks can be used in either lined canals or unlined ditches. They may have a headwall with grooves for flashboards or a slide gate (sluice gate) to control the upstream water level. Booher (1974) recommends that the bottom of permanent check structures be level with the bottom of the canal or ditch so that when open, water can drain from the channel. Booher (1974) further recommends that permanent checks be large enough to keep resistance to flow at a minimum.

Portable checks can be removed when an irrigation is complete and reset at another location along the canal or ditch to irrigate another area of the farm. They consist of canvas, metal, or plastic or rubber sheeting supported by a pipe or wooden crosspiece. The crosspiece is laid across the banks of the channel. When water must flow past the check, a notch is provided in the upper edge of the dam below the crosspiece to serve as a spillway. The elevation of the spillway is adjusted by rolling the material around the crosspiece.

Various types of semiautomatic check structures have been developed. These structures usually include a timer or other automating device that causes the check to open or close when an irrigation is completed. Semiautomatic checks must be manually reset before the next irrigation can begin. Semiautomatic checks that open at the end of an irrigation are called drop-open checks, and those that close at the end of an irrigation are drop-closed checks. With drop-open checks, all checks are set in the closed position manually prior to water being introduced into the canal or ditch. As water enters the channel, it backs up behind the first check. When the irrigation is complete, the first check opens automatically and water flows downstream and backs up behind the next check. A major disadvantage of drop-open checks is that if a check fails to open, lands downstream are not irrigated. With drop-closed checks, all checks are manually set in the open position before water enters the canal or ditch. As water enters, it flows to the downstream end of the channel until the downstream-most area has been irrigated. The downstream-most check then closes automatically and the next area upstream along the canal or ditch is irrigated.

Checks that reset automatically have also been developed. These devices include a programmable controller, and pneumatic or hydraulic cylinders that open, close, and adjust gates. A series of checks are opened and closed sequentially according to the desires of the irrigator (as programmed into the controller) until all areas along the channel have been irrigated. The controller automatically resets the gates at the appropriate time. Irrigation can proceed downstream from the upstream end of the channel or upstream from the downstream end.

(vi) Check-Drops   A check-drop structure functions as both a check and a drop. Flashboards mounted in a headwall provide upstream head control, while an abrupt drop in channel bottom elevation from the upstream to downstream sides of the flashboards allows a nonerosive flow velocity to be maintained in the channel.

7.3.1g Outlets

Outlets are devices for releasing the desired flow of water into distribution laterals, basins, borders, and furrows. They should be easy to operate and should protect from erosion the canal or ditch and the surface of the field at the point of release. Outlets from open channels include turnouts, siphon tubes, and spiles.

(i) Turnouts   Turnouts similar to those in Figure 2.1 are constructed in the banks of farm canals and ditches to provide and control flow to basins, borders, and distribution laterals. They usually have removable flashboards or a circular or rectangular slide gate to regulate flow. Drop-open gates (similar to drop-open checks) are utilized in semiautomatic turnouts. Programmable controllers and slide gates with hydraulic or pneumatic cylinders allow turnouts to be opened, closed, and modulated automatically.

(ii) Siphon Tubes   Siphon tubes are curved aluminum or plastic pipes that are laid over the bank of delivery canals and ditches to deliver water to borders and furrows (see Figure 7.1). Water flows into the tube, is pulled (siphoned) over the bank of the delivery channel, and delivered into borders and furrows when there is sufficient operating head and the tube is properly positioned and full of water (primed). The operating head when the outlet end of the tube is submerged, is the difference in elevation between the water surfaces at the entrance and outlet ends of the tube. For a free-flowing tube, the operating head is the difference in elevation between the water surface at the tubes entrance and the center of its outlet end.

Siphon tubes are available in diameters ranging between 13 and 150 mm (0.5 and 6 in) and lengths of 1.2 to 3.0 m (4 to 10 ft). The discharge of a siphon tube depends on its diameter, length, and inside roughness as well as the number and degree of bends and the operating head. Figure 7.6 can be used to estimate the flow from standard aluminum or plastic siphon tubes.

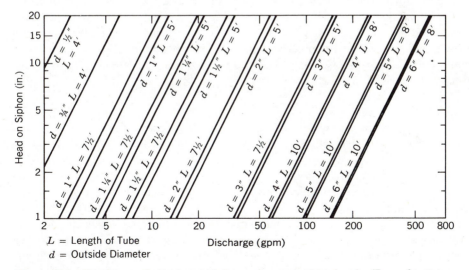

L = Length of Tube
d = Outside Diameter

Figure 7.6   Discharge of aluminum siphon tubes at various heads. (From Section 15, Chapter 5 of the *SCS National Engineering Handbook,* 1984.)

(iii) Spiles Spiles are 25 to 64 mm (1.0 to 2.5 in) diameter pipes that are permanently installed in the bank of level or slightly sloping canals and ditches to deliver water to furrows (see Figure 7.1). When flow is controlled by regulating head (usually with check structures), spiles must be carefully installed so that the same amount of flow is delivered to each basin, border, or furrow. Gates on each spile are also used to regulate flow.

7.3.2   Low-Head Pipelines

Water is often conveyed from the water source and distributed to basins, borders, and furrows in low-head pipelines. These pipelines essentially eliminate evaporation and seepage losses, can be laid on nonuniform grades, and allow water to be conveyed uphill against the land slope. They allow the area that would otherwise be occupied by open ditches/canals to be planted to crops. They also reduce maintenance requirements, make water control easier, and eliminate ditch bank weed problems.

Surface-irrigated farms may have permanent, semiportable, or portable low-head pipelines. In permanent pipelines, water is conveyed from the water source to the fields in buried supply pipes and distributed to basins, borders, and furrows in buried distribution pipes. Semiportable pipelines have buried supply lines and pipes laid on the ground surface to distribute water. In a fully portable system, surface pipes are used for both supply and distribution. Buried pipes are made of either PVC, steel, concrete, wrapped aluminum, or asbestos-cement. PVC, and aluminum pipe in 6.1, 9.1, or 12.2 m (20, 30, or 40 ft) lengths with quick couplers or hose are used for portable surface pipes.

Low-head pipelines have an inlet, one or more outlets, and various structures for surge protection, air relief, flow measurement, head control, and debris and sand removal. Pressure relief, air release, and vacuum relief valves that are used for pressurized pipelines are also used with low-head pipelines. Information presented in Chapter 5 for pressurized pipelines applies to low-head pipelines. Friction loss is computed using Eqs. 5.13, 5.14, 5.15, 5.16, and 5.17, and pipes are selected using the procedures in Section 5.3.2f.

7.3.2a   Inlets

Inlet structures are needed to carry water from the water source into low-head pipelines. Inlets normally include gates or valves for on–off service, flow regulation, and head control; and debris racks to prevent trash and animals from entering the pipeline. They should also have a flow measurement device such as a weir, flume, orifice, or propeller-type meter (flow measurement devices are described in Chapter 8); and a screen to remove weed seeds, moss, etc. When the source water carries excessive amounts of suspended material, inlets should have an enlarged section, called a *sand trap*, to slow the water and allow suspended particles to settle out. Such inlets should be constructed to allow easy cleaning. Because most inlets are open to the atmosphere, they normally provide surge protection and air release.

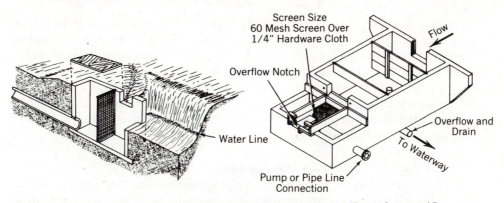

Figure 7.7   Some gravity inlets for buried low-head pipelines. (From Section 15, Chapter 3 of the *SCS National Engineering Handbook*, 1967.)

Inlets may be classified as *gravity inlets* or *pump stands*. Gravity inlets that connect the pipeline directly to the water source are used when the water surface elevation of the source is sufficient to allow gravity flow into the pipeline and to provide the pressure needed at every pipeline outlet. Typical gravity inlets for buried low-head pipelines are shown in Figure 7.7.

Example pump stands for low head pipelines are shown in Figure 7.8. Pump stands must be high enough to provide the pressure needed at every pipe outlet yet low enough to protect the pipeline (by allowing water to overflow when there is excessive inlet pressure). Vent pipes similar to the one in Figure 7.8*b* are used when

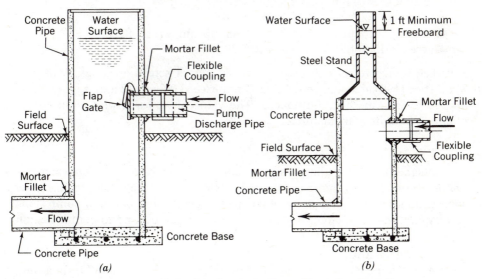

Figure 7.8   Some pump stands for low-head pipelines. (From Portland Cement Association, 1969.)

the head requirements of the pipeline are high (but less than 6 m or 20 ft). These vent pipes should be large enough to limit velocities to 3 m/s (10 ft/s) when the full pump outflow is flowing through them. Similarly, the downward velocity in pump stand barrels should not exceed 0.6 m/s (2 ft/s). There must also be a flexible coupling between the pump and stand to prevent the transmission of vibrations to the stand or pipeline. When the pipeline is higher than the pump the usual practice is to carry the discharge pipe over the top of the pump stand to prevent backflow. Otherwise, a check valve may be required.

7.3.2b Debris Racks and Screens

Trash carried by irrigation water that plugs siphon tubes and spiles as well as pipe gates and valves is removed by placing debris racks on the upstream side of gravity inlets. These racks also prevent animals from entering low-head pipelines. Screens are required to remove finer debris such as weed seeds and moss. Several different types of screens and trash racks for gravity inlets have been developed (Bergstrom, 1961; Bondurant and Kemper, 1985; Courtland et al., 1956; and Pugh and Evans, 1964). Trash is prevented from entering pump stands by placing screens around the inlet of the pump suction line.

Head loss through debris racks and screens is computed with the following equation.

$$h_1 = K_r \frac{V^2}{2g} \tag{7.5}$$

where

h_1 = head loss through trash rack or screen (m, ft);
K_r = head loss coefficient (from Figure 7.9);
V = velocity through trash rack or screen openings (m/s, ft/s);
g = acceleration of gravity (9.81 m/s^2, 32.2 ft/s^2).

V is the average velocity through the trash rack or screen openings. Walker and Skogerboe (1987) recommended that V not exceed 0.6 m/s (2 ft/s). Figure 7.9 also applies to wire mesh fabric screens. The value of A_r for a wire mesh fabric is obtained from its manufacturers.

7.3.2c Sand Traps

A sand trap is a settling basin for removing sand from irrigation water. They are incorporated into pump stands and gravity inlets by placing pipeline inverts some distance above the bottom of the inlet structure. The barrel of pump stands/gravity inlets must be large enough to keep velocities low and to permit cleaning.

7.3.2d Standpipes

A standpipe is a vertical pipe or box extending above ground from a buried pipeline (see Figure 7.10). Standpipes provide surge protection, air release, flow regulation, and head control in buried low-head pipelines. In additional to pump

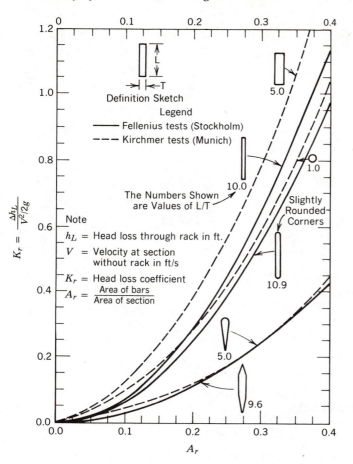

Figure 7.9   Trash rack or screen head loss coefficients for different bar shapes. (From U.S. Army, Corps of Engineers, Waterways Experiment Station, 1959.)

stands, the main types of standpipes are gate stands, overflow stands, and float-valve stands.

   (i) Gate Stands   Gate stands are installed where branch lines take off from the mainline. Each outlet of a gate stand is equipped with a slide gate or gate valve. Gate stands must be large enough to accommodate the gates to be used and to permit access for maintenance and repair. They are used to control flow into branch lines or to increase pressure upstream.

   (ii) Overflow Stands   Overflow stands include an internal baffle that allows a constant pressure to be maintained at each outlet upstream of the stand. An optional slide gate is sometimes included to bypass water through the baffle to obtain lower upstream pressures. Overflow stands also create a drop in the

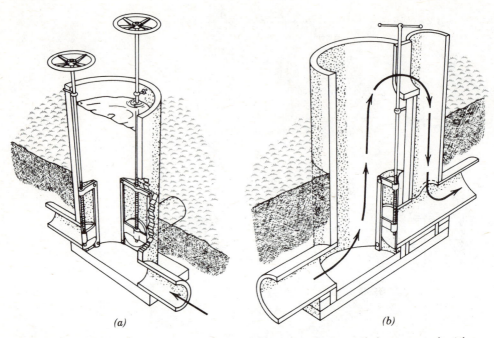

Figure 7.10 Examples of (*a*) a gate stand and (*b*) an overflow stand that are used with low-head pipelines. (From Section 15, Chapter 3 of the *SCS National Engineering Handbook*, 1967.)

hydraulic gradient that limits downstream pipeline pressures. This type of structure is not used in pipelines constructed on flat or very slight slopes.

(iii) *Float-Valve Stands*   The float-valve stand in Figure 7.11 automatically controls the head on the downstream pipeline. The primary component of the structure is a float valve that gradually closes as the water level in the stand rises and opens when the water level drops. These valves are commercially available in sizes ranging from 100 to 600 mm (4 to 24 in) with capacities from 2 to 2000 l/s (300 to 30,000 gpm). Stands should be at least 760 mm (30 in) in diameter and provide 300 to 600 mm (1 to 2 ft) of freeboard.

Float-valve stands are useful on steep slopes and are usually installed at vertical intervals of about 3 m (10 ft). A series of automatic float valves along a pipeline may require careful design to prevent unsable valve action and pressure surges (Kruse, et al., 1980).

7.3.2e   Outlets

Outlets are devices that release water from low-head pipelines into basins, borders, and furrows. For buried pipelines, they consist of a riser pipe and one or more valves to control the flow. Outlets should release water into fields without causing erosion. They may include alfalfa or orchard valves, various types of hydrants, and gated pipe.

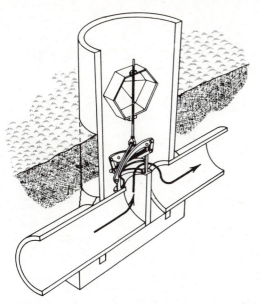

Figure 7.11   Section of a float-valve stand for a low-head pipeline. (From Section 15, Chapter 3 of the *SCS National Engineering Handbook*, 1967.)

   (i)   *Alfalfa Valves*   Alfalfa valves, like the one in Figure 7.12, are used to release relatively large flows into basins, borders, and several furrows (by releasing water into a head ditch or gated pipe). They are mounted on top of a riser 7 to 8 cm (3 in) below the ground surface. Water flows from all sides of an alfalfa valve when it is open.
   The discharge from an alfalfa is computed with the following equation.

$$Q = (K)(C_d)(A)(h)^{0.5} \tag{7.6}$$

$$A = \pi D d$$

where

Q = discharge of valve (l/min, gpm);
C_d = discharge coefficient;
A = area of opening (cm^2, in^2);
D = diameter of valve opening (cm, in);
d = vertical distance of valve opening (cm, in);
h = head loss across the valve, that is, difference between head on upstream side of the valve and the depth of water above the valve (cm, in);
K = unit constant. ($K = 2.66$ for Q in l/min, A in cm^2, and h in cm. $K = 7.22$ for Q in gpm, A in in^2, and h in in.)

For an alfalfa valve C_d is equal to 0.7 (Walker and Skogerboe, 1987). Booher (1974) recommends that the head on the upstream side of the valve be approximately 30 cm (1 ft) above the ground surface. Thus, the size of the valve should be chosen to limit h to 30 cm (1 ft) or less.

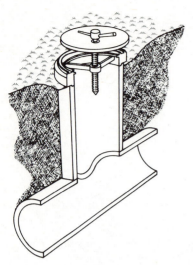

Figure 7.12 Section of an alfalfa valve for a low-head pipeline. (From Section 15, Chapter 3 of the *SCS National Engineering Handbook*, 1967.)

(ii) Orchard Valves An orchard valve is located inside a riser pipe as shown in Figure 7.13. Because the opening size of orchard valves is smaller than the diameter of the riser, they are generally used where smaller flows (than those delivered through alfalfa valves) are required. The capacity of an orchard valve is computed using Eq. 7.6 with $C_d = 0.6$ (Walker and Skogerboe, 1987).

(iii) Open-Pot Hydrant In this type of outlet the riser extends far enough above ground for two or more slide-gate openings to be installed at or slightly above ground level. An orchard valve, located in the riser below the slide gates (see Figure 7.14) is adjusted to maintain the water level 2 to 8 cm (1 to 3 in) above the

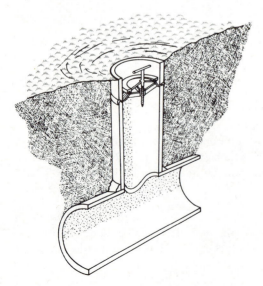

Figure 7.13 Section of an orchard valve for a low-head pipeline. (From Section 15, Chapter 3 of the *SCS National Engineering Handbook*, 1967.)

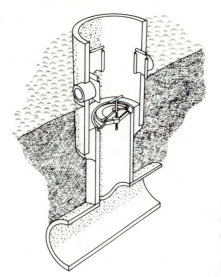

Figure 7.14 Open-pot hydrant with orchard valve on top of a riser pipe and three slide-gate outlets for a low-head pipeline. (From Section 15, Chapter 3 of the *SCS National Engineering Handbook*, 1967.)

slide gates. The slide gates, which are normally designed to limit exit velocities to 0.9 m/s (3 ft/s), are adjusted to provide the desired flow.

(iv) *Pot Hydrants*   In a pot hydrant (see Figure 7.15), the pot is capped, the slide gates are located on the outside of the pot (to allow gate adjustment), and the riser does not contain an orchard valve. Flow is controlled by adjusting pipeline pressure and slide-gate openings. Screw-type valves that break up the jet of water leaving pot hydrants and reduce the erosion hazard are sometimes used in lieu of slide gates. Pot hydrants are used for irrigating orchards and permanent crops

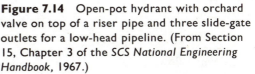

Concrete Cap

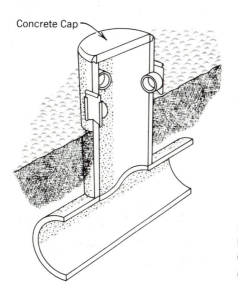

Figure 7.15 Section of a pot hydrant for a low-head pipeline. (From Section 15, Chapter 3 of the *SCS National Engineering Handbook*, 1967.)

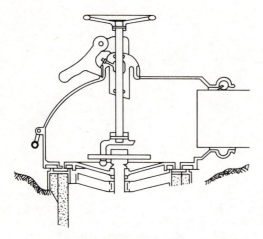

Figure 7.16 Cast aluminum portable hydrant for low-head pipelines.

where small flows are distributed to small furrows. Care must be taken to keep pipeline pressures below 30 to 60 cm (1 to 2 ft) and to use gates that seal tightly when closed.

(v) Overflow Pot Hydrants   These outlets are used on steep slopes. They are a combination of an open pot hydrant and overflow stand.

(vi) Portable Hydrants   A sketch of a portable hydrant made of cast aluminum and galvanized sheet metal are shown in Figures 7.16 and 7.17, respectively. These hydrants are placed over alfalfa valves to connect portable surface pipe to buried supply pipes. They typically have a mechanism for sealing the hydrant to the riser pipe and an exterior handle that connects to the handle of the alfalfa valve. When the hydrant is in place, the exterior handle is utilized to open and close the alfalfa valve. Portable hydrants may have a single outlet,

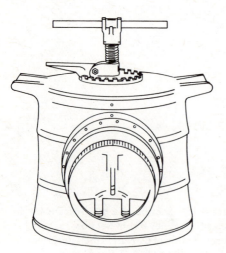

Figure 7.17 Sheet metal portable hydrant for low-head pipelines.

permitting flow in one direction only, or outlets on two sides. A friction or clamp joint is used to connect portable surface pipe to the hydrant.

   (vii) Pneumatic Valves   A pneumatic valve consists of an inflatable O-ring that is mounted in an alfalfa valve. When inflated with air, the tube forms an annular seal between the alfalfa valve seat and lid. Pneumatic valves are used with basins, borders, and furrows (usually in conjunction with gated pipes). Some pneumatic valves can be programmed (by the irrigator) to open and close as necessary to carry out an irrigation schedule.

   (viii) Water-operated Valves   Valves that utilize water for actuation are available for on–off service and modulating discharge into gated pipes and buried distribution laterals. Water from the pipeline is used to close the valve. A battery-powered, timer-activated, 3-way pilot valve is utilized to control valve opening and closing (Hart et al., 1980).

   (ix) Surge-Flow Valves   These devices, which were developed for surge irrigation systems, alternate water between two borders or two groups (sets) of furrows. They consist of a cast aluminum tee with either two independent valves that control flow from each outlet or a single valve that switches flow from one outlet to the other. The inlet of the tee is connected to the supply pipeline either directly or via an existing hydrant.

   Microprocessor-based control systems allow the irrigator to vary cycle times during the irrigation. Electrical power for the microprocessor and to operate the valve is obtained from batteries and/or solar panels.

   (x) Gated Pipe   Gated pipes are portable aluminum or PVC pipes with uniformly spaced outlets for releasing water into furrows (see Figure 7.1). Water flow from each gate is regulated by controlling the size of the gate opening. Gated pipe is usually manufactured in lengths of about 9 m (30 ft) and diameters ranging from 10 to 30 cm (4 to 12 in). Individual pipes are connected to a hydrant or to each other with friction or clamp couplings. A portable plug is installed at the end of the line. Flexible tubing, sometimes called lay-flat tubing, with outlet tubes is another type of gated pipe. Because flexible gated pipe is made of plastic, rubber, or canvas, it can be rolled up for easy moving and storage.

   Gated pipes must be designed to provide adequate operating pressure to each gate. Equations 5.13 through 5.17 are used for friction loss and pressure calculations in gated pipe. The U.S. Soil Conservation Service recommends a Hazen–Williams C of 144 for 9.1 m (30 ft) lengths of aluminum gated pipe and $C = 130$ for PVC pipe with gates. Erosive exit velocities resulting from high heads and small gate openings must be avoided. Butterfly valves and orifices may be installed at pipe joints to dissipate excess pressure or flexible tubes, called socks, can be attached to each gate to minimize erosion. The socks dissipate the energy of the high velocity streams flowing from the gates.

   Surge, pneumatic, and water-operated valves are used with gated pipes to automate furrow irrigation systems.

7.4   The Surface Irrigation Process

Surface irrigation is accomplished by causing water to flow over the land surface. Typically, water is diverted into the field from a supply ditch or pipe and flows behind a distinct wetting front over the soil surface. As flow occurs some water enters (i.e., infiltrates) the soil. Under normal conditions flow continues until the advancing wetting front reaches the opposite end of the field. Water then begins either to leave the field as surface runoff or to be stored on the surface if runoff is prevented by diking the downstream end of the field. Normally, water continues to be diverted into the field until the irrigation requirement has been supplied. After inflow ends, flow across the field usually continues, but the depth of flow decreases, beginning at the upstream end of field. When the depth of flow at the upstream end becomes zero, a recession or drying front is formed. This recession front moves downstream until it reaches the downstream end of the field or it meets a similar front moving upstream from the downstream end of the field. When no water remains on the surface the irrigation is complete.

The complete surface irrigation process, diagrammed in Figure 7.18, is divided into the advance, storage, depletion, and recession phases. These phases are named according to the most noticeable process that occurs during the phase.

7.4.1   Advance Phase

The advance phase begins when water is turned into the field and ends when water reaches the downstream end of the field. During advance a sharply defined water front with water on the inflow side of the front and dry field on the other side moves across the field.

The rate of advance decreases with time as the wetted area behind the water front increases. Typical relationships between the distance that water has advanced

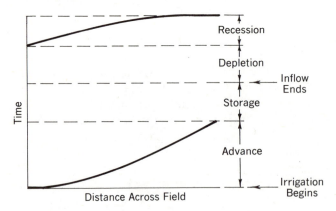

Figure 7.18   The complete surface irrigation process. (From Bassett *et al.*, 1980.)

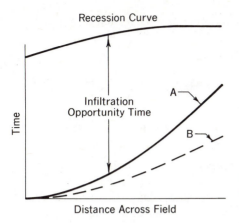
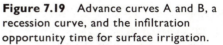

Figure 7.19 Advance curves A and B, a recession curve, and the infiltration opportunity time for surface irrigation.

from the upstream end of a "typical" field and time are illustrated in Figure 7.19. The advance represented by curve A in Figure 7.19 is slower than that represented by curve B. This occurs because either

1. the inflow rate to the field (stream size) is lower,
2. the field slope is flatter,
3. the intake (infiltration) rate of the soil is higher, and/or
4. the hydraulic roughness is greater for curve A than for curve B.

7.4.2 Storage Phase

The storage phase begins when the advance phase ends. It occurs only if inflow to the field continues after water has advanced to the downstream end of the field. The storage phase ends when inflow ends.

7.4.3 Depletion Phase

The depletion phase begins when the storage phase ends (i.e., when inflow ends) and ends when the depth of flow at the inflow end of the field becomes zero.

7.4.4 Recession Phase

Recession begins when the depletion phase ends. A "drying" front moves from the inflow to downstream end of the field. Recession continues until either the front reaches the end of the field or it encounters a receding front moving toward the inflow end of the field. The latter condition seldom occurs.

The position of the front moving from the inflow end of the field versus time during recession is shown in Figure 7.19. Like advance, the rate of recession depends upon the inflow rate, the slope of the field, the infiltration capacity of the soil, and the hydraulic roughness. Recession will tend to be most rapid when the inflow rate is low, the field slope is steep, the infiltration capacity is high, and/or the hydraulic roughness is small.

7.5 The Effectiveness (Quality) of Surface Irrigations

As discussed in Section 2.4.5d, the effectiveness of irrigation qualitatively describes the application efficiency, uniformity, and adequacy of irrigation. The concept of irrigation effectiveness is extremely useful in evaluating designs and management strategies. In the sections that follow, the efficiency, uniformity and adequacy of surface irrigations are discussed. Procedures for estimating these parameters (and hence the effectiveness of irrigation) are presented.

7.5.1 Uniformity

The uniformity of application describes how evenly an irrigation system distributes water over a field. Perfect (100 percent) uniformity means that the entire field receives an equal depth of water. Less than perfect uniformities imply that some areas of a field receive more water than other areas. This may result in "over irrigating" the crop in one portion of the field and under-watering the crop in another portion.

In order for the depth of infiltration to be identical across a field and for uniformities to be high, the infiltration opportunity time must be the same throughout the field. The infiltration opportunity time is the time interval during which water is available to enter the soil between the time it arrives at a point during the advance phase and departs during recession. The infiltration opportunity time is defined graphically in Figure 7.19 as the vertical distance between the advance and recession curves. High uniformities are possible only when this vertical distance is constant across the field. This is normally facilitated by a "flat" advance curve (i.e., by rapid advance of water across the field). Thus, highest uniformities can be expected on fields with steep slopes, low hydraulic roughness, low infiltration capacity soils, and/or when large stream sizes are used.

In furrow irrigation a constant wetted perimeter at every cross section along the furrow is also required for high uniformity. This is virtually impossible to achieve, since the wetted perimeter is dependent upon stream size, and stream size diminishes along the furrow as water is absorbed into the furrow. This difference is most noticeable between the up- and downstream ends of a furrow, and when there is little to no runoff from the end of graded furrows.

The uniformity of application can be evaluated using the Christiansen Uniformity Coefficient (C_u). The coefficient C_u is computed using Eq. 2.13. The following example illustrates a procedure for estimating C_u along an irrigation furrow.

EXAMPLE 7.2 **Computing the Uniformity of Irrigation Along an Irrigation Furrow**

Given:

Station from Head of Furrow	Depth of Infiltration at Each Station (in)
1	4.00
2	3.80
3	3.65
4	3.60
5	3.50
6	3.40

Required:

Uniformity of irrigation, C_u

Solution:

Quantifying parameters in Eq. 2.13.

$$X = (4.00 + 3.80 + 3.65 + 3.60 + 3.50 + 3.40)/6 = 3.66 \text{ in}$$

$$\begin{aligned}\sum |d| = &|4.00 - 3.66| + |3.80 - 3.66| + |3.65 - 3.66| + |3.60 - 3.66| \\ &+ |3.50 - 3.66| + |3.40 - 3.66| \\ = &\; 0.97\end{aligned}$$

$$C_u = \left(1.0 - \frac{0.97}{6(3.66)}\right)100 = 95.6 \text{ percent}$$

7.5.2 Application Efficiency

The application efficiency is defined as the amount of water that is beneficially used by the crop divided by the total amount of water applied. When leaching is neglected, beneficial use equals the amount of water stored in the root zone, and Eq. 2.12 becomes

$$E_a = \frac{RZ}{\forall} 100 \qquad (\text{when } L = 0) \tag{7.7}$$

$$RZ = \frac{D(\theta_{fc} - \theta_i)}{100} = \forall - DP - RO \tag{7.8}$$

$$\forall = \bar{Q}t/A \tag{7.9}$$

where

E_a = efficiency of application (percent);
RZ = amount of water stored in the root zone;
$\forall$ = total water applied;
D = depth of the root zone;

θ_{fc} and θ_i = volumetric water contents in percent at field capacity and prior to
             irrigation, respectively;
$\bar{Q}$ = average stream size during the irrigation;
t = duration of the irrigation;
DP = deep percolation;
RO = runoff.
A = area irrigated.

Runoff is water that flows from the downstream end of the field and is not reused.
      Maximum application efficiency (E_a = 100 percent) occurs when RZ equals
∀. Efficiency E_a is seldom 100 percent unless the field is severely underirrigated,
since some applied water runs off from the end of the field or becomes deep
percolation. Evaporation is usually neglected.
      The amount of runoff and deep percolation (and hence the application
efficiency, E_a) depend on the relationship between the time required for water to
advance to the end of the field (T_a) and the time required for soil at the downstream
end of the field to infiltrate the desired depth of water (T_I). The ratio of T_a/T_I can be
changed by changing either T_a or the desired application depth and thus T_I. The
desired depth cannot exceed the storage capacity of the soil. Time T_a increases
when

1. stream size is decreased,
2. length of run is increased,
3. hydraulic roughness is increased,
4. field slope is decreased, and/or
5. infiltration capacity of the soil is increased.

      Figure 7.20 illustrates how deep percolation, runoff, and application efficien-
cy vary with T_a/T_I. For a typical soil Figure 7.20 indicates that increasing T_a/T_I
decreases runoff while deep percolation increases. Application efficiency E_a increases from 0 to a maximum and then decreases slightly as T_a/T_I becomes larger.
      When stream sizes remain constant during the irrigation of graded borders
and furrows (are not cut back), there is a conflict between obtaining maximum

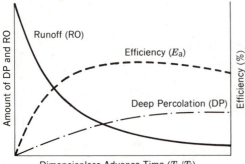

Figure 7.20 Surface runoff (RO),
deep percolation (DP), and
application efficiency (E_a) for a typical
surface irrigated soil. T_a is the time
to advance the end of the field; T_I is
the time to infiltrate the desired
depth of application.

uniformity and efficiency. A fast advance time usually results in a relatively high uniformity and a low efficiency. Conversely, high efficiencies are generally associated with slower advance times and lower uniformities. Thus, surface irrigation usually involves a compromise.

7.5.3 Adequacy of Irrigation

Runoff and deep percolation can be controlled and application efficiency maximized by not filling the soil to field capacity (i.e., by not allowing IOT to exceed T_I anywhere in the field). This practice may cause at least a portion of the field to be inadequately irrigated. The adequacy of irrigation is the percent of the field receiving sufficient water to maintain the quantity and quality of crop production at a "profitable" level. Since this definition requires crop, soil, and market conditions to be specified, alternative concepts have been used to describe the adequacy of irrigation. The storage efficiency, as defined by Eq. 2.16, is often used for surface irrigation.

   The following example problem illustrates a procedure for determining the effectiveness of a surface irrigation.

EXAMPLE 7.3 Computing the Effectiveness of a Surface Irrigation

Given:

- advance data in columns 1 and 2 of the following table
- the following cumulative infiltration—time graph
- furrow spacing = 3.5 ft
- stream size = 6 gpm
- furrow is 513 ft long
- water enters the furrow for 1012 min (16.9 hr)
- 4 in of water is required to fill the root zone to field capacity
- assume that recession is instantaneous when inflow to the furrow ends

Required:
1. E_a
2. E_s
3. C_u

   Solution Steps:
1. compute the infiltration opportunity time and depth of infiltration along the furrow
2. compute average depth of infiltration
3. compute depth of deep percolation
4. compute depth of application
5. compute depth of runoff
6. compute depth stored in the root zone
7. compute E_a
8. compute E_s
9. compute C_u

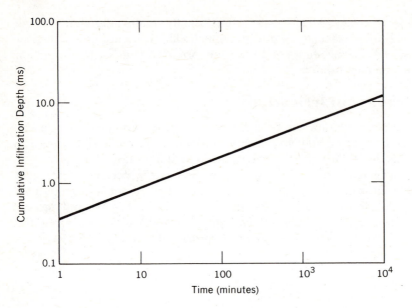

Solution Step 1

(compute IOT, and the depth of infiltration along furrow)

Distance Along Furrow (ft)	Minutes of Application	Advance Time (min)	IOT[a] min	Inches[b] Infiltrated
0	1012	0	1012	4.88
100	1012	22	990	4.84
200	1012	83	929	4.73
300	1012	180	832	4.53
400	1012	314	698	4.24
500	1012	482	530	3.82
513	1012	506	506	3.75

[a] IOT = Infiltration opportunity time = Col. 2 − Col. 3
[b] Inches of infiltration from infiltration curve for the IOT values in column 4

Solution Step 2

(Compute average depth of infiltration)

$$\bar{D} = \frac{\left(\dfrac{4.88 + 4.84}{2}\right)(100) + \left(\dfrac{4.84 + 4.73}{2}\right)(100) + \cdots + \left(\dfrac{3.82 + 3.75}{2}\right)13}{513}$$

$$= 4.52 \text{ in}$$

Solution Step 3
(Compute depth of deep percolation)

$$DP = \bar{D} - \text{depth to fill root zone}$$
$$= 4.52 - 4.0 = 0.52 \text{ in}$$

Solution Step 4
(Compute depth of application)

$$D_a = \frac{(6 \text{ gpm})(1012 \text{ min})}{(513 \text{ ft})(3.5 \text{ ft})} \left(\frac{12 \text{ in}}{\text{ft}}\right)\left(\frac{1 \text{ ft}^3}{7.48 \text{ gal}}\right)$$

$$= 5.43 \text{ in}$$

Solution Step 5
(Compute depth of runoff)

$$RO = D_a - \bar{D}$$
$$= 5.43 - 4.52 = 0.91 \text{ in}$$

Solution Step 6
(Compute depth stored in root zone)

$$D_s = \frac{(4)(400) + \left(\dfrac{4 + 3.82}{2}\right)100 + \left(\dfrac{3.82 + 3.75}{2}\right)13}{513}$$

$$= 3.98 \text{ in}$$

Solution Step 7
(Compute E_a)

$$E_a = 100\left(\frac{D_a - (DP + RO)}{D_a}\right) = 100\left(\frac{5.43 - (0.52 + 0.91)}{5.43}\right)$$

$$= 73.7\%$$

Solution Step 8
(Compute E_s)

$$E_s = 100\left(\frac{D_s}{D_n}\right) = 100\left(\frac{3.98}{4.00}\right) = 99.5 \text{ percent}$$

Solution Step 9
(Compute C_u) Using a procedure similar to Example 7.1

$$C_u = \left(1.0 - \frac{2.65}{5.13(4.52)}\right)100 = 88.6 \text{ percent}$$

7.5.4   Improving the Effectiveness of Surface Irrigation

Land smoothing, reuse of tailwater (i.e., reuse of water that runs off the downstream end of surface irrigated fields), cutback irrigation, and surge-flow irrigation can be employed to improve the effectiveness of surface irrigation. Land smoothing, which involves the use of soil removed from high spots on the land surface to fill depressions, improves effectiveness by increasing the uniformity of application. Tailwater reuse and cutback irrigation increase application efficiency by reducing runoff losses and, where the use of these practices allows larger stream sizes, deep percolation losses. Cutback irrigation is the practice of decreasing (cutting back) inflow to graded furrows after water has advanced to the downstream end of the field. Surge-flow irrigation, which is accomplished by cycling inflow to borders and furrows on and off, can improve both the uniformity of application and the irrigation efficiency in many soils.

7.5.4a   Land Smoothing

Land smoothing is the process of moving soil from high spots on the land surface to low spots to provide a more uniform plane for water flow. Land smoothing usually improves the uniformity of water application within basins, borders, and furrows.

The first step in land smoothing is to determine the depth and composition of the soil profile throughout the field. This will indicate the extent to which topsoil can be removed without reducing crop production. Cuts that expose cobblestones and other undesirable material can permanently reduce crop production. Exposure of the subsoil frequently, however, presents no problem, or may require only the application of nitrogen fertilizer (Marr, 1965).

Land smoothing requires an accurate topographic map to establish the location of surface drainways, supply lines, roads, and fields. When there are abrupt changes in slope and/or elevation it may be desirable and less expensive to subdivide the field into parcels of land and smooth each subunit separately (rather than smoothing the entire field). When subdividing the area, an effort to maintain rectangular-shaped fields should be made, since irregularly shaped fields are difficult to irrigate and cultivate.

Once field layout is complete a grid should be established in each subunit as per Example 7.4. The elevation at each grid point and either the eyeball or least-squares method can then be used to determine the grade across and along the field that minimizes the amount of material to be excavated and transported. The amount of material that must be cut or filled at each grid point is then computed. Because of compaction by machinery in fill areas and other factors, it may be necessary to adjust the elevation of the entire plane to obtain a ratio of the sum of the cuts to the sum of the fills of 1.2 to 1.5.

The following example demonstrates the eyeball and least-squares methods for making land smoothing calculations. A procedure for adjusting the elevation of the graded plane to obtain a ratio of cuts to fills of 1.2 to 1.5 is also included.

EXAMPLE 7.4   Land Smoothing Calculations

Given:

|← 100′ →| →| 50′ |← ↓

8.5[1]	9.2	9.8	10.3	50′ ↑
6.7	8.2	8.6	8.8		↑ N
				100′
7.0	7.8	7.4	8.3	↓

[1] All grid elevations are in ft.

Required:
- grade in north-south and west-east directions

- elevation at each grid point after the land has been smoothed and the elevation of the planes has been adjusted so that the ratio of cuts to fills is 1.2 to 1.5

- total volume of excavation

Solution:

	1	2	3	4		
A	8.5	9.2	9.8	10.3	37.8	9.45
B	6.7	8.2	8.6	8.8	32.3	8.08
C	7.0	7.8	7.4	8.3	30.5	7.63
	22.2	25.2	25.8	27.4	100.60	
	7.40	8.40	8.60	9.13		8.38

Method I. "eyeball"

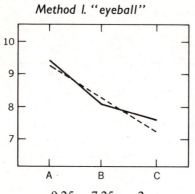

$$\text{Slope} = \frac{9.25 - 7.25}{200'} = \frac{2}{200} = 1 \text{ percent}$$

8.5	9.20	9.2	9.20	9.8	9.20	10.3	9.20
	f0.70		0		c0.60		c1.10
6.7	8.20	8.2	8.20	8.6	8.20	8.8	8.20
	f1.50		0		c0.40		c0.60
7.0	7.20	7.8	7.20	7.4	7.20	8.3	7.20
	f0.20		c0.60		c0.20		c1.10

$\sum c = 4.60$

$\sum f = 2.40$

$$\frac{\sum c}{\sum f} = \frac{4.60}{2.40} = 1.92 \qquad \text{too high}$$

reduce ratio by raising plane 0.1 ft.

8.5	9.30	9.2	9.30	9.8	9.30	10.3	9.30
	f0.80		f0.10		c0.50		c1.00
6.7	8.30	8.2	8.30	8.6	8.30	8.8	8.30
	f1.60		f0.10		c0.30		c0.50
7.0	7.30	7.8	7.30	7.4	7.30	8.3	7.30
	f0.30		c0.50		c0.10		c1.00

$\sum c = 3.90$

$\sum f = 2.90$

$$\frac{\sum c}{f} = \frac{3.90}{2.90} = 1.34 \qquad \text{OK}$$

Volume of excavation $= (\sum c)(100)(100)/27$
$$= 1444.4 \text{ yd}^3$$

Method 2.   Least-Squares-Method
Equation 7.10 is used to compute the grade in either the north-south or west-east directions.

$$G_{ns} \text{ or } G_{we} = \frac{\sum (SH) - \dfrac{(\sum S)(\sum H)}{n}}{\sum(S)^2 - \dfrac{(\sum S)^2}{n}} (100) \qquad (7.10)$$

where

G_{we} = slope of the line that best fits the points that represent the average land slope in a west to east direction across the field;

G_{ns} = slope of the line that best fits the points that represent the average land slope in a north to south direction across the field;

$\sum (SH)$ = the sum of the products of the station distance and elevation of each grid point;

$(\sum S)(\sum H)$ = the product of the sums of the station distances and the elevations of grid point;

n = the number of grid points;

$\sum (S)^2$ = the sum of the squares of the station distances of each grid point;

$(\sum S)^2$ = the square of the sum of the station distance of each grid point.

For NS direction
$$\sum (SH) = 100(9.45) + 200(8.08) + 300(7.63) = 4850$$
$$\sum S = 100 + 200 + 300 = 600$$
$$\sum H = 9.45 + 8.08 + 7.63 = 25.16$$
$$\sum (S)^2 = 100^2 + 200^2 + 300^2 = 140,000$$

$$G_{ns} = \left(\frac{4850 - \dfrac{(600)(25.16)}{3}}{140,000 - \dfrac{600^2}{3}} \right)(100) = -0.91 \text{ percent}$$

The negative sign indicates that the elevation decreases from north to south.

For WE direction

$$\sum (SH) = 100(7.40) + 200(8.40) + 300(8.60) + 400(9.13) = 8652$$
$$\sum S = 100 + 200 + 300 + 400 = 1000$$
$$\sum H = 7.40 + 8.40 + 8.60 + 9.13 = 33.53$$
$$\sum (S)^2 = 100^2 + 200^2 + 300^2 + 400^2 = 300,000$$

$$G_{we} = \left(\frac{8652 - \dfrac{1000(33.53)}{4}}{300,000 - \dfrac{1000^2}{4}} \right)(100) = 0.54 \text{ percent}$$

The positive sign indicates that the elevation increases from west to east.

Once the grades have been determined, a procedure similar to the "eyeball" method is followed to determine cuts/fills at each grid point, balance the cuts to fills, and calculate the volume of excavation.

The computer program in Appendix K performs land smoothing calculations using the least squares method.

7.5.4b   Tailwater Reuse

Tailwater systems similar to the one diagrammed in Figure 7.21 collect runoff water and make it available for reuse. Reuse systems usually include collection ditches or diked areas at the lower end of the field, an open channel or pipe drain that directs the collected water to a storage reservoir and a means of returning the collected water to the same field or delivering it to another field. The return/delivery system includes pumps pipelines and/or open channels. The design of various types of reuse systems is presented in Hart et al., (1980).

Runoff generated during an irrigation should be used for subsequent sets on that field or to irrigate other fields (Hart et al., 1980). Runoff water should be

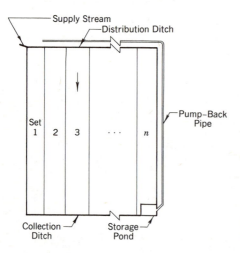

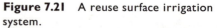

Figure 7.21   A reuse surface irrigation system.

collected and used to shorten the time that primary source water is used during other sets (to reduce the amount of primary source water used) or to decrease the primary source flowrate used in other sets. It can also be utilized to provide all the water for smaller sets.

Runoff should be reused during the set in which it is generated only to supply flow during the second stage of cutback irrigation. Water is obtained entirely from the primary source until inflow is cutback. After cutback, only runoff water is used. Other schemes for reusing water during the same irrigation in which it is generated, such as starting new streams or steadily reducing the primary source flowrate as runoff flow constantly increases, have proved unsatisfactory and are not recommended (Hart et al., 1980).

7.5.4c   Cutback Irrigation

Cutback irrigation involves decreasing (cutting back) the rate of inflow to graded furrows during an irrigation to reduce runoff losses and improve application efficiency. The basic strategy of cutback irrigation is to reduce runoff from a field by matching the rate at which water enters the soil to the average infiltration capacity of the soil along the length of the furrow. Runoff is eliminated when the rate at which inflow decays with time exactly coincides with the decay of the average infiltration rate with time for the entire length of field. Inflow is usually cutback in discrete steps (rather than continuously). In Figure 7.22 runoffs for irrigation without cutback, one cutback, and multiple cutbacks are compared. This figure indicates that runoff is reduced by increasing the number of cutbacks. Because of high labor requirements, inflows are usually only cutback once or twice per irrigation. Automated surface irrigation systems allow more frequent cutbacks and result in improved application efficiency.

Cutback irrigation requires the installation of valves in supply pipes or the construction of level bays with spile outlets along the side of open ditches. Such systems may be automatically or manually controlled. Manual cutback systems require more labor than noncutback systems.

Implementation of cutback irrigation when inflow to the field is constant can result in the generation of significant quantities of excess inflow water. For example, if a constant flow of 2000 l/min (530 gpm) of water from a water user organization canal is being turned into the field to provide flow during the advance phase of a cutback irrigation, and inflow is cutback to 1200 l/min (320 gpm), what does one do with the extra 800 l/min (180 gpm)? One solution is store the excess water in a reservoir. During the next irrigation, the higher stream sizes used prior to cutback are obtained by pumping from the reservoir. After cutback, excess delivery inflow replenishes the reservoir. Another solution is a split-set technique where the total set or field segment is divided into two parts. The first half of the set is irrigated with the entire delivery flow until water runs off the field. The entire irrigation delivery is then directed onto the other half of the field segment for the same length of time. Water is then reintroduced into the first field segment so that the entire delivery flow is distributed across the total set for the remainder of the irrigation.

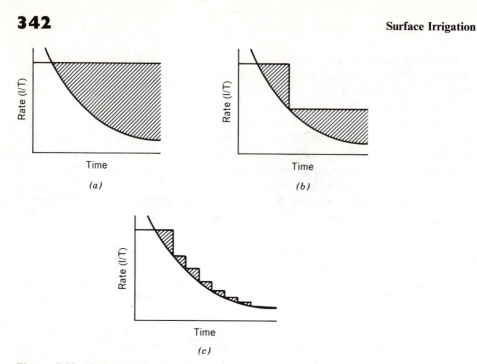

Figure 7.22 Relationships between inflow rate (to the field) and time with and without cutback irrigation. The cross-hatched areas are the amount of runoff from the field. (*a*) No cutback. (*b*) One cutback. (*c*) Several cutbacks.

7.5.4d Cablegation

Cablegation is a semiautomated concept for achieving cutbacks in furrow systems. This system, diagrammed in Figure 7.23, consists of a gated pipe with the gates located near the top of the pipe and laid on a slope along the head end of the field. This pipe serves as both a conveyance and distribution pipe. A movable plug restrained by a cable in the pipe obstructs flow through the pipe and causes flow from the gates into a furrow. The other end of the cable is attached to a reel at the pipe inlet. During the irrigation, the cable is unrolled from the reel and the plug moves slowly downstream. New gates begin to discharge water as flow from some of the upstream gates end. The rate of gate outflow steadily decreases from the plug toward the inlet end of the pipe because of the increasing elevation in that direction. Thus, inflow to a furrow begins at a maximum rate when the plug is adjacent to the furrow and steadily decreases as the plug moves downstream. Kincaid (1984) has developed dimensionless design relationships for cablegation systems.

7.5.4e Surge Flow Irrigation

Surge flow irrigation involves intermittent (rather than continuous) inflow to furrows. It is accomplished by alternating periods of constant, nonzero inflow to furrows with rest periods of zero inflow. The duration of time between successive

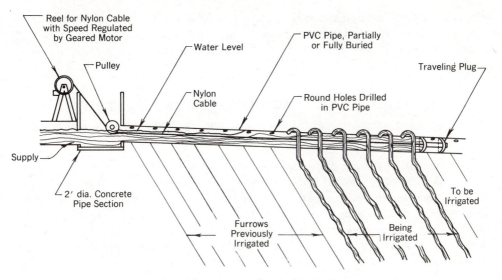

Figure 7.23 Schematic of a cablegation surface irrigation system.

inflow periods, called the *cycle time*, is chosen so that several on–off cycles are required to complete the advance phase of the irrigation. During the advance phase, the duration of rest periods is normally long enough for most, if not all, water to infiltrate before the next inflow period begins. The ratio of on to off times is the cycle ratio.

In many soils, surging increases the rate of advance during inflow periods because of reduced surface roughness and lower infiltration rates in the previously wetted portion of the field. Reasons for the lower infiltration rates include greater infiltration times, the break down of soil aggregates, the entrapment of air in the soil, and the development of surface seals (Walker and Skogerboe, 1987).

More rapid advance makes higher frequency, lighter surface irrigations possible, improves the uniformity of the irrigation, and allows higher application efficiencies to be achieved. Proper management is required to reduce deep percolation and runoff losses and to achieve higher application efficiencies. Initial experience with surge irrigation indicates that when advance is complete, runoff can be reduced by adjusting cycle times so that inflow ends when water has advanced about 75 percent of the distance across the field. Initial experience also indicates that higher uniformities result when cycle times are varied during advance so that about the same length of previously unwetted field is wet during each inflow period. It has also been found that surge irrigation reduces both spatial and temporal differences in advance rates (Walker and Skogerboe, 1987).

Furrows are surge irrigated using the surge-flow valves (described in Section 7.3.2e) and gated pipe. The valves switch water back and forth between two sets of furrows. Although the cycle time can vary during an irrigation, the cycle ratio must remain constant (i.e., the on-time must equal the off-time) for there to be continuous flow through the surge-flow valve. Research to develop information and procedures for designing and operating surge irrigation systems is being conducted. The use of surge irrigation with borders is also being investigated.

7.6   Design of Surface Irrigation Systems

Surface irrigation system design involves developing specifications for economically and technically feasible layouts of application, delivery, and drainage facilities. As discussed in Section 2.4.4, this is an iterative process that begins when all the necessary data have been assembled, a suitable water source identified, and the design daily irrigation requirement (DDIR) determined. The designs of basin, border, and furrow systems with constant stream sizes are discussed in the following sections.

7.6.1   Infiltration Data for Surface System Design

The relationship between cumulative infiltration depth and time is a prerequisite to surface system design. This relationship can be developed from the infiltration rate versus time relationship for the soil. Equation 5.12 is often used to represent this relationship for surface irrigated soils.

The relationship between cumulative infiltration depth and time is obtained by integrating Eq. 5.12 with respect to time. The following equation results.

$$F = ct^d \qquad\qquad\qquad (7.11a)$$

where

$$c = \frac{a}{b + 1}$$
$$d = b + 1$$

F = cumulative infiltration depth (mm, in)

The U.S. Soil Conservation Service uses the following equation to relate cumulative infiltration depth and time in minutes.

$$F = et^f + g \qquad\qquad\qquad (7.11b)$$

where

t = time since infiltration began (min);
e, f, g = parameters from Tables 7.6 and 7.7.

The Soil Conservation Service (SCS) has classified major soils into groups called intake families and determined values of $e, f,$ and g for each family. Tables 7.6 and 7.7 list $e, f,$ and g values for several SCS intake families. The values in Table 7.6 apply to basins and borders, while those in Table 7.7 are used for furrows.

Table 7.6   Values of e, f, and g in Eq. 7.11b for Basins and Borders

Intake Family	e (mm)	e (in)	f	g (mm)	g (in)
0.1	0.6198	0.0244	0.661	6.985	0.275
0.3	0.9347	0.0368	0.721	6.985	0.275
0.5	1.1862	0.0467	0.756	6.985	0.275
1.0	1.7805	0.0701	0.785	6.985	0.275
1.5	2.2835	0.0899	0.799	6.985	0.275
2.0	2.7534	0.1084	0.808	6.985	0.275
3.0	3.6500	0.1437	0.816	6.895	0.275
4.0	4.4450	0.1750	0.823	6.895	0.275

Source: Soil Conservation Service, "Border Irrigation," *SCS National Engineering Handbook* (1974), Chapter 4, Section 15, U.S. Department of Agriculture, Washington, D.C.

Table 7.7   Values of e, f, and g in Eq. 7.11b for Furrows

Intake Family	e (mm)	e (in)	f	g (mm)	g (in)
0.05	0.5334	0.0210	0.6180	6.985	0.275
0.10	0.6198	0.0244	0.6610	6.985	0.275
0.15	0.7110	0.0276	0.6834	6.985	0.275
0.20	0.7772	0.0306	0.6988	6.985	0.275
0.25	0.8534	0.0336	0.7107	6.985	0.275
0.30	0.9246	0.0364	0.7204	6.985	0.275
0.35	0.9957	0.0392	0.7285	6.985	0.275
0.40	1.064	0.0419	0.7356	6.985	0.275
0.45	1.130	0.0445	0.7419	6.985	0.275
0.50	1.196	0.0471	0.7475	6.985	0.275
0.60	1.321	0.0520	0.7572	6.985	0.275
0.70	1.443	0.0568	0.7656	6.985	0.275
0.80	1.560	0.0614	0.7728	6.895	0.275
0.90	1.674	0.0659	0.7792	6.895	0.275
1.00	1.786	0.0703	0.785	6.985	0.275
1.50	2.284	0.0899	0.799	6.985	0.275
2.00	2.753	0.1084	0.808	6.985	0.275

Source: Soil Conservation Service, "Furrow Irrigation," *SCS National Engineering Handbook* (1984), Chapter 5, Section 15, U.S. Department of Agriculture, Washington, D.C.

It is best to conduct field tests to obtain infiltration data for design. Infiltration data are collected with ring infiltometers, the ponding method, the blocked furrow technique, the inflow/outflow method, and recirculating infiltrometers. Infiltration characteristics can also be determined from advance data using the two-point method of Elliott and Walker (1982). Infiltration data for designing basins and borders are usually obtained with the ring infiltrometer, ponding, or two-point methods. Either the blocked furrow, inflow/outflow, recirculating infiltrometer, or two-point method are normally used to obtain infiltration data for designing furrow systems.

7.6.1a Ring Infiltrometers

A typical *single-ring infiltrometer* consists of a metal cylinder that is driven into the soil (see Figure 7.24*a*). Merriam et al., (1980) recommend that the cylinder have a diameter of 250 mm (10 in) or more, be 400 mm (16 in) tall, have a wall thickness of 1.5 mm (1/16 in), and be driven at least 150 mm (6 in) into the soil.

Several rings should be used to determine the infiltration characteristics of the field. During a test, enough water to fill each ring to a datum point (usually equal to the desired depth of irrigation) is added and the water level measured periodically. Both the distance the water level has dropped and the time of measurement should be recorded for each ring. These data are plotted on log-log paper as shown in Figure 7.25 and a line fit through the data points for each ring. The lines will rarely coincide because of intrafield variations in soil (Merriam et al., 1980). After careful examination of the lines, a single line that best represents the infiltration characteristics of the field is determined.

Double-ring infiltrometers (Figure 7.24*b*) are sometimes used to improve the accuracy of the infiltration data collected. *Double-ring infiltrometers* consist of two

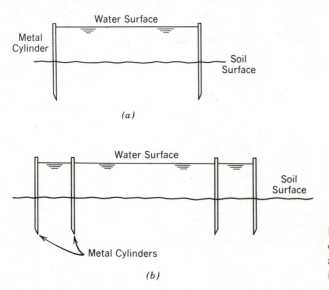

(a)

(b)

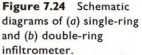

Figure 7.24 Schematic diagrams of (*a*) single-ring and (*b*) double-ring infiltrometer.

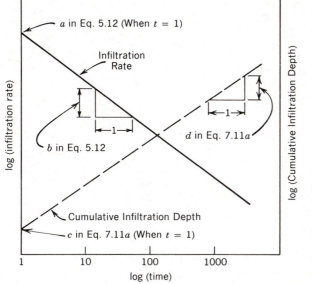

Figure 7.25 Log-log plots of infiltration rate versus time and cumulative infiltration depth versus time for a typical soil. The definitions of a and b in Eq. 5.12 and c and d in Eq. 7.11a are illustrated.

Labels in figure:
a in Eq. 5.12 (When $t = 1$)
Infiltration Rate
log (infiltration rate)
b in Eq. 5.12
d in Eq. 7.11a
log (Cumulative Infiltration Depth)
Cumulative Infiltration Depth
c in Eq. 7.11a (When $t = 1$)
log (time)

concentric rings (rather than a single ring). The outer ring provides a buffer of water that minimizes lateral movement of water from the inner ring when the wetting front has penetrated below the bottom of the rings.

Because ring infiltration data may not agree with the actual amount of infiltration that occurs during irrigation, they are often adjusted utilizing measured inflow, runoff, advance, and recession data. Example 7.5 illustrates a procedure for obtaining an adjusted cumulative infiltration relationship for basin and border system design. In this example, inflow and runoff data are utilized to estimate the average depth of infiltration, while the average infiltration opportunity time is computed from advance and recession data. The adjusted cumulative infiltration curve is obtained by shifting the original ring infiltration curve to have a value equal to the average depth of infiltration (calculated from inflow–runoff data) when t equals the average infiltration opportunity time (determined from advance and recession data). The slope of the adjusted curve equals that of the original ring infiltration relationship. Values of c and d in Eq. 7.11a for the adjusted relationship are determined by noting that c equals F when t is 1 and that d is the slope of the relationship.

EXAMPLE 7.5 Adjusting Infiltration Data Using Measured Advance, Stream size, and Inflow Time Data

Given:
- stream size $= 200$ l/min/m of width
- inflow time $= 38$ min
- the advance and recession data in the following solution table
- the following cumulative infiltration curve

Required:
• adjusted cumulative infiltration relationship

Solution:

Solution Steps
1. compute the average depth of infiltration
2. compute average infiltration opportunity time from advance data
3. adjust the cumulative infiltration curve

Solution Step 1
(Compute average infiltration depth)

$$\text{Infiltration depth} = \left(\frac{200\ 1}{\min\ m}\right)\left(\frac{38\ \min}{90\ m}\right)\left(\frac{1000\ mm}{m}\right)\left(\frac{1\ m^3}{1000\ 1}\right)$$

$$= 84\ mm$$

Solution Step 2
(Compute average IOT)

Distance Along Basin (m)	Advance Time (min)	Recession Time (min)[a]	IOT (min)
0	0	128	128
15	3	128	125
30	7	128	121
45	12	128	116
60	17	128	111
75	22	128	106
90	28	128	100

[a] Recession is instantaneous.

Average IOT = 115 min

Solution Step 3
(Adjusting cumulative infiltration curve)

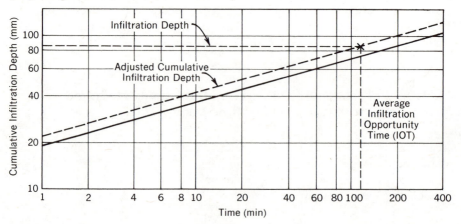

7.6.1b   The Ponding Method

The principles and procedures of this method of measuring infiltration are similar to the ring infiltrometer method. Earth dikes are used (rather than metal rings) to pond water. A trench that is made impermeable can be constructed to limit lateral subsurface flow and improve the accuracy of the data collected. The use of two concentric ponds, *double ponds,* is an alternate technique for minimizing lateral subsurface flow. The ability to use a larger ground surface area is the primary advantage of the ponding method over ring infiltrometers. Infiltration rates from single ponds without impermeable trenches are generally higher than the true infiltration rate.

7.6.1c   The Blocked Furrow Method

Because a large portion of infiltration flows laterally from furrows, data obtained with ring infiltrometers and the ponding method (where flow is vertical) are not recommended for furrow system design. The blocked furrow method is similar to the ponding method.

The *blocked furrow method* involves installing cutoff plates at the up- and downstream ends of test sections in three adjacent furrows. Infiltration is measured in the center furrow by measuring the volume of water required to maintain a constant water level. The furrows on each side of the test furrow provide a buffer that improves accuracy. The depth of infiltration is computed by dividing measured infiltration volumes by the product of furrow spacing times the length of the test section.

7.6.1d   The Inflow/Outflow Method

In the inflow/outflow method, the infiltration rate is determined by measuring the rates of flow into and out of a section of furrow. When the depth of flow in the furrow is changing slowly, the infiltration rate equals the difference between the inflow and outflow rates. The infiltration rate versus time relationship for the furrow is obtained by simultaneously measuring inflow and outflow with flumes or weir plates several times during the irrigation. These data are plotted on log-log paper as shown in Figure 7.25. Values of a and b are determined by noting that a equals f when t is 1 and that b is the slope of the relationship.

The cumulative depth of infiltration versus time relationship needed for design is obtained by substituting a and b into Eq. 7.11a. This relationship can be adjusted to more accurately represent furrow infiltration using the procedure described in the previous section.

This version of the inflow/outflow method must be carefully applied to obtain accurate and reliable infiltration data. When flow measuring points are close to one another, it is often difficult to detect differences in flow because of normal measurement errors. When the points are far apart, large differences in infiltration opportunity time along the furrow can limit the usefulness of the data.

7.6.1e   Recirculating Infiltrometers

A *recirculating infiltrometer* (diagrammed in Figure 7.26) is an inflow/outflow device. Water from the source tank is added to the constant-head device as needed

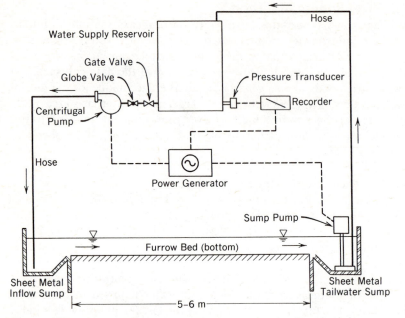

Figure 7.26 Schematic of a recirculating furrow of infiltrometer.

to maintain a constant flow from the orifice. Water that flows through the test section (without infiltrating) collects in the sump at the downstream end of the section and is pumped back to the supply tank. The water level in the supply tank is monitored during the test to determine infiltration rate and cumulative infiltration depth versus time relationships. Either a water stage recorder or an electronic water level sensor and data logger are utilized to mechanize recirculating infiltrometers.

7.6.1f Two-Point Method

Although it was developed for furrows, the two-point method can also be used to obtain infiltration data for designing basin and border systems. The two-point method, which is based upon Eq. 7.17, uses a power function of the following form to represent the relationship between the distance that water has advanced across a basin, border, or furrow and time.

$$X = pt^r \qquad (7.12)$$

$$r = \frac{\ln 2}{\ln T_a - \ln T_{0.5a}} \qquad (7.12a)$$

where

X = distance that water has advanced across the basin, border, or furrow (m, ft);
t = time since the start of advance (min);
p, r = fitted parameters;
T_a = time to advance the full length of the field (min);
$T_{0.5a}$ = time to advance half the length of the field (min).

Measured advance data are also used to evaluate K and a in the following equation for computing cumulative infiltration.

$$F = kt^a + \frac{f_0 t}{K} \tag{7.13}$$

where

$$a = \frac{\ln V_a + \ln V_{0.5a}}{\ln T_a - \ln T_{0.5a}} \tag{7.13a}$$

$$V_a = \frac{QT_a}{KL} - 0.77A_0 - \frac{f_0 T_a}{K(r+1)} \tag{7.13b}$$

$$V_{0.5a} = \frac{2QT_{0.5a}}{KL} - 0.77A_0 - \frac{f_0 T_{0.5a}}{K(r+1)} \tag{7.13c}$$

$$\sigma_z = \frac{a + r(1-a) + 1}{(1+a)(1+r)} \tag{7.13d}$$

$$k = \frac{V_a}{\sigma_z T_a^a} \tag{7.13e}$$

where

F = cumulative infiltration (m³/m, ft³/ft);
Q = streamsize (l/min, gpm);
t = time from the start of the irrigation (min);
f_0 = final infiltration rate (l/min/m, gpm/ft);
L = length of the furrow (m, ft);
A_0 = cross-sectional area of flow at furrow inlet (m², ft²);
r = exponent in the advance function (see Eq. 7.12);
K = unit constant ($K = 1000$ for Q in l/min, A_0 in m², and f_0 in l/min/m.
    $K = 7.48$ for Q in gpm, A_0 in ft², and f_0 in gpm/ft.).

A derivation of the previous equations is presented in Elliott and Walker (1982) and Walker and Skogerboe (1987).

The values of a and K in Eq. 7.13 are evaluated using the following procedure.

1. Conduct field experiments to determine T_a, $T_{0.5a}$, f_0, and A_0.
2. Compute r using Eq. 7.12a.
3. Compute V_a and $V_{0.5a}$ using Eqs. 7.13b and 7.13c, respectively.
4. Compute a using Eq. 7.13a.
5. Compute σz using Eq. 7.13d.
6. Compute k using Eq. 7.13e.

Steps 2 through 6 of this procedure are illustrated in the following example problem. Techniques for measuring T_a, $T_{0.5a}$, f_0, and A_0 are discussed in sections following the example. The computer program in Appendix M also performs these calculations.

EXAMPLE 7.6 Determining Infiltration Characteristics Using the Two-Point Method

Given:
- information from Example 7.5
- the border is 15 m wide and 90 m long
- $A_0 = 0.45$ m^2
- $f_0 = 1.75$ l/min/m
- $Q = 3000$ l/min

Required:
1. r in Eq. 7.12
2. K and a in Eq. 7.13
3. cumulative infiltration depth in millimeters after 200 min

Solution:
1. determining r using Eq. 7.12

$$r = \frac{\ln 2}{\ln 28 - \ln 12} = 0.818$$

2. determining k and a

$$V_a = \frac{(3000)(28)}{(1000)(90)} - 0.77(0.45) - \frac{(1.75)(28)}{(1000)(1.818)} = 0.560$$

$$V_{0.5a} = \frac{2(3000)(12)}{(1000)(90)} - 0.77(0.45) - \frac{(1.75)(12)}{(1000)(1.818)} = 0.442$$

$$a = \frac{\ln 0.560 - \ln 0.442}{\ln 28 - \ln 12} = 0.279$$

$$\sigma_z = \frac{0.279 + 0.818(1 - 0.279) + 1}{(1.279)(1.818)} = 0.804$$

$$k = \frac{0.560}{(0.804)(28)^{0.279}} = 0.275$$

3. determining F for 200 minutes

$$F = 0.275(200)^{0.279} + \frac{(1.75)(200)}{1000} = 1.556 \text{ m}^3/\text{m}$$

$$\text{Depth in cm} = \left(\frac{1.556 \text{ m}^3/\text{m}}{15 \text{ m}}\right)\left(\frac{100 \text{ cm}}{\text{m}}\right) = 10.4 \text{ cm}$$

(i) Determining T_a and $T_{0.5a}$ These parameters are measured in the field by placing at least 10 stakes, spaced no more than 30 m (100 ft) apart along test basins, borders, or furrows and recording the time when water first reaches each stake during advance. In basins and borders, this may require considerable judgement because of irregular advance rates across the width of the basin or border. In such cases, a rectangular grid of stakes (rather than a single row of stakes) is sometimes employed to improve the quality of the data collected.

Basins, borders, and furrows should be monitored during tests to estimate the maximum nonerosive stream size and to determine the maximum depth of flow. Flow depth data are useful in determining furrow sizes and ridge heights for basins and borders. In addition, the extent of lateral wetting from furrows should be measured to estimate furrow spacing (this requires that the design depth of irrigation be applied during the test). Recession data should also be collected for borders by recording the time when water disappears at each stake.

Tests should be replicated, if possible, to account for intrafield differences in soil texture and structure. Vehicle and/or implement traffic through the field may cause large differences in soil structure between wheel track and nonwheel track furrows. It is desirable to repeat tests in the same basin, border, or furrow to evaluate differences in advance rates between irrigations. Advance rates are normally slower in freshly cultivated soils than in soils that have not been cultivated between irrigations.

(ii) Determining the Final Infiltration Rate, f_0   The final infiltration rate is determined by continuing ring infiltrometer, ponding, blocked furrow, inflow/outflow, and recirculating infiltrometer tests long enough for the infiltration rate to become constant. This constant rate is, by definition, f_0. Values of the basic infiltration rate from Table 5.3 can also be used to approximate f_0.

Because of the effect that lateral flow below ring infiltrometers and infiltration ponds has on infiltration measurements, double ring infiltrometers and double ponds or ponds with impermeable trenches should be used to determine f_0. Elliott and Walker (1982) suggest the inflow/outflow method for determining f_0 for furrows.

(iii) Determining the Upstream Flow Area, A_0 A_0 can be measured with a rillmeter (see Figure 7.27) or estimated using the Manning equation (Eq. 7.2). Recommended values of the Manning roughness coefficient, n, are given in Table 7.8. The computer programs in Appendices I, L, and M use the Manning equation to determine A_0.

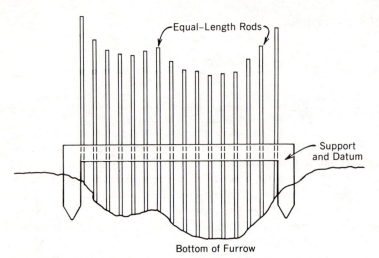

Figure 7.27 A rillmeter for measuring the distance from the datum to the bottom of the furrow (rill). The cross-sectional area of the furrow is computed from these distances.

7.6.2 The Use of Mathematical Models in Surface System Design

Mathematical models are being used more and more in surface system design. They are used to make a variety of calculations, including the position of wetting and drying fronts during advance and recession, and the uniformity and efficiency of irrigation. They improve the quality of the design by allowing the designer to consider several combinations of stream size, field length, and slope (rather than just a few). In addition, spatial and temporal changes in soil characteristics between and within irrigations can be more thoroughly evaluated.

Several techniques with varying levels of sophistication have been developed to mathematically model surface irrigation processes. Most of these approaches are

Table 7.8 Recommended Values of the Manning Roughness Coefficient, *n*

Condition	*n*
Previously irrigated and smooth soil	0.02
Freshly tilled soil	0.04
Dense growth obstructs water movement	0.15

Source: W. R. Walker and G. V. Skogerboe, *Theory and Practice of Surface Irrigation* (1987), Prentice-Hall, Inc., Englewood, Cliffs, N.J.

based upon the fundamental principles of conservation of mass, momentum, and/ or energy. A more complete discussion of these techniques is found in Bassett et al., (1980) and Walker and Skogerboe (1987).

For the spatially varied and unsteady flow that occurs during surface irrigations, application of the conservation of mass principle yields the following equation (called the *continuity equation*).

$$\frac{\partial Q}{\partial x} + \frac{\partial A}{\partial t} + I = 0 \tag{7.14}$$

where

A = cross-sectional area of flow;
Q = flowrate;
I = infiltration rate per unit length;
t = time;
x = distance along direction of flow from inlet.

Application of Newton's second law to surface irrigation systems yields the following equation (called the *momentum equation*).

$$\left(\frac{1}{Ag}\right)\left(\frac{\partial Q}{\partial t}\right) + \left(\frac{2Q}{A^2g}\right)\left(\frac{\partial Q}{\partial x}\right) + (1 - Fr^2)\left(\frac{\partial y}{\partial x}\right) = S_0 - S_f \tag{7.15}$$

where

$$Fr = \frac{Q^2 T}{A^3 g}$$

A = cross-sectional area of flow;
Q = stream size;
Y = depth of flow;
g = acceleration due to gravity;
t = time;
x = distance in direction flow;
T = top width of flow cross section;
Fr = Froude number;
S_0 = slope of basin, border, or furrow;
S_f = slope of energy grade line.

Equations 7.14 and 7.15, called the Saint-Venant equations, are derived in Chow (1959). Computer models that use the Saint-Venant equations to simulate surface irrigations are called *hydrodynamic models*.

Hydrodynamic models that accurately simulate border irrigation have been developed by Bassett (1972), Kincaid et al., (1972), Bassett and Fitzsimmons (1976), and Katapodes and Strelkoff (1977a). Souza (1981) and Haie (1984) have modeled furrow irrigations using the hydrodynamic approach. Hydrodynamic models are complex and expensive to use.

Strelkoff and Katapodes (1977) developed a simpler and faster surface irrigation modeling approach that requires less computer storage (than the hydrodynamic approach) by observing that the acceleration and inertial terms in Eq. 7.15 can be neglected (since flow velocities during most surface irrigations are small). When the acceleration and inertial terms are neglected Eq. 7.15 reduces to

$$\frac{\partial y}{\partial x} = S_0 - S_f \tag{7.16}$$

Surface irrigation models based on Eqs. 7.14 and 7.16 are called zero-inertia models. The *zero-inertia* approach has been verified against field data for basins, borders, and furrows (Clemmens and Fangmeier, 1978; Clemmens, 1979; Elliott and Walker, 1982). Design methodologies for level basins (Clemmens and Strelkoff, 1979) and for sloping borders and basins (Strelkoff and Clemmens, 1981) have been developed using the zero-inertia approach.

In the *kinematic-wave* approach of mathematically simulating advance and recession, the momentum equation (Eq. 7.15) is replaced by a uniform flow equation such as the Manning, Chezy, or Darcy-Weisbach equation. Because of the uniform flow assumption, kinematic-wave models are sometimes called "uniform depth," "uniform flow," or "normal-depth" models. The uniform flow assumption greatly simplifies the analysis, but limits the use of the approach to free draining, graded borders, and furrows (since basins, diked borders and furrows, and dead-level fields cannot be represented mathematically).

When compared to solutions obtained with the zero-inertia and hydrodynamic models, kinematic-wave simulations for borders were closer to hydrodynamic solutions at large times and advance distances (Katapodes and Strelkoff, 1977). These comparisons also showed better agreement between kinematic-wave and hydrodynamic solutions for steeper slopes.

Another important approach to surface irrigation modeling is the *volume-balance* method. This approach has been proved with field data and is the basis for procedures used by the U.S. Soil Conservation Service to design level and graded borders. The volume-balance approach also accurately simulates furrow irrigations.

The volume balance approach is based on the following form of the continuity equation.

$$Qt = V_y(t) + V_z(t) \qquad \text{for } t \leq T_a \tag{7.17}$$

where

Q = stream size;
t = time;
T_a = time required for water to advance across the length of the field;
$V_y(t)$ = volume of water on soil surface at time, t;
$V_z(t)$ = volume of water that has infiltrated at time, t.

This equation states that the inflow volume equals the sum of the volumes of water stored on the soil surface and infiltrating the soil during the time period t.

The volume of water on the soil surface is determined by integrating the flow area over the advance distance, X.

$$V_y(t) = \int_0^x A(x, t)\, dx = \bar{A}X = 0.77(A_0)(X) \tag{7.18}$$

where

$$
\begin{aligned}
X &= \text{advance distance;} \\
A(x, t) &= \text{cross-sectional area of flow;} \\
x &= \text{distance from the field inlet;} \\
t &= \text{elapsed time since irrigation began;} \\
\bar{A} &= \text{average cross-sectional area of flow over the advance distance } X; \\
A_0 &= \text{cross-sectional area of flow at the field inlet.}
\end{aligned}
$$

The Manning equation is often used to estimate $\bar{A}$ for the stream size, Q.

The amount of infiltration is given by

$$V_z(t) = \int_0^x Z(x, t - T_x)\, dx \tag{7.19}$$

where

$$
\begin{aligned}
Z(x, t - T_x) &= \text{depth of infiltration at a distance } x \text{ from the inlet at time } t; \\
T_x &= \text{time required for water to advance a distance } x.
\end{aligned}
$$

In Eq. 7.19, a unit width is implied for basins and borders. For furrow systems, the right side of Eq. 7.19 must be multiplied by the furrow spacing to obtain a volume.

A convenient form of Eq. 7.17 results when advance is represented by Eq. 7.12 and $V_z(t)$ is computed with Eq. 7.13. After setting X equal to the length of the field, L, and t equal to T_a, this equation is:

$$\frac{(Q)(T_a)}{K} - 0.77(A_0)(L) - \sigma_z(L)(k)(T_a)^a - \sigma \frac{(f_0)(T_a)(L)}{K} = 0 \tag{7.20}$$

$$\sigma = \frac{1}{1 + r} \tag{7.20a}$$

where Q, T_a, A_0, L, σ_z, k, a, and r are as defined for Eqs. 7.12 and 7.13. A derivation of Eq. 7.20 is given in Walker and Skogerboe (1987).

Equation 7.20 is used to determine the time, T_a, required for water to advance across the length of the field. Because r in Eqs. 7.13b, 7.13c, and 7.13d depends on the stream size, Q, Eqs. 7.20 and 7.12a must be solved simultaneously. The computer program in Appendix L performs the iterative calculations necessary to solve these equations. This program also returns values for $T_{0.5a}$ (called THALF in the program), and p and r in Eq. 7.12. The program is based on the procedures used to solve the following example problem.

EXAMPLE 7.7 Computing T_a and $T_{0.5a}$, and p and r in Eq. 7.12

Given:
- $S = 0.10$ percent
- the following information from Example 7.6

$$L = 90 \text{ m}$$
$$W = 15 \text{ m}$$
$$k = 0.275$$
$$a = 0.278$$
$$f_0 = 1.75 \text{ l/min/m}$$

Required:
1. T_a and $T_{0.5a}$ for $Q = 2000$ l/min
2. p and r in Eq. 7.12

Solution:
a. T_a and $T_{0.5a}$ are determined by solving Eqs. 7.12a and 7.20 simultaneously. This involves the following steps.

1. Determine A_0 (using the program in Appendix I).
2. Select an initial value for r.
3. Compute σ_z with Eq. 7.13d.
4. Compute σ with Eq. 7.20a.
5. Select an initial value for T_a.
6. Solve Eq. 7.20. Does the absolute value of the result exceed 0.005?
   Yes—an acceptable value of T_a has been found. Go to Step 7.
   No—select a new value for T_a and repeat Step 6.
7. Select a value for $T_{0.5a}$.
8. Solve Eq. 7.20 with $L = L/2$. Does the absolute value of the result exceed 0.005?
   Yes—an acceptable value of $T_{0.5a}$ has been found. Go to Step 9.
   No—select a new value for $T_{0.5a}$ and repeat Step 8.
9. Compute r' with Eq. 7.12a. Is the absolute value of $r - r'$ less than 0.001?
   Yes—an acceptable value of r has been found. Go to Step 10.
   No—set $r = r'$ and repeat Steps 5 through 9.
10. Substitute $X = L$ and $t = T_a$ into Eq. 7.12 and solve for p.

Step 1. Determining A_0

$n = 0.02$ from Table 7.8
Depth of flow $= 0.0195$ m   (computed with the computer program in Appendix I)
$A_0 = (0.0195)(15 \text{ m}) = 0.293 \text{ m}^2$.

Step 2. Setting the initial value of r

$r = 0.700$

Step 3.   Computing σ_z with Eq. 7.13d

$$\sigma_z = \frac{0.279 + 0.7(1 - 0.279) + 1.0}{(1 + 0.279)(1 + 0.7)} = 0.820$$

Step 4.   Computing σ with Eq. 7.20a

$$\sigma = \frac{1}{1 + 0.7} = 0.588$$

Step 5.   Selecting an initial value of T_a

$$T_a = \frac{L}{V} = \frac{(A_0)(L)}{Q}$$

$$= \frac{(0.293)(90)(1000 \text{ l/m})}{2000}$$

$$= 13.2 \text{ min}$$

Because the bisection method is used in the program to solve Eq. 7.20, the initial value of T_a must be greater than or equal to the actual T_a. Therefore, the approximate T_a computed above is multiplied by 5 to ensure that the initial "guess" exceeds the actual T_a.

Initial $T_a = (5)(13.2) = 66$ min.

Step 6.   Solving Equation 7.20 for T_a

$$\frac{(2000)(66)}{1000} - 0.77(0.293)(90) - (0.820)(90)(0.275)(66)^{0.279}$$

$$- \frac{(0.588)(1.75)(90)(66)}{1000} = 40.26$$

Since $40.26 > 0.005$, select a new value for T_a and repeat this step.

After several iterations $T_a = 40.54$ min

Step 7.   Selecting an initial value for $T_{0.5a}$

Following reasoning outlined in Step 5, set the initial value of $T_{0.5a}$ equal to T_a.

Initial $T_{0.5a} = 40.54$ mins

Step 8.   Solving Eq. 7.20 for $T_{0.5a}$

The procedures used in this step are identical to those used in Step 6 (except $X = L/2$ and $t = T_{0.5a}$).

After several iterations $T_{0.5a} = 16.57$ min

Step 9.   Computing r' with Eq. 7.12a

$$r' = \frac{\ln 2}{\ln 40.56 - \ln 16.57} = 0.774$$

Since the absolute value or $r - r'$ exceeds 0.001, Steps 5 through 9 must be repeated. After several iterations

$$r = 0.776$$
$$T_a = 39.94 \text{ min}$$
$$T_{0.5a} = 16.36 \text{ min}$$

Step 10.   Computing p in Eq. 7.12

$$p = \frac{90}{(39.94)^{0.776}} = 5.138$$

7.6.3   Design of Level Basins

As described in Section 7.2.1, level-basin irrigation involves dividing the irrigated area into level units bounded by temporary or semipermanent dikes or ridges. Temporary ridges may be used for a single irrigation or for an entire cropping season. Semipermanent ridges are used for perennial crops and for annual crops such as rice that are grown on the same land year after year.

As with other irrigation systems, the design of level-basin systems is an iterative process. It involves adjusting system layouts, inflow times, streamsizes, basin dimensions, and the number of basins irrigated per set until the desired blend of efficiency, uniformity, adequacy, convenience of operation, and cost is achieved. Design procedures for level basins can also be used for level borders and furrows.

7.6.3a   Layout

Example layouts for basin irrigation systems are shown in Figure 7.28. It is desirable that the long axis of basins be perpendicular to supply channels and pipelines. This maximizes the spacing between supply lines/channels. Drains are sometimes placed at the ends of basins midway between the supply lines (as shown in Figure 7.28a) to remove excess water resulting from overirrigation or heavy rainfall. Tier layouts, like the one in Figure 7.28b, allow a single supply channel or pipeline outlet to serve several basins. Key factors affecting system layout include water source location, field topography, and basin dimensions.

(i) Water Source Location   When possible, the water source should be located so that all basins on the farm can be irrigated via gravity. It is also desirable that the water source be near the center of the irrigated area to minimize the size of supply channels and/or pipelines.

(ii) Terrain   The general topography of the land influences basin shape. Rectangular-shaped basins can be constructed on uniformly sloping lands, while basins are often irregularly shaped on undulating terrain (since ridges must follow

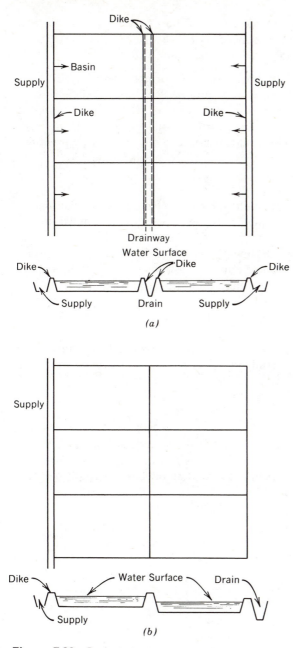

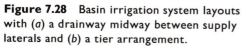

Figure 7.28   Basin irrigation system layouts with (*a*) a drainway midway between supply laterals and (*b*) a tier arrangement.

the land surface contours). Lands with steep natural slopes are terraced (benched) to obtain level benches on which to construct the basins. The banks between adjacent terraces are sometimes stabilized with masonry walls.

   (iii) *Basin Dimensions* Basin dimensions are normally determined by the infiltration characteristics of the soil and the stream size. Relatively small basins are required on soils with high infiltration capacities, such as sands, even when large stream sizes are available. Basins on finer textured soils can be small or large depending on the stream size. Where local experience is not available, a preliminary estimate of basin size can be made using the guidelines in Table 7.9.

7.6.3b   Land Smoothing
   Land smoothing improves the uniformity and efficiency of irrigation by eliminating high and low areas within basins that cause uneven infiltration of ponded water. It also reduces labor requirements and greatly facilitates the layout of supply channels, roadways, and drainways by permitting rectangular- rather than odd-shaped basins. Basins are sometimes constructed with small, sloping channels to allow drainage and flow through them (in tier arrangements). Land smoothing is discussed in Section 7.5.4a.

7.6.3c   Irrigation Time
   Irrigation time is the infiltration opportunity time required to infiltrate the desired depth of irrigation. It is determined by substituting the desired irrigation depth into either Eq. 7.11a, 7.11b, or 7.13. Values for the various coefficients and exponents in these equations are best evaluated from field data.

7.6.3d   Stream Size
   In level basin irrigation, stream sizes should be as large as possible to maximize application efficiency and uniformity, but small enough to not to cause "excessive" erosion. Using large stream sizes that flood basins rapidly reduces

Table 7.9 Suggested Basin Areas for Different Soil Types and Rates of Water Flow

Soil Type	Unit Areas in ha/l/s
Sand	0.067
Sandy loam	0.20
Clay loam	0.40
Clay	0.67

Source: W. E. Hart, H. G. Collins, G. Woodward, and A. S. Humpherys, "Design and Operation of Gravity or Surface Irrigation Systems." In *Design and Operation of Farm Irrigation Systems*, M. E. Jensen (Ed.), copyright © 1980 by ASAE. p. 511. Reprinted by permission of ASAE.

differences in infiltration opportunity times across the basin and normally decreases deep percolation losses. Field trials should be conducted to determine the maximum stream size that does not cause erosion to be "excessive" or basin ridges to be overtopped. The design stream size must not exceed this value.

7.6.3e Inflow Time

The time that water flows into basins, called the *inflow time*, is usually selected to allow the desired irrigation depth to be applied at the far end of the basin. When there is no runoff, basin inflow time is the sum of the advance time and the time required to deliver the volume of water needed to provide the desired depth of irrigation. This relationship is:

$$T_i = T_a + \frac{(D)(A_B)}{(K)(Q)} \tag{7.21}$$

where

T_i = inflow time (min);
T_a = time to advance across the basin (min);
D = desired depth of irrigation (mm, in);
A_B = area of basin (m², ft²);
Q = stream size (l/min, m³/s, ft³/s);
K = unit constant. ($K = 1.0$ for D in mm, A in m², and Q in l/min. $K = 60000$ for D in mm, A in m², and Q in m³/s. $K = 720$ for D in in, A in ft², and Q in ft³/s.)

Time T_a can be determined from measured infiltration data with Eq. 7.20. The cross-sectional area at the upstream end of level basins, A_0, can be estimated with the Manning equation by assuming that the slope, S_0, equals the upstream flow depth divided by the basin length.

The application efficiency is usually the criterion used to judge basin designs. The following equation is used to compute application efficiency.

$$E_a = \frac{(D)(A_B)(100)}{(K)(Q)(T_i)} \tag{7.22}$$

where

E_a = application efficiency in percent;
K = unit constant from Eq. 7.21.

When the maximum safe stream size (with respect to erosion) does not result in an acceptable application efficiency, it may be necessary to reduce the size of the basin or introduce water into the basin at more than one location (i.e., increase the number of inlets so that the total inflow rate to the basin is increased as the inflow rate at each inlet is reduced).

7.6.3f   Ridge Dimensions

The crown width of a basin should be at least as wide as the ridge is high. Basin ridges should have a settled height that equals or exceeds the maximum depth of flow plus 5 cm (2 in).

7.6.3g   Number of Basins Irrigated per Set

The following equation is used to compute the number of basins irrigated per set.

$$N_B = \frac{(N_T)(T_i)(\text{DDIR})(E_a)}{(144,000)(D)}$$

(7.23)

where

N_B = number of basins irrigated per set;
N_T = total number basins being irrigated;
T_i = inflow time (min);
DDIR = design daily irrigation requirement (mm/day, in/day);
E_a = application efficiency (percent);
D = desired depth of irrigation (mm, in).

7.6.3h   Delivery System

Delivery system design begins when the desired combination of irrigation time, stream size, inflow time, basin dimensions, and the number of basins irrigated per set have been identified. The procedures outlined in Sections 7.3.1 and 7.3.2 are used for open channels and low-head pipelines, respectively.

7.6.4   Design of Graded Borders

In border irrigation, fields are subdivided into relatively long, narrow stripes of uniform width called borders. Earth ridges are constructed along the edge of each border to guide a sheet flow of water down the sloping long axis of the border. The surface of each border is smoothed to minimize cross-slope (between adjacent ridges) and to eliminate furrows and other depressions that concentrate the flow and reduce the uniformity of application. The lower ends of borders are normally open, but can be diked to prevent runoff. Borders are also described in Section 7.2.3.

The design of graded (sloping) borders is an iterative process that involves adjusting system layout, land slopes, inflow times, border length, stream sizes, and the number of borders irrigated per set until the desired combination of uniformity, efficiency, adequacy, and convenience of operation are achieved. Level borders are designed using the procedures in Section 7.6.3.

7.6.4a   Layout

Border irrigation systems are laid out so that a supply channel or pipeline delivers water to the upper end of each border. In addition, it is desirable that border ridges be constructed parallel to a field boundary to make irrigation and other cultural operations simpler and more convenient. In smaller fields or fields with low infiltration capacity soils, borders often extend across the full length of the field. In larger fields or where high infiltration capacity soils are being irrigated, it is frequently necessary to have two or more borders across the length of the field to obtain acceptable application efficiency and uniformity. A drainway across the lower end of adjacent borders is often installed to carry away runoff water.

Border system layout is also affected by the location of the water source and the length and width of the borders. These effects are described in the following sections.

(i) *Water Source Location*   As with other surface irrigation systems the water source should be located so that all borders can be irrigated via gravity. In addition, the water source should be as near the center of the irrigated area as possible to minimize the size of supply channels or pipelines.

(ii) *Border Length*   Borders should be as long as possible to reduce labor requirements and system cost, but short enough to retain reasonable application efficiency and uniformity. Long borders reduce labor requirements and system cost by minimizing the number of borders and the length of delivery channels and pipelines. Application efficiencies and uniformities are normally higher on shorter borders (than on longer ones) because smaller advance times are possible.

Soil type is an extremely important factor affecting border length. On soils with low infiltration capacities, acceptable application efficiencies and uniformities are possible on borders up to 800 m (2600 ft) long (Booher, 1974). Booher also observes that it may be necessary to limit length to 100 m (300 ft) or less on high infiltration capacity soils. To obtain uniform water distribution in fields with highly variable soils, Booher (1974) recommends that border lengths be adjusted so that only soils with similar infiltration capacities are included in a border. Typical border lengths for different soils are given in Table 7.10. These recommendations

Table 7.10   Typical Border Lengths for Different Soils

Soil	Typical Border Length in (m)	(ft)
Clay	180–350	600–1150
Clay loam	90–300	300–1000
Sandy loam	90–250	300–800
Loamy sand	75–150	250–500
Sand	60–90	200–300

Source: L. J. Booher, *Surface Irrigation* (1974), Food and Agriculture Organization of the United Nations, Rome, Italy, 160 pp.

Note: Values are based on experience in the southwestern United States.

are based on experience in the southwestern United States (Booher, 1974) and should only be used as guidelines for planning field trials (to measure infiltration and recession) and for developing initial layout concepts.

(iii) Border Width   Borders must be wide enough to accommodate at least one pass of the farm equipment to be used. It is, however, desirable that they be wide enough for an even number of passes. In borders with zero cross-slope between adjacent ridges, the maximum border width depends on the ability of water to spread laterally from the inlet of the border to the ridges. The amount of lateral spread is related to the slope in the direction of flow. Table 7.11 lists maximum border width as a function of the slope in the direction of flow.

When leveling the field in the direction perpendicular to the ridges is not practical (because of insufficient soil depth or economics, for example), Booher (1974) recommends that the elevation difference between the ridges of a border not exceed 3 cm (1 in) and/or that the difference in ground surface elevation between the up- and downhill sides of a ridge not exceed 6 cm (2 in). Thus, borders should not be wider than 9 m (30 ft) on 1.0 percent cross-slopes (9 cm/0.01 = 9 m). An alternative solution would be to grade the entire field to a uniform 1.0 percent slope (rather than grading each border). The maximum border width in this case would be 3 m or 10 ft (3 cm/0.01 = 3 m). A maximum width of 6 m or 20 ft (6 cm/0.01 = 6 m) could be used for borders with zero cross-slope on fields with a 1.0 percent cross-slope. These border widths should be compared to the appropriate value in Table 7.9. Border width should not exceed the smallest of these values.

7.6.4b   Land Smoothing

Land smoothing improves the application uniformity of border irrigation by eliminating furrows and other depressions that concentrate the flow. Although borders with zero cross-slope (from ridge to ridge) are preferred because higher uniformities are possible, borders with cross-slopes are sometimes constructed on

Table 7.11   **Recommended Maximum Border Widths for Different Slopes in the Direction of Flow**

Slope in Percent	Maximum Border Width in	
	(m)	(ft)
level	60	200
0.0–0.1	35	120
0.1–0.5	20	60
0.5–1.0	15	50
1.0–2.0	12	40
2.0–4.0	9	30
4.0–6.0	6	20

Source: Soil Conservation Service, "Border Irrigation, *SCS National Engineering Handbook* (1974), Chapter 4, Section 15, U.S. Department of Agriculture, Washington, D.C.

lands with severe natural terrain. As discussed in the previous section, the width of borders, in such cases, is selected so that the elevation difference between ridges does not exceed 3 cm (1 in) and/or the difference in ground surface elevation between the up- and downhill sides of a ridge is not more than 6 cm (2 in). Land smoothing calculations may be required for each border.

The slopes with the smallest excavation will be used, at least, initially. It may be necessary to adjust these slopes if they are not reasonable or to improve system performance (see Section 7.6.4d). Booher (1974) recommends minimum slopes of 0.2 to 0.3 percent and maximum slopes of 2 percent for sandy loams and up to 7 percent for pastured clay soils with water-stable aggregates. Land smoothing is discussed in Section 7.5.4a.

7.6.4c   Irrigation Time

The irrigation time for border irrigation is determined by substituting the desired depth of irrigation into either Eq. 7.11a, 7.11b, or 7.13. It is best to conduct field tests to obtain the data needed to evaluate the fitted parameters in the infiltration equations.

7.6.4d   Stream Size

The design stream size for a graded border must be small enough to be nonerosive and large enough to adequately spread water across the width of the border. The design stream size must also result in rates of advance and recession that are essentially equal.

The maximum nonerosive stream size for nonsod forming crops, such as alfalfa and small grains, is estimated with the following equation.

$$Q_{max} = \frac{K}{S_0^{-0.75}} \tag{7.24}$$

where

Q_{max} = maximum nonerosive stream size (m^3/s/m, ft^3/s/ft);
S_0 = slope of the graded border (percent);
K = unit constant. ($K = 1.765(10)^{-4}$ for Q_{max} in m^3/s/m. $K = 1.899(10)^{-3}$ for Q_{max} in ft^3/s/ft.)

For well-established, dense sod crops, Q_{max} computed with Eq. 7.24 is doubled.

Stream size selection must consider the effect of the depletion and recession phases as well as the storage and advance phases. The duration of the depletion phase, often called the lag time, for graded borders can be estimated with the following equations:

$$T_L = \frac{n^{1.2}(Q/K)^{0.2}}{120\left(S_0 + \left(\dfrac{0.0094n(Q/K)^{0.175}}{T_I^{0.88}S_0^{0.5}}\right)\right)^{1.60}} \qquad \text{for} \quad S \le 0.4 \text{ percent} \tag{7.25a}$$

$$T_L = \frac{n^{1.2}(Q/K)^{0.2}}{120S_0^{1.6}} \qquad \text{for} \quad S > 0.4 \text{ percent} \tag{7.25b}$$

where

T_L = lag time (min);
  n = Manning roughness factor from Table 7.8;
  Q = stream size (m^3/s/m, ft^3/s/ft);
S_0 = border slope (m/m, ft/ft);
  T_I = infiltration opportunity for desired application depth (min);
  K = unit constant. ($K = 1.0$ for Q in m^3/s/m. $K = 10.76$ for Q in ft^3/s/ft.)

Field tests to determine the position of the recession front as a function of time are recommended. Normally, recession data for only a single stream size (that is approximately equal to the expected stream size) is needed, since recession rates are not usually extremely sensitive to stream size (Merriam et al., 1980).

Values of r and p in Eq. 7.12, and T_a and $T_{0.5a}$, which can be obtained using the procedure demonstrated in Example 7.7, are used with measured recession data to compute the application efficiency and uniformity, deep percolation, and runoff (see Example 7.3). Equation 7.22 can also be used to determine E_a when estimates of uniformity, deep percolation, and runoff are not needed. The design stream size normally maximizes application efficiency.

Application efficiency can be modified by adjusting border length and/or slope. Such changes should be considered when the maximum efficiency for a layout is unacceptably low. In such cases it will be necessary to adjust border length, slope, and stream size until an acceptable application efficiency is obtained.

7.6.4e Inflow Time

As with other surface irrigation systems, the inflow time for borders is normally selected to allow the desired depth of irrigation to be applied at the far end of the border. The inflow time is estimated by assuming that the advance and recession curves are parallel. Thus, the inflow time is given by

$$T_i = T_I - T_L \qquad\qquad (7.26)$$

where

T_i = inflow time (min);
T_I = infiltration opportunity time for desired application depth (min);
T_a = time to advance across the border (min);
T_L = lag time (min).

More convenient inflow times may be obtained by changing the depth of water applied per irrigation. Either Eq. 7.11a, 7.11b, or 7.13 can be used to estimate the time needed to infiltrate a specified depth of water.

7.6.4f Ridge Dimensions

The height of border ridges must be at least 3 cm (1 in) greater than the maximum depth of flow (Booher, 1974). Maximum flow depth information should be collected in the field. When these data are not available, one of the following equations can be used to estimate the depth of flow at the upstream end of a border.

$$d = K_1 T_L^{3/16} Q^{9/16} n^{3/8} \qquad \text{for} \quad S_0 \leq 0.4 \text{ percent} \qquad (7.27a)$$

$$d = K_2 Q^{0.6} n^{0.6} S_0^{-0.3} \qquad \text{for} \quad S_0 > 0.4 \text{ percent} \qquad (7.27b)$$

where

d = normal depth of flow at upstream end of the border (mm, in);
T_L = lag time (min);
Q = stream size (m^3/s/m, ft^3/s/ft);
n = Manning roughness factor from Table 7.8;
S_0 = slope of the border (m/m, ft/ft);
K_1 = unit constant. (K_1 = 2454 for Q in m^3/s/m and d in mm. K_1 = 25.4 for Q in ft^3/s/ft and d in in.)
K_2 = unit constant. (K_2 = 1000 for Q in m^3/s/m and d in mm. K_2 = 9.46 for Q in ft^3/s/ft and d in in.)

The sides of border ridges must be stable when wet. Ridges constructed in clay soils may have base widths of only 60 cm (2 ft), while those installed in loose sandy soils may be up to 2.4 m (8 ft) wide (Booher, 1974).

7.6.4g   Number of Borders Irrigated per Set
The number of borders to be irrigated during a set is calculated with Eq. 7.23. The number of borders irrigated per set and the total number of borders in the field, respectively, are N_B and N_T.

7.6.4h   Delivery System
Delivery channels and pipelines are designed when the desired blend of irrigation time, slope, border length, stream size, and the number of borders irrigated per set have been determined. The procedures in Sections 7.3.1 and 7.3.2 are utilized to design the delivery system.

7.6.5 Design of Furrow Systems

In furrow irrigation, water is distributed across fields in small, evenly spaced, shallow, usually sloping channels. In contrast to basin and border irrigation, furrows do not wet the entire soil surface. Efficient irrigation therefore depends on the lateral movement of water from the furrows. Most crops, except those such as rice that are grown in ponded water, can be irrigated with furrows. Furrow systems are also described in Section 7.2.3.

Furrow system design is an iterative process similar to those used for basin and border system design. System layout, furrow slope, inflow and irrigation times, furrow length, stream size, and the number of furrows irrigated per set are adjusted until the desired combination of efficiency, uniformity, adequacy, and convenience of operation are obtained. The basin irrigation design procedures described in Section 7.6.3 are used to design systems with level impoundment-type furrows.

7.6.5a Layout

Furrow systems should be laid out so that furrows are parallel to a field boundary and have a delivery channel or pipeline at their upper end. In small fields or fields with low infiltration capacity soils, furrows often extend across the full field length. In larger fields or where soils with high infiltration capacities are being irrigated, it may be necesssary to have furrow lengths equal to an even fraction of the total field length. Drainways at the lower end of graded furrows are normally needed to carry away runoff water.

Steep natural terrains are sometimes bench terraced to obtain mildly sloping areas where furrows can be installed. In irregular terrain, graded furrows that generally follow field contours can be used. Field and vegetable crops growing on slopes up to 8 to 10 percent can be irrigated with contour furrows (Booher, 1974). The effect of water source location and furrow length on system layout are discussed in the following sections.

(i) Water Source Location   When possible, the water source should be located so that the entire field can be irrigated via gravity. It is also desirable that the water source be near the center of the irrigated area to minimize the size of delivery channels and pipelines.

(ii) Furrow Length   Where agriculture is mechanized, furrows should be as long as possible to reduce labor requirements and system cost, but short enough to retain reasonable application efficiency and uniformity. However, when labor is plentiful or inexpensive and/or the water supply is limited, short furrows may be most suitable (since application efficiency and uniformity normally increase as furrow length decreases). Short furrows will also be desirable when a variety of crops are to be grown in a small area.

Relatively short furrows are required on sands and other soils with rapid infiltration characteristics and low water holding capacities. Because furrow lengths can normally be increased as average depths of application become larger, furrows can be much longer for deep rooted crops on clay soils than for shallow rooted crops grown in sandy soils.

Booher (1974) recommends that furrow lengths be adjusted according to soil type and slope in fields with large soil and slope variations. Each furrow should be confined to similar soils and slopes.

Table 7.12 lists maximum lengths of cultivated furrows suggested by Booher (1974) for different soils, slopes, and depths of water to be applied. These values should be used as guidelines for planning field trials and developing initial layout concepts.

7.6.5b Land Smoothing

Land smoothing improves the application efficiency and uniformity of furrow irrigation by providing a constant slope along the direction of flow. Field slopes that minimize excavation should be used in the initial stages of system design when they are reasonable. It may be necessary to modify these slopes later in the design process to improve the effectiveness of irrigation and/or the convenience of system operation.

Table 7.12 Suggested Maximum Lengths for Cultivated Furrows for Different Soils, Slopes, and Depths to be Applied

A.   Lengths in meters, depths in centimeters

Furrow Slope	Average Depth of Water Applied (cm)											
	7.5	15	22.5	30	5	10	15	20	5	7.5	10	12.5
	Clays				Loams				Sands			
Percent	. . . .*Meters*.											
0.05	300	400	400	400	120	270	400	400	60	90	150	190
0.1	340	440	470	500	180	340	440	470	90	120	190	220
0.2	370	470	530	620	220	370	470	530	120	190	250	300
0.3	400	500	620	800	280	400	500	600	150	220	280	400
0.5	400	500	560	750	280	370	470	530	120	190	250	300
1.0	280	400	500	600	250	300	370	470	90	150	220	250
1.5	250	340	430	500	220	280	340	400	80	120	190	220
2.0	220	270	340	400	180	250	300	340	60	90	150	190

B.   Lengths in feet; depths in inches

Furrow slope	Average Depth of Water Applied (in)											
	3	6	9	12	2	4	6	8	2	3	4	5
	Clays				Loams				Sands			
Percent	. . . .*Feet*.											
0.05	1 000	1 300	1 300	1 300	400	900	1 300	1 300	200	300	500	600
0.1	1 100	1 400	1 500	1 600	600	1 100	1 400	1 500	300	400	600	700
0.2	1 200	1 500	1 700	2 000	700	1 200	1 500	1 700	400	600	800	1 000
0.3	1 300	1 600	2 000	2 600	900	1 300	1 600	1 900	500	700	900	1 300
0.5	1 300	1 600	1 800	2 400	900	1 200	1 500	1 700	400	600	800	1 000
1.0	900	1 300	1 600	1 900	800	1 000	1 200	1 500	300	500	700	800
1.5	800	1 100	1 400	1 600	700	900	1 100	1 300	250	400	600	700
2.0	700	900	1 100	1 300	600	800	1 000	1 100	200	300	500	600

Source: L. J. Booher, *Surface Irrigation* (1974), Food and Agriculture Organization of the United Nations, Rome, Italy, 160 pp.

Hart et al. (1980) recommend that furrow slopes not exceed 1.0 percent except in arid areas where rainfall-induced erosion is not a hazard. In such areas furrow slopes of up to 3.0 percent may be acceptable. In humid areas furrow slopes usually should not be more than 0.3 to 0.5 percent. A minimum slope of 0.03 to 0.05 percent is required in humid areas to assure adequate surface drainage. Land smoothing is discussed in Section 7.5.4a.

7.6.5c Irrigation Time

The irrigation time for furrow irrigation is determined by substituting the desired depth of irrigation into either Eq. 7.11a, 7.11b, or 7.13. Values for the parameters in these equations can be quantified using field data from blocked furrow, inflow/outflow, recirculating infiltrometer, or two-point method tests.

7.6.5d Stream Size

In furrow systems, application uniformity generally increases as stream sizes become larger. Maximum efficiency, however, may not be associated with the largest stream sizes in furrow systems without cutback or runoff reuse. In such systems, application efficiency is reduced by the large runoff losses that occur with large stream sizes. Thus, furrow stream sizes must be carefully selected to obtain the desired blend of irrigation effectiveness and convenience of operation.

Furrow stream sizes must not exceed the maximum nonerosive stream size determined in field trials. The following equation provides guidance in selecting stream sizes for field trials.

$$Q_{max} = \frac{K}{S_0} \tag{7.28}$$

where

Q_{max} = maximum nonerosive stream size (l/min, gpm);
S_0 = furrow slope in the direction of flow (percent);
K = unit constant. ($K = 40$ for Q_{max} in l/min. $K = 10$ for Q_{max} in gpm.)

Furrow streamsizes are sometimes selected on the basis of the one-fourth rule. This rule states that the time required for water to advance through a furrow should be one-fourth of the irrigation time. Example 7.8 illustrates the use of the one-fourth rule to determine the design streamsize.

EXAMPLE 7.8 Selecting Furrow Stream Sizes Using the One-Fourth Rule

Given:
• measured advance data for stream sizes of 1, 3, and 5 m^3/h plotted in the figure below
• irrigation time = 360 min
• field length is approximately 120 m

Required:
• design stream size for the time of advance to be one-fourth of the irrigation time

Solution:

Solution Steps
1. compute desired advance time, T_a
2. Use T_a and advance data to estimate stream size

Solution Step 1

(Compute T_a)

$$T_a = \frac{T_I}{4} = \frac{360}{4} = 90 \text{ min}$$

Solution Step 2
(Determine design stream size)

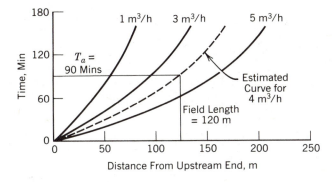

From the above figure the design stream size is approximately 4 m³/h.

Since large runoff losses may result, cutback irrigation or tailwater reuse is recommended when the one-fourth rule is followed. In situations where cutback irrigation or tailwater reuse is not used, it may be desirable to use a smaller stream size (than would be used if the one-fourth rule were applied) to slow advance, decrease runoff, and improve application efficiency.

When the one-fourth rule is not used, the design stream size for a furrow is determined from rate of advance data for different stream sizes (the depletion and recession phases are normally short relative to the advance and storage phases and can usually be neglected). The procedure demonstrated in Example 7.7 can be used to generate the required advance data. The stream size that results in the highest application efficiency is normally selected as the design stream size. When advance data are available, application efficiency and uniformity, deep percolation, and runoff can be determined with the procedures demonstrated in Example 7.3 or with the computer program in Appendix J. Application efficiency can also be computed with Eq. 7.22 (when estimates of uniformity, deep percolation, and runoff are not required).

Application efficiency can be modified by adjusting furrow length and/or slope. Such changes should be considered when the maximum efficiency for a layout is unacceptably low. In such cases it will be necessary to adjust furrow length, slope, and/or stream size until an acceptable application efficiency is obtained.

7.6.5e Inflow Time

The inflow time for furrow irrigation is normally selected so that the desired depth of irrigation is applied at the far end of the furrow. The design inflow time is determined by assuming that the duration of the depletion and recession phases is negligible. The following equation is usually used to compute the inflow time for furrow irrigation systems.

$$T_i = T_a + T_l \tag{7.29}$$

where

T_i = inflow time (min);
T_a = time to advance across the field (min);
T_l = irrigation time (min).

More convenient inflow times may be obtained by adjusting the depth of water applied per irrigation. This is accomplished using either Eq. 7.11a, 7.11b, or 7.13.

7.6.5f Furrow Spacing

The distance between furrows should be based on optimum crop spacing, modified if necessary to obtain adequate lateral wetting and accommodate farm equipment. A standard furrow spacing is often used for a number of different crops that make use of the same farm equipment. This eliminates changing the spacing of the tool attachments when the equipment is moved from one crop to another.

The lateral movement of water from furrows in soils with uniform profiles depends primarily on the texture of the soil. Typical wetting patterns for sands, loams, and clays are illustrated in Figure 7.29. Sandy soils that tend to have a vertical wetted pattern should have closer furrow spacing than clay or loam soils.

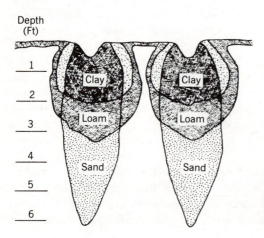

Figure 7.29 Wetting patterns in furrows of differing soil types. (From Section 15, Chapter 5 of the *SCS National Engineering Handbook*, 1984.)

Soils with nonuniform profiles will generally have greater lateral movement of water than soils laying above less permeable layers or above abrupt changes in soil texture.

In most situations, the primary consideration in selecting furrow spacing is to place them close enough so that soil between adjacent furrows is wet before water moves beyond the depth of the root zone. Wetting patterns can be determined by digging a trench across a furrow after the desired depth of irrigation has been applied.

Sometimes furrows are placed so that a dry area remains between adjacent furrows or groups of furrows. The resulting wetting pattern is similar to that obtained with "line-source" trickle irrigation. Furrows in orchards, vineyards, and hopyards are often spaced in this way to conserve water and control weeds.

7.6.5g Furrow Shape and Size

Most furrows are V-shaped, 10 to 20 cm (4 to 8 in) deep, and 20 to 30 cm (8 to 12 in) wide at the top. Shallower furrows 10 to 15 cm (4 to 6 in) deep are often used for shallow rooted crops especially during germination, while broad-based, U-shaped furrows with bottom widths of 15 to 20 cm (6 to 8 in) are sometimes used in low infiltration capacity soils. Shallow furrows require that the fields be carefully graded to a uniform slope and that furrows have a uniform depth and shape along the length of the field to prevent overtopping. Broad-based furrows promote infiltration by increasing the wetted perimeter over which infiltration occurs. Deep furrows allow at least a portion of the root zone to be above the furrow bottom, thus minimizing the adverse effects caused by waterlogging of the root zone in high rainfall areas where water may accumulate and remain for extended periods of time. On the other hand, deep furrows require a large volume of water to adequately irrigate the upper portion of the root zone—an important point where crops have shallow roots due to compaction below the furrow bottom.

Furrow shapes are normally modified by water flow. On steep slopes narrower channels tend to form, whereas on flatter slopes broader channels usually result. These tendencies are greater on sandy soils than on clay soils.

7.6.5h Number of Furrows Irrigated per Set

Equation 7.23 is used to determine the number of furrows irrigated during a set. N_B and N_T are the number of furrows irrigated per set and the total number of furrows in the field, respectively.

7.6.4i Delivery System

Delivery channels and pipelines are designed when the desired blend of inflow time, furrow slope, furrow length, stream size, and number of furrows irrigated per set has been determined. The procedures in Sections 7.3.1 and 7.3.2 are utilized to design the delivery system.

Homework Problems

7.1 A canal is being designed to convey water from a lake to a 50-ha irrigated farm. The farm irrigation system is designed to provide a design daily irrigation requirement of 8 mm/day and to operate 24 h day. Determine the required capacity of
a. An unlined ditch, and
b. A lined channel.

7.2 Design a triangular-shaped unlined ditch to convey water to the farm in Problem 7.1. The average slope along the route of the ditch is approximately 0.35 percent and the soil is a noncolloidal silt loam.

7.3 Repeat Problem 7.2 for a trapezoidal-shaped unlined ditch.

7.4 Repeat Problem 7.2 for a rectangular-shaped concrete lined canal.

*7.5 Use the computer program in Appendix I to solve Problem 7.2.

*7.6 Use the computer program in Appendix I to solve Problem 7.3.

*7.7 Use the computer program in Appendix I to solve Problem 7.4.

7.8 A division box similar to the one in Figure 7.3*b* is being used to deliver 2000 and 3000 l/min, respectively, into two outlet channels downstream of the structure. Determine the weir crest elevation for each outlet of the division box. The water surface in the box is 0.75 m above the bottom of the structure and each outlet is 0.6 m wide. The discharge of a weir is given by the following equation:

$$Q = 110287(W - 0.2\,h)h^{3/2}$$

where

Q = discharge (l/min);
W = width of weir (m);
h = head above weir crest (m).

7.9 In Problem 7.8, 10000 l/min is delivered to the division box. Determine the flow into each outlet channel if the weir crest elevations of the outlets are 0.6 and 0.7 m above the bottom of the division box, respectively.

7.10 Determine the flow through a 25 mm (1 in) diameter, 1.5 m (5 ft) long aluminum siphon tube when the difference between the water surface elevations in the head ditch and the furrow is 23 cm (9 in).

---

* Indicates that a computer program will facilitate the solution of the problems so marked.

7.11 A trash rack is placed in a 1 m wide, concrete-lined rectangular channel that has a normal depth of flow of 0.5 m. The channel has a slope of 0.5 percent. The bars of the trash rack have square corners and are 4 mm thick by 20 mm wide. There is a 2-cm opening between adjacent bars. Estimate the head loss across the trash rack.

7.12 Determine the discharge from a 15-cm-diameter alfalfa valve that is opened 3 cm. The head in the pipeline is 30-cm.

7.13 Repeat Problem 7.12 for a 15-cm-diameter orchard valve.

7.14 Determine
   a. the uniformity of application,
   b. the application efficiency, and
   c. the storage efficiency

for 200-m-long furrows spaced 75 cm apart. A stream size of 40 l/min enters the furrows for 24 hours. Recession is instantaneous and begins 2 hours after inflow ends. The readily available water holding capacity is 15 cm and advance and infiltration are described by the following relationships.

$$X = 4.5 \, t^{0.56}$$

$$I = 10.0 \, t^{0.40}$$

where

X = distance from upstream end of the field that water has advanced (m);
t = time (min);
I = cumulative depth of infiltration (mm).

*7.15 Solve Problem 7.14 using the computer program in Appendix J.

7.16 Smooth the following field for surface irrigation. Determine:
   a. the grades in the north–south and west–east directions, and
   b. the total volume of excavation.

		10.0	11.2	11.5	10.8
N		9.8	10.5	10.8	10.7
W	E	9.9	10.6	11.0	10.9
	S	10.1	9.8	11.3	11.0

The elevations are in meters and the points are spaced in a 20-m-square grid.

*7.17 Use the computer program in Appendix K to solve Problem 7.16.

7.18 Use the following advance and inflow/outflow information to determine an adjusted cumulative infiltration function for a 250-m-long furrow. The spacing between furrows is 75 cm. Inflow and outflow measurements were made at the upstream end of the furrow and at a station 50 m further downstream. There was 19.37 m^3 of runoff from each furrow.

Distance Along Furrow (m)	Advance Time (min)	Recession Time (min)
0	0	1000
25	4	1000
50	8	1000
75	14	1000
100	22	1000
125	31	1000
150	42	1000
175	55	1000
200	72	1000
225	91	1000
250	115	1000

Time Since Inflow Began (min)	Inflow Rate (l/min)	Outflow Rate (l/min)
10	36.00	15.01
20	36.00	28.49
40	36.00	29.63
70	36.00	31.61
130	36.00	32.29
210	36.00	33.33
310	36.00	33.41
410	36.00	33.49
510	36.00	33.85
910	36.00	34.39

7.19 Estimate the expected stream size for a 12-ha basin constructed in a sandy loam soil.

7.20 Determine the application efficiency for the basin in Example 7.7 for a desired application depth of 15 cm. Assume that the basin is level (i.e., that $S_0 = 0$).

7.21 Using information from Example 7.7, determine the application efficiency for a graded border with a slope of 0.1 percent. The desired application depth is 15 cm.

7.22 Determine the ridge height for the graded border in Problem 7.21.

7.23 Using the infiltration data from Problem 7.18 determine the irrigation time required to apply 100 mm of irrigation water.

7.24 Estimate the maximum nonerosive stream size for a furrow with a 0.5 percent slope.

7.25 Use the one-fourth rule and the advance and adjusted infiltration data from Problem 7.18 plus the following additional advance data to determine
a. the irrigation time,
b. the stream size, and
c. the inflow time

for a furrow irrigation system. The readily available water holding capacity is 48 mm. Is the resulting stream size acceptable if the field slope is 0.40 percent?

Distance From Upstream End of Furrow in m	Advance Time in Minutes for	
	$Q = 18$ l/min	$Q = 24$ l/min
0	0	0
25	4	4
50	11	10
75	21	17
100	36	28
125	58	42
150	89	61
175	133	86
200	196	118
225	283	161
250	404	216

7.26 Repeat Problem 7.25 for a readily available water holding capacity of 67 mm.

*7.27 Use the computer program in Appendix J to determine
a. the application efficiency,
b. the Christiansen uniformity coefficient, and
c. the storage efficiency

for the stream size, application depth, and inflow time calculated in Problem 7.25.

*7.28 Use data from Problems 7.18 and 7.25 to determine the stream size that maximizes application efficiency. Also determine the required irrigation and inflow times.

*7.29 Use the computer program in Appendix L and the advance and stream size data from Problem 7.18 to determine the application efficiency for a stream size of 30 l/min. The slope of the field is 0.4 percent and the desired application depth is 10 cm.

References

Aisenbrey, A. J., Jr., R. B. Hayes, H. J. Warren, D. L. Winsett, and R. B. Young (1978). Design of small canal structures. U.S. Dept. of the Interior Bureau of Reclamation, Denver, Colo., 435 pp.

Bassett, D. L. (1972). Mathematical model of the water advance in border irrigation. *Trans. ASAE*, **15**(6), pp. 992–995.

Bassett, D. L., D. D. Fangmeier, and T. Strelkoff (1980). Hydraulics of surface irrigation. In *Design and Operation of Farm Irrigation Systems*, M. E. Jensen (Ed.), ASAE Monograph 3, St. Joseph, MI, pp. 447–498.

Bassett, D. L., and D. W. Fitzsimmons (1976). Simulating overland flow in border irrigation. *Trans. ASAE*, **19**(4), pp. 674–680.

Benami, A., and A. Ofen (1984). Irrigation engineering: Sprinkler, trickle, and surface irrigation; Principles, design, and agricultural practices. Irrigation Engineering Scientific Publications, Bet Dagan, Volcani Center, Israel, 257 pp.

Bergstrom, W. (1961). Weed seed screens for irrigation systems. Pacific Northwest Cooperative Extension Service (Wash., Oreg., Idah.), Extension Publication Bulletin 43, 7 pp.

Bishop, A. A., M. E. Jensen, and W. A. Hall (1967). Surface irrigation systems. In *Irrigation of Agricultural Lands*, R. M. Hagen, H. R. Haise, and T. W. Edminster (Ed.), ASA Monograph 11, pp. 865–884.

Bondurant, J. A., and W. D. Kemper (1985). Self-cleaning, non-powered trash screens for small irrigation flows. *Trans. ASAE*, **28**(1), pp. 113–117.

Booher, L. J. (1974). Surface irrigation. Food and Agriculture Organization of the United Nations, Rome, Italy, 160 pp.

Bos, M. G., J. A. Replogle, and A. J. Clemmens (1984). *Flow Measuring Flumes for Open Channel Systems*. Wiley, New York, 312 pp.

Chow, V. T. (1959). *Open Channel Hydraulics*. McGraw-Hill, New York, 680 pp.

Clemmens, A. J. (1979). Verification of the zero-inertia model for surface irrigation. *Trans. ASAE*, **22**(6), pp. 1306–1309.

Clemmens, A. J., and D. D. Fangmeier (1978). Discussion of Strelkoff and Katapodes (1977). *J. Irrigation and Drainage Division*, American Society of Civil Engineers, **104**(IR3), pp. 337–339.

Clemmens, A. J., and T. Strelkoff (1979). Dimensionless advance for level-basin irrigation. *J. Irrigation and Drainage Devision*, American Society of Civil Engineers, **105**(3), pp. 259–273.

Couthard, T. L., J. C. Wilcox, and H. O. Lacy (1956). Screening irrigation water. Agricultural Engineering Division Bulletin A.E. 6, University of British Columbia, Vancouver, 14 pp.

Dedrick, A. R., L. J. Erie, and A. J. Clemmens (1982). Level-basin irrigation. In *Advances in Irrigation*, Vol. 1, D. Hillel (Ed.), Academic Press, New York, pp. 105–145.

Elliott, R. L. (1981). Zero-inertia furrow irrigation modeling applied to the derivation of infiltration parameters. Ph.D dissertation, Colorado State Univ., Ft. Collins, 181 pp.

Elliott, R. L., and W. R. Walker (1982). Field evaluation of furrow infiltration and advance functions. *Trans. ASAE*, **25**(2), pp. 396–400.

Haie, N. (1984). Hydrodynamic simulation of continuous and surged surface flow. Ph.D dissertation, Utah State Univ., Logan, 147 pp.

Hart, W. E., H. G. Collins, G. Woodward, and A. S. Humpherys (1980). Design and operation of gravity or surface irrigation systems. In *Design and Operation of Farm*

Irrigation Systems, M. E. Jensen (Ed.), ASAE Monograph 3, St. Joseph, MI, pp. 501–580.

Katapodes, N. D., and T. Strelkoff (1977a). Hydrodynamics of border irrigation-complete model. *J. Irrigation and Drainage Division*, American Society of Civil Engineers, **103**(IR3), pp. 309–324.

Katapodes, N. D., and T. Strelkoff (1977b). Dimensionless solution of border irrigation advance. *J. Irrigation and Drainage Division*, American Society of Civil Engineers, **103**(IR4), pp. 401–407.

Kincaid, D. C., D. E. Heermann, and E. A. Kruse (1972). Hydrodynamics of border irrigation. *Trans. ASAE*, **15**(4), pp. 674–680.

Kincaid, D. C. (1984). Cablegation: V. Dimensionless design relationships. *Trans. ASAE*, **27**(3), pp. 769–772, 778.

Kraatz, D. B. (1977). Irrigation canal lining. FAO Land and Water Development Series No. 1, Food and Agriculture Organization of the United Nations, Rome, Italy, 199 pp.

Kraatz, D. B., and I. K. Mahajan (1975). Small hydraulic structures. Irrigation and Drainage Paper No. 26, Parts 1 and 2, Food and Agriculture Organization of the United Nations, Rome, Italy, 407 pp. and 293 pp., respectively.

Kruse, E. G., A. S. Humpherys, and E. J. Pope (1980). Farm water distribution systems. In *Design and Operation of Farm Irrigation Systems*, M. E. Jensen (Ed.), ASAE Monograph 4, St. Joseph, MI, pp. 394–443.

Marr, J. C. (1965). Grading land for surface irrigation. Circular 438, California Agricultural Experiment Station Extension Service, Univ. of California, Davis, 55 pp.

Merriam, J. L., M. N. Shearer, and C. M. Burt (1980). Evaluating irrigation systems and practices. In *Design and Operation of Farm Irrigation Systems*, M. E. Jensen (Ed.), ASAE Monograph 4, St. Joseph, MI, pp. 721–760.

Peri, G., D. I. Norum, and G. V. Skogerboe (1979). Evaluation and improvement of basin irrigation. Water Management Technical Report 49B, Water Management Research Project, Engineering Research Center, Colorado State Univ. Ft. Collins, 179 pp.

Peri, G., D. I. Norum, and G. V. Skokerboe (1979). Evaluation and improvement of border irrigation. Water Management Technical Report 49C, Water Management Research Project, Engineering Research Center, Colorado State Univ., Ft. Collins, 105 pp.

Pugh, W. J., and N. A. Evans (1964). Weed seed and trash screens for irrigation water. Colorado Agricultural Experiment Station Bulletin 522-S, Colorado State Univ., Fort Collins.

Schwab, G. O., R. K. Frevert, T. W. Edminster, and K. K. Barnes (1981), *Soil and Water Conservation Engineering*. Wiley, New York, 525 pp.

Souza, F. (1981). Non-linear hydrodynamic model of furrow irrigation. Ph.D dissertation, Univ. of California-Davis, Davis, 172 pp.

Soil Conservation Service (1974). Border irrigation, Chapter 4, Section 15 of the *SCS National Engineering Handbook*, U.S. Department of Agriculture, Washington, D.C.

Soil Conservation Service (1984). Furrow irrigation, Chapter 5, Section 15 of the *SCS National Engineering Handbook*, U.S. Department of Agriculture, Washington, D.C.

Strelkoff, T., and A. J. Clemmens (1981). Dimensionless stream advance in sloping borders. *J. Irrigation and Drainage Division*, American Society of Civil Engineers, **107**(IR4), pp. 361–382.

Strelkoff, T., and N. D. Katapodes (1977). Border irrigation hydraulics with zero-inertia. *J. Irrigation and Drainage Division*, American Society of Civil Engineers, **103**(IR3), pp. 325–342.

Walker, W. R., and G. V. Skogerboe (1987). *Theory and Practice of Surface Irrigation.* Prentice-Hall, Englewood Cliffs, N.J.

U.S. Army (1959). Hydraulic design criteria. Corps of Engineers, Waterways Experiment Station, Vicksburg, Miss.

U.S. Department of the Interior (1978). Report on the water conservation opportunities study. U.S. Bureau of Reclamation and Bureau of Indian Affairs, Washington, D.C.

8

Flow Measurement

8.1 Introduction

Effective irrigation management and maintenance programs require a knowledge of the amount and rate of water use. Data describing the volume of water use and/or volumetric flowrate (volume per unit time) are invaluable in controlling water application and detecting changes in performance due to system malfunction or maintenance problems. Table 8.1 lists some possible causes of changes in irrigation system flow.

All irrigation systems should include a measuring device for determining system flow. Provision should also be included for measuring flow at locations where it is divided between and within fields and at other critical points within irrigation systems.

The primary purpose of this chapter is to introduce the major devices and methods of measuring flow in irrigation systems. Devices and procedures for making flow measurements in open channels and pipelines are presented. Equations for estimating flow and designing measuring devices for specific applications are included.

8.2 Flow Volume and Volumetric Flowrate

The volume of water use and the volumetric flowrate describe flow in irrigation systems. The relationship between these two quantities is

$$V = \frac{Q \Delta t}{K} \tag{8.1}$$

Table 8.1   Some Possible Causes of Changes in Irrigation System Flow

Increased Flow

Improperly adjusted gates, valves, checks
Pipeline leaks and breaks
Pressure downstream of pressure regulators is too high
Worn or oversize sprinkler nozzles, emission devices, etc.
System on too long (as indicated by higher than expected volumes of flow)

Decreased Flow

Improperly adjusted gates, valves, checks
Clogged sprinklers, emission devices, screens, filters, etc.
Pump wear
Pressure downstream of pressure regulators too low
Existence of entrapped air in the system
System not on long enough (as indicated by lower than expected volumes of flow)

Source: L. G. James and W. M. Shannon, "Flow Measurement and System Maintenance." In *Trickle Irrigation for Crop Production*, F. S. Nakayama and D. A. Bucks (Eds.), copyright © 1986, Elsevier Science Publishing Co., Inc., p. 280. Reprinted by permission of Elsevier Science Publishing Co., Inc. and authors.

Table 8.2   Common Units of Flow Volume and Flowrate and Unit Conversions Used in Irrigation

Volume

1 cubic meter	= 1000 liters
1 liter	= 1000 cubic centimeters
1 cubic meter	= 35.31 cubic feet
1 cubic meter	= 264.2 gallons
1 acre-foot	= 1233.3 cubic meters
1 acre-foot	= 43560 cubic feet
1 acre-foot	= 12 acre-inches
1 acre-foot	= 325,829 gallons

Flowrate

1 l/min	= 60 l/s
1 cubic meter/s	= 1000 l/s
1 l/s	= 15.85 gallons/minute
1 cubic meter/s	= 35.31 cubic feet/s
1 gallon/min	= 3.785 l/min
1 gallon/h	= 3.785 l/h
1 cubic foot/s	= 448.8 gallons/min
1 cubic foot/s	= 1.9835 acre-feet/day
1 acre-inch/h	= 452.6 gallons/min
1 acre-inch/h	= 1.0083 cubic feet/s

where

V = volume of flow (m³, ac-ft);

Q = volumetric flowrate (l/s, gpm);

Δt = time interval (min)

K = unit constant. ($K = 16.67$ for V in m³ and Q in l/s. $K = 325,829$ for V in acre-ft and Q in gpm.)

    The primary units of volume in the SI and English unit systems are the cubic meter and the acre-foot, respectively. Liters per second (l/s) and gallons per minute (gpm) are, respectively, the primary SI and English units for volumetric flowrate (American Society of Agricultural Engineers, 1986). Table 8.2 lists several other units of volume and flowrate as well as factors for converting from one unit to another.

    The following example illustrates the use of the factors in Table 8.2 to convert from one unit of volume/flowrate to another.

EXAMPLE 8.1   Flowrate and Volume Unit Conversions

Given:

$V = 2530$ cubic meters

Required:
(a V in acre-feet
(b) V in liters

Solution:

(a) V in acre-feet $= \dfrac{2530}{1233.5} = 2.05$ ac-ft

(b) V in liters $= (2530 \text{ m}^3)(1000 \text{ l/m}^3) = 2,530,000 \text{ l}$

8.3   Flow Measurement in Open Channels

There are several methods of measuring flow in open irrigation channels. Flowrate is measured using either the volumetric, velocity-area, control section, or dilution methods. Flow volume can be determined from flowrate measurements using Eq. 8.1.

8.3.1   Volumetric Method

In this method, the time required for the flow to fill a container of known volume is measured. The flowrate is determined by dividing the volume of the container by

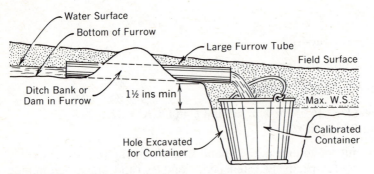

Figure 8.1 Volumetric measurement of outflow from an irrigation furrow.

the time required to fill it. An installation for measuring furrow flows with the volumetric method is shown in Figure 8.1. The volumetric method is also used to measure the discharge of individual sprinklers and trickle emission devices.

The accuracy of volumetric measurements depends on the size of the container relative to the accuracy of the timing device. For example, to achieve an accuracy of 1 percent or less with a timing device that is accurate to within 0.2 seconds, the container must be large enough so that at least 20 seconds is required to fill it. Similarly a filling time of 10 seconds is required for 2 percent accuracy and a 4-second filling time is needed for 5 percent accuracy.

8.3.2   Velocity-Area Method

The velocity-area approach involves measuring the velocity and cross sectional area of the channel and using the following equation to compute the flowrate.

$$Q = KVA \tag{8.2}$$

where

Q = flowrate (l/s, gpm);

V = the average flow velocity (m/s, ft/s);

A = cross sectional area normal to the flow (m^2, ft^2);

K = unit constant. ($K = 1000$ for Q in l/s, V in m/s, and A in m^2. $K = 448.8$ for Q in gpm, V in ft/s, and A in ft^2.)

8.3.2a   Area

In wide and/or irregularly shaped channels the cross section (normal to flow) is often divided into several segments as shown in Figure 8.2. The area of each segment is the product of the width of the segment and its average depth. The areas of the segments are summed to determine the total cross-sectional area.

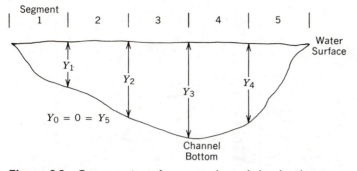

Figure 8.2   Cross section of an open channel that has been subdivided into five segments with equal widths. Y_i is the depth of flow at location i.

8.3.2b   Velocity

Determination of the velocity is difficult, since there are normally significant variations in velocity within irrigation channel cross sections. The float method of measuring velocity can be used when high accuracy is not required or when costly installations are not warranted. A current meter (see Figure 8.3) is used when more accurate measurements are required. Velocities can also be measured with chemical tracers.

(i) The Float Method   This method requires a straight section of channel that has a uniform cross section and is free of surface disturbances and cross currents. Float measurements should be made on windless days to avoid wind-induced deflections of the float.

For best results, a string or tape should be stretched across the beginning, midpoint, and end of the channel section. These locations must be far enough apart to allow accurate travel time measurements. The float should be released a sufficient distance upstream for it to attain the stream velocity before it enters the test section. The times when the float passes each station should be noted and recorded. Wide channels are often divided into segments and a float used to determine the velocity of each segment.

The average velocity of the stream (or stream segment) is computed with the following equation.

$$V = (CF)(V_f) \qquad\qquad (8.3)$$

where

V = average velocity of the stream (or stream segment);

CF = velocity correction factor from Table 8.3;

V_f = float velocity.

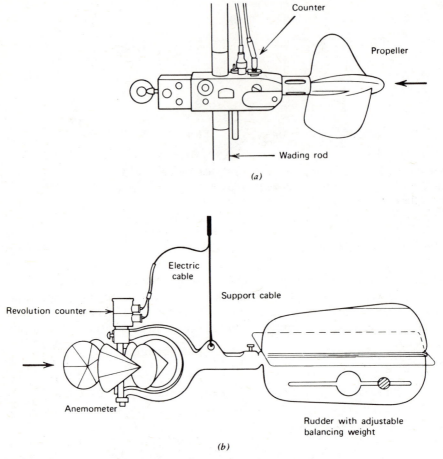

Figure 8.3 (*a*) Propeller- and (*b*) Price-type current meters.

(ii) Current Meters   Two types of current meters are shown in Figure 8.3. When immersed in moving water, the meter's impeller revolves at a speed that is proportional to the water velocity. The water velocity is determined by noting the time required for a certain number of revolutions and using a meter calibration curve that relates the speed of meter rotation to water velocity.

Current meters are used to measure velocity at several depths and positions across a channel. Standard procedure is to divide a cross section into segments (as previously described) and use a current meter to determine the average velocity in each segment from either multiple-depth, two-depth, or single-depth measurements.

The *multiple-depth method* involves measuring the velocity at several closely spaced points from the bottom of the channel to the water surface in each segment. If there is equal vertical spacing between measuring points, the mean velocity

Table 8.3   Correction Factor for Eq. 8.3

Average Flow Depth		
(m)	(ft)	CF
0.3	1.0	0.66
0.6	2.0	0.68
0.9	3.0	0.70
1.2	4.0	0.72
1.5	5.0	0.74
1.8	6.0	0.76
2.7	9.0	0.77
3.7	12.0	0.78
4.6	15.0	0.79
≥ 6.1	≥ 20.0	0.80

Source: U.S. Bureau of Reclamation, *Water Measurement Manual* (1975), Department of the Interior, U.S. Government Printing Office, Washington, D.C., 327 pp.

approximates the average of the measured velocities. The multiple-depth method is more accurate than either the two-depth or single-depth methods.

In the *two-depth method*, the velocity is determined at 0.2 and 0.8 of the depth in each channel segment. The average of these two measurements approximates the mean velocity for ordinary conditions.

In the *single-depth method*, the velocity is determined at a point 0.6 of the stream depth below the water surface. This method is generally employed for shallow depths where use of the two-depth method is difficult. The single-depth method is used for flow depths of 0.3 m (1 ft) or less.

(iii) *Tracer Methods*   In this method, the time required for a "charge" of a chemical tracer to travel through a test section of known length is measured and used to compute the velocity, by dividing the test section length by the travel time. Dye tracers, such as fluorescein or potassium permanganate, are detected visually. When the tracer is salt (NaCl), electrodes placed at the ends of the test section are used to determine when the tracer enters and leaves the test section.

8.3.3   Control Sections

This method involves the use of natural or constructed/installed control sections with stable depth of flow (stage) versus discharge (volumetric flowrate) relationships. Discharge is determined from measurements of stage using the stage-discharge relationship for the control section. Stage-discharge relationships are determined by field or laboratory calibration and analytically using the conservation of energy principle and boundary layer theory (Bos et al., 1984).

Natural controls are cross sections or channel reaches where the channel cross section, slope, and roughness remain relatively constant and there is no tailwater. In situations where a suitable natural control is not available, flow measuring devices including thin-plate weirs, flumes, and orifices are constructed/ installed to provide a stable stage-discharge relationship. Hydraulic structures with definite stage-discharge relationships, such as culverts and sluice gates, can also be used to accurately measure flow in open irrigation channels.

8.3.3a   Natural Controls

Stage versus discharge relationships for natural controls, called *rating curves*, are usually developed in the field for a wide range of flows using the velocity-area method. Rating curve shifts due to such things as sediment accumulation must be considered to make accurate flow measurements. Once the rating curve has been established, a continuous record of discharge can be obtained by making continuous water stage measurements. The U.S. Bureau of Reclamation *Water Measurement Manual* (1975) gives procedures for developing rating curves using the velocity-area method and for making measurements of water stage.

8.3.3b   Thin-Plate Weirs

Thin-plate weirs, also called *sharp-crested weirs*, consist of a smooth, vertical, flat plate installed across the channel and perpendicular to flow. The plate obstructs flow causing water to back up behind the weir plate and to flow over the weir crest as shown in Figure 8.4. The distance from the bottom of the channel to the weir crest is the crest height. The depth of flow over the weir crest (measured at a specified distance upstream of the weir plate) is called the head. The overflowing sheet of water is known as the nappe.

Thin-plate weirs are most accurate when the nappe completely springs free of the upstream edge of the weir crest and air is able to pass freely around the nappe.

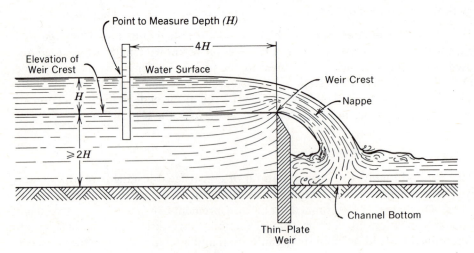

Figure 8.4 Profile of a thin-plane weir.

A head of at least 6 cm (2 in) and a crest thickness of no more than 1 or 2 mm (0.03 or 0.08 in) are required for water to spring free of the weir crest (U.S. Bureau of Reclamation, 1975). Knife-edged crests are not recommended because they are easily damaged by debris, sediment, and rust pitting (Ackers et al., 1978). Recommended crest geometry for different types of thin-plate weirs is shown in Figure 8.5. Downstream water levels that do not permit free aeration around the nappe should be avoided because they reduce measurement accuracy and require additional measurements and calculations.

The positions of the weir crest relative to the bottom of the approach section can also affect the performance of thin-plate weirs. The U.S. Bureau of Reclamation (1975) recommends that the crest height of a thin-plate weir be at least twice the head on the weir. Otherwise, the contraction of the nappe is not complete vertically and standard tables and equations relating head and discharge are not valid.

While easy to construct/install and convenient to use, thin-plate weirs may not always be suitable. They are not recommended when flow contains sediment and debris, because measurement accuracy is reduced by deposition in the approach channel. In addition, they cannot be used in channels on flat grades where the required difference in elevation between the water levels on the up- and downstream side of the weir is not available.

The crest of thin-plate weirs may extend across the full width of the channel or be notched. The most common notch shapes are rectangular, triangular, trapezoidal.

(i) Full-width Weirs   Full-width weirs are called *suppressed rectangular weirs* because their sides are coincident with the sides of the approach channel and no lateral contraction of the nappe is possible. Provision for ventilating the nappe must be provided to prevent the development of a zone of low pressure below the nappe and the tendency for the nappe to cling to the crest (Ackers et al., 1978). The discharge of a full-width thin-plate weir can be computed with the following equation:

$$Q = KbH^{3/2} \quad \text{for} \quad H < P/2 \tag{8.4}$$

$$H > 6 \text{ cm (0.2 ft)}$$

$$P > 0.3 \text{ m (1 ft)}$$

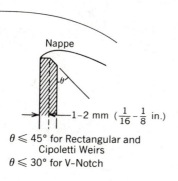

$\theta \leqslant 45°$ for Rectangular and
  Cipoletti Weirs
$\theta \leqslant 30°$ for V-Notch

Figure 8.5   Thin-plate weir section.

where

Q = discharge (l/s, gpm, cfs);
b = crest width (m, ft);
H = difference between the crest and the water surface at a point upstream
   from the weir a distance of four times the maximum head on the crest
   (m, ft);
P = crest height (m, ft);
K = unit constant. (K = 1838 for Q in l/s and b and H in m. K = 1495 for Q in
   gpm and b and H in ft, and K = 3.33 for Q in cfs and b and H in ft.)

Equation 8.4 applies when the approach velocity is negligible. The following equation applies when the approach velocity is appreciable and cannot be neglected.

$$Q = Kb((H + h_a)^{3/2} - h_a^{3/2}) \qquad (8.5)$$

where

h_a = velocity head of approach flow (m, ft).

   (ii) Contracted Rectangular Weirs   Figure 8.6 shows a contracted rectangular weir. To be considered contracted, the sides of the weir notch must be far enough from the sides of the channel to allow full horizontal contraction of the nappe. This requires that the distance from the sides of the weir to the sides to the approach channel be at least twice the head but not less than 0.3 m (1 ft) (U.S. Bureau of Reclamation, 1975). The nappe must be fully contracted vertically as well. The following equation relates discharge to the upstream head.

$$Q = K(b - 0.2H)H^{3/2} \qquad (8.6)$$

where Q, b, H, and K are as defined for Eq. 8.4.

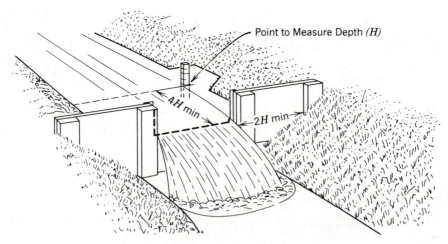

Figure 8.6   Rectangular weir with end contractions.

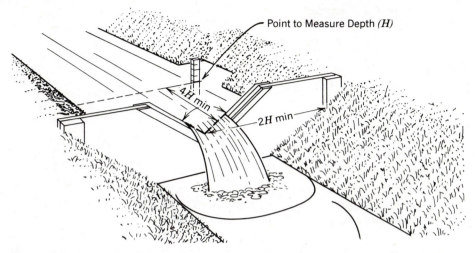

Point to Measure Depth *(H)*

4H min

2H min

Figure 8.7 V-notch, thin-plate weir.

Equation 8.6 applies when the approach velocity is negligible. The following equation applies when the approach velocity cannot be neglected.

$$Q = K(b - 0.2H)((H + h_a)^{3/2} - h_a^{3/2})$$ (8.7)

where all terms are as defined for Eqs. 8.4 and 8.5.

(iii) Triangular (V-Notch) Weirs The triangular or *V*-notch thin-plate weir is an accurate flow-measuring device, particularly for flows less than 30 l/s (about 450 gpm) and is as accurate as other types of thin-plate weirs for flows from 30 to 300 l/s (450 to 4500 gpm) (U.S. Bureau of Reclamation, 1975). A fully contracted thin-plate triangular weir is diagrammed in Figure 8.7.

To operate properly triangular weirs should be installed so that the minimum distance from the channel bank to the weir edge is at least twice the head on the weir. In addition, the distance from the bottom of the approach channel to the point of the weir notch should also be at least twice the head on the weir (U.S. Bureau of Reclamation, 1975).

The head-discharge relationship for a fully (both horizontally and vertically) contracted triangular thin-plate weir is

$$Q = K_1 C_d \tan\left(\frac{\theta}{2}\right) h_e^{5/2}$$ (8.8)

$$h_e = H + K_2 k_h$$

where

Q = discharge (l/s, gpm);
C_d = coefficient of discharge from Figure 8.8;
θ = notch angle (degrees);
h_e = effective head (m, ft);

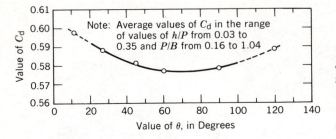

B = Width of the approach channel
P = Height of notch vertex above bottom of channel

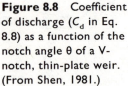

Figure 8.8 Coefficient of discharge (C_d in Eq. 8.8) as a function of the notch angle θ of a V-notch, thin-plate weir. (From Shen, 1981.)

H = difference between the crest and the water surface at a point upstream from the weir a distance of four times the maximum head on the crest (m, ft);

k_h = constant from Figure 8.9 (m, ft);

K_1 = unit constant (K_1 = 2362 for Q in l/s, and h, H, and k in m;
       K_1 = 1920 for Q in gpm, and h, H, and k in ft);

K_2 = unit constant (K_2 = 3.28 for H in m. K_2 = 1.00 for H in ft).

(iv) *Trapezoidal Weirs*   One type of trapezoidal-shaped thin-plate weir is the Cipolletti weir. The notch of this fully contracted weir inclines outward (toward the sides of the approach channel) at a slope of 1 horizontal to 4 vertical. As with other thin-plate weirs, the height of the weir crest above the bottom of the approach channel should be at least twice the head over the crest, and the distance from the sides of the notch to the sides of the channel should also be at least twice

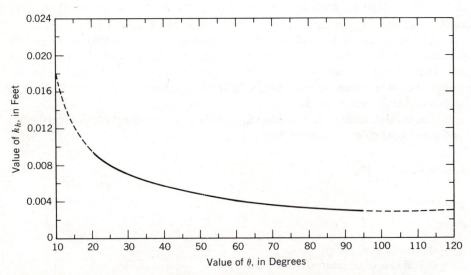

Figure 8.9   Value of k_h (in Eq. 8.8) as a function of the notch angle θ of a V-notch, thin-plate weir. (From Shen, 1981.)

the head (U.S. Bureau of Reclamation, 1975). Cipolletti weirs should not be used for heads less than about 6 cm (2 in), nor for heads greater than 1/3 the crest length.

Equation 8.9 describes the head–discharge relationship for Cipolletti weirs:

$$Q = KbH^{3/2} \tag{8.9}$$

where

Q, b, and H are as defined for Eq. 8.4
K = unit constant ($K = 1859$ for Q in l/s, and b and H in m. $K = 1511$ for Q in gpm, and b and H in ft. $K = 3.367$ for Q in cfs and b and H in ft).

Equation 8.9 is used when the velocity of approach is negligible. When the approach velocity cannot be neglected the following equation is used:

$$Q = Kb(H + 1.5h_a)^{3/2} \tag{8.10}$$

where all terms are as defined for Eqs. 8.4 and 8.5.

8.3.3c Flumes

A *flume* is a specially shaped channel section that is constructed or installed in open channels to obtain a stable stage-discharge relationship for flow measurement. Flumes have a converging inlet section that directs flow into a section with a level or sloping bed and constricted width called a throat. The inlet section serves as a transition between the channel and the throat. Downstream of the throat is a diverging outlet section that returns flow to the channel. The throat acts as a control and creates a unique relationship between the water level in the converging section and flowrate through the flume (provided the water level in the diverging section of the flume is low enough not to affect the stage in the converging section).

The principal advantages of flumes are that they cause small friction losses and are not very sensitive to velocity of approach. In addition, most flumes are not subject to deposition of silt and debris because of relatively high flow velocities through them.

Flumes are classified as either short- or long-throated depending on the length of the throat section relative to the upstream head. In long-throated flumes, where the length of the throat section is at least twice the upstream head (measured above the flume floor), the throat is sufficiently long for an essentially hydrostatic pressure distribution to develop. This allows the head-discharge relationship to be derived analytically and gives flume designers the freedom to vary flume dimensions to satisfy the specific requirements of a particular site.

Because the throats of short-throated flumes are too short for the development of a hydrostatic pressure distribution within them, head-discharge relationships are determined by field or laboratory calibration. Short-throated flumes are not, therefore, normally "custom" designed according to site requirements, as are long-throated flumes. Instead, the designer must select the standard short-throated flume design that most nearly meets the requirements of the site.

Short-throated flumes including Parshall, cutthroat, WSC, HS, and various trapezoidal flumes are used to measure flow in open irrigation channels.

Long-throated flumes with elevated floor sills and without width contractions are also used in measuring irrigation flows.

(i) The Parshall Flume   Parshall flumes, like the one in Figure 8.10, are used extensively in the United States to measure irrigation flows. Flows ranging from 0.3 to 85,000 l/s (4.5 gpm to 3000 cfs) can be accurately measured with properly sized and constructed/installed Parshall flumes. Parshall flumes may be built of wood, concrete, galvanized sheet metal, or other materials. Large flumes are usually constructed on site, while smaller flumes can be purchased as prefabricated units that are installed in the channel. Dimensions and ranges of flow for various standard Parshall flumes as well as design procedures and information can be found in the U.S. Bureau of Reclamation *Water Measurement Manual* (1975).

The primary advantage of Parshall flumes is their ability to provide accurate flow measurements over a wide range of flows with a minimum of head loss (U.S. Bureau of Reclamation, 1975). This allows their use in relatively shallow channels with flat grades. The main disadvantages of Parshall flumes are their relatively large size (Walker and Skogerboe, 1987) and the accurate workmanship required for satisfactory performance (U.S. Bureau of Reclamation, 1975).

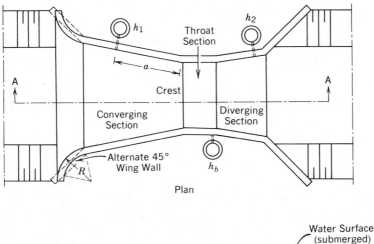

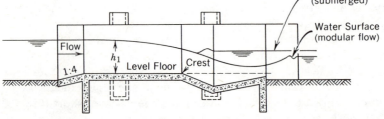

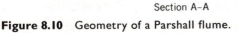

Figure 8.10   Geometry of a Parshall flume.

The rate of flow through a Parshall flume may or may not be affected by the flow depth downstream of the throat. Free-flow conditions exist when the tailwater depth is not high enough to affect flow. A flume is considered to be submerged when the tailwater depth is sufficient to affect flow and the submergence ratio, S, computed with Eq. 8.11, exceeds the transition submergence S_t. Table 8.4 lists values of S_t for different standard Parshall flume designs. Equation 8.11 is

$$S = \frac{h_d}{h_u}$$ (8.11)

where

S = submergence ratio;
h_d = head downstream of throat (m, ft);
h_u = head upstream of throat (m, ft).

Table 8.4 Free-Flow and Submerged-Flow Coefficients and Exponents for Parshall Flumes

W	C_f for Q in (l/s)	C_f for Q in (cfs)	C_s for Q in (l/s)	C_s for Q in (cfs)	n_f	n_s	S_t
1 in	9.57	0.338	8.47	0.299	1.55	1.000	0.56
2	19.14	0.676	17.33	0.612	1.55	1.000	0.61
3	28.09	0.992	25.91	0.915	1.55	1.000	0.64
6	58.34	2.06	47.01	1.66	1.58	1.080	0.55
9	86.94	3.07	71.08	2.51	1.53	1.060	0.63
12	113.28	4.00	88.08	3.11	1.52	1.080	0.62
18	169.92	6.00	125.17	4.42	1.54	1.115	0.64
24	226.56	8.00	168.22	5.94	1.55	1.140	0.66
30	283.20	10.00	204.47	7.22	1.555	1.150	0.67
3 ft	339.84	12.00	243.55	8.60	1.56	1.160	0.68
4	453.12	16.00	314.35	11.10	1.57	1.185	0.70
5	566.40	20.00	383.74	13.55	1.58	1.205	0.72
6	679.68	24.00	448.87	15.85	1.59	1.230	0.74
7	792.96	28.00	514.01	18.15	1.60	1.250	0.76
8	906.24	32.00	577.73	20.40	1.60	1.260	0.78
10	1136.48	40.13	702.05	24.79	1.59	1.275	0.80
12	1345.20	47.50	830.91	29.34	1.59	1.275	0.80
15	1658.42	58.56	1024.33	36.17	1.59	1.275	0.80
20	2180.64	77.00	1346.90	47.56	1.59	1.275	0.80
25	2702.86	95.44	1669.46	58.95	1.59	1.275	0.80
30	3225.08	113.88	1992.03	70.34	1.59	1.275	0.80
40	4269.24	150.75	2636.88	93.11	1.59	1.275	0.80
50	5313.68	187.63	3282.00	115.89	1.59	1.275	0.80

Source: G. V. Skogerboe, M. L. Hyatt, J. D. England, and J. R. Johnson, Design and Calibration of Submerged Open Channel Flow Measurement Structures: Part 2, Parshall Flumes (1967). Rep. WG31-2, Utah Water Research Laboratory, College of Engineering, Utah State Univ., Logan, May.

The following head-discharge equation is used for free-flowing Parshall flumes (i.e., when the tailwater does not affect the head upstream of the throat):

$$Q = C_f(KH)^{n_f} \qquad \text{for} \quad S \leq S_t \tag{8.12}$$

where

Q = discharge (l/s, cfs);
C_f = coefficient from Table 8.4;
H = head (see Figure 8.10) (m, ft);
n_f = flow exponent from Table 8.4;
K = unit constant (K = 3.28 for H in m. K = 1.00 for H in ft);
S_t = transition submergence from Table 8.4.

When the flume is submerged (i.e., when $S > S_t$) the following equation applies.

$$Q = \frac{C_s(K(h_u - h_d))^{n_f}}{[-(\log S + C)]^{n_s}} \qquad \text{for} \quad S > S_t \tag{8.13}$$

where

h_u = upstream head (m, ft);
h_d = dowstream head (m, ft);
C_s = coefficient from Table 8.4;
n_s = exponent from Table 8.4;
C = 0.0044 for Parshall flumes;
K = unit constant from Eq 8.12.

(ii) Cutthroat Flumes Cutthroat flumes have a rectangular cross section, a level floor, a uniformly converging inlet section, and a uniformly expanding outlet section. The throat occurs at the intersection of the inlet and outlet sections. A plane view of a cutthroat flume and construction details are given in Figure 8.11.

Because cutthroat flumes have level floors, they are easier and more economical to build and install than Parshall flumes. The level floor also allows them to be placed directly on channel beds and to be installed inside of concrete-lined channels (Walker and Skogerboe, 1987).

Equations 8.12 and 8.13 with appropriate values of n_f, n_s, and S_t from Figure 8.12 and $C = 0$ are used to relate head and discharge for free- and submerged flow conditions. The following equations are used to determine C_f and C_s.

$$C_f = K K_f W^{1.025} \tag{8.14}$$

$$C_s = K K_s W^{1.025} \tag{8.15}$$

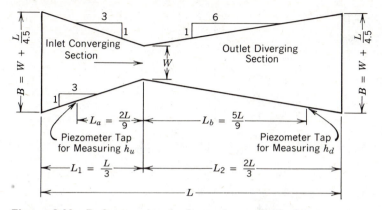

Figure 8.11 Definition sketch of a cutthroat flume. *Source:* G. V. Skogerboe, R. S. Bennett, and W. R. Walker, "Generalized Discharge Relations for Cutthroat Flumes", copyright © 1972 by the American Society of Civil Engineers, p. 570. Reprinted by permission of ASCE from the Journal IR/ASCE, (Dec. 1972) Vol 98, No. 4, p. 581.

where

W = throat width (ft);
K_f = coefficient from Figure 8.12;
K_s = coefficient from Figure 8.12.
K = unit constant ($K = 28.31$ for Q and H in Eqs. 8.12 and 8.13 in l/s and m, respectively. $K = 1.00$ for Q and H in Eqs. 8.12 and 8.13 in cfs and ft, respectively).

(iii) Trapezoidal Flumes A typical trapezoidal flume has approach, converging, throat, diverging, and exit sections as shown in Figure 8.13. The American Society of Agricultural Engineers (ASAE) has published dimensions and calibration information for several standard trapezoidal flumes (ASAE Standard S359.1, 1982). Dimensions keyed to Figure 8.13 and free-flow discharge equations for four of these flumes are listed in Tables 8.5 and 8.6. Flumes 1 and 2 are designed for use with concrete-lined channel sections. Flumes 3 and 4 are recommended primarily for use in unlined channels (ASAE Standard S359.1, 1982). Flume 4 has a zero-width throat and is intended for measuring small flows, such as in individual furrows (ASAE Standard S359.1, 1982).

Table 8.7 lists discharge correction factors for submerged flow for flumes 1, 2, and 3. The correction procedure involves measuring h and using it to compute the free-flow discharge, Q (with the appropriate equation from Table 8.7) and the submergence, S. S is computed with Eq. 8.11.

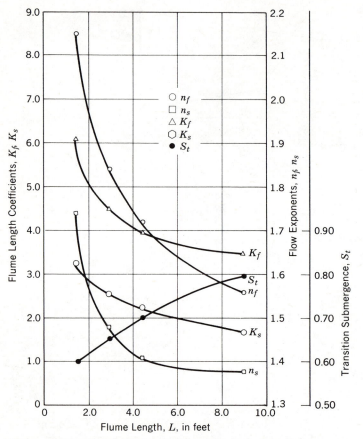

Figure 8.12 Generalized free-flow and submerged-flow coefficients (K_f and K_s), exponents (n_f and n_s), and transition submergence (S_t) for cutthroat flumes. These data are used in Equations 8.12, 8.13, 8.14, and 8.15. *Source*: G. V. Skogerboe, R. S. Bennett, and W. R. Walker, "Generalized Discharge Relations for Cutthroat Flumes," copyright © 1972 by the American Society of Civil Engineers, p. 581. Reprinted by permission of ASCE from the Journal *IR/ASCE*, (Dec. 1972), Vol 98, No. 4, p. 581.

The submergence correction factor is then determined using Table 8.7. The submerged flow discharge is calculated with the following equation.

$$Q_s = K_{sf} Q \tag{8.16}$$

where

Q_s = submerged discharge;
K_{sf} = submergence correction factor from Table 8.7;
Q = free-flow discharge.

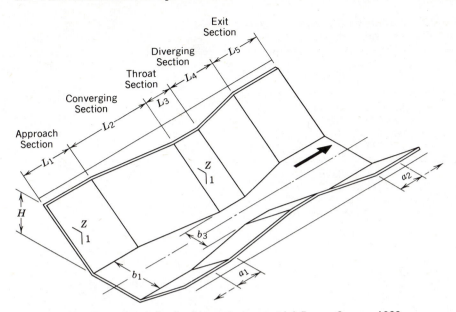

Figure 8.13 Typical standard calibrated trapezoidal flume. *Source: 1982 Yearbook*, copyright © 1982 by ASAE, p. 542. Reprinted by permission of ASAE.

    (iv) WSC Flume   The WSC flume, developed at Washington State College (now Washington State University), is a trapezoidal flume with a 60° V-notch throat. Although standard dimensions and calibration curves for three different flume sizes were developed, only the smallest of the three designs is considered here. Layout and dimensions for this flume are presented in Figure 8.14. This flume is designed to measure flows ranging from 4 to 100 l/min (1 to 26 gpm).

    Trout (1986) developed the following head-discharge relationship for a fiberglass version of the WSC flume manufactured by the Powlus Company.

$$Q = K(h_u - 0.15)^{2.63} \tag{8.17}$$

Table 8.5 **Dimensions of Standard Calibrated Trapezoidal Measuring Flumes, Expressed in Feet**[a]

No.	b_1	b_3	Z	H	L_1	L_2	L_3	L_4	L_5	a_1[b]	b_2[b]
1	1.00	0.40	1.00	1.333	1.25	1.422	1.00	1.422	0.50	0.146	0.0625
2	2.00	1.00	1.25	3.00	2.00	3.00	2.50	2.00	1.00	0.50	0.50
3	1.33	0.67	0.58	2.312	1.50	1.326	1.667	1.326	0.50	0.25	0.25
4	0.167	—	0.58	0.562	0.583	0.578	0.583	0.0578	0.25	0.125	—

Source: American Society of Agricultural Engineers, *1982-1983 Agricultural Engineers Yearbook* (1982), St. Joseph, MI Copyright © by ASAE, p. 543. Reprinted by permission of ASAE.
[a] Dimensions apply to Figure 8.13.
[b] Distance from edge of section to point of depth measurement h_u and h_d.

Table 8.6 Discharge Equations of Standard Calibrated Trapezoidal Measuring Flumes

Flume No.	Equation	h_u Range (ft)	Q Range (cfs)
1	$Q = 3.23h_u^{2.5} + 0.63h_u^{1.5} + 0.05$	0.20–1.20	0.05–5.96
2	$Q = 4.27h_u^{2.5} + 1.67h_u^{1.5} + 0.19$	0.30–2.70	0.54–58.8
3	$Q = 1.46h_u^{2.5} + 2.22h_u^{1.5}$	0.20–2.20	0.24–17.4
4	$Q = 1.55h_u^{2.58}$	0.15–0.50	0.012–0.26

Source: American Society of Agricultural Engineers, *1982-1983 Agricultural Engineers Yearbook* (1982), St. Joseph, MI. Copyright © 1982 by ASAE, p. 543. Reprinted by permission of ASAE.

Note: The units of Q and h are cfs and feet, respectively.

where

Q = discharge (l/min, gpm);

h_u = upstream head measured along the sloping side of the flume inlet section (mm);

K = unit constant. ($K = 5.43(10)^{-4}$ for Q in l/min. $K = 1.43(10)^{-4}$ for Q in gpm. h_u is in mm in both cases.)

(v) (HS) H Flumes These relatively inexpensive and easily constructed flumes are useful for measuring flows up to 23 l/s (365 gpm) (Withers and Vipond, 1980). It is important that HS flumes be carefully constructed to maximum tolerance of ± 3 mm ($\pm\frac{1}{8}$ in) for acceptable accuracy (Withers and Vipond, 1980). Layout information for HS flumes is given in Figure 8.15.

Table 8.7 Submergence Correction Factor for Standard Calibrated Trapezoidal Flume I, 2, and 3 in Tables 8.5 and 8.6

Submergence Ratio, S	Submergence Correction Factor, K_{sf}	
	Flumes 1 and 2	Flume 3
0.70	0.993	1.000
0.75	0.984	1.000
0.80	0.970	0.996
0.85	0.945	0.988
0.90	0.902	0.972
0.92	0.875	0.964
0.94	0.838	0.953
0.95	0.815	0.946

Source: American Society of Agricultural Engineers, *1982-1983 Agricultural Engineers Yearbook* (1982), St. Joseph, MI. Copyright © 1982 by ASAE, p. 543. Reprinted by permission of ASAE.

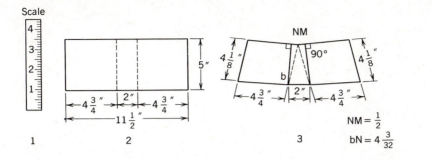

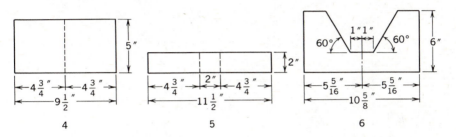

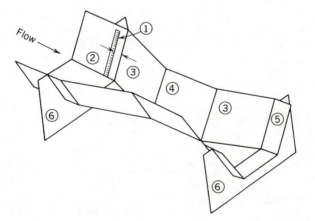

Figure 8.14 Layout and dimension (in inches) of the WSC V-notch flume. (From Section 15, Chapter 9 of the *SCS National Engineering Handbook*, 1962.)

Head and free-flow discharge are related by the following equation (Bos, 1976).

$$\log Q = A + B \log h_u + C(\log h_u)^2 \tag{8.18}$$

where

Q = free flow discharge (m³/s);
h_u = head (m);
A, B, C = coefficients from Table 8.8.

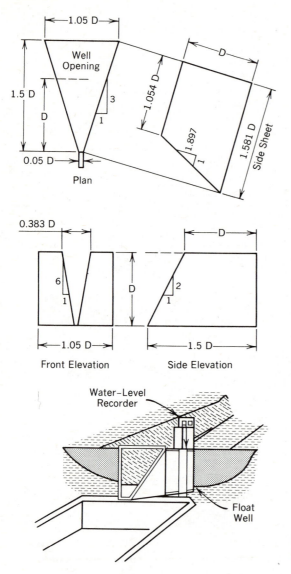

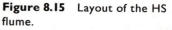

Figure 8.15 Layout of the HS flume.

(vi) *Long-Throated Flumes* In the long-throated flumes considered here, the control section is formed by an upward-sloping ramp (in the direction of flow) and a level sill. A long-throated flume formed by installing a ramp and sill in a trapezoidal channel is shown in Figure 8.16a. Long-throated flumes can be constructed, without altering the geometry of the channel cross section, on the bottom of straight, uniform reaches of lined channels whose length is at least ten times their width. In unlined channels, permanent, usually concrete channel sections containing the ramp and sill are constructed. Prefabricated, portable sheet metal sections with ramps and sills are available for small unlined channels.

Table 8.8   Coefficients in Eq. 8.18 for Four HS Flumes

Flume Depth, D (cm)	(in)	A	B	C
12.2	4.80	−0.4361	2.5151	0.1379
18.3	7.20	−0.4430	2.4908	0.1657
24.4	9.61	−0.4410	2.4571	0.1762
30.5	12.00	−0.4382	2.4193	0.1790

Source: M. G. Bos (Ed.), *Discharge Measurement Structures* (1976), Publication 20, International Institute for Land Reclamation and Improvement, Wageningen, the Netherlands.

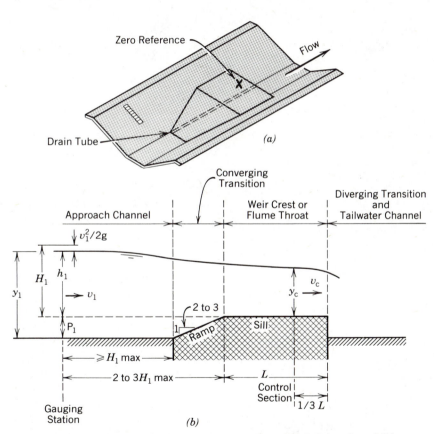

Figure 8.16   (*a*) A long-throated flume in a trapezoidal channel, and (*b*) a definition sketch for long-throated flumes. *Source*: M. G. Bos, J. A. Replogle, and A. J. Clemmens, *Flow Measuring Flumes for Open Channel Systems*, copyright © 1984 by John Wiley & Sons, Inc., New York, pp. 37, 66. Reprinted by permission of John Wiley & Sons, Inc.

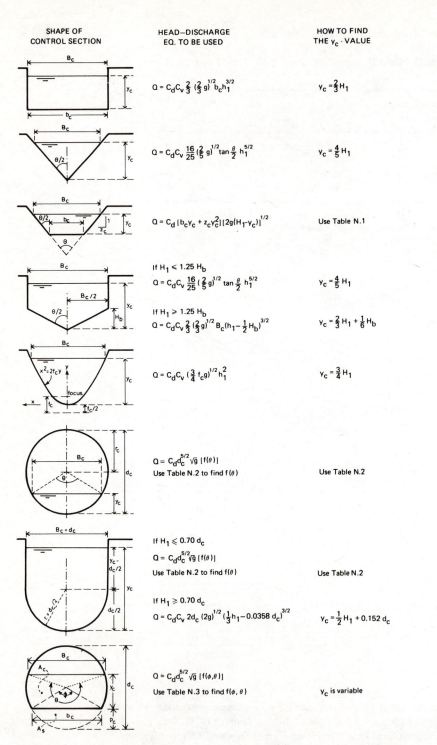

SHAPE OF CONTROL SECTION	HEAD–DISCHARGE EQ. TO BE USED	HOW TO FIND THE y_c · VALUE
B_c, y_c, b_c	$Q = C_d C_v \frac{2}{3} (\frac{2}{3} g)^{1/2} b_c h_1^{3/2}$	$y_c = \frac{2}{3} H_1$
B_c, $\theta/2$, y_c	$Q = C_d C_v \frac{16}{25} (\frac{2}{5} g)^{1/2} \tan\frac{\theta}{2} h_1^{5/2}$	$y_c = \frac{4}{5} H_1$
B_c, $\theta/2$, b_c, z_c, θ, y_c	$Q = C_d [b_c y_c + z_c y_c^2][2g(H_1 - y_c)]^{1/2}$	Use Table N.1
B_c, $B_c/2$, $\theta/2$, H_b, y_c	If $H_1 \leqslant 1.25 H_b$ $Q = C_d C_v \frac{16}{25} (\frac{2}{5} g)^{1/2} \tan\frac{\theta}{2} h_1^{5/2}$ If $H_1 \geqslant 1.25 H_b$ $Q = C_d C_v \frac{2}{3} (\frac{2}{3} g)^{1/2} B_c (h_1 - \frac{1}{2} H_b)^{3/2}$	$y_c = \frac{4}{5} H_1$ $y_c = \frac{2}{3} H_1 + \frac{1}{6} H_b$
B_c, $x^2 = 2f_c y$, focus, x, f_c, $\frac{1}{2}f_c/2$, y_c	$Q = C_d C_v (\frac{3}{4} f_c g)^{1/2} h_1^2$	$y_c = \frac{3}{4} H_1$
B_c, r_c, d_c, θ, y_c	$Q = C_d d_c^{5/2} \sqrt{g} [f(\theta)]$ Use Table N.2 to find $f(\theta)$	Use Table N.2
$B_c = d_c$, $y_c - d_c/2$, $r = d_c/2$, $d_c/2$, y_c	If $H_1 \leqslant 0.70 d_c$ $Q = C_d d_c^{5/2} \sqrt{g} [f(\theta)]$ Use Table N.2 to find $f(\theta)$ If $H_1 \geqslant 0.70 d_c$ $Q = C_d C_v 2d_c (2g)^{1/2} (\frac{1}{3} h_1 - 0.0358 d_c)^{3/2}$	Use Table N.2 $y_c = \frac{1}{2} H_1 + 0.152 d_c$
B_c, A_c, d_c, y_c, ϕ, θ, b_c, P_c, A_s'	$Q = C_d d_c^{5/2} \sqrt{g} [f(\phi, \theta)]$ Use Table N.3 to find $f(\phi, \theta)$	y_c is variable

Figure 8.17 Head–discharge for long-throated flumes. *Source*: M. G. Bos, J. A. Replogle, and A. J. Clemmens, *Flow Measuring Flumes for Open Channel Systems*, copyright © 1984 by John Wiley & Sons, Inc., New York, p. 214. Reprinted by permission of John Wiley & Sons, Inc.

406

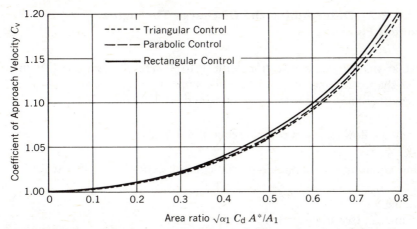

Figure 8.18 C_v (in the equations in Figure 8.17) as a function of the ratio of $\sqrt{\alpha}\,C_d A^*/A_1$. (From Bos et al., 1984.) *Source:* M. G. Bos, J. A. Replogle, and A. J. Clemmens, *Flow Measuring Flumes for Open Channel Systems,* copyright © 1984 by John Wiley & Sons, Inc., New York, p. 212. Reprinted by permission of John Wiley & Sons, Inc.

Portable sheet metal devices for making temporary measurements in lined channels have also been designed (Bos et al., 1984).

One major advantage of these long-throated flumes is that they can be used to accurately measure a large range of flows in several different channel shapes, including rectangular, triangular, trapezoidal, circular, and parabolic. In addition, head–discharge relationships that are accurate to within 2 percent can be developed analytically (Ackers et al., 1978; Bos et al., 1984). Other important attributes include minimal head loss through them and their ability to successfully pass floating debris and sediment. They also tend to be the most economical of all structures for accurately measuring flow (Bos et al., 1984).

Bos et al. (1984) present head–discharge equations that were derived analytically for long-throated flumes with ramps and sills in rectangular, triangular, trapezoidal, parabolic, circular, and truncated V (see Figure 8.17) cross sections. These equations are presented in Figure 8.17.

Figure 8.18 gives C_v as a function of $\sqrt{\alpha_1}\,C_d A^*/A_1$, where α_1 is the velocity distribution coefficient, C_d is the coefficient of discharge (Eq. 8.19), A_1 is the cross-sectional area of flow at the gauging station where h_1 is measured, and A^* equals the imaginary area of flow at the control section if the water depth would equal h_1. Bos et al. (1984) recommend an α_1 value of 1.04.

C_d is determined with the following equation.

$$C_d = \left(\frac{H_1}{L} - 0.07\right)^{0.018} \qquad \text{for} \quad 0.1 < \frac{H_1}{L} < 1.0 \tag{8.19}$$

where
C_d = coefficient of discharge;
H_1 = total head upstream of the sill (m, ft);
L = length of the horizontal portion of the sill (m, ft).

408

EXAMPLE 8.2   Determining the Discharge Through a Longthroated Flume

Given:
- a triangular-shaped control section with 1 to 1 side slopes ($z_c = 1.0$) in a triangular-shaped channel with 1.5 to 1 side slopes ($z = 1.5$)
- the sill length (L) and height (P_1) are 1.0 m and 0.15 m, respectively

Required:
- discharge for a flow depth in control section of 0.20 m

Solution:
Use the flow equation for triangular-shaped control sections from Figure 8.17.

Solution Steps
1. Determine θ
2. Determine H_1 (see Figure 8.16)
3. Determine C_d using Eq. 8.19
4. Set $V_1 = 0$
5. Determine h_1 (see Figure 8.16)
6. Determine A^*
7. Determine Y_1 (see Figure 8.16)
8. Determine A_1
9. Determine C_v from Figure 8.18
10. Determine Q using the appropriate equation from Figure 8.17
11. Determine V_1^*
12. If $V_1^* = V_1$, a solution has been found. Otherwise, set $V_1 = V_1^*$ and repeat steps 5 through 12.

Step 1
$$\theta = \tan(1/1) = 45.0°$$

Step 2
$$H_1 = (5/4)Y_c = (5/4)(0.20 \text{ m}) = 0.25 \text{ m}$$

Step 3
$$C_d = (0.25/1.00 - 0.07)^{0.018} = 0.970$$

Step 4
$$V_1 = 0$$

Step 5
$$h_1 = 0.25 - \frac{1.04(0)^2}{2(9.81)} = 0.25 \text{ m}$$

Step 6
$$A^* = zh_1^2 = (1.0)(0.25)^2 = 0.0625 \text{ m}^2$$

Step 7
$$Y_1 = h_1 + P_1 = 0.25 + 0.15 = 0.40 \text{ m}$$

Step 8

$$A_1 = zY_1^2 = (1.5)(0.40)^2 = 0.24 \text{ m}^2$$

Step 9

$$(\alpha_1)^{0.5}(C_d)\left(\frac{A^*}{A_1}\right) = (1.04)^{0.5}(0.970)\left(\frac{0.0625}{0.24}\right) = 0.26$$

$C_v = 1.02$ from Figure 8.18

Step 10

$$Q = (0.970)(1.02)(16/25)((2/5)(9.81))^{1/2}\tan(45)(0.25)^{5/2}$$
$$= 3.92(10)^{-2} \text{ m}^3/\text{s}$$

Step 11

$$V_1^* = Q/A_1 = 3.92(10)^{-2}/0.24 = 0.16 \text{ m/s}$$

Step 12

Since $V_1^* \neq V_1$, set $V_1 = V_1^*$ and return to step 5.

Step 5

$$h_1 = 0.249 \text{ m}$$

Step 6

$$A^* = 0.062 \text{ m}^2$$

Step 7

$$Y_1 = 0.399 \text{ m}$$

Step 8

$$A_1 = 0.239 \text{ m}^2$$

Step 9

$$C_v = 1.02$$

Step 10

$$Q = 3.88(10)^{-2} \text{ m}^3/\text{s}$$

Step 11

$$V_1^* = 0.16 \text{ m/s}$$

Step 12

Have a solution since $V_1^* = V_1 = 0.16$ m/s.

The solution is $Q = 3.88(10)^{-2} \text{ m}^3/\text{s}$.

8.3.3d Orifices

Orifices for measuring irrigation water are usually either circular or rectangular in shape and placed in vertical surfaces that are perpendicular to flow. Orifices are generally located near the bottom of the channel so that they are fully submerged (i.e., the downstream water level is well above the top of the orifice). This minimizes the difference between the up- and downstream water levels and allows fully submerged orifices to be used in situations where there is insufficient fall for a weir and when the use of a flume is not justified (U.S. Bureau of Reclamation, 1975).

Free-flow orifices that discharge into the air are not widely used in open channels because of the large difference in head across them (U.S. Bureau of Reclamation, 1975). Thin-plate weirs are normally used in lieu of free-flow orifices.

An orifice is partially submerged when the downstream water level is between the top and bottom of its opening. Partial submergence should be avoided because of the difficulty in making head measurements (U.S. Bureau of Reclamation, 1975).

Fully submerged orifices may be either contracted or suppressed. In a *contracted orifice*, the opening is far enough from the walls of the approach channel or other disturbing surfaces to allow the full development of a *vena contracta*. This requires that the distance from the orifice edges to the bounding edges of the channel be at least twice the smallest dimension of the orifice (U.S. Bureau of Reclamation, 1975). A *suppressed orifice* is one whose perimeter partly or fully coincides with the sides of the approach channel or other surfaces that would eliminate or reduce contraction.

For true orifice flow, the water level upstream of the orifice must always be well above the top edge of the orifice. If the upsteam water surface drops below the top of the opening, the orifice performs as a weir and does not follow the laws of orifice discharge.

Submerged orifices are normally installed to have a negligible velocity of approach. This requires that the area of the water prism 6 to 9 m (20 to 30 ft) upstream of the orifice be at least 8 times the area of the orifice (U.S. Bureau of Reclamation, 1975).

The following equation relates head and discharge for contracted, fully submerged orifices when the velocity of approach is negligible:

$$Q = 0.61KAH^{1/2} \tag{8.20}$$

where

Q = orifice discharge (l/s, gpm);
A = area of orifice opening (cm^2, in^2);
H = head (m, ft);
K = unit constant. (K = 0.443 for Q in l/s, A in cm^2, and H in m. K = 25.0 for Q in gpm, A in in^2, and H in ft.)

When the velocity of approach becomes appreciable, the following formula should be used to compute orifice discharge.

$$Q = 0.61KA(H + h_a)^{1/2} \tag{8.21}$$

where, h_a is the velocity head in the approach channel in meters or feet.

The following equation should be used for suppressed, submerged orifices:

$$Q = 0.61K(1 + 0.15r)A(H + h_a)^{1/2} \tag{8.22}$$

where r is the ratio of the suppressed portion of the perimeter to the total perimeter of the orifice. When the velocity of approach is negligible $h_a = 0$.

8.3.4 Dilution Methods

The *dilution method* involves injecting a known amount of a chemical, fluorescent, or radioactive tracer into the flow and measuring its dilution after it has flowed far enough downstream to mix completely with the water and produce a uniform concentration. No measurements of area or distance are required, since the total flow is determined directly. The dilution method is used to measure the discharge in small channels where a limited number of measurements is required and other methods would be impractical or too expensive (Charlton, 1978). Additional information on the dilution method is available in the U.S. Bureau of Reclamation *Water Measurement Manual* (1975).

Discharge is computed using the following equation.

$$Q = q\left(\frac{C_1 - C_2}{C_2 - C_0}\right) \tag{8.23}$$

where

Q = discharge (volume/time);
q = rate at which tracer is added to the flow (volume/time);
C_1 = concentration of tracer solution added to the flow (mass/volume);
C_2 = concentration of flow after tracer is uniformly mixed with the flow (mass/volume);
C_0 = concentration of flow upstream of tracer injection point (mass/volume).

8.4 Flow Measurement in Pipelines

Several types of devices can be used to measure the volume of flow and/or volumetric flowrate in pipelines. These devices may be classified as differential pressure, rotating mechanical, bypass, ultrasonic, or insertion meters.

8.4.1 Differential Pressure Flowmeters

Venturi tubes, orifice plates, and elbow meters are the main types of differential pressure flowmeters used in irrigation pipelines. Differential pressure flowmeters create a pressure difference that is proportional to the square of the volumetric flowrate. The pressure difference is normally created by causing flow to pass through a contraction. Manometers, Bourdon gauges, or pressure transducers are normally utilized to measure the pressure difference.

8.4.1a Venturi Tubes

A Venturi tube is diagrammed in Figure 8.19. The pressure drop between the inlet and throat is created as water passes through the throat. In the section downstream of the throat the gradual increase in cross-sectional area causes the

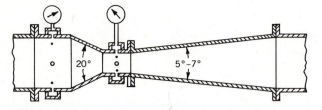

Figure 8.19 Venturi flowmeter. (From James and Shannon, 1986.) *Source*: L. G. James and W. M. Shannon, "Flow Measurement and System Maintenance." In *Trickle Irrigation for Crop Production,* F. S. Nakayama and D. A. Bucks, copyright © 1986 by Elsevier Science Publishing Co., Inc., p. 284. Reprinted by permission of Elsevier Science Publishing Co., Inc., and authors.

velocity to decrease and the pressure to increase. The pressure drop between the Venturi's inlet and throat is related to the volumetric flowrate. This relationship is

$$Q = \frac{Cd^2 K (P_1 - P_2)^{1/2}}{[1 - (d/D)^2]^{1/2}} \qquad (8.24)$$

where

Q = discharge (l/min, gpm);
C = flow coefficient;
D = diameter of upstream section (cm, in);
d = diameter of contraction (cm, in);
P_1 = pressure in upstream section (kPa, psi);
P_2 = pressure in contraction (kPa, psi);
K = unit constant. (K = 6.66 for Q in l/min, d and D in cm, and P_1 and P_2 in kPa. K = 29.86 for Q in gpm, d and D in in, and P_1 and P_2 in psi.)

The flow coefficient C varies with Reynolds number. For Venturi tubes that have Reynolds numbers that exceed 200,000, C equals 0.98. Venturi tubes for chemical injection can sometimes be used to measure flow.

8.4.1b Orifice Plates

An *orifice* consists of a thin plate with a square-edged hole that is clamped between flanges in a pipe. Most orifice plates have a circular-shaped hole concentric with the pipe as in Figure 8.20. The principle of operation of an orifice is

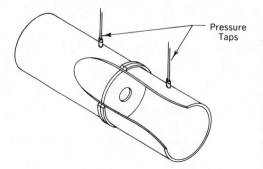

Pressure Taps

Figure 8.20 Orifice flowmeter. *Source*: L. G. James and W. M. Shannon, "Flow Measurement and System Maintenance." In *Trickle Irrigation for Crop Production,* F. S. Nakayama and D. A. Bucks, copyright © 1986 by Elsevier Science Publishing Co., Inc., p. 284. Reprinted by permission of Elsevier Science Publishing Co., Inc., and authors.

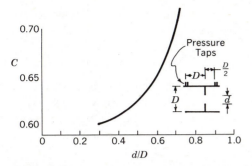

Figure 8.21 Flow coefficient (C in Eq. 8.24) for square-edged circular-shaped orifices as a function of the ratio of orifice opening to inside diameter for Reynolds numbers greater than 100,000. *Source*: L. G. James and W. M. Shannon, "Flow Measurement and System Maintenance." In *Trickle Irrigation for Crop Production*, F. S. Nakayama and D. A. Bucks (Eds.), copyright © 1986 by Elsevier Science Publishing Co., Inc., p. 285. Reprinted by permission of Elsevier Science Publishing Co., Inc. and authors.

similar to a venturi tube. Increased velocity in the orifice creates a pressure drop between the up- and downstream sides of the orifice plate. The pressure drop across an orifice normally exceeds that for a venturi tube with the same d to D ratio. Orifices are, however, normally less expensive than venturi tubes.

Equation 8.24 can be used to compute the volumetric flowrate through an orifice plate for various pressure drops. The flow coefficient for square-edged circular-shaped concentric orifices varies with the ratio of d to D and Reynolds number. Figure 8.21 can be used to obtain values of C when the Reynolds number, computed by substituting the orifice diameter into Eq. 6.10, exceeds 100,000 and pressure taps are located 1.0 and 0.5 pipe diameters up- and downstream of the orifice plate.

Eccentric and chord orifice plates, like those in Figure 8.22, are recommended for use with sediment-laden waters. These orifices prevent sediment accumulation in the pipe but are less accurate than concentric orifices.

Choosing the diameter of the contraction in Venturi tubes and orifice plates is extremely important. It is desirable that the diameter of the contraction, d, be

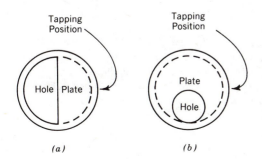

Figure 8.22 (*a*) Chord and (*b*) eccentric orifice plates. *Source*: L. G. James and W. M. Shannon, "Flow Measurement and System Maintenance." In *Trickle Irrigation for Crop Production*, F. S. Nakayama and D. A. Bucks (Eds.), copyright © 1986 by Elsevier Science Publishing Co., Inc., p. 286. Reprinted by permission of Elsevier Science Publishing Co., Inc. and authors.

large enough to minimize head loss. However, if d is too large (relative to D), accurate measurement of the pressure difference is difficult.

8.4.1c   Elbow Meters

Another important type of differential pressure flowmeter is an *elbow meter*. Pressure differences between the outside and inside walls of an elbow are related to volumetric flowrate. Equation 8.25 is used to compute volumetric flowrate when the pressure difference and cross-sectional area of the elbow are known.

$$Q = C_e K A (P_o - P_i)^{1/2} \qquad\qquad (8.25)$$

where

Q = discharge (l/min, gpm);
C_e = elbow meter flow coefficient from Figure 8.23;
A = cross-sectional area of elbow (cm^2, in^2);
P_o = pressure on outside of elbow (kPa, psi);
P_i = pressure on inside of elbow (kPa, psi);
K = unit constant. (K = 8.49 for Q in l/min, A in cm^2, and P_o and P_i in kPa.
    K = 38.02 for Q in gpm, A in in^2, and P_o and P_i in psi.)

As shown in Figure 8.23, the elbow meter flow coefficient, C_e, ranges between 0.63 and 0.83, depending on size, shape, and type of elbow.

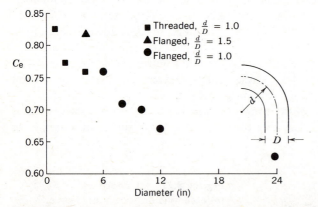

Figure 8.23   Flow coefficient (C_e in Eq. 8.25) for different diameter elbow meters for Reynolds numbers greater than 100,000. *Source*: L. G. James and W. M. Shannon, "Flow Measurement and System Maintenance." In *Trickle Irrigation for Crop Production*, F. S. Nakayama and D. A. Bucks (Eds.), copyright © 1986 by Elsevier Science Publishing Co., Inc., p. 287. Reprinted by permission of Elsevier Science Publishing Co., Inc., and authors.

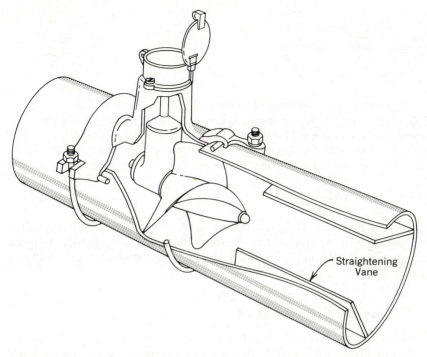

Figure 8.24 A propeller meter installed in a section of pipe. (From Section 15, Chapter 9 of the *SCS National Engineering Handbook*, 1962.)

8.4.2 Rotating Mechanical Flowmeters

There are many types of *rotating mechanical flowmeters* used in pipelines. These flowmeters normally have a rotor that revolves at a speed roughly proportional to the flowrate, and a device for recording and displaying the total volume of flow and/or volumetric flowrate. The rotor may be a propeller (as shown in Figure 8.24) or axial-flow turbine, or a vane-wheel with the flow impinging tangentially at one or more points.

Calibration tests are usually needed to accurately relate rotor revolutions to flow. The lowest flowrate that can be accurately measured by a rotating mechanical flowmeter depends on the amount of bearing friction that can be tolerated while the occurrence of cavitation often establishes the largest flowrate that can be measured. Head loss through most rotating mechanical flowmeters is moderate.

8.4.3 Bypass Flowmeters

A *bypass* or *shunt meter* is another type of flowmeter used in irrigation pipelines. As shown in Figure 8.25, a bypass meter is an orifice or other differential pressure device with a small mechanical flowmeter across the pressure taps (rather than a

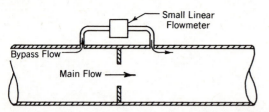

Figure 8.25 Schematic of a bypass meter. *Source*: L. G. James and W. M. Shannon, "Flow Measurement and System Maintenance." In *Trickle Irrigation for Crop Production*, F. S. Nakayama and D. A. Bucks (Eds.), copyright © 1986 by Elsevier Science Publishing Co., Inc., p. 287. Reprinted by permission of Elsevier Science Publishing Co., Inc. and authors.

pressure measuring device). The relationship between volumetric flowrate in the main pipe and flow in the bypass line is essentially linear for properly designed bypass flowmeters. Both volume of flow and volumetric flowrate can usually be obtained.

8.4.4   Ultrasonic Flowmeters

Various types of devices that use beams of ultrasound to measure flow velocity and hence volumetric flowrate are called *ultrasonic flowmeters* (the electronic circuitry required by ultrasonic flowmeters usually also provide the volume of flow). Some of these devices require the presence of suspended particles, air bubbles, and/or fluid turbulence to measure flow, while others do not. Ultrasonic meters do not obstruct flow and thus cause no loss of pressure and do not have mechanical parts that wear.

Because ultrasonic beams will travel through the wall of a pipe, ultrasonic flowmeters can be either portable or built-in. Portable models have the transmitter and receiver mounted in a housing that is clamped onto the outside of the pipe. This eliminates the inconvenience and expense of breaking open the pipe for flowmeter installation. With built-in ultrasonic flowmeters the transmitter and receiver are factory mounted in a short section of pipe. Built-in meters are installed in the pipeline and cannot be conveniently moved to another location. Built-in ultrasonic flowmeters are more accurate than clamp-on units because the relative position of the transmitter and receiver is fixed and the cross-sectional area of the flowmeter sections is precisely known.

8.4.4a   Single-path Diagonal-beam Ultrasonic Flowmeters

The *single-path diagonal-beam meter* is one of the earliest and most widely used types of ultrasonic flowmeters. This type of device transmits two ultrasonic signals diagonally across the pipe, as shown in Figure 8.26. One of the signals travels downstream and the other upstream. The difference in travel times of the two beams is related to the flow velocity in the pipe. Because this relationship is normally obtained by calibration and since the flow velocity along only a single

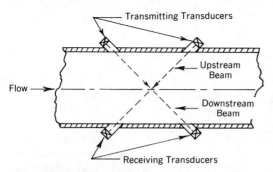

Figure 8.26 Principle of a single-path, diagonal-beam, ultrasonic flowmeter. *Source*: L. G. James and W. M. Shannon, "Flow Measurement and System Maintenance." In *Trickle Irrigation for Crop Production*, F. S. Nakayama and D. A. Bucks (Eds.), copyright © 1986 by Elsevier Science Publishing Co., Inc., p. 288. Reprinted by permission of Elsevier Science Publishing Co., Inc. and authors.

line across the pipe is used, flowmeter accuracy is highest when the actual velocity profile within the pipe is similar to the one that existed when the flowmeter was calibrated. Both built-in and clamp-on models of single-path diagonal-beam ultrasonic flowmeters are available.

8.4.4b Multichordal Diagonal Ultrasonic Flowmeters

The accuracy of single-path diagonal-beam meters can be improved by using a *multichordal diagonal-beam* arrangement similar to that in Figure 8.27. Flowmeter accuracy is less sensitive to the velocity profile within the pipe, since several beams of ultrasound travel diagonally across the pipe rather than just one. This type of ultrasonic flowmeter does, however, require more complicated circuitry for signal processing. Because of the close tolerances needed for high accuracy, multichordal meters are available only as built-in units.

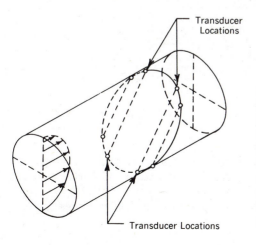

Figure 8.27 A four-chordal diagonal-beam ultrasonic flowmeter. *Source*: L. G. James and W. M. Shannon, "Flow Measurement and System Maintenance." In *Trickle Irrigation for Crop Production*, F. S. Nakayama and D. A. Bucks (Eds.), copyright © 1986 by Elsevier Science Publishing Co., Inc., p. 289. Reprinted by permission of Elsevier Science Publishing Co., Inc., and authors.

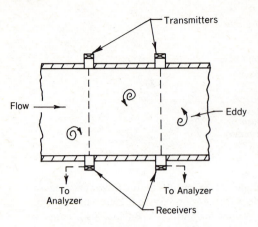

Figure 8.28 Principle of the cross-correction ultrasonic flowmeter. *Source*: L. G. James and W. M. Shannon, "Flow Measurement and System Maintenance." In *Trickle Irrigation for Crop Production*, F. S. Nakayama and D. A. Bucks (Eds.), copyright © 1986 by Elsevier Science Publishing Co., Inc., p. 289. Reprinted by permission of Elsevier Science Publishing Co., Inc. and authors.

8.4.4c Cross-correlation Ultrasonic Flowmeters

Cross-correlation ultrasonic flowmeters employ two transverse beams of ultrasound, one located a short distance upstream of the other, as shown in Figure 8.28, to measure flow velocity within a pipe. The volumetric flowrate is calculated from the time required for flow discontinuities such as aggregations of suspended particles, air bubbles, or fluid turbulence (eddies) to pass from the up- to downstream beams. Like single-path diagonal-beam meters, the accuracy of cross-correlation meters is extremely sensitive to deviations of the actual velocity profile from the one that existed during calibration. Cross-correlation meters require the presence of flow discontinuities and are well suited to clamp-on operation.

8.4.4d Doppler-effect Ultrasonic Flowmeters

A *Doppler-effect ultrasonic meter* (see Figure 8.29) measures the velocity of suspended particles or small air bubbles being carried by the flow. The suspended particles and air bubbles reflect some of the ultrasonic signals that are transmitted into the flow to the receiver. Transmitted and reflected signals are compared and

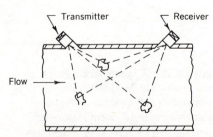

Figure 8.29 Principle of the Doppler-effect ultrasonic flowmeter. *Source*: L. G. James and W. M. Shannon, "Flow Measurement and System Maintenance." In *Trickle Irrigation for Crop Production*, F. S. Nakayama and D. A. Bucks (Eds.), copyright © 1986 by Elsevier Science Publishing Co., Inc., p. 290. Reprinted by permission of Elsevier Science Publishing Co., Inc. and authors.

the volumetric flowrate determined. In addition to being sensitive to the velocity profile within the pipe, Doppler-effect flowmeter readings are affected by changes in the velocity of sound in the water caused by temperature and density variations. Like cross-correlation meters, Doppler-effect meters are well suited to clamp-on operation.

8.4.5 Insertion Flowmeters

The main insertion meters are the Pitot tube and integrating-type Pitot tubes.

8.4.5a Pitot Tubes

A *Pitot tube* (see Figure 8.30) consists of two small-diameter concentric tubes pointing directly upstream. The inner tube measures the total flow energy (kinetic plus potential energy), while the outer tube senses only potential (pressure) energy. Because water in the mouth of the inner tube is brought to rest and its kinetic energy converted to potential (pressure) energy, the difference in pressure between the inner and outer tubes equals the kinetic energy of flow. The velocity of flow can be computed from this pressure difference using the following equation.

$$V = CKh^{1/2} \tag{8.26}$$

where

V = velocity of flow (m/s, ft/s);
C = flow coefficient (C for a well-designed Pitot tube is approximately 1.00);
h = head difference (cm, in);
K = unit constant. ($K = 0.443$ for V in m/s, and h in cm. $K = 2.315$ for V in ft/s, and h in in.)

When the velocity is known, Eq. 8.2 is used to compute the volumetric flowrate.

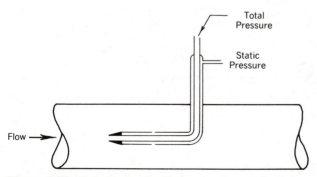

Figure 8.30 Schematic of a Pitot tube. *Source:* L. G. James and W. M. Shannon, "Flow Measurement and System Maintenance." In *Trickle Irrigation for Crop Production,* F. S. Nakayama and D. A. Bucks (Eds.), copyright © 1986 by Elsevier Science Publishing Co., Inc., p. 290. Reprinted by permission of Elsevier Science Publishing Co., Inc. and authors.

Because Pitot tubes like the one in Figure 8.30 measure the velocity at a single point and since the velocity varies across a pipe, it is usually necessary to measure velocity at several locations within a pipe to accurately determine V. When lower accuracy is acceptable and the velocity profile is symmetric V can be obtained from a single velocity measurement at a point located three quarters the pipe radius from the pipe center.

8.4.5b   Integrating Pitot Tubes

Several *insertion-type flowmeters* similar to the one in Figure 8.31 are available commercially. These meters typically have tubes with several strategically located, upstream-facing holes to "average" the total energy (kinetic plus potential energy) across the pipe. They also have another tube for sensing the potential energy (static pressure). The differential pressure (total − static pressure) obtained with these types of meters can be used in Eqs. 8.2 and 8.26 to determine the volumetric flowrate. The flow coefficient, C, in Eq. 8.26 must, however, be determined by calibration.

8.4.6   Flowmeter Installation

Flowmeters perform best when velocity profiles are symmetric and flow does not rotate. Asymmetrical velocity profiles and flow rotation are caused by bends, valves, and other fittings that significantly disturb flow. It is usually recommended, therefore, that flowmeters be installed in long, straight sections of pipe free of fittings that distort flow. There must be enough straight pipe on each side of the flowmeter to prevent upstream and/or downstream disturbances from affecting flowmeter performance.

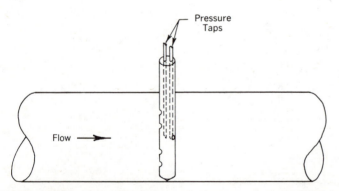

Figure 8.31   A commercially available insertion-type flowmeter. *Source*: L. G. James and W. M. Shannon, "Flow Measurement and System Maintenance." In *Trickle Irrigation for Crop Production*, F. S. Nakayama and D. A. Bucks (Eds.), copyright © 1986 by Elsevier Science Publishing Co., Inc., p. 292. Reprinted by permission of Elsevier Science Publishing Co., Inc. and authors.

Some flowmeters are more sensitive to flow disturbances than others. The length of straight pipe required upstream of flowmeters varies from 5 to 50 pipe diameters. The minimum length of straight pipe required downstream of flowmeters is 5 to 10 pipe diameters. Thus, flowmeters require anywhere from 12 to 60 pipe diameters of straight pipe free of fittings that distort flow for best performance.

When an adequate length of straight pipe is not available, flowmeter accuracy can frequently be improved by in-place calibration and/or by installing a flow straightener. In-place calibration involves determining the head versus discharge relationship for the flowmeter after it has been installed. This procedure allows the effect of flow distortions to be included in the head–discharge relationship.

Four different flow-straightener designs with varying abilities to correct velocity profile distortions and reduce flow rotation are shown in Figure 8.32. Tube bundle straighteners effectively reduce flow rotation, adequately correct asymmetric velocity profiles, but create substantial pressure loss. The AMCA and etiole straighteners remove flow rotation and have negligible head loss across them, but do not correct velocity profile distortions. Perforated plate straighteners are easily

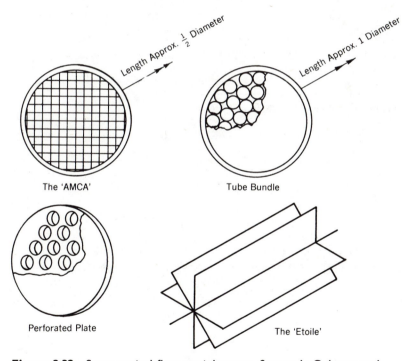

Figure 8.32 Some typical flow straighteners. *Source*: L. G. James and W. M. Shannon, "Flow Measurement and System Maintenance." In *Trickle Irrigation for Crop Production*, F. S. Nakayama and D. A. Bucks (Eds.), copyright © 1986 by Elsevier Science Publishing Co., Inc., p. 293. Reprinted by permission of Elsevier Science Publishing Co., Inc. and authors.

installed between two flanges and effectively correct distorted velocity profiles, but do not reduce flow rotation. Pressure loss across a perforated plate-type straightener can be excessive.

Homework Problems

8.1 Determine the volume of water applied to a surface-irrigated field during a 24-hour set for an inflow rate of 350 gpm.

8.2 Determine the volume of water applied in Problem 8.1 in cubic meters. What is the inflow rate in l/s?

8.3 It requires 45.3 seconds to fill a 5-gallon container. Determine the volumetric flowrate in
(a) gallons per minute (gpm), and
(b) liters per second (l/s).

8.4 Determine the percent accuracy to which the flowrate can be determined in Problem 8.3 if the stopwatch being used can be read to the nearest 0.1 second.

8.5 Use the data from the following table to determine the cross-sectional area perpendicular to flow in a 5-m-wide irrigation channel.

Distance from Near Bank (m)	Depth of Flow (m)
1.0	0.72
2.0	1.14
3.0	1.22
4.0	0.68

8.6 The float method was used to measure the surface velocity of flow in each segment of the channel in Problem 8.5. The test section was 20 m long. The following table lists the travel times measured for each channel segment.

Channel Segment (m)	Travel Time (s)
0–1.0	19.8
1.0–2.0	17.7
2.0–3.0	16.3
3.0–4.0	16.9
4.0–5.0	18.9

Determine the average velocity of flow in the channel.

8.7 Determine the volumetric flowrate of the irrigation channel in Problems 8.5 and 8.6.

8.8 Develop a theoretical rating curve for a long, straight reach of 2-m-wide rectangular shaped, concrete-lined channel with a uniform slope of 0.01 percent. The channel is 1.0 m deep.

8.9 Determine the maximum flowrate for the channel in Problem 8.8 if a full-width weir with a crest height of 0.67 m is installed in the channel.

8.10 Determine the approach velocity when the channel (with the full-width weir) in Problems 8.8 and 8.9 is flowing at full capacity. Is the approach velocity large enough to significantly affect the discharge?

8.11 Determine the maximum flowrate for the channel in Problem 8.8 if a contracted rectangular weir with a crest width and height of 0.4 and 0.67 m, respectively, is installed in the channel. Is the approach velocity significant?

8.12 Determine the maximum flowrate for the channel in Problem 8.8 if a 90° V-notch weir with a crest height of 0.67 m is installed in the channel.

8.13 Determine the free-flow discharge through a standard design Parshall flume with a 24-inch throat width for an upstream head of 2.0 feet.

8.14 Repeat Problem 8.13 for a downstream head of
(a) 0.8 ft, and
(b) 1.8 ft.

8.15 Determine the free-flow discharge through a standard design cutthroat flume with a throat width of 24 inches and a length of 8 feet for an upstream head of 2.0 feet.

8.16 Repeat Problem 8.15 for a downstream head of
(a) 0.8 ft, and
(b) 1.8 ft.

8.17 The upstream head on a small flume is 0.3 ft. Determine the free flow discharge in l/s for:
(a) ASAE trapezoidal flume 4,
(b) The WSC flume,
(c) A 12.2-cm-deep-type HS flume, and
(d) A 5-cm-diameter submerged orifice (with a negligible approach velocity).

8.18 A long-throated flume with a sill length (L) and height (P_1) of 1.0 and 0.2 m, respectively, is installed in a 2.0 m wide by 1.0 m deep rectangular-shaped concrete-lined channel. Determine the discharge for $H_1 = 0.5$ m. What is the sill referenced head in the control section?

8.19 A Venturi tube with a 4-inch-diameter throat is installed in a 10-inch-diameter pipe. The pressure in the throat of the Venturi tube is 10 psi lower than upstream pressure. Determine the flowrate in the pipeline.

8.20 A 4-inch-diameter orifice is installed in a 10-inch-diameter pipeline. The pressure difference across the orifice is 10 psi. Determine the flowrate through the pipeline.

8.21 A 4-inch-diameter elbow with a 3-inch radius of curvature (measured as in Figure 8.23) is being used to measure the flowrate in a pipeline. Determine the flowrate if the pressure difference between the inside and outside of the elbow is 10 psi.

8.22 A Pitot tube (like the one in Figure 8.30) is used to measure the discharge of a 10-inch-diameter pipe. The difference in pressure between the inner and outer tubes was measured at five equally spaced locations across the pipe. The following table lists the measured pressure difference for each of these locations. Determine the flowrate through the pipeline.

Measurement Location percent of Pipe Diameter	Pressure (in)
5.0	0.48
15.0	2.42
25.0	5.20
35.0	7.64
45.0	11.00

References

Ackers, P., W. R. White, J. A. Perkins, and A. J. M. Harrison (1978). *Weirs and Flumes for Flow Measurement.* Wiley, New York, 327 pp.

American Society of Agricultural Engineers (1982). *1982–1983 Agricultural Engineers Yearbook.* St. Joseph, MI.

Bos, M. G. (Ed.) (1976). *Discharge Measurement Structures.* Publication 20, International Institute for Land Reclamation and Improvement, Wageningen, the Netherlands.

Bos, M. G. (1985). *Long-throated Flumes and Broad-crested Weirs.* Martinus Nijhoff/Dr W. Junk Publishers, Boston, Mass., 141 pp.

Bos, M. G., J. A. Replogle, and A. J. Clemmens (1984). *Flow Measuring Flumes for Open Channel Systems.* Wiley, New York, 321 pp.

Brakensiek, D. L., H. B. Osborn, and W. J. Rawls (1979). Field manual for research in agricultural hydrology. Agriculture Handbook 224, Science and Education Administration, U.S. Department of Agriculture, U.S. Government Printing Office, Washington D.C., 547 pp.

Charlton, F. G. (1978). Measuring flow in open channels: A review of methods. Report 75, Construction Industry Research and Information Association, London, 160 pp.

Lansford, W. M. (1936). The use of an elbow in a pipeline for determining the flow in a pipe. Engineering Experiment Station Bulletin 289, Univ. of Illinois, Urbana, 36 pp.

Hayward, A. T. J. (1979). *Flowmeters: A Basic Guide and Source-Book for Users.* Wiley, New York, 197 pp.

Replogle, J. A., and M. G. Bos (1982). Flow measurement flumes: Application to irrigation water management. In *Advances in Irrigation,* Vol. 1, Academic Press, New York, pp. 147–217.

Roberson, J. A., and C. T. Crowe (1980). *Engineering Fluid Mechanics.* Houghton Mifflin, Boston, Mass., 661 pp.

Shen, J. (1981). Discharge characteristics of triangular-notch thin-plate weirs. U.S. Geological Survey Water-Supply Paper 1617–B, U.S. Goverment Printing Office, Washington, D.C., 45 pp.

Simon, A. L. (1976). *Practical Hydraulics.* Wiley, New York, 306 pp.

Skogerboe, G. V., M. L. Hyatt, J. D. England, and J. R. Johnson (1967). Design and calibration of submerged open channel flow measurement structures: Part 2, Parshall flumes. Rep. WG 31-3, Utah Water Research Laboratory, College of Engineering, Utah State Univ., Logan, May.

Skogerboe, G. V., R. S. Bennett, and W. R. Walker (1972). Generalized discharge relations for cutthroat flumes. *J. Irrigation Drainage Division,* ASCE, Vol. 98, No. IR4, pp. 569–583.

Soil Conservation Service (1962). Measurement of irrigation water, Chapter 9, Section 15, of the *SCS National Engineering Handbook,* U.S. Department of Agriculture, Washington, D.C., 72 pp.

Streeter, W. L. (1966). *Fluid Mechanics.* McGraw-Hill, New York, 705 pp.

Trout, T. (1986). Installation and use of the Powlus furrow flume. Personal communication, May 1986.

U.S. Department of the Interior (1975). *Water Measurement Manual.* U.S. Bureau of Reclamation, U.S. Government Printing Office, Washington, D.C. 327 pp.

Vennard, J. K., and R. L. Street (1974). *Elementary Fluid Mechanics.* Wiley, New York, 740 pp.

Walker, W. R., and G. V. Skogerboe (1987). *The Theory and Practice of Surface Irrigation.* Prentice-Hall, Inc., Englewood Cliffs, N.J.

Withers, B., and S. Vipond (1980). *Irrigation Design and Practice.* Cornell Univ. Press, Ithaca, N.Y., 306 pp.

Appendix A

Extraterrestrial Radiation

Monthly values of extraterrestrial radiation (R_a) for the northern and southern hemispheres are given in Table A1. Equations A1 through A6 can be used to estimate daily values of R_a for computer application.

$$R_a = 1.26714(h_{d0}/r_{ve}^2)\left[h_s \frac{\pi}{180} \sin(\Phi) \sin(\delta) + \cos(\Phi) \cos(\delta) \sin(h_s) \right] \tag{A1}$$

$$h_{d0} = 12.126 - 1.85191(10)^{-3} \text{ ABS}(\Phi) + 7.61048(10)^{-5}(\Phi)^2 \tag{A2}$$

$$r_{ve} = 0.98387 - 1.11403(10)^{-4}(J) + 5.2774(10)^{-6}(J)^2 - 2.68285(10)^{-8}(J)^3$$
$$+ 3.61634(10)^{-11}(J)^4 \tag{A3}$$

$$h_s = \cos^{-1}(-\tan \Phi \tan \delta) \tag{A4}$$

$$\delta = \frac{180}{\pi}(0.006918 - 0.399912 \cos \theta + 0.070257 \sin \theta - 0.006758 \cos 2\theta$$
$$+ 0.000907 \sin 2\theta - 0.002697 \cos 3\theta - 0.001480 \sin 3\theta) \tag{A5}$$

$$\theta = 0.986(J - 1) \tag{A6}$$

where,

R_a = extraterrestrial radiation (mm/day);
h_{d0} = daytime hours at zero declination (h);
r_{ve} = radius vector of earth;
h_s = sunrise to sunset hour angle (degrees);
Φ = location latitude (degrees) (Φ is positive for north latitudes and negative for south latitudes);
δ = declination of the sun (degrees);
J = days from Jan. 1 (e.g., $J = 1$ for Jan. 1, $J = 2$ for Jan. 2, ..., $J = 365$ for Dec. 31);
θ = day of year expressed in degrees (i.e., $\theta = 0°$ is Jan. 1, $\theta = 90°$ is Apr. 2, $\theta = 180°$ is July 2, ...).

Table A1 Extraterrestrial Radiation (R_a) Expressed in Equivalent Evaporation in mm/day

				Northern Hemisphere														*Southern Hemisphere*						
Jan	Feb	Mar	Apr	May	June	July	Aug	Sept	Oct	Nov	Dec	Lat	Jan	Feb	Mar	Apr	May	June	July	Aug	Sept	Oct	Nov	Dec
3.8	6.1	9.4	12.7	15.8	17.1	16.4	14.1	10.9	7.4	4.5	3.2	50°	17.5	14.7	10.9	7.0	4.2	3.1	3.5	5.5	8.9	12.9	16.5	18.2
4.3	6.6	9.8	13.0	15.9	17.2	16.5	14.3	11.2	7.8	5.0	3.7	48	17.6	14.9	11.2	7.5	4.7	3.5	4.0	6.0	9.3	13.2	16.6	18.2
4.9	7.1	10.2	13.3	16.0	17.2	16.6	14.5	11.5	8.3	5.5	4.3	46	17.7	15.1	11.5	7.9	5.2	4.0	4.4	6.5	9.7	13.4	16.7	18.3
5.3	7.6	10.6	13.7	16.1	17.2	16.6	14.7	11.9	8.7	6.0	4.7	44	17.8	15.3	11.9	8.4	5.7	4.4	4.9	6.9	10.2	13.7	16.7	18.3
5.9	8.1	11.0	14.0	16.2	17.3	16.7	15.0	12.2	9.1	6.5	5.2	42	17.8	15.5	12.2	8.8	6.1	4.9	5.4	7.4	10.6	14.0	16.8	18.3
6.4	8.6	11.4	14.3	16.4	17.3	16.7	15.2	12.5	9.6	7.0	5.7	40	17.9	15.7	12.5	9.2	6.6	5.3	5.9	7.9	11.0	14.2	16.9	18.3
6.9	9.0	11.8	14.5	16.4	17.2	16.7	15.3	12.8	10.0	7.5	6.1	38	17.9	15.8	12.8	9.6	7.1	5.8	6.3	8.3	11.4	14.4	17.0	18.3
7.4	9.4	12.1	14.7	16.4	17.2	16.7	15.4	13.1	10.6	8.0	6.6	36	17.9	16.0	13.2	10.1	7.5	6.3	6.8	8.8	11.7	14.6	17.0	18.2
7.9	9.8	12.4	14.8	16.5	17.1	16.8	15.5	13.4	10.8	8.5	7.2	34	17.8	16.1	13.5	10.5	8.0	6.8	7.2	9.2	12.0	14.9	17.1	18.2
8.3	10.2	12.8	15.0	16.5	17.0	16.8	15.6	13.6	11.2	9.0	7.8	32	17.8	16.2	13.8	10.9	8.5	7.3	7.7	9.6	12.4	15.1	17.2	18.1
8.8	10.7	13.1	15.2	16.5	17.0	16.8	15.7	13.9	11.6	9.5	8.3	30	17.8	16.4	14.0	11.3	8.9	7.8	8.1	10.1	12.7	15.3	17.3	18.1
9.3	11.1	13.4	15.3	16.5	16.8	16.7	15.7	14.1	12.0	9.9	8.8	28	17.7	16.4	14.3	11.6	9.3	8.2	8.6	10.4	13.0	15.4	17.2	17.9
9.8	11.5	13.7	15.3	16.4	16.7	16.6	15.7	14.3	12.3	10.3	9.3	26	17.6	16.4	14.4	12.0	9.7	8.7	9.1	10.9	13.2	15.5	17.2	17.8
10.2	11.9	13.9	15.4	16.4	16.6	16.5	15.8	14.5	12.6	10.7	9.7	24	17.5	16.5	14.6	12.3	10.2	9.1	9.5	11.2	13.4	15.6	17.1	17.7
10.7	12.3	14.2	15.5	16.3	16.4	16.4	15.8	14.6	13.0	11.1	10.2	22	17.4	16.5	14.8	12.6	10.6	9.6	10.0	11.6	13.7	15.7	17.0	17.5
11.2	12.7	14.4	15.6	16.3	16.4	16.3	15.9	14.8	13.3	11.6	10.7	20	17.3	16.5	15.0	13.0	11.0	10.0	10.4	12.0	13.9	15.8	17.0	17.4
11.6	13.0	14.6	15.6	16.1	16.1	16.1	15.8	14.9	13.6	12.0	11.1	18	17.1	16.5	15.1	13.2	11.4	10.4	10.8	12.3	14.1	15.8	16.8	17.1
12.0	13.3	14.7	15.6	16.0	15.9	15.9	15.7	15.0	13.9	12.4	11.6	16	16.9	16.4	15.2	13.5	11.7	10.8	11.2	12.6	14.3	15.8	16.7	16.8
12.4	13.6	14.9	15.7	15.8	15.7	15.7	15.7	15.1	14.1	12.8	12.0	14	16.7	16.4	15.3	13.7	12.1	11.2	11.6	12.9	14.5	15.8	16.5	16.6
12.8	13.9	15.1	15.7	15.7	15.5	15.5	15.6	15.2	14.4	13.3	12.5	12	16.6	16.3	15.4	14.0	12.5	11.6	12.0	13.2	14.7	15.8	16.4	16.5
13.2	14.2	15.3	15.5	15.5	15.3	15.3	15.5	15.3	14.7	13.6	12.9	10	16.4	16.3	15.5	14.2	12.8	12.0	12.4	13.5	14.8	15.9	16.2	16.2
13.6	14.5	15.3	15.6	15.3	15.0	15.1	15.4	15.3	14.8	13.9	13.3	8	16.1	16.1	15.5	14.4	13.1	12.4	12.7	13.7	14.9	15.8	16.0	16.0
13.9	14.8	15.4	15.4	15.1	14.7	14.9	15.2	15.3	15.0	14.2	13.7	6	15.8	16.0	15.6	14.7	13.4	12.8	13.1	14.0	15.0	15.7	15.8	15.7
14.3	15.0	15.5	15.5	14.9	14.4	14.6	15.1	15.3	15.1	14.5	14.1	4	15.5	15.8	15.6	14.9	13.8	13.2	13.4	14.3	15.1	15.6	15.5	15.4
14.7	15.3	15.6	15.3	14.6	14.2	14.3	14.9	15.3	15.3	14.8	14.4	2	15.3	15.7	15.7	15.1	14.1	13.5	13.7	14.5	15.2	15.5	15.3	15.1
15.0	15.5	15.7	15.3	14.4	13.9	14.1	14.8	15.3	15.4	15.1	14.8	0	15.0	15.5	15.7	15.3	14.4	13.9	14.1	14.8	15.3	15.4	15.1	14.8

Source: J. Doorenbos and W. O. Pruitt, "Guidelines for Predicting Crop Water Reclamation," Irrigation and Drainage Paper 24 (1977), FAO, United Nations, 144 pp.

Appendix B

Computing daytime hours and percentage of total daytime hours

$$h_{di} = \frac{(h_{si})(h_{d0i})}{90} \tag{B1}$$

where,

h_{di} = daytime hours for day i;

h_{si} = sunrise or sunset hour angle (degrees) for day i (see Appendix A for equation for computing h_s);

h_{d0i} = daytime hours at zero declination for day i (see Appendix A for equation for computing h_{d0}).

$$P_i = \left(\frac{h_{di}}{h_a}\right)100 \tag{B2}$$

where,

P_i = percent of total annual daytime occurring during day i;

h_a = total daytime hours per year.

$$h_a = \sum_{i=1}^{365} h_{di} \tag{B3}$$

Appendix C

K_{SCS} Values for Equation 1.21

Table C1 K_{SCS} Values for Eq. 1.21 Only

Crop	Jan.	Feb.	Mar.	Apr.	May	June	July	Aug.	Sept.	Oct.	Nov.	Dec.	Earliest Growth Date (spring)	Latest Growth Date (fall)
Alfalfa	0.63	0.73	0.86	0.99	1.08	1.13	1.11	1.06	0.99	0.91	0.78	0.64	50°	28° Frost
Avacados	0.27	0.42	0.58	0.70	0.78	0.81	0.77	0.71	0.63	0.54	0.43	0.30	—	—
Citrus	0.63	0.66	0.68	0.70	0.71	0.71	0.71	0.71	0.70	0.68	0.67	0.64	—	—
Grapes	0.20	0.24	0.33	0.50	0.71	0.80	0.80	0.76	0.61	0.50	0.35	0.23	55°	50°
Deciduous orchards with cover crop	0.63	0.73	0.86	0.98	1.09	1.13	1.11	1.06	0.99	0.90	0.78	0.66	50°	45°
Deciduous orchards without cover crop	0.17	0.25	0.40	0.63	0.88	0.96	0.95	0.82	0.54	0.30	0.19	0.15	50°	45°
Pasture grass	0.49	0.57	0.73	0.85	0.90	0.92	0.92	0.91	0.87	0.79	0.67	0.55	45°	45°
Walnuts	0.09	0.13	0.23	0.44	0.69	0.92	0.98	0.88	0.69	0.49	0.31	0.15	—	—

Source: Soil Conservation Service (SCS), "Irrigation Water Requirements," *Tech. Release 21*, USDA-SCS (1970), 88 pp.

Table C2 K_{SCS} **Values for Annual Crops for Eq. 1.21 Only**

Crop	Percent of Growing Season											Earliest Planting Date (spring)	Latest Growth Date (fall)
	0	10	20	30	40	50	60	70	80	90	100		
Dry beans	0.50	0.59	0.72	0.90	1.04	1.11	1.12	1.02	0.89	0.74	0.60	60°	32° frost
Snap beans	0.50	0.55	0.61	0.69	0.79	0.89	0.98	1.05	1.10	1.12	1.12	—	28° frost
Sugar beets	0.49	0.49	0.61	0.78	0.96	1.10	1.21	1.25	1.22	1.14	1.04	28° frost	32° frost
Corn (grain)	0.44	0.49	0.58	0.71	0.92	1.05	1.08	1.06	1.00	0.93	0.85	55°	—
Corn (silage)	0.44	0.47	0.54	0.64	0.80	0.98	1.06	1.08	1.06	1.02	0.96	55°	—
Corn (sweet)	0.44	0.50	0.58	0.72	0.92	1.05	1.08	1.08	1.06	1.03	1.00	—	—
Cotton	0.20	0.25	0.34	0.50	0.79	0.96	1.02	0.94	0.80	0.66	0.49	62°	32° frost
Spring grain	0.28	0.46	0.71	0.94	1.14	1.31	1.27	1.04	0.69	0.30	0	45°	32° frost
Melons & cantaloupes	0.44	0.48	0.56	0.66	0.76	0.82	0.81	0.78	0.74	0.71	0.68	—	—
Peas	0.50	0.59	0.72	0.89	1.03	1.11	1.12	1.10	1.06	1.01	0.96	—	—
Potato (Irish)	0.33	0.40	0.50	0.72	0.96	1.18	1.31	1.37	1.36	1.30	1.23	60°	32° frost
Grain sorghum	0.30	0.38	0.60	0.83	1.02	1.08	1.00	0.88	0.76	0.65	0.56	—	—
Soybeans	0.19	0.26	0.33	0.41	0.54	0.73	0.91	1.02	0.93	0.77	0.66	—	—
Tomatoes	0.44	0.46	0.48	0.56	0.76	0.96	1.03	0.99	0.90	0.80	0.69	—	—
Small vegetables	0.29	0.40	0.56	0.69	0.78	0.82	0.82	0.80	0.72	0.58	0.38	—	—
Winter wheat (fall)	0.28	0.34	0.45	0.58	0.72	0.85	0.97	1.08	1.18	1.28	1.36	45°	—
(spring)	1.36	1.36	1.35	1.34	1.33	1.30	1.18	0.91	0.57	0.23	0	45°	—

Source: Soil Conservation Service (SCS), "Irrigation Water Requirements," Tech. Release 21, USDA-SCS (1970), 88 pp.

Appendix D

Daily Climatic Data for Prosser, Wash., USA

Month/Day	Precipitation (mm)	Pan Evaporation[a] (mm)	Temperature in °C (max)	Temperature in °C (min)	Relative Humidity (max)	Relative Humidity (min)	R_s (mm)	Wind (m/s)
4/1	0.0	4.1	16.1	8.9	98.0	30.0	7.2	3.0
4/2	0.0	3.0	12.2	7.2	77.0	31.0	3.2	3.0
4/3	0.0	2.8	10.6	0.0	94.0	47.0	7.4	3.1
4/4	0.0	3.0	12.2	0.0	80.0	31.0	6.4	2.9
4/5	1.3	3.0	10.0	−1.1	85.0	31.0	6.8	2.7
4/6	2.5	0.8	13.3	0.6	98.0	38.0	3.5	2.3
4/7	0.0	0.5	6.1	−1.1	94.0	69.0	5.7	1.8
4/8	0.0	3.3	10.0	−2.2	96.0	43.0	8.6	1.8
4/9	0.0	5.1	13.9	0.6	76.0	29.0	9.1	2.8
4/10	0.0	0.5	17.8	2.8	76.0	28.0	5.4	1.6
4/11	0.0	0.5	20.6	8.3	98.0	29.0	5.0	3.1
4/12	14.0	0.3	19.4	5.6	84.0	52.0	4.9	4.5
4/13	0.0	3.8	13.3	6.7	80.0	47.0	6.4	3.3
4/14	0.0	4.3	13.3	1.1	81.0	46.0	6.7	4.2
4/15	0.0	2.5	8.9	−2.8	84.0	38.0	7.3	3.5
4/16	0.0	14.5	12.2	−2.8	93.0	35.0	9.0	1.8
4/17	0.0	1.3	13.9	1.1	75.0	25.0	8.9	2.4
4/18	0.0	1.3	14.4	−3.3	70.0	33.0	10.0	4.7
4/19	0.0	1.8	11.7	−3.3	76.0	30.0	10.1	3.7
4/20	0.0	5.8	13.3	−3.3	64.0	20.0	10.1	1.7
4/21	0.0	5.8	15.6	0.6	58.0	17.0	10.0	1.8
4/22	0.0	6.6	20.0	2.8	64.0	20.0	10.0	1.7
4/23	0.0	6.4	25.6	8.9	62.0	22.0	10.0	1.5
4/24	0.0	6.6	24.4	0.6	76.0	20.0	10.2	2.9
4/25	0.0	6.9	17.2	1.1	68.0	28.0	10.1	2.0
4/26	0.0	6.9	17.8	2.2	67.0	26.0	10.1	1.8
4/27	0.0	5.8	21.1	3.9	81.0	30.0	9.1	1.5
4/28	8.4	0.8	22.8	6.1	98.0	28.0	9.3	3.0
4/29	0.0	7.1	13.9	−1.1	83.0	30.0	10.1	4.5
4/30	0.0	4.8	16.1	3.9	75.0	33.0	8.7	1.4
Totals/Means	26.2	119.9	15.3	1.7	80.4	32.9	239.3	2.7

Note: 46.25°N latitude, 275 m above mean sea level

[a] $Kp = 0.80$. (ET_0 for grass reference crop $= 0.80 \times$ pan evaporation).

433

Appendix D (Cont.)

Month/Day	Precipitation (mm)	Pan Evaporation[a] (mm)	Temperature in °C (max)	(min)	Relative Humidity (max)	(min)	R_s (mm)	Wind (m/s)
5/1	0.0	3.8	21.7	6.1	87.0	36.0	7.2	1.2
5/2	0.0	3.8	23.3	3.9	93.0	28.0	7.8	1.2
5/3	0.0	4.3	18.9	2.8	86.0	23.0	8.1	4.1
5/4	0.0	6.4	13.3	−2.2	86.0	23.0	10.4	4.3
5/5	0.0	6.6	16.7	0.0	70.0	24.0	11.0	2.3
5/6	0.0	6.1	20.6	6.7	75.0	27.0	9.7	1.8
5/7	0.0	8.4	24.4	7.8	74.0	30.0	10.4	3.0
5/8	0.0	7.1	17.8	0.6	91.0	32.0	8.5	4.0
5/9	0.0	7.1	17.8	3.3	84.0	36.0	8.2	2.2
5/10	0.0	7.4	18.9	5.6	91.0	45.0	8.2	2.9
5/11	0.0	5.8	18.3	2.2	94.0	34.0	9.8	1.8
5/12	0.0	6.6	22.8	4.4	82.0	34.0	10.6	1.8
5/13	0.0	7.1	23.9	4.4	94.0	36.0	8.0	2.1
5/14	0.0	4.1	22.2	6.1	93.0	47.0	8.5	1.1
5/15	0.0	5.8	23.9	5.6	96.0	44.0	11.1	1.2
5/16	0.0	6.1	25.6	8.3	83.0	29.0	10.6	1.1
5/17	0.0	6.1	28.3	10.0	98.0	30.0	3.7	2.1
5/18	1.3	1.8	15.6	0.6	96.0	52.0	10.3	2.0
5/19	0.0	0.7	19.4	1.1	89.0	26.0	11.7	2.6
5/20	0.0	6.1	22.8	5.6	83.0	29.0	9.0	1.5
5/21	0.0	6.4	26.7	11.7	73.0	34.0	11.3	1.2
5/22	0.0	7.9	30.0	12.8	93.0	29.0	10.0	1.2
5/23	0.0	7.9	26.7	5.6	74.0	33.0	10.5	2.9
5/24	0.0	7.6	26.7	10.6	77.0	29.0	11.6	1.3
5/25	0.0	7.9	30.0	12.8	80.0	34.0	10.9	1.3
5/26	0.0	11.2	30.0	6.1	74.0	27.0	6.5	3.0
5/27	0.0	7.1	17.8	4.4	99.0	35.0	7.3	3.7
5/28	5.8	2.5	15.0	6.1	99.0	56.0	8.2	2.0
5/29	0.0	6.4	21.1	6.1	96.0	33.0	10.7	1.7
5/30	0.0	6.6	23.9	5.0	83.0	30.0	11.3	1.5
5/31	0.0	6.6	24.4	6.1	80.0	26.0	10.1	1.9
Totals/Means	7.1	189.3	22.2	5.5	86.2	33.3	291.2	2.1

[a] $Kp = 0.80$.

Month/Day	Precipitation (mm)	Pan Evaporation[a] (mm)	Temperature in °C		Relative Humidity		R_s (mm)	Wind (m/s)
			(max)	(min)	(max)	(min)		
6/1	0.0	6.4	27.2	8.9	89.0	32.0	9.1	1.4
6/2	0.0	5.3	23.3	2.8	100.0	46.0	11.9	2.2
6/3	0.0	8.6	23.9	7.2	83.0	24.0	10.4	2.5
6/4	0.0	6.6	22.2	4.4	93.0	30.0	8.1	1.7
6/5	0.0	11.4	21.1	4.4	94.0	42.0	8.8	3.0
6/6	0.0	11.4	22.2	3.3	92.0	34.0	10.1	2.7
6/7	0.0	11.7	24.4	2.8	92.0	36.0	11.2	2.7
6/8	0.0	5.3	25.0	3.3	84.0	34.0	11.6	2.7
6/9	0.0	7.9	30.6	11.1	72.0	24.0	11.9	2.7
6/10	0.0	9.1	31.1	11.7	72.0	22.0	12.0	2.7
6/11	0.0	9.1	32.8	12.2	74.0	23.0	10.6	2.7
6/12	0.0	5.6	32.8	15.6	94.0	25.0	6.0	1.7
6/13	0.0	5.8	25.6	11.7	97.0	52.0	5.0	2.0
6/14	0.0	5.6	26.7	11.7	93.0	57.0	11.8	1.0
6/15	0.0	7.1	31.1	12.2	90.0	38.0	11.7	1.1
6/16	0.0	8.6	34.4	14.4	86.0	30.0	12.1	1.2
6/17	0.0	10.2	35.0	15.6	80.0	28.0	11.5	1.3
6/18	0.0	12.7	35.0	17.8	52.0	30.0	12.2	2.6
6/19	0.0	10.7	35.0	15.6	65.0	24.0	11.7	2.0
6/20	0.0	10.7	39.4	18.3	78.0	28.0	10.2	1.2
6/21	0.0	10.7	38.3	17.8	85.0	31.0	7.1	1.5
6/22	0.0	6.9	32.2	13.3	88.0	48.0	11.8	2.0
6/23	0.0	10.7	33.9	12.8	77.0	30.0	11.9	2.0
6/24	0.0	9.4	35.0	14.4	83.0	30.0	11.1	1.2
6/25	0.0	8.1	35.6	18.9	87.0	38.0	8.9	1.4
6/26	0.0	7.9	36.1	17.8	94.0	44.0	7.7	1.4
6/27	0.0	7.9	33.9	13.9	98.0	40.0	8.6	2.4
6/28	21.1	7.9	27.2	13.3	94.0	33.0	2.7	1.7
6/29	1.8	1.0	18.3	13.3	98.0	32.0	5.4	0.6
6/30	0.0	3.0	24.4	14.4	98.0	60.0	7.7	0.9
Totals/Means	22.9	243.3	29.8	11.8	86.1	34.8	291.1	1.9

[a] $Kp = 0.80$.

Appendix D (Cont.)

Month/Day	Precipitation (mm)	Pan Evaporation[a] (mm)	Temperature in °C		Relative Humidity		R_s (mm)	Wind (m/s)
			(max)	(min)	(max)	(min)		
7/1	0.0	3.3	27.8	12.2	93.0	42.0	9.0	1.7
7/2	0.0	3.3	28.9	8.9	90.0	33.0	7.3	2.5
7/3	0.0	10.2	23.3	6.7	95.0	46.0	12.3	1.9
7/4	0.0	7.1	23.3	6.7	98.0	34.0	9.8	2.7
7/5	0.0	7.4	23.3	10.0	88.0	33.0	11.9	2.0
7/6	0.0	7.4	25.0	7.8	96.0	32.0	10.7	1.9
7/7	0.0	7.4	26.7	13.3	87.0	32.0	9.1	1.5
7/8	0.0	7.9	29.4	11.7	93.0	36.0	9.3	1.3
7/9	0.0	7.9	28.3	12.2	84.0	33.0	11.5	1.2
7/10	0.0	5.8	30.6	12.2	84.0	26.0	11.5	1.4
7/11	0.0	7.9	33.3	15.0	80.0	24.0	10.2	1.2
7/12	0.0	8.4	36.1	15.6	92.0	27.0	11.4	1.5
7/13	0.0	8.4	33.9	15.6	85.0	27.0	7.7	1.2
7/14	0.0	9.4	31.7	8.9	86.0	43.0	11.1	3.1
7/15	0.0	8.4	22.2	5.0	94.0	33.0	5.7	3.8
7/16	0.0	10.4	21.1	5.6	98.0	46.0	11.5	1.9
7/17	0.0	5.8	25.0	11.7	85.0	36.0	10.9	1.5
7/18	0.0	7.6	30.6	13.3	87.0	32.0	10.9	1.3
7/19	0.0	7.9	33.9	12.8	88.0	33.0	11.4	1.5
7/20	0.0	7.9	35.0	15.0	92.0	32.0	10.0	1.6
7/21	0.0	8.9	35.0	12.2	81.0	25.0	11.2	1.8
7/22	0.0	9.1	28.3	6.1	93.0	29.0	11.8	2.6
7/23	0.0	10.2	26.1	8.3	88.0	36.0	11.4	1.3
7/24	0.0	9.1	30.0	11.7	74.0	32.0	11.5	1.4
7/25	0.0	9.4	34.4	14.4	70.0	28.0	11.1	1.5
7/26	0.0	9.4	37.2	17.8	65.0	26.0	10.0	1.8
7/27	0.0	9.4	36.7	17.2	77.0	27.0	9.9	2.2
7/28	0.0	11.2	38.9	22.2	75.0	30.0	10.3	1.6
7/29	0.0	10.2	40.0	18.9	81.0	33.0	10.3	1.1
7/30	0.0	8.6	40.6	18.9	81.0	36.0	5.3	1.3
7/31	3.0	11.7	33.9	13.3	98.0	48.0	10.6	2.0
Totals/Means	3.0	256.8	30.7	12.3	86.4	33.2	316.6	1.8

[a] Kp = 0.80.

Month/Day	Precipitation (mm)	Pan Evaporation[a] (mm)	Temperature in °C		Relative Humidity		R_s (mm)	Wind (m/s)
			(max)	(min)	(max)	(min)		
8/1	0.0	12.2	35.0	7.8	98.0	32.0	11.0	1.5
8/2	0.0	12.2	27.2	7.2	96.0	34.0	10.1	3.1
8/3	0.0	7.4	25.0	11.7	95.0	42.0	8.4	1.5
8/4	0.0	7.1	25.0	10.6	87.0	40.0	10.6	1.6
8/5	0.0	7.1	28.9	12.2	92.0	32.0	10.4	1.2
8/6	0.0	6.4	32.2	13.9	95.0	34.0	10.3	1.2
8/7	0.0	8.4	32.8	17.8	80.0	37.0	9.3	1.7
8/8	0.0	8.4	38.3	23.9	80.0	27.0	5.5	1.7
8/9	0.0	8.6	36.1	18.3	90.0	38.0	7.9	1.1
8/10	23.4	3.0	35.6	11.1	100.0	38.0	6.2	1.3
8/11	0.0	4.6	27.8	8.3	100.0	44.0	8.8	1.4
8/12	0.0	7.6	25.6	9.4	91.0	38.0	10.2	2.6
8/13	0.0	8.1	28.3	9.4	99.0	35.0	8.6	1.5
8/14	0.0	4.3	27.2	8.9	98.0	40.0	9.3	2.0
8/15	0.0	4.3	22.8	9.4	97.0	37.0	9.4	1.1
8/16	0.0	4.6	27.8	11.7	78.0	36.0	9.7	1.2
8/17	0.0	7.6	30.6	11.1	95.0	28.0	9.7	1.9
8/18	0.0	5.8	30.6	12.8	90.0	34.0	9.9	1.2
8/19	0.0	7.4	32.2	14.4	77.0	31.0	9.5	1.3
8/20	0.0	4.3	28.9	18.9	89.0	40.0	7.8	1.4
8/21	0.0	7.6	36.1	18.3	84.0	35.0	8.9	1.2
8/22	0.0	7.6	35.6	13.3	80.0	36.0	9.5	1.2
8/23	0.0	7.4	33.9	16.7	70.0	30.0	9.5	1.5
8/24	0.0	8.1	33.3	16.7	69.0	34.0	9.6	1.6
8/25	0.0	8.4	33.9	15.6	81.0	32.0	9.3	1.4
8/26	0.0	7.1	36.1	13.3	85.0	28.0	9.2	1.0
8/27	0.0	6.9	35.0	11.1	96.0	31.0	9.5	1.2
8/28	0.0	5.1	31.1	6.7	97.0	22.0	9.2	2.0
8/29	0.0	5.1	28.3	13.3	97.0	30.0	5.2	1.4
8/30	0.0	5.1	24.4	9.4	98.0	44.0	6.1	1.5
8/31	0.0	6.9	24.4	8.3	93.0	34.0	8.8	1.5
Totals/Means	23.4	214.6	30.6	12.6	89.6	34.6	277.2	1.5

[a] Kp = 0.80.

Appendix D (Cont.)

Month/Day	Precipitation (mm)	Pan Evaporation[a] (mm)	Temperature in °C (max)	Temperature in °C (min)	Relative Humidity (max)	Relative Humidity (min)	R_s (mm)	Wind (m/s)
9/1	0.0	5.6	26.7	11.7	88.0	34.0	8.5	1.4
9/2	0.0	6.1	31.1	13.9	90.0	34.0	7.2	1.2
9/3	0.0	3.8	32.2	17.2	89.0	38.0	4.9	1.2
9/4	0.0	5.3	29.4	8.9	98.0	46.0	8.6	2.6
9/5	0.0	5.1	26.1	9.4	98.0	42.0	8.4	1.5
9/6	0.0	5.3	27.2	10.0	96.0	36.0	7.8	1.2
9/7	0.0	5.3	30.6	12.2	90.0	34.0	8.1	1.2
9/8	0.0	5.1	29.4	12.2	97.0	38.0	8.3	1.0
9/9	0.0	6.1	32.2	12.8	79.0	35.0	3.3	1.2
9/10	0.0	0.3	26.1	3.9	98.0	54.0	7.7	2.1
9/11	0.0	3.0	19.4	4.4	98.0	35.0	3.6	2.8
9/12	0.0	3.0	15.6	8.3	98.0	61.0	7.1	1.1
9/13	1.3	3.0	18.9	8.9	80.0	47.0	7.4	1.8
9/14	0.0	6.9	20.0	6.7	70.0	30.0	7.8	3.4
9/15	0.0	6.9	16.1	4.4	74.0	35.0	7.8	3.3
9/16	0.0	5.1	20.6	4.4	97.0	36.0	7.7	1.5
9/17	0.0	4.6	23.3	8.9	86.0	32.0	7.5	1.4
9/18	0.0	0.3	28.3	10.0	81.0	32.0	6.8	1.2
9/19	0.0	0.3	28.3	13.9	98.0	30.0	3.3	2.5
9/20	14.7	0.3	22.2	10.0	99.0	70.0	5.7	1.6
9/21	0.0	3.3	18.3	6.1	97.0	52.0	6.8	1.7
9/22	0.0	3.0	22.2	6.7	97.0	44.0	7.0	1.1
9/23	0.0	3.8	25.0	7.8	94.0	34.0	6.9	1.2
9/24	0.0	4.3	26.7	12.8	96.0	31.0	3.2	1.2
9/25	0.0	2.0	22.8	12.2	98.0	63.0	3.0	1.2
9/26	0.0	2.0	19.4	6.7	97.0	56.0	3.4	1.7
9/27	12.7	2.3	16.7	6.1	99.0	57.0	5.6	0.9
9/28	0.0	4.3	19.4	5.6	99.0	57.0	3.8	3.1
9/29	0.3	4.3	17.2	6.1	99.0	57.0	5.3	3.4
9/30	0.0	4.3	18.3	4.4	99.0	57.0	6.3	1.8
Totals/Means	29.0	115.1	23.7	8.9	92.6	43.6	188.8	1.8

[a] $Kp = 0.80$.

Month/Day	Precipitation (mm)	Pan Evaporation[a] (mm)	Temperature in °C (max)	(min)	Relative Humidity (max)	(min)	R_s (mm)	Wind (m/s)
10/1	0.0	3.8	17.8	3.9	99.0	57.0	6.3	1.4
10/2	0.0	2.5	22.8	7.8	99.0	57.0	5.2	1.2
10/3	0.0	2.8	20.6	3.3	99.0	57.0	6.0	2.2
10/4	1.3	2.5	17.8	6.7	100.0	41.0	5.3	1.2
10/5	0.0	3.0	17.2	2.8	77.0	44.0	5.7	1.4
10/6	0.0	4.1	17.8	8.3	96.0	56.0	1.8	2.0
10/7	1.8	0.8	12.8	2.2	91.0	46.0	4.1	2.0
10/8	0.0	2.5	13.3	2.8	98.0	68.0	3.1	2.6
10/9	0.0	1.8	13.3	3.3	99.0	43.0	5.5	1.0
10/10	0.0	1.5	16.7	2.2	97.0	46.0	5.5	1.0
10/11	0.5	1.8	18.9	3.3	98.0	40.0	5.4	1.2
10/12	0.0	2.5	20.0	4.4	98.0	45.0	5.3	1.1
10/13	0.0	2.8	21.7	5.6	98.0	45.0	4.9	1.0
10/14	0.0	2.3	21.1	5.6	94.0	45.0	5.0	1.0
10/15	0.0	2.3	21.1	7.8	94.0	52.0	3.6	1.1
10/16	0.0	3.0	20.6	8.9	83.0	40.0	4.4	1.1
10/17	0.0	3.3	20.0	1.7	94.0	40.0	3.1	3.5
10/18	0.0	3.3	13.3	−2.2	99.0	42.0	5.1	2.6
10/19	0.0	2.0	11.7	−4.4	97.0	40.0	5.0	1.3
10/20	0.0	1.8	13.3	2.8	92.0	48.0	3.8	1.0
10/21	0.0	1.0	11.7	1.1	98.0	71.0	1.6	1.0
10/22	5.3	4.8	7.8	1.1	97.0	70.0	2.4	1.1
10/23	0.0	1.0	16.7	8.3	97.0	66.0	3.0	1.4
10/24	0.0	10.2	16.7	5.6	98.0	64.0	4.2	1.2
10/25	12.2	1.0	17.8	5.6	97.0	61.0	3.3	1.1
10/26	2.5	2.3	16.7	3.9	95.0	45.0	4.2	2.7
10/27	0.0	3.3	12.2	1.1	96.0	42.0	4.2	3.7
10/28	0.0	1.5	15.0	2.2	97.0	76.0	1.0	1.1
10/29	27.4	1.0	8.9	2.2	94.0	68.0	2.2	1.5
10/30	0.0	1.0	7.8	3.3	97.0	55.0	3.3	0.9
10/31	0.0	1.0	12.8	0.0	97.0	55.0	1.9	1.2
Totals/Means	51.1	78.7	16.0	3.6	95.6	52.4	125.5	1.5

[a] Kp = 0.80.

Appendix E

Minor Loss Coefficients

Use the Equation $M_l = kv^2/2g$ Unless Otherwise Indicated.

① Perpendicular square entrance:

$$k = 0.50 \qquad \text{if edge is sharp}$$

② Perpendicular rounded entrance

$R/d =$	0.05	0.1	0.2	0.3	0.4
$k \ =$	0.25	0.17	0.08	0.05	0.04

③ Perpendicular reentrant entrance

$$k = 0.8$$

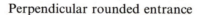

④ Additional loss due to skewed entrance

$$k = 0.505 + 0.303 \sin \alpha + 0.226 \sin^2 \alpha$$

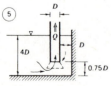

⑤ Suction pipe in sump with conical mouthpiece

$$M_l = D + \frac{5.6Q}{\sqrt{2g}\,D^{1.5}} - \frac{v^2}{2g}$$

Without mouthpiece

$$M_l = 0.53D + \frac{4Q}{\sqrt{2g}\,D^{1.5}} - \frac{v^2}{2g}$$

Width of sump shown: $3.5D$

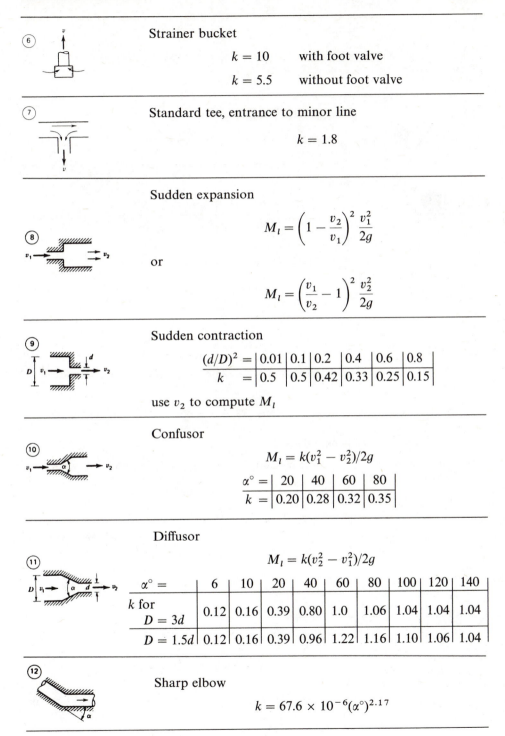

⑥ Strainer bucket

$$k = 10 \quad \text{with foot valve}$$

$$k = 5.5 \quad \text{without foot valve}$$

⑦ Standard tee, entrance to minor line

$$k = 1.8$$

⑧ Sudden expansion

$$M_l = \left(1 - \frac{v_2}{v_1}\right)^2 \frac{v_1^2}{2g}$$

or

$$M_l = \left(\frac{v_1}{v_2} - 1\right)^2 \frac{v_2^2}{2g}$$

⑨ Sudden contraction

$(d/D)^2 =$	0.01	0.1	0.2	0.4	0.6	0.8
$k \quad =$	0.5	0.5	0.42	0.33	0.25	0.15

use v_2 to compute M_l

⑩ Confusor

$$M_l = k(v_1^2 - v_2^2)/2g$$

$\alpha° =$	20	40	60	80
$k =$	0.20	0.28	0.32	0.35

⑪ Diffusor

$$M_l = k(v_2^2 - v_1^2)/2g$$

$\alpha° =$	6	10	20	40	60	80	100	120	140
k for $D = 3d$	0.12	0.16	0.39	0.80	1.0	1.06	1.04	1.04	1.04
$D = 1.5d$	0.12	0.16	0.39	0.96	1.22	1.16	1.10	1.06	1.04

⑫ Sharp elbow

$$k = 67.6 \times 10^{-6}(\alpha°)^{2.17}$$

Bends

$$k = (0.13 + 1.85(r/R)^{3.5})\sqrt{\alpha°/180°}$$

Close return bend

$$k = 2.2$$

Gate valve

$e/D =$	0	1/4	3/8	1/2	5/8	3/4	7/8
k =	0.15	0.26	0.81	2.06	5.52	17.0	97.8

Globe valve

$$k = 10 \quad \text{when fully open}$$

Rotary valve

$\alpha° =$	5	10	20	30	40	50	60	70	80
k =	0.05	0.29	1.56	5.47	17.3	52.6	206	485	∞

Check valves

Swing type $k = 2.5$ when fully open

Ball type $k = 70.0$

Lift type $k = 12.0$

Angle valve

$$k = 5.0 \qquad \text{if fully open}$$

Segment gate in rectangular conduit

$$k = 0.8 + 1.3\left[\left(\frac{1}{n}\right) - n\right]^2$$

where $n = \varphi/\varphi_0 =$ the rate of opening with respect to the central angle.

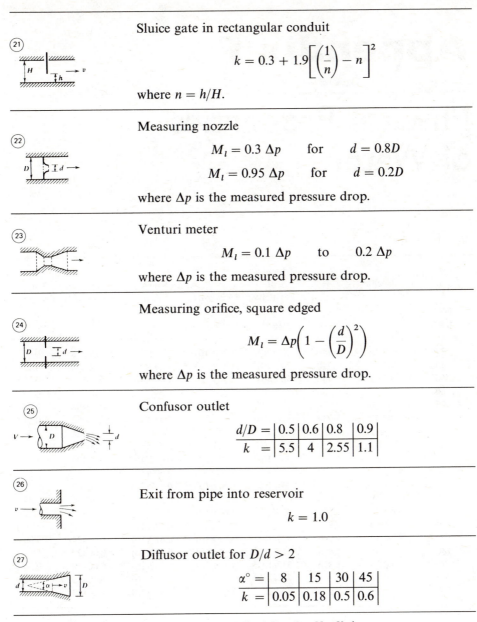

Sluice gate in rectangular conduit

$$k = 0.3 + 1.9\left[\left(\frac{1}{n}\right) - n\right]^2$$

where $n = h/H$.

Measuring nozzle

$$M_l = 0.3\,\Delta p \qquad \text{for} \qquad d = 0.8D$$

$$M_l = 0.95\,\Delta p \qquad \text{for} \qquad d = 0.2D$$

where Δp is the measured pressure drop.

Venturi meter

$$M_l = 0.1\,\Delta p \qquad \text{to} \qquad 0.2\,\Delta p$$

where Δp is the measured pressure drop.

Measuring orifice, square edged

$$M_l = \Delta p\left(1 - \left(\frac{d}{D}\right)^2\right)$$

where Δp is the measured pressure drop.

Confusor outlet

d/D =	0.5	0.6	0.8	0.9
k =	5.5	4	2.55	1.1

Exit from pipe into reservoir

$$k = 1.0$$

Diffusor outlet for $D/d > 2$

$\alpha°$ =	8	15	30	45
k =	0.05	0.18	0.5	0.6

Source: A. L. Simon, *Practical Hydraulics* (1976), John Wiley & Sons, Inc., New York.

Appendix F

Physical Properties of Water

English Units

Temperature, (°F)	Specific Weight,[a] γ, (lb/ft^3)	Density,[a] ρ, (slug/ft^3)	Modulus[b] of Elasticity,[c] E (psi)	Viscosity,[a] μ (lb-s/ft^2)	Kinematic Viscosity,[a] ν (ft^3/s)	Surface[a] Tension,[d] σ, (lb/ft)	Vapor Pressure,[e] ρ_v, (psia)
32	62.42	1.940	$287(10)^3$	$3.746(10)^{-5}$	$1.931(10)^{-5}$	0.00518	0.09
40	62.43	1.940	$296(10)^3$	$3.229(10)^{-5}$	$1.664(10)^{-5}$	0.00614	0.12
50	62.41	1.940	$305(10)^3$	$2.735(10)^{-5}$	$1.410(10)^{-5}$	0.00509	0.18
60	62.37	1.938	$313(10)^3$	$2.359(10)^{-5}$	$1.217(10)^{-5}$	0.00504	0.26
70	62.30	1.936	$319(10)^3$	$2.050(10)^{-5}$	$1.059(10)^{-5}$	0.00498	0.36
80	62.22	1.934	$324(10)^3$	$1.799(10)^{-5}$	$0.930(10)^{-5}$	0.00492	0.51
90	62.11	1.931	$328(10)^3$	$1.595(10)^{-5}$	$0.826(10)^{-5}$	0.00486	0.70
100	62.00	1.927	$331(10)^3$	$1.424(10)^{-5}$	$0.739(10)^{-5}$	0.00480	0.95
110	61.86	1.923	$332(10)^3$	$1.284(10)^{-5}$	$0.667(10)^{-5}$	0.00473	1.27
120	61.71	1.918	$332(10)^3$	$1.168(10)^{-5}$	$0.609(10)^{-5}$	0.00467	1.69
130	61.55	1.913	$331(10)^3$	$1.069(10)^{-5}$	$0.558(10)^{-5}$	0.00460	2.22
140	61.38	1.908	$330(10)^3$	$0.981(10)^{-5}$	$0.514(10)^{-5}$	0.00454	2.89
150	61.20	1.902	$328(10)^3$	$0.905(10)^{-5}$	$0.476(10)^{-5}$	0.00447	3.72
160	61.00	1.896	$326(10)^3$	$0.838(10)^{-5}$	$0.442(10)^{-5}$	0.00441	4.74
170	60.80	1.890	$322(10)^3$	$0.780(10)^{-5}$	$0.413(10)^{-5}$	0.00434	5.99
180	60.58	1.883	$318(10)^3$	$0.726(10)^{-5}$	$0.385(10)^{-5}$	0.00427	7.51
190	60.36	1.876	$313(10)^3$	$0.678(10)^{-5}$	$0.362(10)^{-5}$	0.00420	9.34
200	60.12	1.868	$308(10)^3$	$0.637(10)^{-5}$	$0.341(10)^{-5}$	0.00413	11.52
212	59.83	1.860	$300(10)^3$	$0.593(10)^{-5}$	$0.319(10)^{-5}$	0.00404	14.70

S.I. Units[f]

Temperature, (°C)	Specific Weight,[a] γ, (kN/m²)	Density, ρ, (kg/m³)[g]	Modulus of Elasticity,[c] E (kN/m²)	Viscosity, μ (N-s/m²)	Kinematic Viscosity, ν (m²/s)	Surface Tension,[d] σ, (N/m)	Vapor Pressure, ρ_v, (kN/m²)
0	9.805	999.8	$1.98(10)^6$	$1.781(10)^{-3}$	$1.785(10)^{-6}$	0.0756	0.61
5	9.807	1000.0	$2.05(10)^6$	$1.518(10)^{-3}$	$1.519(10)^{-6}$	0.0749	0.87
10	9.804	999.7	$2.10(10)^6$	$1.307(10)^{-3}$	$1.306(10)^{-6}$	0.0742	1.23
15	9.798	999.1	$2.15(10)^6$	$1.139(10)^{-3}$	$1.139(10)^{-6}$	0.0735	1.70
20	9.789	998.2	$2.17(10)^6$	$1.002(10)^{-3}$	$1.003(10)^{-6}$	0.0728	2.34
25	9.777	997.0	$2.22(10)^6$	$0.890(10)^{-3}$	$0.893(10)^{-6}$	0.0720	3.17
30	9.764	995.7	$2.25(10)^6$	$0.798(10)^{-3}$	$0.800(10)^{-6}$	0.0712	4.24
40	9.730	992.2	$2.28(10)^6$	$0.653(10)^{-3}$	$0.658(10)^{-6}$	0.0696	7.38
50	9.689	988.0	$2.29(10)^6$	$0.547(10)^{-3}$	$0.553(10)^{-6}$	0.0679	12.33
60	9.642	983.2	$2.28(10)^6$	$0.466(10)^{-3}$	$0.474(10)^{-6}$	0.0662	19.92
70	9.589	977.8	$2.25(10)^6$	$0.404(10)^{-3}$	$0.413(10)^{-6}$	0.0644	31.16
80	9.530	971.8	$2.20(10)^6$	$0.354(10)^{-3}$	$0.364(10)^{-6}$	0.0626	47.34
90	9.466	965.3	$2.14(10)^6$	$0.315(10)^{-3}$	$0.326(10)^{-6}$	0.0608	70.10
100	9.399	958.4	$2.07(10)^6$	$0.282(10)^{-3}$	$0.294(10)^{-6}$	0.0589	101.33

[a] From "Hydraulic Models," *A.S.C.E. Manual of Engineering Practice*, No. 25, ASCE, 1942.

[b] Approximate values averaged from many sources.

[c] At atmospheric pressure.

[d] In contact with air.

[e] From J. H. Keenan and F. G. Keyes, *Thermodynamic Properties of Steam*, Wiley, 1936.

[f] Compiled from many sources including interpolation of English Units table above. *Handbook of Chemistry and Physics*, 54th Ed., The CRC Press, 1973, and *Handbook of Tables for Applied Engineering Science*, The Chemical Rubber Co., 1970.

[g] $1000 \text{ kg/m}^3 = 1 \text{ g/cm}^3$.

Appendix G

FORTRAN Code for an Irrigation Scheduling Model

```
      WRITE(5,*)'*******************************************************'
      WRITE(5,*)'* THIS PROGRAM USES EITHER THE WRIGHT-JENSEN PENMAN, *'
      WRITE(5,*)'* DOORENBOS-PRUITT PENMAN, OR THE PAN EVAPORATION    *'
      WRITE(5,*)'* METHOD OF COMPUTING ETO TO SCHEDULE IRRIGATIONS.   *'
      WRITE(5,*)'*     DATA REQUIRED:                                 *'
      WRITE(5,*)'*        (A)  FC, PWP, SAT, DRZ FOR SOIL             *'
      WRITE(5,*)'*        (B)  INITIAL SOIL MOISTURE CONTENT          *'
      WRITE(5,*)'*        (C)  CROP MAD                               *'
      WRITE(5,*)'*        (D)  DAILY CLIMATIC DATA  (FROM AN EXTERNAL *'
      WRITE(5,*)'*             DATA FILE SIMILAR TO APPENDIX D -- SEE *'
      WRITE(5,*)'*             PAGE 453 FOR MORE INFORMATION)         *'
      WRITE(5,*)'*     OUTPUT INCLUDES:                               *'
      WRITE(5,*)'*        (A)  DAILY ET, EFF RAIN, IRRI, DP, RO, DIR  *'
      WRITE(5,*)'*        (B)  DAILY WATER CONTENT                    *'
      WRITE(5,*)'*        (C)  CUMULATIVE ET, IRRI, DP, RO    LGJ386  *'
      WRITE(5,*)'*******************************************************'
C
C IN THIS PROGRAM THE DEVICE NUMBERS FOR EXTERNAL DATA FILES, THE
C TERMINAL (SCREEN), AND THE PRINTER ARE 4, 5, AND 6, RESPECTIVELY
C
      INTEGER DBEG,YEAR,DEND,M(214),D(214),DAY,DA(13),UNIT
      REAL KP,LAT,MAD,KC,IRRI,IRRTOT,ITOTAL,IRRIFC
      REAL EP(214),P(214),TMAX(214),TMIN(214),RHMAX(214),RS(214),
     &RHMIN(214),U(214),X(214,10),KARR(13),D1(12)
      CHARACTER MONTH$(12)*9,A$*1,B$*1,CHAR*66,LUNIT*2,DUNIT*6,HUNIT*6
   10 WRITE(5,*)'    '
      WRITE(5,*)'EITHER SI OR ENGLISH UNITS MAY BE USED.  ENTER: '
      WRITE(5,*)'    1    FOR SI UNITS '
      WRITE(5,*)'    2    FOR ENGLISH UNITS '
      READ(5,11) UNIT
   11 FORMAT(I1)
      IF (UNIT.EQ.1) THEN
```

```
          LUNIT='MM'
          DUNIT='MM/DAY'
          HUNIT='METERS'
        ELSE
          LUNIT='IN'
          DUNIT='IN/DAY'
          HUNIT='FEET'
        END IF
        WRITE(5,*)'                    '
        LAT=0
        WRITE(5,*)' ENTER THE DESIRED METHOD OF COMPUTING ETO:  '
        WRITE(5,*)'          1    WRIGHT-JENSEN PENMAN             '
        WRITE(5,*)'          2    DOORENBOS-PRUITT PENMAN          '
        WRITE(5,*)'          3    PAN EVAPORATION                  '
        READ(5,*) METHOD
        CHG=0
139 CONTINUE
        IF (METHOD.EQ.3) THEN
          WRITE(5,*)'ENTER THE PAN COEFFICIENT "KP"  '
          READ(5,*) KP
        END IF
        IF (CHG.EQ.12) GO TO 499
        IF (METHOD.EQ.2) THEN
109     CONTINUE
        WRITE(5,*)'ENTER THE STATION LATITUDE (USE "+" FOR NORTH LATITU
   &DES, "-" FOR SOUTH LATITUDES) '
          READ(5,*) LAT
          HDO=12.126-.00185191*ABS(LAT)+.0000761048*LAT**2
          IF (CHG.EQ.9) GO TO 499
        END IF
 29 CONTINUE
        WRITE(5,*)'ENTER POROSITY "POR"  (% BY VOLUME) '
        READ(5,*) POR
        IF(CHG.EQ.1.) GO TO 499
 39 CONTINUE
        WRITE(5,*)'ENTER FIELD CAPACITY "FC"  (% BY VOLUME) '
        READ(5,*) FC
        IF(CHG.EQ.2.) GO TO 499
 49 CONTINUE
        WRITE(5,*)'ENTER PERMANENT WILTING POINT "PWP" (% BY VOLUME) '
        READ(5,*) PWP
        IF(CHG.EQ.3.) GO TO 499
 59 CONTINUE
        WRITE(5,*)'ENTER MAXIMUM ALLOWABLE DEFICIENCY "MAD"  (% ) '
        READ(5,*) MAD
        IF(CHG.EQ.4.) GO TO 499
 69 CONTINUE
```

```
IF (UNIT.EQ.1) THEN
   WRITE(5,*)'ENTER DEPTH OF THE ROOT ZONE "DRZ"  (MM) '
   READ(5,*) DRZ
ELSE
   WRITE(5,*)'ENTER DEPTH OF THE ROOT ZONE "DRZ" (INCHES) '
   READ(5,*) DRZ
   DRZ=DRZ*25.4
END IF
IF(CHG.EQ.5) GO TO 499
WCC=MAD*PWP/100.+FC*(1.-MAD/100.)
DAY=0
IRRTOT=0
IRRI=0
DPTOT=0
DP=0
ROTOT=0
RO=0
PETOT=0
ETTOT=0
D1(1)=31
D1(2)=28
D1(3)=31
D1(4)=30
D1(5)=31
D1(6)=30
D1(7)=31
D1(8)=31
D1(9)=30
D1(10)=31
D1(11)=30
D1(12)=31
79 CONTINUE
IF (UNIT.EQ.1) THEN
   WRITE(5,*)'ENTER THE STATION ELEVATION IN METERS '
   READ(5,*) H
ELSE
   WRITE(5,*)'ENTER THE STATION ELEVATION IN FEET '
   READ(5,*) H
   H=H/3.281
END IF
IF(CHG.EQ.6) GO TO 499
PA=1013.-0.1152*H+0.00000544*H**2
89 CONTINUE
WRITE(5,*)'     ENTER DATE OF THE BEGINNING OF THE GROWING SEASON
&(MONTH,DAY,YEAR) '
READ(5,*) MBEG,DBEG,YEAR
IF(CHG.EQ.7) GO TO 499
```

```
    129 CONTINUE
        WRITE(5,*)'ENTER THE TYPE OF KC FUNCTION: '
        WRITE(5,*)'    1    CONSTANT (FOR ENTIRE IRRIGATION SEASON) '
        WRITE(5,*)'    2    MONTHLY '
        WRITE(5,*)'    3    GROWTH STAGE (FROM TABLE 1.4) '
        READ(5,*) KCCODE
     99 CONTINUE
        IF (KCCODE.NE.3) THEN
           WRITE(5,*)'ENTER DATE THAT THE GROWING SEASON ENDS (MONTH,DAY)'
           READ(5,*) MEND,DEND
           JEND=0
           IF (MEND.EQ.4) THEN
             JEND=JEND+DEND
           ELSE
             N=MEND-1
             DO 45 I=4,N
               JEND=JEND+D1(I)
     45      CONTINUE
             JEND=JEND+DEND
           END IF
        END IF
        IF (CHG.EQ.8) GO TO 499
        IF (KCCODE.EQ.1) THEN
           WRITE(5,*)'ENTER KC FOR ENTIRE IRRIGATION SEASON '
           READ(5,*) KC
        END IF
C
C    ------   ENTERING MONTHLY "KC" VALUES FOR THE GROWING SEASON -----
C
        MONTH$(1)='JANUARY'
        MONTH$(2)='FEBRUARY'
        MONTH$(3)='MARCH'
        MONTH$(4)='APRIL'
        MONTH$(5)='MAY'
        MONTH$(6)='JUNE'
        MONTH$(7)='JULY'
        MONTH$(8)='AUGUST'
        MONTH$(9)='SEPTEMBER'
        MONTH$(10)='OCTOBER'
        MONTH$(11)='NOVEMBER'
        MONTH$(12)='DECEMBER'
        DO 43 I=1,12
     43 KARR(I)=0.0
        IF (KCCODE.EQ.2) THEN
           J1=MBEG
           IF (DBEG.LT.15) THEN
              J1=J1-1
```

```
              COUNT=D1(J1)-(15-DBEG)
          ELSE
              COUNT=DBEG-15
          END IF
          K=MEND
          IF (DEND.GT.15) K=K+1
          DA(1)=D1(J1)-COUNT+1
          L=1
          DO 100 I=J1,K
          L=L+1
          DA(L)=DA(L-1)+D1(I)
   91     FORMAT(A9)
          WRITE(5,90)'ENTER KC FOR ',MONTH$(I)
   90     FORMAT(A13,A9)
          READ(5,*) KARR(L-1)
          JBEG=J1
  100     CONTINUE
          L=1
C
C  ------ ENTERING GROWTH STAGE "KC" INFORMATION  ------
C
      ELSE IF (KCCODE.EQ.3) THEN
          WRITE(5,*)'ENTER KC AND # OF DAYS IN STAGE 1  (KC, # OF DAYS)'
          READ(5,*) KARR(1),DA(1)
          WRITE(5,*)'ENTER THE # OF DAYS IN STAGE 2'
          READ(5,*) DA(2)
          DA(2)=DA(2)+DA(1)
          WRITE(5,*)'ENTER KC AND # OF DAYS IN STAGE 3  (KC, # OF DAYS)'
          READ(5,*) KARR(3),DA(3)
          DA(3)=DA(3)+DA(2)
          WRITE(5,*)'ENTER KC AT THE END OF STAGE 4 AND THE # OF DAYS IN
     &STAGE 4  (KC, # OF DAYS)'
          READ(5,*) KARR(4),DA(4)
          DA(4)=DA(4)+DA(3)
          DUM=0
          DO 140 I=1,MBEG-1
  140     DUM=DUM+D1(I)
          JBEG=DUM+DBEG
          TARGET=JBEG+DA(4)
          DO 150 I=MBEG,12
              DUM=DUM+D1(I)
              IF (DUM.GT.TARGET) GO TO 200
              OLD=DUM
  150     CONTINUE
  200     MEND=I
          DEND=TARGET-OLD-1
          JEND=OLD+DEND-90
```

```
        END IF
        IF (CHG.EQ.11) GO TO 499
    119 CONTINUE
        WRITE(5,*)'ENTER INITIAL SOIL MOISTURE CONTENT  (% BY VOLUME) '
        READ(5,*) WC
C
C   ------  ECHO CHECKING INPUT DATA  -------
C
    499 I=5
    500 WRITE(I,*)'   '
        IF (METHOD.EQ.1) THEN
        WRITE(I,*)'                    WRIGHT-JENSEN PENMAN METHOD'
        ELSE IF (METHOD.EQ.2) THEN
        WRITE(I,*)'                 DOORENBOS-PRUITT PENMAN METHOD'
        ELSE IF (METHOD.EQ.3) THEN
        WRITE(I,*)'                    PAN EVAPORATION METHOD'
        END IF
        WRITE(I,*)'                             FOR'
        WRITE(I,*)'                    IRRIGATION SCHEDULING'
        WRITE(I,40) YEAR
        WRITE(I,*)'   '
     40 FORMAT(22X,I4,'  CLIMATIC DATA')
        WRITE(I,*)'    '
        IF (UNIT.EQ.1) THEN
          WRITE(I,30)'1. POROSITY = ',POR,' %      6. ELEVATION = ',H,HUN
       &IT
        ELSE
          WRITE(I,30)'1. POROSITY = ',POR,' %      6. ELEVATION = ',H*3.28
       &1,HUNIT
        END IF
        WRITE(I,32)'2. FC        = ',FC,' %      7. STARTING DATE: ',MBEG,
       &'/',DBEG,'/',YEAR
        WRITE(I,34)'3. PWP       = ',PWP,'%      8. ENDING DATE: ',MEND,
       &'/',DEND
        WRITE(I,30)'4. MAD       = ',MAD,' %      9. LATITUDE  = ',LAT,
       &'DEGREES'
        IF (UNIT.EQ.1) THEN
          WRITE(I,36)'5. DRZ       = ',DRZ,' ',LUNIT,'    10. INITIAL WATE
       &R CONTENT = ',WC,' %'
        ELSE
          WRITE(I,36)'5. DRZ       = ',DRZ/25.4,' ',LUNIT,'    10. INITIAL
       & WATER CONTENT = ',WC,' %'
        END IF
        WRITE(I,*)'    '
        WRITE(I,*)'    '
        IF (KCCODE.EQ.1) THEN
          WRITE(I,31)' 11. KC = ',KC,' FOR ENTIRE PERIOD.'
```

```
 31    FORMAT(4X,A10,F4.2,A19)
       ELSE IF (KCCODE.EQ.2) THEN
         WRITE(I,*)'11. A DIFFERENT KC FOR EACH MONTH WILL BE USED.'
         WRITE(I,*)'                        '
         DO 33 N=1,8
           M1=N-1+MBEG
           IF(KARR(N).GT.0.01) WRITE(I,35)' KC FOR ',MONTH$(M1),' = ',
      &KARR(N)
 33    CONTINUE
 35 FORMAT(8X,A8,A9,A3,F4.2)
       ELSE
         WRITE(I,*)'11.  A DIFFERENT KC FOR EACH GROWTH STAGE WILL BE USED
      &.'
         WRITE(I,7)'THERE ARE ',DA(1),' DAYS IN GROWTH STAGE #1,          KC
      &     = ',KARR(1)
         WRITE(I,8)'THERE ARE ',DA(2)-DA(1),' DAYS IN GROWTH STAGE #2'
         WRITE(I,7)'THERE ARE ',DA(3)-DA(2),' DAYS IN GROWTH STAGE #3,
      &   KC = ',KARR(3)
         WRITE(I,7)'THERE ARE ',DA(4)-DA(3),' DAYS IN GROWTH STAGE #4, FI
      &NAL KC = ',KARR(4)
       END IF
  7 FORMAT(5X,A10,I2,A37,F4.2)
  8 FORMAT(5X,A10,I2,A24)
       IF (METHOD.EQ.3) THEN
         WRITE(I,*)'       '
         WRITE(I,37)'12. THE PAN COEFFICIENT(KP) = ',KP
 37    FORMAT(A30,F4.2)
       END IF
 93 CONTINUE
 30 FORMAT(1X,A15,F6.1,A23,F9.2,1X,A6)
 32 FORMAT(1X,A15,F6.1,A26,I2,A1,I2,A1,I4)
 34 FORMAT(1X,A15,F6.1,A25,I2,A1,I2)
 36 FORMAT(1X,A15,F6.1,A2,A2,A31,F6.2,A3)
       WRITE(5,*)' '
       IF (I.EQ.6) GOTO 601
C
C ------- CORRECTING INPUT ERRORS -------
C
       WRITE(5,*)' ANY CHANGES?  (Y/N) '
       READ(5,422) B$
 422 FORMAT(A1)
       IF (B$.EQ.'Y') THEN
         WRITE(5,*)'WHICH NUMBER WOULD YOU LIKE TO CHANGE?'
         READ(5,*) CHG
         IF(CHG.EQ.1) GO TO 29
         IF(CHG.EQ.2) GO TO 39
         IF(CHG.EQ.3) GO TO 49
```

```
         IF(CHG.EQ.4) GO TO 59
         IF(CHG.EQ.5) GO TO 69
         IF(CHG.EQ.6) GO TO 79
         IF(CHG.EQ.7) GO TO 89
        IF(CHG.EQ.8) GO TO 99
         IF(CHG.EQ.9) GO TO 109
         IF(CHG.EQ.10) GO TO 119
         IF(CHG.EQ.11) GO TO 129
        IF(CHG.EQ.12) GO TO 139
      ELSE
        I=6
        GOTO 500
      END IF
  601 WC1=WC
C
C   ---   READING CLIMATIC DATA FROM AN EXTERNAL DATA FILE   ---
C
C       (THIS DATA FILE IS SIMILAR TO APPENDIX D)
C       CLIMATIC DATA SHOULD HAVE THE FOLLOWING UNITS:
C           TEMPERATURES IN DEGREES C
C           RELATIVE HUMIDITIES IN PERCENT
C           PRECIPIATION AND PAN EVAPORATION IN MM
C           SOLAR RADIATION IN MM OF WATER
C           WIND SPEED IN M/S
C
C   ------------------------------------------------------------
C
      DO 300 I=1,214
         READ(4,*) (X(I,J), J=1,10)
  300 CONTINUE
      DO 350 I=1,214
        M(I)=INT(X(I,1))
        D(I)=INT(X(I,2))
        P(I)=X(I,3)
        EP(I)=X(I,4)
        TMAX(I)=X(I,5)
        TMIN(I)=X(I,6)
        RHMAX(I)=X(I,7)
        RHMIN(I)=X(I,8)
        RS(I)=X(I,9)
        U(I)=X(I,10)
  350 CONTINUE
      WRITE(6,*)'          '
      WRITE(6,*)'          '
      CHAR=' DATE    ET    PE    WC    IRRI   DP    RO   AVE DIR'
      WRITE(6,360) CHAR
      WRITE(6,370)LUNIT,LUNIT,'% BY VOL',LUNIT,LUNIT,LUNIT,DUNIT
```

```
      CHAR='-----------------------------------------------------------'
      WRITE(6,360) CHAR
  360 FORMAT(3X,A60)
  370 FORMAT(13X,A2,5X,A2,3X,A8,3X,A2,5X,A2,4X,A2,4X,A6)
      I=0
      D3=0.
      J=90
      DAY=0
C
C  -------- MAIN CALCULATION LOOP  -----------
C
  400 I=I+1
      J=J+1
      IF (I.GT.214) GOTO 450
      IF (M(I).LT.MBEG) GOTO 400
      IF (M(I).GT.MBEG) GOTO 405
      IF (D(I).LT.DBEG) GOTO 400
  405 IF (I.GT.JEND) GOTO 450
      DAY=DAY+1
      D3=D3+1.
  410 WCI=WC
C
C  ------- DETERMINING MONTHLY "KC" VALUES  -------
C
      IF (KCCODE.EQ.2) THEN
         IF (D3.GT.DA(L))THEN
            L=L+1
            COUNT=0
         ELSE
            KC=KARR(L)+(KARR(L+1)-KARR(L))/D1(JBEG+L-1)*COUNT
            COUNT=COUNT+1
         END IF
C
C  ------- DETERMINING GROWTH STAGE "KC" VALUES  -------
C
      ELSE IF (KCCODE.EQ.3) THEN
         IF (D3.LE.DA(1)) THEN
            KC=KARR(1)
         ELSE IF (D3.LE.DA(2)) THEN
            KC=KARR(1)+(KARR(3)-KARR(1))/(DA(2)-DA(1))*(D3-DA(1))
         ELSE IF (D3.LE.DA(3)) THEN
            KC=KARR(3)
         ELSE
            KC=KARR(3)-(KARR(3)-KARR(4))/(DA(4)-DA(3))*(D3-DA(3))
         END IF
      END IF
C
```

```
C  ------   COMPUTE ETO  ------
C
      IF (METHOD.EQ.1) CALL WJP(TMAX,TMIN,RHMAX,RHMIN,J,RS,U,ET,I,PA)
      IF (METHOD.EQ.2) CALL DPP(TMAX,TMIN,RHMAX,RHMIN,J,LAT,HDO,RS,U,ET,
     &I,PA)
      IF (METHOD.EQ.3) CALL PAN(EP,KP,ET,I,LAT)
      ET=KC*ET
      PETOT=PETOT+P(I)
      ETTOT=ETTOT+ET
      WC=WCI-100.*(ET-P(I))/DRZ
C
C  -----   COMPUTES RO (ASSUMING THAT RO OCCURS WHEN ROOT ZONE  -----
C  -----   REACHES SATURATION)                                  -----
C
      IF (WC.GE.FC) THEN
         IF (WC.LE.POR) THEN
            DP=(WC-FC)/100.*DRZ
            DPTOT=DPTOT+DP
         ELSE
C
C  -----   COMPUTES DEEP PERCOLATION FROM ROOT ZONE  ------
C
            DP=(POR-FC)/100.*DRZ
            DPTOT=DPTOT+DP
            RO=(WC-POR)/100.*DRZ
            ROTOT=ROTOT+RO
         END IF
         WC=FC
      END IF
      C=25.4
C
C  -----   OUTPUTTING DAILY VALUES  -----
C
  430 IF (P(I).GT.0.0) THEN
         IF (IRRI.GT.0) THEN
            IF (UNIT.EQ.1) THEN
              WRITE(6,432) M(I),'/',D(I),ET,P(I),WC,IRRI,DP,RO,ETBAR
            ELSE
              WRITE(6,432) M(I),'/',D(I),ET/C,P(I)/C,WC,IRRI/C,DP/C,RO/C,
     &ETBAR/C
            END IF
         ELSE
            IF (UNIT.EQ.1) THEN
              WRITE(6,434) M(I),'/',D(I),ET,P(I),WC,DP,RO
            ELSE
              WRITE(6,434) M(I),'/',D(I),ET/C,P(I)/C,WC,DP/C,RO/C
            END IF
```

```
            END IF
        ELSE IF (IRRI.GT.0) THEN
            IF (UNIT.EQ.1) THEN
              WRITE(6,436) M(I),'/',D(I),ET,WC,IRRI,DP,RO,ETBAR
            ELSE
              WRITE(6,436) M(I),'/',D(I),ET/C,WC,IRRI/C,DP/C,RO/C,ETBAR/C
            END IF
          ELSE
            IF (UNIT.EQ.1) THEN
              WRITE(6,438) M(I),'/',D(I),ET,WC
            ELSE
              WRITE(6,438) M(I),'/',D(I),ET/C,WC
            END IF
        END IF
  432 FORMAT(4X,I2,A,I2,2F7.2,1X,5F7.2)
  434 FORMAT(4X,I2,A,I2,F7.2,F7.2,1X,F7.2,7X,2F7.2)
  436 FORMAT(4X,I2,A,I2,F7.2,8X,5F7.2)
  438 FORMAT(4X,I2,A,I2,F7.2,8X,F7.2)
      DP=0.
      RO=0.
C
C  -----  DETERMINING AMOUNT OF IRRIGATION  ------
C
      IF (WC.GE.WCC) IRRI=0
      IF (WC.GT.WCC+1.) THEN
         GO TO 400
      END IF
      IRRI=(FC-WC)/100.*DRZ
      ETBAR=IRRI/DAY
      DAY=0
      IRRTOT=IRRTOT+IRRI
      WC=FC
      GO TO 400
  450 IRRIFC=(WC1-WC)/100.*DRZ
      ITOTAL=IRRTOT+IRRIFC
      WRITE(6,*)'        '
C
C  ------  OUTPUTTING TOTALS  -------
C
      IF (UNIT.EQ.1) THEN
         WRITE(6,460)' TOTALS ',ETTOT,PETOT,ITOTAL,'*',DPTOT,ROTOT
      ELSE
         WRITE(6,460)' TOTALS ',ETTOT/C,PETOT/C,ITOTAL/C,'*',DPTOT/C,
     &ROTOT/C
      END IF
  460 FORMAT(A7,2X,F7.2,F7.2,8X,F7.2,A1,F6.2,F7.2)
      WRITE(6,*)'      '
```

```
      WRITE(6,*)'    '
      IF (UNIT.EQ.1) THEN
      WRITE(6,470)' *  THIS INCLUDES ',IRRIFC,LUNIT,' NEEDED TO FILL THE
     & SOIL'
      ELSE
      WRITE(6,470)' *  THIS INCLUDES ',IRRIFC/C  ,LUNIT,' NEEDED TO FILL
     & THE SOIL'
      END IF
      WRITE(6,480)'   FROM ',WC,'% TO THE INITIAL WC OF ',WC1,'%'
  470 FORMAT(A18,F6.2,1X,A2,A24)
  480 FORMAT(A9,F6.2,A24,F6.2,A2)
      WRITE(6,*)'    '
      STOP
      END
C
C
C     SUBROUTINE FOR THE WRIGHT-JENSEN PENMAN METHOD
C
C
      SUBROUTINE WJP(TMAX,TMIN,RHMAX,RHMIN,J,RS,U,ET,I,PA)
      REAL TMAX(214),TMIN(214),RHMIN(214),RHMAX(214),RS(214),U(214),LAT
      LAT=0.0
      TA=(TMAX(I)+TMIN(I))/2.
      ESA=EXP((19.08*TA+429.4)/(TA+237.3))
      RH=(RHMAX(I)+RHMIN(I))/2.
      EA=ESA*RH/100.
      RSO=-17.01116+0.426114*J-0.00172917*J**2.+0.00000184803*J**3.
      CONST=SIN(RAD(0.01745*J-0.26175))
      A1=0.325+0.045*CONST
      QN=0.77*RS(I)-(0.000000001*((TMAX(I)+273.16)**4+(TMIN(I)+273.16)**
     &4)*(A1-0.044*EA**0.5)*(-0.18+1.22*RS(I)/RSO))
      EA=(0.197+0.261*U(I))*(ESA-EA)
      DELTA=4098.*ESA/(TA+237.3)**2.
      GAMMA=1615.*PA/(2490000.-2130.*TA)
      ET=(DELTA*QN+GAMMA*EA)/(DELTA+GAMMA)
      RETURN
      END
C
C
C     SUBROUTINE FOR THE DOORENBOS-PRUITT PENMAN METHOD
C
C
      SUBROUTINE DPP(TMAX,TMIN,RHMAX,RHMIN,J,LAT,HDO,RS,U,ET,I,PA)
      REAL TMAX(214),TMIN(214),RHMIN(214),RHMAX(214),RS(214),U(214),LAT
      TA=(TMAX(I)+TMIN(I))/2.
      ESA=EXP((19.08*TA+429.4)/(TA+237.3))
      RH=(RHMAX(I)+RHMIN(I))/2.
```

```
      EA=ESA*RH/100.
      THETA=0.986*(J-1)
      DEL=.006918-.399912*COS(RAD(THETA))+.070257*SIN(RAD(THETA))-.00675
     &8*COS(RAD(2*THETA))
      DEL=DEL+.000907*SIN(RAD(2*THETA))-.002697*COS(RAD(3*THETA))-.00148
     &*SIN(RAD(3*THETA))
      DEL=180./3.14159*DEL
      HS=ACOS(-(TAN(RAD(LAT))*TAN(RAD(DEL))))
      RVE=.98387-.000111403*J+.0000052774*J**2.-2.68285E-8*J**3+3.61634E
     &-11*J**4
      RA=HS*SIN(RAD(LAT))*SIN(RAD(DEL))+COS(RAD(LAT))*COS(RAD(DEL))*SIN(
     &RAD(HS))
      RA=1.26714*(HDO/RVE**2.)*RA
      QN=.75*RS(I)-(2.E-9*(TA+273.16)**4.*(.34-.044*EA**.5)*(-.35+1.8*RS
     &(I)/RA))
      EA=(.27+.2333*U(I))*(ESA-EA)
      DELTA=4098.*ESA/(TA+237.3)**2.
      GAMMA=1615.*PA/(2490000.-2130.*TA)
      ET=(DELTA*QN+GAMMA*EA)/(DELTA+GAMMA)
      RETURN
      END
C
C
C     PAN EVAPORATION SUBROUTINE
C
C
      SUBROUTINE PAN(EP,KP,ET,I,LAT)
      REAL KP,EP(214)
      LAT=0
      ET=KP*EP(I)
      RETURN
      END
C
C
C     DEGREE TO RADIAN CONVERSION FUNCTION
C
C
      FUNCTION RAD(X)
      REAL LAT
      RAD=X/57.29577952
      RETURN
      END
```

Appendix H

FORTRAN Code for a Pipeline Design Program

```
      WRITE (5,*) '***************************************************'
      WRITE (5,*) '*                                                 *'
      WRITE (5,*) '* PIPELINE DESIGN PROGRAM (187-LGJ)               *'
      WRITE (5,*) '* PROGRAM COMPUTES THE PRESSURE DISTRIBUTION ALONG *'
      WRITE (5,*) '*      (A) SPRINKLE/TRICKLE LATERALS WITH UP TO    *'
      WRITE (5,*) '*          300 SPRINKLERS/EMISSION POINTS AND AS   *'
      WRITE (5,*) '*          MANY AS 3 DIFFERENT DIAMETERS           *'
      WRITE (5,*) '*      (B) SUBMAINS WITH UP TO 300 LATERALS AND AS *'
      WRITE (5,*) '*          MANY AS 10 DIFFERENT DIAMETERS          *'
      WRITE (5,*) '*      (C) MAINLINES WITH UP TO 300 SUBMAINS AND AS *'
      WRITE (5,*) '*          MANY AS 10 DIFFERENT DIAMETERS          *'
      WRITE (5,*) '*                                                 *'
      WRITE (5,*) '*      DATA WILL BE REQUESTED AS NEEDED           *'
      WRITE (5,*) '***************************************************'
C
C  IN THIS PROGRAM THE DEVICE NUMBERS FOR THE TERMINAL (SCREEN) AND
C  PRINTER ARE 5 AND 6, RESPECTIVELY.
C
      DIMENSION E1(300),P1(300),PD(300),H1(300),Q0(300),DL(300),ED(10)
      DIMENSION Q1(300),V1(300),D0(300),DELX(300),CL(300),NCP(300)
      CHARACTER UP$*4,Y$*12,ID$*50,A$*1,B$*1,C$*1,D$*1,E$*1,NO*2,YES*3
      CHARACTER DUNTS$*2,QUNTS$*5,PUNTS$*3,VUNTS$*6,LUNTS$*2,T$*1
      CHARACTER BARB$*3,P$*8,PT$*8,F$*1,R$*1
      REAL LL(300),LD(10),K,K1,K2,K3,L9,L
      INTEGER PIPTYP,UNTS,BL,PRES,ED,UPBD,DFLAG,EFLAG,PFLAG,QFLAG,SC1
      DATA A$,B$,C$,D$,E$,F$,NO,YES/'A','B','C','D','E','F','NO','YES'/
      DO 170 I=1,300
      CL(I)=0
  170 CONTINUE
      WRITE (5,*) 'TO START TYPE ANY CHARACTER'
      READ (5,180) R$
  180 FORMAT (A1)
```

459

```
      WRITE (5,*) 'EITHER SI OR ENGLISH UNITS MAYBE USED. ENTER:'
      WRITE (5,*) '     1  TO USE SI UNITS'
      WRITE (5,*) '     2  TO USE ENGLISH UNITS.'
      READ (5,*) UNTS
      IF (UNTS.EQ.2) GOTO 360
      LUNTS$='M'
      QUNTS$='L/MIN'
      DUNTS$='MM'
      PUNTS$='KPA'
      VUNTS$='M/S'
      K=9.81
      K1=572888/9.81
      K2=0.06479
      K3=16.666
      GOTO 450
  360 LUNTS$='FT'
      QUNTS$='GPM'
      DUNTS$='INS'
      PUNTS$='PSI'
      VUNTS$='FT/S'
      K=2.31
      K1=1/2.31
      K2=29
      K3=.3208
  450 E2=0
      WRITE (5,*) ' TYPE OF PIPELINE:'
      WRITE (5,*) '     1  SPRINKLE LATERAL'
      WRITE (5,*) '     2  TRICKLE LATERAL'
      WRITE (5,*) '     3  SUBMAIN'
      WRITE (5,*) '     4  MAIN LINE'
      WRITE (5,*) ' '
      WRITE (5,*) 'ENTER 1, 2, 3, OR 4 '
      READ (5,*) PIPTYP
      IF (PIPTYP.NE.2) GOTO 580
      IF (UNTS.EQ.2) GOTO 570
      QUNTS$='L/HR'
      GOTO 580
  570 QUNTS$='GPH'
  580 PFLAG=0
      MFLAG=0
      EFLAG=0
      DFLAG=0
      IF (PIPTYP.GT.2) GOTO 4850
      WRITE (5,*) 'ENTER LATERAL ID INFO (UP TO 50 CHARACTERS) '
      READ (5,650) ID$
  650 FORMAT (A50)
      WRITE (6,660) ID$
```

```
 660 FORMAT (6X,A50)
     BL=0
     SED=1
     IF (PIPTYP.EQ.1) GOTO 910
     WRITE (5,*) 'ENTER NUMBER OF EMISSION POINTS ALONG LATERAL'
     READ (5,*) N
     WRITE (5,*) 'ENTER NUMBER OF EMISSION DEVICES PER EMISSION POINT'
     READ (5,*) NED
     IF (NED.EQ.1) GOTO 780
     WRITE (5,760) 'ENTER THE SPACING BETWEEN EMISSION DEVICES IN ',LUN
    $TS$
 760 FORMAT (A46,A2)
     READ (5,*) SED
 780 WRITE (5,*) 'ARE THERE BARB LOSSES? ENTER YES OR NO.'
     READ (5,790) BARB$
 790 FORMAT (A3)
     IF (BARB$.EQ.'NO ') GOTO 880
     IF (BARB$.NE.'YES') GOTO 780
 830 WRITE (5,*) 'USING FIGURE 6.11, IS THE SIZE OF THE IN-LINE BARB'
     WRITE (5,*) 'LARGE, STANDARD, OR SMALL? ENTER:'
     WRITE (5,*) '        1   FOR LARGE '
     WRITE (5,*) '        2   FOR STANDARD'
     WRITE (5,*) '        3   FOR SMALL'
     READ (5,*) BL
     IF (BL.LT.1) GOTO 830
     IF (BL.GT.3) GOTO 830
 880 WRITE (5,890) 'ENTER DISTANCE TO FIRST EMISSION DEVICE IN ',LUNTS$
 890 FORMAT (A43,A2)
     GOTO 960
 910 NED=1
 920 WRITE (5,*) 'ENTER NUMBER OF SPRINKLERS ALONG LATERAL'
     READ (5,*) N
     IF (N.LT.1) GOTO 920
     IF (N.GT.300) GOTO 920
     WRITE (5,950) 'ENTER DISTANCE TO FIRST SPRINKLER IN ',LUNTS$
 950 FORMAT (A37,A2)
 960 READ (5,*) D1
     IF (PIPTYP.EQ.2) GOTO 990
     WRITE (5,980) 'ENTER SPRINKLER SPACING IN ',LUNTS$
 980 FORMAT (A27,A2)
     GOTO 1010
 990 WRITE (5,1000) 'ENTER EMISSION POINT SPACING IN ',LUNTS$
1000 FORMAT (A32,A2)
1010 READ (5,*) D2
1020 IF (PIPTYP.EQ.2) GOTO 1050
     WRITE (5,*) 'ARE FLOW CONTROL NOZZLES TO BE USED? ENTER YES OR NO.
    $'
```

```
        GOTO 1060
1050 WRITE (5,*) 'ARE PRESSURE COMPENSATING EMISSION DEVICES TO BE USED
    $? ENTER YES OR NO.'
1060 READ (5,1061) R$
1061 FORMAT (A1)
     IF (R$.EQ.'N') GOTO 1250
     IF (R$.NE.'Y') GOTO 1020
     IF (PIPTYP.EQ.2) GOTO 1080
     WRITE (5,1070) 'ENTER FLOW CONTROL NOZZLE DISCHARGE RATE IN ',QUNT
    $S$
1070 FORMAT (A44,A5)
     GOTO 1100
1080 WRITE (5,1090) 'ENTER PRES COMPENSATING EMISSION DEVICE DISCHARGE
    $RATE IN ',QUNTS$
1090 FORMAT (A58,A5)
1100 READ (5,*) Q5
     IF (PIPTYP.EQ.2) GOTO 1180
     WRITE (5,*) 'ENTER MINIMUM PRESSURE REQUIRED FOR PROPER FLOW CONTR
    $OL NOZZLE'
     IF (UNTS.EQ.2) GOTO 1150
     WRITE (5,1140) 'OPERATION IN ',PUNTS$,' NOTE: THIS IS USUALLY 300
    $KPA.'
1140 FORMAT (A13,A3,A30)
     GOTO 1210
1150 WRITE (5,1160) 'OPERATION IN ',PUNTS$,'. NOTE: THIS IS USUALLY 40
    $PSI.'
1160 FORMAT (A13,A3,A31)
     GOTO 1210
1180 WRITE (5,*) 'ENTER MINIMUM PRESSURE REQUIRED FOR PROPER PRESSURE'
     WRITE (5,1200) 'COMPENSATING EMISSION DEVICE OPERATION IN ',PUNTS$
1200 FORMAT (A38,1X,A3)
1210 READ (5,*) P7
     B=0
     A=Q5/P7**.5
     GOTO 1400
1250 IF (PIPTYP.EQ.2) GOTO 1370
     WRITE (5,1270) 'ENTER THE DIAMETER OF THE SPRINKLER RANGE NOZZLE I
    $N ',DUNTS$
1270 FORMAT (A52,A2)
     WRITE (5,*) ' (TO ENTER 5/32 ins INPUT 5,32; TO ENTER 1.5 INS OR M
    $M INPUT 1.5,1)'
     READ (5,*) D3,D4
     IF (D4.GT.0) GOTO 1300
     D4=1
1300 WRITE (5,1310) 'ENTER THE DIAMETER OF THE SPRINKLER SPREADER NOZZL
    $E IN ',DUNTS$
1310 FORMAT (A55,A4)
```

```
        WRITE (5,*) '(ENTER 0,1 WHEN THERE IS NO SPREADER NOZZLE)'
        READ (5,*) D5,D6
        IF (D6.LE.0) D6=1.0
        A=K2*((D3/D4)**2+(D5/D6)**2)
        B=.5
        GOTO 1410
1370 WRITE (5,1380) 'ENTER A AND B IN THE EQUATION ',QUNTS$,'= A*P**B
    $ WHERE P IS IN ',PUNTS$
1380 FORMAT (A30,A5,A25,A3)
        READ (5,*) A,B
1400 IF (PIPTYP.NE.2) GOTO 1410
        A=A/60.
1410 IF (DFLAG.GT.0) GOTO 2100
        DFLAG=1
1430 WRITE (5,*) 'THERE MAYBE AS MANY AS 3 DIFFERENT DIAMETERS ALONG TH
    $E LENGTH OF A'
        WRITE (5,*) 'LATERAL. ENTER THE NUMBER OF DIFFERENT PIPE DIAMETERS
    $ ALONG THE LATERAL.'
1460 READ (5,*) NDIA
        ED(1)=N*NED
        ED(2)=N*NED
        IF (NDIA.EQ.1) GOTO 1530
        IF (NDIA.LT.4) GOTO 1570
        WRITE (5,*) 'NDIA > 3, ENTER ANOTHER VALUE THAT IS LESS THAN OR EQ
    $UAL TO 3.'
        GOTO 1460
1530 WRITE (5,1540) 'ENTER LATERAL DIAMETER IN ',DUNTS$
1540 FORMAT (A26,A2)
        READ (5,*) LD(1)
        GOTO 1920
1570 IF (NDIA.EQ.3) GOTO 1670
1580 IF (PIPTYP.EQ.2) GOTO 1600
        WRITE (5,*) 'WITH THE UPSTREAM MOST SPRINKLER AS NUMBER 1, ENTER'
        WRITE (5,*) 'THE SPRINKLER NUMBER WHERE THE DIAMETER CHANGES.'
        GOTO 1620
1600 WRITE (5,*) 'WITH THE UPSTREAM MOST EMISSION POINT AS NUMBER 1, EN
    $TER THE'
        WRITE (5,*) 'EMISSION POINT NUMBER WHERE THE DIAMETER CHANGES.'
1620 READ (5,*) ED(1)
        IF (ED(1).LE.N) GOTO 1820
        IF (PIPTYP.EQ.2) GOTO 1660
        WRITE (5,*) 'THE SPR# ENTERED > THE# OF SPKS ALONG THE LATERAL.
    $TRY AGAIN.'
        GOTO 1580
1660 WRITE (5,*) 'THE EM PT# ENTERED > THE TOTAL# OF EM PTS ALONG THE
    $ LATERAL. TRY AGAIN.'
        GOTO 1580
```

```
1670 IF (PIPTYP.EQ.2) GOTO 1680
     WRITE (5,*) 'WITH THE UPSTREAM MOST SPRINKLER AS NUMBER 1, ENTER
    $NUMBER 1, ENTER THE'
     WRITE (5,*) 'SPRINKLER NUMBERS WHERE THE DIAMETER CHANGES OCCUR.'
     GOTO 1710
1680 WRITE (5,*) 'WITH THE UPSTREAM MOST EMISSION POINT AS NUMBER 1, EN
    $TER THE'
     WRITE (5,*) 'EMISSION POINT NUMBERS WHERE THE DIAMETER CHANGES OCC
    $UR'
1710 READ (5,*) ED(1),ED(2)
     IF (ED(1).LT.ED(2)) GOTO 1760
     NDUM=ED(1)
     ED(1)=ED(2)
     ED(2)=NDUM
1760 IF (ED(2).LT.N) GOTO 1820
     IF (PIPTYP.EQ.2) GOTO 1790
     WRITE(5,*) 'ONE OF THE SPR #S ENTERED > THE TOTAL # OF SPRS.'
     WRITE(5,*) 'ENTER THE SPR #S WHERE THE LATERAL DIAMETER CHANGES.'
     GOTO 1710
1790 WRITE(5,*)'ONE OF THE EM PT #S ENTERED > THE TOTAL # OF EM PTS.'
     WRITE(5,*)'ENTER THE EM PT #S WHERE THE LATERAL DIAMETER CHANGES
    $.'
     GOTO 1710
1820 DO 1870 I=1,NDIA-1
     IF (PIPTYP.EQ.2) GOTO 1850
     WRITE (5,1840) 'ENTER THE DIAMETER UPSTREAM OF SPRINKLER ',ED(I),'
    $ IN ',DUNTS$
1840 FORMAT (A47,I3,A4,A2)
     GOTO 1860
1850 WRITE (5,1840) 'ENTER THE DIAMETER UPSTREAM OF EMISSION DEVICE ',E
    $D(I),' IN ',DUNTS$
1860 READ (5,*) LD(I)
1870 CONTINUE
     IF (PIPTYP.EQ.2) GOTO 1900
     WRITE (5,1890) 'ENTER THE DIAMETER DOWNSTREAM OF SPRINKLER ',ED(ND
    $IA-1),' IN ',DUNTS$
1890 FORMAT (A49,I3,A4,A2)
     GOTO 1910
1900 WRITE (5,1890) 'ENTER THE DIAMETER DOWNSTREAM OF EMISSION DEVICE '
    $,ED(NDIA-1),' IN ',DUNTS$
1910 READ (5,*) LD(NDIA)
1920 DOO=LD(1)
     DO 1970 I=1,N*NED
     IF (I.GT.ED(1)) DOO=LD(2)
     IF (I.GT.ED(2)) DOO=LD(3)
1960 DO(I)=DOO
1970 CONTINUE
```

```
      IF (PIPTYP.EQ.1) GOTO 2100
      IF (BL.EQ.0) GOTO 2100
C
C  --------  BARB LOSSES  -----------
C
      DO 2090 I=1,N*NED
      IF (UNTS.EQ.2) GOTO 2060
      IF (BL.NE.1) GOTO 2000
      CL(I)=30.91*DO(I)**(-1.935)
      GOTO 2090
 2000 IF (BL.NE.2) GOTO 2010
      CL(I)=23.03*DO(I)**(-1.947)
      GOTO 2090
 2010 CL(I)=19.47*DO(I)**(-2.018)
      GOTO 2090
 2060 IF (BL.NE.1) GOTO 2070
      CL(I)=.194*DO(I)**(-1.935)
      GOTO 2090
 2070 IF (BL.NE.2) GOTO 2080
      CL(I)=.139*DO(I)**(-1.947)
      GOTO 2090
 2080 CL(I)=.0934*DO(I)**(-2.018)
 2090 CONTINUE
 2100 IF (MFLAG.GT.0) GOTO 2130
 2110 CALL MATL (C,Y$)
      MFLAG=1
 2130 IF (PFLAG.GT.0) GOTO 2280
 2140 WRITE (5,*) 'THE PRESSURE AT EITHER THE UPSTREAM OR DOWNSTREAM'
      WRITE (5,*) 'END OF THE LATERAL IS REQUIRED. ENTER:'
      WRITE (5,*) '    1  TO INPUT A DOWNSTREAM PRESSURE'
      WRITE (5,*) '    2  TO INPUT AN UPSTREAM PRESSURE'
      READ (5,*) PRES
 2190 IF (PRES.EQ.2) GOTO 2250
      IF (PIPTYP.EQ.2) GOTO 2220
      WRITE (5,2210) 'ENTER PRESSURE IN ',PUNTS$,' AT DOWNSTREAM MOST SP
     $RINKLER'
 2210 FORMAT (A18,A3,A29)
      GOTO 2240
 2220 WRITE (5,2230) 'ENTER PRESSURE IN ',PUNTS$,' AT DOWNSTREAM MOST EM
     $ISSION DEVICE'
 2230 FORMAT (A18,A3,A35)
 2240 READ (5,*) P10
      GOTO 2270
 2250 WRITE (5,2210) 'ENTER PRESSURE IN ',PUNTS$,' AT UPSTREAM END OF LA
     $TERAL'
      READ (5,*) P10
 2270 PFLAG=1
```

```
2280 IF (EFLAG.GT.0) GOTO 2760
     WRITE (5,2300) 'ENTER ELEVATION OF SUBMAIN AT UPSTREAM END OF LATE
     $RAL IN ',LUNTS$
2300 FORMAT (A57,A2)
     READ (5,*) E10
     EFLAG=1
2310 IF (PIPTYP.EQ.2) GOTO 2350
     WRITE (5,*) 'YOU MAY ENTER THE ELEVATION AT EACH SPRINKLER OF THEY
     $ CAN'
     GOTO 2360
2350 WRITE (5,*) 'YOU MAY ENTER THE ELEVATION AT EACH EMISSION DEVICE O
     $R THEY CAN'
2360 WRITE (5,*) 'BE CALCULATED FROM USER SUPPLIED SLOPE DATA. INPUT "E
     $"'
     IF (PIPTYP.EQ.2) GOTO 2380
     WRITE (5,*) 'IF YOU WISH TO ENTER THE ELEVATION AT EACH SPRINKLER,
     $ "S"'
     WRITE (5,*) 'TO COMPUTE ELEVATION AT EACH SPRINKLER FROM SLOPE DAT
     $A'
     GOTO 2390
2380 WRITE (5,*) 'IF YOU WISH TO ENTER THE ELEVATION AT EACH EMISSION D
     $EVICE, "S"'
     WRITE (5,*) 'TO COMPUTE ELEVATION AT EACH EMISSION DEVICE FROM SLO
     $PE DATA'
2390 READ (5,1061) R$
     IF (R$.EQ.'S') GOTO 2510
     IF (R$.NE.'E') GOTO 2310
     DO 2490 I=1,N*NED
     IF (PIPTYP.EQ.2) GOTO 2460
     WRITE (5,2450) 'ELEVATION IN ',LUNTS$,' OF SPRINKLER ',I,' IS'
2450 FORMAT (A13,A2,A14,I3,A3)
     GOTO 2480
2460 WRITE (5,2470) 'ELEVATION IN ',LUNTS$,' OF EMISSION POINT ',I,' IS
     $'
2470 FORMAT (A13,A2,A19,I3,A4)
2480 READ (5,*) E1(I)
2490 CONTINUE
     GOTO 2960
2510 WRITE (5,*) 'YOU MAY SPECIFY TWO SLOPES ALONG THE PIPELINE, + FOR'
     WRITE (5,*) 'DOWNHILL SLOPES, - FOR UPHILL SLOPES (IN THE DIRECTIO
     $N OF FLOW) '
     WRITE (5,*) 'ENTER SLOPE(S) IN        $UPSTREAM END '
     READ (5,*) S1,S2
     L9=D1+(N-1)*D2+FLOAT(NED-1)*SED
     IF (S1.NE.S2) GOTO 2590
     L=L9
     GOTO 2620
```

```
2590 WRITE (5,2600) 'ENTER THE LENGTH OF SLOPE S1 IN ',LUNTS$,' (S1 MUS
    $T BE > ',D1,' AND < ',L9,' )'
2600 FORMAT (A32,A2,A15,F4.0,A7,F5.0,A2)
     READ (5,*) L
2620 SL=S1
     X=0
     DELX(1)=D1
     EI1=E10
     I1=0
     DO 2750 I=1,N
     DO 2730 J=1,NED
     X=X+DELX(I1+1)
     IF (X.GT.L) SL=S2
     I1=I1+1
     E1(I1)=EI1-SL/100.*DELX(I1)
     EI1=E1(I1)
     DELX(I1+1)=SED
2730 CONTINUE
     IF (I.GE.N) GOTO 2750
     DELX(I1+1)=D2
2750 CONTINUE
2760 IF (PRES.EQ.2) GOTO 2960
C
C ------- CALCULATION LOOP FOR KNOWN DOWNSTREAM PRESSURE ---------
C
     I1=N*NED
     P1(I1)=P10
     Q3=0
     DO 2940 I=1,I1
     J=I1+1-I
     IF (P1(J).LT.0) GOTO 4660
     IF (B.GT.0) GOTO 2890
     Q0(J)=Q5
     IF (PIPTYP.NE.2) GOTO 2870
     Q0(J)=Q5/60.
2870 IF (P1(J).GE.P7) GOTO 2900
     Q0(J)=A*P1(J)**.5
     GOTO 2900
2890 Q0(J)=A*P1(J)**B
2900 Q3=Q3+Q0(J)
     H1(J)=K1*(DELX(J)+CL(J))*(.285*C)**(-1.852)*Q3**1.85/D0(J)**4.87
     IF (J.EQ.1) GOTO 2950
     P1(J-1)=P1(J)+H1(J)+(E1(J)-E1(J-1))/K
2940 CONTINUE
2950 P10=P1(J)+H1(J)+(E1(J)-E10)/K
     GOTO 3750
C    ------- CALCULATION LOOP FOR KNOWN UPSTREAM PRESSURE   -------
```

```
C
C            THE PRESSURE AND DISCHARGE AT EACH SPRINKLER/EMISSION
C            DEVICE ARE DETERMINED VIA AN ITERATIVE TECHNIQUE. THE
C            SOLUTION BEGINS BY GUESSING A TOTAL Q FOR THE LATERAL
C            AND USING IT TO COMPUTE THE PRESSURE AND DISCHARGE OF
C            EACH SPRINKLER/EMISSION DEVICE. THE DISCHARGES ARE
C            SUMMED AND COMPARED TO THE INITIAL Q. THE PROCESS IS
C            REPEATED UNTIL THE Q COMPUTED DURING AN ITERATION IS
C            WITHIN 1 PERCENT OF THE Q USED TO BEGIN THE ITERATION.
C
C      ------------------------------------------------------------------
C
 2960 K5=0
      QDUM=0
      PI1=P10
      EI1=E10
C
C -----    ESTIMATE MAX Q BY NEGLECTING FRICTION LOSSESS    ------
C
      DO 3110 I=1,N*NED
      P1(I)=PI1-(E1(I)-EI1)/K
      PI1=P1(I)
      EI1=E1(I)
      IF (P1(I).LT.0) GOTO 4660
      IF (B.GT.0) GOTO 3090
      Q0(I)=Q5
      IF (PIPTYP.NE.2) GOTO 3070
      Q0(I)=Q5/60.
 3070 IF (P1(I).GE.P7) GOTO 3100
      Q0(I)=A*P1(I)**.5
      GOTO 3100
 3090 Q0(I)=A*P1(I)**B
 3100 QDUM=QDUM+Q0(I)
 3110 CONTINUE
C
C -----    ESTIMATE MIN Q BY USING MAX Q TO COMPUTE FRICTION LOSSESS -----
C
      Q6=QDUM
      DO 3190 I=1,N*NED
      H1(I)=K1*(DELX(I)+CL(I))*(.285*C)**(-1.852)*Q6**1.85/D0(I)**4.87
      Q6=Q6-Q0(I)
 3190 CONTINUE
      Q2=0
      PI1=P10
      EI1=E10
      DO 3310 I=1,N*NED
      P1(I)=PI1-H1(I)-(E1(I)-EI1)/K
```

```
        PI1=P1(I)
        EI1=E1(I)
        IF (P1(I).LT.0) GOTO 4660
        IF (B.GT.0) GOTO 3290
        Q0(I)=Q5
        IF (PIPTYP.NE.2) GOTO 3250
        Q0(I)=Q5/60.
3250 IF (P1(I).GE.P7) GOTO 3300
        Q0(I)=A*P1(I)**.5
        GOTO 3300
3290 Q0(I)=A*P1(I)**B
3300 Q2=Q2+Q0(I)
3310 CONTINUE
3320 Q3=(QDUM+Q2)/2
        Q7=.1*Q3
        IF (ABS (QDUM-Q2).LE.Q7) GOTO 3690
3350 Q6=Q3
        Q7=.01*Q6
        PI1=P10
        EI1=E10
        DO 3530 I=1,N*NED
        H1(I)=K1*(DELX(I)+CL(I))*(.285*C)**(-1.852)*Q6**1.85/D0(I)**4.87
        P1(I)=PI1-H1(I)-(E1(I)-EI1)/K
        PI1=P1(I)
        EI1=E1(I)
        IF (P1(I).LT.0) GOTO 4660
        IF (B.GT.0) GOTO 3480
        Q0(I)=Q5
        IF (PIPTYP.NE.2) GOTO 3450
        Q0(I)=Q5/60.
3450 IF (PIPTYP.GE.P7) GOTO 3490
        Q0(I)=A*P1(I)**.5
        GOTO 3490
3480 Q0(I)=A*P1(I)**B
3490 Q6=Q6-Q0(I)
        IF (Q6.GT.0) GOTO 3530
        Q3=Q2
        GOTO 3350
3530 CONTINUE
        IF (ABS (Q0(N*NED)-Q6).LE.Q7) GOTO 3690
        K5=K5+1
        IF (K5.LT.20) GOTO 3640
        WRITE (5,*) '*************************************************'
        WRITE (5,*) '*                                               *'
        WRITE (5,*) '* AN ANSWER HAS NOT BEEN OBTAINED IN 20 ITERATIONS *'
        WRITE (5,*) '* SEE THE INSTRUCTOR.                           *'
        WRITE (5,*) '*                                               *'
```

```
          WRITE (5,*) '****************************************************'
          GOTO 4830
 3640 IF (QO(N*NED)-Q6.GE.0) GOTO 3670
          QDUM=Q3
          GOTO 3320
 3670 Q2=Q3
          GOTO 3320
 3690 Q3=0
          DO 3730 I=1,N*NED
          Q3=Q3+QO(I)
 3730 CONTINUE
C
C  ----------  PRINT INPUT DATA AND RESULTS  ------------
C
 3750 WRITE (6,*) ' '
          IF (NDIA.GT.1) GOTO 3780
          WRITE (6,3850) 'LATERAL DIAMETER              = ',LD(1),DUNTS$
          GOTO 3860
 3780 DO 3820 I=1,NDIA-1
          IF (PIPTYP.EQ.2) GOTO 3810
          WRITE (6,3800) 'LAT DIA UPSTRM OF SPR      ',ED(I),' = ',LD(I),DU
      $NTS$
 3800 FORMAT (5X,A28,I3,A3,F5.2,1X,A4)
          GOTO 3820
 3810 WRITE (6,3800) 'LAT DIA UPSTRM OF EM DEV   ',ED(I),' = ',LD(I),DU
      $NTS$
 3820 CONTINUE
          IF (PIPTYP.EQ.2) GOTO 3840
          WRITE (6,3800) 'LAT DIA DOWNSTRM OF SPR    ',ED(NDIA-1),' = ',LD(N
      $DIA),DUNTS$
          GOTO 3860
 3840 WRITE (6,3800) 'LAT DIA DOWNSTRM OF EM DEV ',ED(NDIA-1),' = ',LD(N
      $DIA),DUNTS$
 3860 IF (PIPTYP.EQ.2) GOTO 3870
          WRITE (6,3850) 'DISTANCE TO FIRST SPRINKLER     = ',D1,LUNTS$
          WRITE (6,3850) 'SPRINKLER SPACING               = ',D2,LUNTS$
 3850 FORMAT (5X,A34,F5.2,1X,A2)
          GOTO 3900
 3870 WRITE (6,3850) 'DISTANCE TO FIRST EMIS DEVICE   = ',D1,LUNTS$
          WRITE (6,3850) 'EMISSION POINT SPACING          = ',D2,LUNTS$
 3900 IF (B.GT.0) GOTO 3970
          IF (UNTS.EQ.1) GOTO 3950
          IF (PIPTYP.EQ.2) GOTO 3920
          WRITE (6,3850) 'FLOW CONTROL NOZZLE GPM         = ',Q5,'    '
          GOTO 3970
 3920 WRITE (6,3850) 'PRESSURE COMPENSATING GPH        = ',Q5,'    '
          GOTO 3980
```

```
3950 IF (PIPTYP.EQ.2) GOTO 3960
     WRITE (6,3850) 'FLOW CONTROL NOZZLE L/MIN       = ',Q5,'     '
3970 WRITE (6,3990) QUNTS$,' = ',A,'*P**',B
     GOTO 4000
3960 WRITE (6,3850) 'PRESSURE COMPENSATING L/HR      = ',Q5,'     '
3980 WRITE (6,3990) QUNTS$,' = ',A*60,'*P**',B
3990 FORMAT (31X,A5,A3,F5.2,A4,F3.2)
4000 WRITE (6,4010) 'PIPE MATERIAL                   = ',Y$
4010 FORMAT (5X,A34,A12)
     WRITE (6,*) ' '
     WRITE (6,*) ' '
     IF (PIPTYP.EQ.2) GOTO 4030
     WRITE (6,*) '    SPRINKLER      ELEVATION     PRESSURE      FLOW
    $'
     GOTO 4040
4030 WRITE (6,*) ' EMISSION DEV      ELEVATION     PRESSURE      FLOW
    $'
4040 WRITE (6,4070) '      NUMBER',LUNTS$,PUNTS$,QUNTS$
4070 FORMAT (2X,A11,10X,A2,10X,A3,10X,A5)
     WRITE (6,*) ' '
     IF (PIPTYP.EQ.2) GOTO 4130
     WRITE (6,4110) 'MAINLINE',E10,P10,Q3,QUNTS$
4110 FORMAT (5X,A8,1X,3(F13.2),1X,A5)
     GOTO 4150
4130 IF (UNTS.EQ.2) GOTO 4140
     WRITE (6,4110) 'MAINLINE',E10,P10,Q3,'L/MIN'
     GOTO 4150
4140 WRITE (6,4110) 'MAINLINE',E10,P10,Q3,'GPM'
4150 DO 4190 I=1,N*NED,NED
     IF (PIPTYP.EQ.2) GOTO 4170
     WRITE (6,4180) '   SPK',I,E1(I),P1(I),Q0(I)
     GOTO 4190
4170 WRITE (6,4180) 'EM DEV',I,E1(I),P1(I),Q0(I)*60
4180 FORMAT (5X,A6,I3,3(F13.2))
4190 CONTINUE
     Q9=Q3/FLOAT(N*NED)
     Q7=Q3
     Q8=0
     DO 4280 I=1,N*NED
     IF (Q0(I).GT.Q7) GOTO 4260
     Q7=Q0(I)
4260 IF (Q0(I).LT.Q8) GOTO 4280
     Q8=Q0(I)
4280 CONTINUE
     WRITE (6,*) ' '
     WRITE (6,*) ' '
     Q6=(Q8-Q7)*100./Q9
```

```
       IF (PIPTYP.EQ.2) GOTO 4360
       WRITE (6,4370) 'AVE SPRINKLER DISCHARGE       = ',Q9,QUNTS$
       WRITE (6,4370) 'MIN SPRINKLER DISCHARGE       = ',Q7,QUNTS$
       GOTO 4390
 4360  WRITE (6,4370) 'AVE DISCHARGE PER EMISSION PT = ',Q9*60,QUNTS$
 4370  FORMAT (5X,A33,F6.2,1X,A6)
       WRITE (6,4370) 'MIN DISCHARGE PER EMISSION PT = ',Q7*60,QUNTS$
 4390  WRITE (6,4370) '      WRITE (5,*) ' '
       WRITE (5,*) '***********************************************'
       WRITE (5,*) '*                                             *'
       WRITE (5,*) '*    YOU MAY RERUN THE PROGRAM FOR:           *'
       WRITE (5,*) '*         (A) ANOTHER LATERAL DIAMETER        *'
       IF (PIPTYP.EQ.2) GOTO 4400
       WRITE (5,*) '*         (B) ANOTHER SPRINKLER               *'
       GOTO 4410
 4400  WRITE (5,*) '*         (B) ANOTHER EMISSION DEVICE         *'
 4410  WRITE (5,*) '*         (C) ANOTHER OPERATING PRESSURE      *'
       WRITE (5,*) '*         (D) ANOTHER PIPE MATERIAL           *'
       WRITE (5,*) '*         (E) ANOTHER LATERAL                 *'
       WRITE (5,*) '*         (F) SUBMAIN SELECTION               *'
       WRITE (5,*) '*      OR (G) STOP                            *'
       WRITE (5,*) '*                                             *'
       WRITE (5,*) '***********************************************'
       WRITE (5,*) 'ENTER A,B,C,D,E,F OR G'
       READ (5,1061) T$
       IF (T$.EQ.'A') GOTO 1430
       IF (T$.EQ.'B') GOTO 1020
       IF (T$.EQ.'C') GOTO 2190
       IF (T$.EQ.'D') GOTO 2110
       IF (T$.EQ.'E') GOTO 580
       IF (T$.EQ.'F') GOTO 4850
 4650  WRITE (5,*) '***********  THE RUN IS COMPLETE  ***************'
       STOP
 4660  WRITE (5,*) ' '
       WRITE (5,*) ' '
       WRITE (5,*) '***********************************************'
       WRITE (5,*) '*                                             *'
       WRITE (5,*) '* THE PRESSURE ALONG THE PIPELINE IS < 0. YOU MAY: *'
       WRITE (5,*) '*         (A) INCREASE THE UPSTREAM PRESSURE     *'
       WRITE (5,*) '*         (B) INCREASE THE PIPE DIAMETER         *'
       WRITE (5,*) '*         (C) USE A DIFFERENT PIPE MATERIAL      *'
       WRITE (5,*) '*           (D) OR STOP.                         *'
       WRITE (5,*) '*                                             *'
       WRITE (5,*) '***********************************************'
       WRITE (5,*) ' ENTER A,B,C OR D '
       READ (5,1061) T$
       E2=E10
```

```
        IF (T$.EQ.'A') GOTO 2140
        IF (T$.EQ.'B') GOTO 1430
        IF (T$.EQ.'C') GOTO 2110
 4830 WRITE (5,*) '**** THE RUN HAS BEEN TERMINATED ****'
        STOP
C
C
C ------- SUBMAINS -- MAINS ---------
C
C
 4850 IF (PIPTYP.EQ.4) GOTO 4980
        PT$='LATERAL'
        P$='SUBMAIN'
        WRITE (5,*) 'THE SUBMAINS UPSTREAM END IS A:'
        WRITE (5,*) '        (1) PUMP'
        WRITE (5,*) '        (2) MAINLINE'
        WRITE (5,*) 'ENTER 1 OR 2'
        READ (5,*) UPBD
        IF (UPBD.EQ.2) GOTO 4960
        UP$='PUMP'
        GOTO 5010
 4960 UP$='MAIN'
        GOTO 5010
 4980 PT$='SUBMAIN'
        UP$='PUMP'
        P$='MAINLINE'
 5010 WRITE (5,5040) 'ENTER ',P$,' ID INFO (UP TO 50 CHARACTERS) '
 5040 FORMAT (A6,A8,A31)
        READ (5,650) ID$
        WRITE (6,660) ID$
        DFLAG=0
        MFLAG=0
        QFLAG=0
        EFLAG=0
        WRITE (5,5120) 'ENTER THE NUMBER OF POINTS ALONG THE ',P$,' WHERE
      $ ',PT$,'S MAY BE CONNECTED.'
 5120 FORMAT (A37,A8,A7,A7,A19)
        READ (5,*) NCPS
        WRITE (5,5150) 'ENTER THE DISTANCE BETWEEN THE ',UP$,' AND THE FIR
      $ST CONNECTION POINT IN ',LUNTS$
 5150 FORMAT (A31,A4,A35,A4)
        READ (5,*) SCP1
        WRITE (5,5180) 'ENTER THE DISTANCE (SPACING) BETWEEN CONNECTION PO
      $INTS IN ',LUNTS$
 5180 FORMAT (A58,A4)
        READ (5,*) SCP
 5200 WRITE (5,5210) 'ENTER NUMBER OF ',PT$,'S OPERATING SIMULTANEOUSLY'
```

```
5210 FORMAT (A16,A7,A26)
     READ (5,*) N
     IF (N.LE.300) GOTO 5260
     WRITE (5,*) 'THE VALUE ENTERED EXCEEDS 300. ENTER A SMALLER VALUE.
     $'
     GOTO 5200
5260 IF (N.GT.1) GOTO 5540
     WRITE (5,*) 'WITH THE UPSTREAM MOST CONNECTION POINT AS NUMBER 1,
     $ENTER THE'
     WRITE (5,5290) 'CONNECTION POINT NUMBER WHERE THE ',PT$,' IS CONNE
     $CTED'
5290 FORMAT (A34,A7,A14)
5300 READ (5,*) NCP(1)
     IF (NCP(1).LE.NCPS) GOTO 5340
     CALL OVER (NCPS)
     GOTO 5300
5340 LL(1)=SCP1+SCP*FLOAT(NCP(1)-1)
     WRITE (5,5360) 'ENTER THE ',PT$,' DISCHARGE IN ',QUNTS$
5360 FORMAT (A10,A7,A14,A5)
     READ (5,*) Q0(1)
     WRITE (5,5390) 'ENTER DIAMETER OF THE ',P$,' IN ',DUNTS$
5390 FORMAT (A22,A8,A4,A4)
     READ (5,*) DL(1)
     Q1(1)=Q0(1)
     WRITE (5,5430) 'ENTER ELEVATION IN ',LUNTS$,' AT ',PT$,' CONNECTIO
     $N POINT NUMBER ',NCP(1)
5430 FORMAT (A19,A2,A4,A7,A25,I3)
     READ (5,*) E1(1)
     WRITE (5,5460) 'ENTER THE ELEVATION IN ',LUNTS$,' AT THE ',UP$
5460 FORMAT (A27,A2,A8,A4)
     READ (5,*) E10
     WRITE (5,5490) 'ENTER THE DESIRED PRESSURE IN ',PUNTS$,' AT THE ',
     $PT$
5490 FORMAT (A30,A3,A8,A7)
     READ (5,*) PD(1)
     EFLAG=1
     DFLAG=1
     IF (MFLAG.GT.0) GOTO 6430
5540 MFLAG=1
5550 CALL MATL (C,Y$)
     IF (QFLAG.GT.0) GOTO 7260
     QFLAG=1
     IF (N.EQ.1) GOTO 7260
5590 WRITE (5,5600) 'DO ',PT$,'S OPERATE ON BOTH SIDES OF THE ',P$,' ?'
5600 FORMAT (A3,A7,A31,A8,A3)
     WRITE (5,*) 'ENTER YES OR NO.'
     READ (5,1061) R$
```

```
      NS=1
      IF (R$.EQ.'N') GOTO 5700
      IF (R$.NE.'Y') GOTO 5590
      NS=2
      DUM=FLOAT(N)/2.
      DUM1=INT(DUM+.1)
      DUM2=INT (DUM1*2)
      IF (N.LE.DUM2) GOTO 5710
      WRITE (5,5690) 'THERE IS ONE UNPAIRED ',PT$,'. IT IS ASSUMED TO BE
     $ THE DOWNSTREAM MOST ',PT$,'.'
5690  FORMAT (A22,A7,A42,A7,A1)
5700  WRITE (5,*) ' '
5710  WRITE (5,5720) ' WITH THE UPSTREAM MOST ',PT$,' AS NUMBER 1 AND TH
     $E UPSTREAM MOST'
5720  FORMAT (A24,A7,A34)
      WRITE (5,*) 'CONNECTION POINT AS NUMBER 1, ENTER THE CONNECTION PO
     $INT NUMBER WHERE'
      DO 5900 I=1,N,NS
      IF (NS.EQ.1) GOTO 5770
      IF (I.LT.N) GOTO 5830
5770  WRITE (5,5780) PT$,I,' IS CONNECTED.'
5780  FORMAT (1X,A7,I4,A14)
5790  READ (5,*) NCP(I)
      IF (NCP(I).LE.NCPS) GOTO 5900
      CALL OVER (NCPS)
      GOTO 5790
5830  WRITE (5,5840) PT$,'S',I,' AND ',I+1,' ARE CONNECTED.'
5840  FORMAT (1X,A7,A2,I3,A5,I3,A15)
5850  READ (5,*) NCP(I)
      IF (NCP(I).LE.NCPS) GOTO 5890
      CALL OVER (NCPS)
      GOTO 5850
5890  NCP(I+1)=NCP(I)
5900  CONTINUE
      IF (DFLAG.GT.0) GOTO 7290
      WRITE (5,5930) PT$,' DISCHARGE IS:'
5930  FORMAT (A7,A14)
      WRITE (5,5950) '    (1) THE SAME FOR ALL ',PT$,'S'
5950  FORMAT (A25,A7,A)
      WRITE (5,5970) '    (2) DIFFERENT FOR EACH ',PT$
5970  FORMAT (A27,A7)
      WRITE (5,*) 'ENTER 1 OR 2 '
      READ (5,*) SL1
      IF (SL1.GT.1) GOTO 6210
      WRITE (5,6020) 'ENTER DISCHARGE FOR THE ',PT$,'S IN ',QUNTS$
6020  FORMAT (A24,A7,A5,A5)
      READ (5,*) Q0(N)
```

```
         DO 6060 I=1,N
         Q0(I)=Q0(N)
 6060 CONTINUE
         Q9=0
         DO 6190 I=1,N,NS
         J=N-I+1
         IF (NS.EQ.1) GOTO 6130
         IF (N.GT.DUM2) GOTO 6120
         J=J-1
 6120 IF (J.LT.N) GOTO 6160
 6130 Q9=Q9+Q0(J)
         Q1(J)=Q9
         GOTO 6190
 6160 Q9=Q9+Q0(J)+Q0(J+1)
         Q1(J)=Q9
         Q1(J+1)=Q1(J)
 6190 CONTINUE
         GOTO 6430
 6210 WRITE (5,6220) 'WITH THE UPSTREAM MOST ',PT$,' AS NUMBER 1,'
 6220 FORMAT (A23,A7,A13)
         WRITE (5,6240) 'ENTER THE DISCHARGE IN ',QUNTS$,' OF:'
 6240 FORMAT (A23,A5,A4)
         DO 6290 I=1,N
         WRITE (5,6270) PT$,'# ',I
 6270 FORMAT (10X,A7,A2,I2)
         READ (5,*) Q0(I)
 6290 CONTINUE
         Q9=0
         DO 6420 I=1,N,NS
         J=N-I+1
         IF (NS.EQ.1) GOTO 6360
         IF (N.GT.DUM2) GOTO 6350
         J=J-1
 6350 IF (J.LT.N) GOTO 6390
 6360 Q9=Q9+Q0(J)
         Q1(J)=Q9
         GOTO 6420
 6390 Q9=Q9+Q0(J)+Q0(J+1)
         Q1(J)=Q9
         Q1(J+1)=Q1(J)
 6420 CONTINUE
 6430 IF (DFLAG.GT.1) GOTO 7290
 6440 DFLAG=1
         WRITE (5,6460) ' THERE MAY BE AS MANY AS 10 DIFFERENT DIAMETERS AL
        $ONG THE ',P$
 6460 FORMAT (A58,A8)
         WRITE (5,*) 'ENTER THE NUMBER OF DIFFERENT PIPE DIAMETERS'
```

```
6480 READ (5,*) NDIA
     IF (NDIA.LE.10) GOTO 6530
     WRITE (5,*) 'THE VALUE ENTERED EXCEEDS 10 (THE MAXIMUM# OF DIAMET
     $ERS ALLOWED). TRY AGAIN.'
     GOTO 6480
6530 DO 6550 I=1,NDIA
     ED(I)=NCPS
6550 CONTINUE
     IF (NDIA.GT.1) GOTO 6630
     WRITE (5,6590) 'ENTER THE DIAMETER IN ',DUNTS$,' OF THE ',P$
6590 FORMAT (A22,A3,A8,A8)
     READ (5,*) DL(1)
     LL(1)=SCP1+FLOAT(NCPS-1)*SCP
     GOTO 6930
6630 IF (NDIA.GT.2) GOTO 6700
     WRITE (5,*) 'WITH THE UPSTREAM MOST CONNECTION POINT AS NUMBER 1,
     $ENTER'
     WRITE (5,*) 'THE CONNECTION POINT NUMBER WHERE THE DIAMETER CHANGE
     $S.'
6660 READ (5,*) ED(1)
     IF (ED(1).LE.NCPS) GOTO 6800
     CALL OVER (NCPS)
     GOTO 6660
6700 WRITE (5,6710) 'STARTING AT THE UPSTREAM END OF THE ',P$,', ENTER'
6710 FORMAT (A36,A8,A7)
     DO 6790 I=1,NDIA-1
     WRITE (5,6740) 'THE CONNECTION POINT NUMBER WHERE DIAMETER CHANGE
     $',I,' OCCURS'
6740 FORMAT (A50,I3,A7)
6750 READ (5,*) ED(I)
     IF (ED(I).LE.NCPS) GOTO 6790
     CALL OVER (NCPS)
     GOTO 6750
6790 CONTINUE
6800 DO 6880 I=1,NDIA-1
     WRITE (5,6820) 'ENTER THE PIPE DIAMETER UPSTREAM OF CONNECTION POI
     $NT ',ED(I),' IN ',DUNTS$
6820 FORMAT (A53,I3,A4,A3)
     READ (5,*) DL(I)
     IF (I.GT.1) GOTO 6870
     LL(I)=SCP1+FLOAT(ED(I)-1)*SCP
     GOTO 6880
6870 LL(I)=(ED(I)-ED(I-1))*SCP
6880 CONTINUE
     WRITE (5,6900) 'ENTER THE PIPE DIAMETER DOWNSTREAM OF CONNECTION P
     $OINT ',ED(NDIA-1),' IN ',DUNTS$
6900 FORMAT (A55,I3,A4,A3)
```

```
           READ (5,*) DL(NDIA)
           LL(NDIA)=FLOAT(NCPS-ED(NDIA-1))*SCP
6930 IF (EFLAG.GT.0) GOTO 7290
           EFLAG=1
           WRITE (5,6960) 'ENTER THE ELEVATION IN ',LUNTS$,' OF THE ',UP$
6960 FORMAT (A23,A2,A8,A4)
           READ (5,*) E10
           WRITE (5,6220) 'WITH THE UPSTREAM MOST ',PT$,' AS NUMBER 1,'
           WRITE (5,7000) 'ENTER THE ELEVATION IN ',LUNTS$,' OF THE UPSTREAM
     $END OF:'
7000 FORMAT (A23,A2,A24)
           DO 7110 I=1,N,NS
           IF (NS.EQ.1) GOTO 7040
           IF (I.LT.N) GOTO 7070
7040 WRITE (5,6270) PT$,'#',I
7050 READ (5,*) E1(I)
           GOTO 7110
7070 WRITE (5,7080) PT$,'S ',I,' AND ',I+1
7080 FORMAT (A7,A2,I3,A5,I3)
           READ (5,*) E1(I)
           E1(I+1)=E1(I)
7110 CONTINUE
           WRITE (5,6220) 'WITH THE UPSTREAM MOST ',PT$,' AS NUMBER 1,'
           WRITE (5,7140) 'ENTER THE DESIRED PRESSURE IN ',PUNTS$,' AT UPSTRE
     $AM END OF:'
7140 FORMAT (A30,A3,A20)
           DO 7250 I=1,N,NS
           IF (NS.EQ.1) GOTO 7190
           IF (I.LT.N) GOTO 7220
7190 WRITE (5,6270) PT$,'#',I
           READ (5,*) PD(I)
           GOTO 7250
7220 WRITE (5,7080) PT$,'S ',I,' AND ',I+1
           READ (5,*) PD(I)
           PD(I+1)=PD(I)
7250 CONTINUE
7260 NS1=NS
           HDUM=0
           I=N
7290 PIPLEN=SCP1
           J1=1
           J2=1
           DIA=DL(1)
           DO 7550 J=1,NCPS
           IF (J.GE.NCP(J2)) GOTO 7430
           IF (J.GE.ED(J1)) GOTO 7380
           PIPLEN=PIPLEN+SCP
```

```
          GOTO 7550
7380 H1(J2)=K1*PIPLEN*(.285*C)**(-1.852)*Q1(J2)**1.85/DIA**4.87
     HDUM=HDUM+H1(J2)
     J1=J1+1
     DIA=DL(J1)
     GOTO 7530
7430 H1(J2)=K1*PIPLEN*(.285*C)**(-1.852)*Q1(J2)**1.85/DIA**4.87
     H1(J2)=H1(J2)+HDUM
     HDUM=0
     D0(J2)=DIA
     IF (J2.EQ.N) GOTO 7560
     IF (NS1.EQ.1) GOTO 7510
     H1(J2+1)=H1(J2)
     D0(J2+1)=DIA
7510 J2=J2+NS1
     IF (J2.GT.N) GOTO 7560
7530 IF (J2.LT.N) GOTO 7540
     NS1=1
7540 PIPLEN=SCP
7550 CONTINUE
7560 PDEF=0
     P1(N)=PD(N)
     I=N
7590 V1(I)=K3*Q1(I)/(.785*D0(I)**2)
     NS1=NS
     IF (N.LE.DUM2) GOTO 7630
     IF (I.EQ.N) NS1=1
7630 P1(I-NS1)=P1(I)-H1(I)+(E1(I)-E1(I-NS1))/K
     IF (P1(I-NS1).LT.0) GOTO 8610
     IF (P1(I-NS1).GT.PD(I-NS1)) GOTO 7680
     DUM=PD(I-NS1)-P1(I-NS1)
     IF (DUM.GE.PDEF) PDEF=DUM
7680 IF (NS1.EQ.1) GOTO 7710
     V1(I-1)=V1(I)
     P1(I-1)=P1(I)
7710 I=I-NS1
     IF (I.GT.0) GOTO 7590
7740 P10=P1(1)-H1(1)+(E1(1)-E10)/K
C
C  ------------ PRINT INPUT DATA AND RESULTS --------------
C
     WRITE (6,*) ' '
     WRITE (6,*) ' '
     WRITE (6,*) '      INPUT DATA '
     WRITE (6,*) ' '
     WRITE (6,7800) 'REACH   DIAMETER     LENGTH    MATERIAL'
     WRITE (6,7810) '#',DUNTS$,LUNTS$
```

```
 7800 FORMAT (20X,A38)
 7810 FORMAT (22X,A1,7X,A3,10X,A2)
      WRITE (6,*) ' '
      DO 7860 I=1,NDIA
      WRITE (6,7850) I,DL(I),LL(I),Y$
 7850 FORMAT (20X,I3,F11.2,F13.1,3X,A12)
 7860 CONTINUE
      WRITE (6,*) ' '
      WRITE (6,*) ' '
      WRITE (6,7900) PT$,'CONNECTION',LUNTS$,' FROM    DISCHARGE'
 7900 FORMAT (20X,A7,2X,A10,2X,A2,A17)
      WRITE (6,7920) '#        POINT #',UP$,QUNTS$
 7920 FORMAT (23X,A14,4X,A4,8X,A5)
      WRITE (6,*) ' '
      DO 7980 I=1,N
      L=SCP1+FLOAT(NCP(I)-1)*SCP
      WRITE (6,7970) I,NCP(I),L,Q0(I)
 7970 FORMAT (22X,I3,I9,3X,F10.1,F11.1)
 7980 CONTINUE
      WRITE (6,*) ' '
      WRITE (6,*) ' '
      WRITE (6,*) ' '
      WRITE (6,*) '        OUTPUT DATA'
      WRITE (6,*) ' '
      WRITE (6,8080) PT$,'ELEVATION    PRESSURE',PUNTS$,'FLOW    VELOCITY'
      WRITE (6,8070) 'NUMBER',LUNTS$,'DESIRED ACTUAL',QUNTS$,VUNTS$
      WRITE (6,*) ' '
 8070 FORMAT (10X,A6,7X,A2,6X,A14,4X,A5,4X,A4)
 8080 FORMAT (10X,A7,3X,A20,1X,A4,4X,A15)
      WRITE (6,8100) UP$,E10,P10,Q1(1)
 8100 FORMAT (12X,A4,F11.1,9X,2(F9.1))
      DO 8200 I=1,N,NS
      IF (NS.EQ.1) GOTO 8150
      IF (N.LE.DUM2) GOTO 8180
      IF (I.LT.N) GOTO 8180
 8150 WRITE (6,8160) I,E1(I),PD(I),P1(I),Q1(I),V1(I)
 8160 FORMAT (12X,I3,F12.1,3(F9.1),F8.2)
      GOTO 8200
 8180 WRITE (6,8190) I,',',I+1,E1(I),PD(I),P1(I),Q1(I),V1(I)
 8190 FORMAT (10X,I3,A1,I3,1X,4(F9.1),F8.2)
 8200 CONTINUE
      WRITE (6,*) ' '
      WRITE (6,*) ' '
      IF (PDEF.EQ.0) GOTO 8260
      WRITE (6,8250) 'THE PRESSURE AT THE ',UP$,' MUST BE INCREASED TO '
     $,P10+PDEF,PUNTS$
 8250 FORMAT (6X,A20,A4,A22,F6.1,1X,A3)
```

```
8260 WRITE (5,*) '****************************************************'
     WRITE (5,*) '*                                                  *'
     WRITE (5,*) '*        YOU MAY RERUN THE PROGRAM FOR:            *'
     WRITE (5,*) '*            (A) OTHER PIPE DIAMETERS              *'
     WRITE (5,*) '*            (B) ANOTHER MATERIAL                  *'
8350 IF (PIPTYP.EQ.3) GOTO 8390
     WRITE (5,*) '*            (C) ANOTHER MAINLINE                  *'
     WRITE (5,*) '*        OR (D) STOP                               *'
     GOTO 8420
8390 WRITE (5,*) '*            (C) ANOTHER SUBMAIN                   *'
     WRITE (5,*) '*            (D) MAINLINE SELECTION                *'
     WRITE (5,*) '*        OR (E) STOP                               *'
8420 WRITE (5,*) '*                                                  *'
     WRITE (5,*) '****************************************************'
     IF (PIPTYP.EQ.3) GOTO 8490
     WRITE (5,*) 'ENTER A,B,C, OR D'
     GOTO 8500
8490 WRITE (5,*) 'ENTER A,B,C,D OR E'
8500 READ (5,1061) T$
     IF (T$.EQ.'A') GOTO 6440
     IF (T$.EQ.'B') GOTO 5550
     IF (PIPTYP.EQ.3) GOTO 8570
     IF (T$.EQ.'C') GOTO 4850
     STOP
8570 IF (T$.EQ.'C') GOTO 5010
     IF (T$.EQ.'E') GOTO 8670
     PIPTYP=4
     GOTO 4980
8610 WRITE (5,*) '****************************************************'
     WRITE (5,*) '*                                                  *'
     WRITE (5,*) '* BECAUSE THE PRESSURE ALONG THE PIPELINE IS < 0   *'
     WRITE (5,*) '* THE PROGRAM IS BEING STOPPED. SEE THE INSTRUCTOR.*'
     WRITE (5,*) '*                                                  *'
     WRITE (5,*) '****************************************************'
8670 STOP
     END
C
C
C  ----------     PIPE MATERIAL     ----------
C
C
     SUBROUTINE MATL (C,Y$)
     CHARACTER Y$*12
     WRITE (5,*) 'ENTER 1,2,3, OR 4 FOR MATERIALS LISTED BELOW OR'
     WRITE (5,*) 'HAZEN-WILLIAMS COEFFICIENT FOR MATERIALS NOT LISTED.'
     WRITE (5,*) '    (1) ALUMINUM'
     WRITE (5,*) '    (2) POLYETHYLENE'
```

```
      WRITE (5,*) '      (3) STEEL'
      WRITE (5,*) '      (4) PVC'
      WRITE (5,*) 'ENTER 1, 2, 3, 4, OR H-W C '
      READ (5,*) Cl
      IF (Cl.NE.1) GOTO 8820
      C=135
      Y$='ALUMINUM'
      GOTO 8970
8820  IF (Cl.NE.2) GOTO 8860
      C=155
      Y$='POLYETHYLENE'
      GOTO 8970
8860  IF (Cl.NE.3) GOTO 8900
      C=120
      Y$='STEEL'
      GOTO 8970
8900  IF (Cl.NE.4) GOTO 8940
      C=150
      Y$='PVC'
      GOTO 8970
8940  C=Cl
      WRITE (5,*) 'ENTER MATERIAL NAME (up to 12 characters) '
      READ (5,8960) Y$
8960  FORMAT (A12)
8970  RETURN
      END
      SUBROUTINE OVER (NCPS)
      WRITE (5,*) 'THE CONNECTION POINT NUMBER ENTERED EXCEEDS THE TOTAL
     $ NUMBER OF'
      WRITE (5,*) 'CONNECTION POINTS. THE CONNECTION POINT NUMBER CAN NO
     $T EXCEED'
      WRITE (5,9010) NCPS,'. TRY AGAIN.'
9010  FORMAT (I3,A12)
      RETURN
      END
```

Appendix I

FORTRAN Code for an Open Channel Design Program

```
      WRITE (5,*) '****************************************************'
      WRITE (5,*) '*                                                *'
      WRITE (5,*) '*   THIS PROGRAM PERFORMS ITERATIVE CALCULATIONS *'
      WRITE (5,*) '*   FOR DESIGNING RECTANGULAR, TRIANGULAR, AND    *'
      WRITE (5,*) '*   TRAPEZOIDAL SHAPED OPEN CHANNELS  (LGJ 187)   *'
      WRITE (5,*) '*                                                *'
      WRITE (5,*) '****************************************************'
C
C  IN THIS PROGRAM THE DEVICE NUMBERS FOR THE TERMINAL (SCREEN)
C  AND PRINTER ARE 5 AND 6, RESPECTIVELY
C
      CHARACTER A$*1,LUNTS$*2,QUNTS$*5,VUNTS$*4
      COMMON LUNTS$,QUNTS$,VUNTS$
      INTEGER SHAPE,UNTS
      REAL K,K1,N
      WRITE (5,*)
      WRITE (5,*) 'PRESS ANY CHARACTER KEY TO START PROGRAM'
      READ (5,5) A$
    5 FORMAT (A1)
      WRITE (5,*) ' '
      WRITE (5,*) 'EITHER SI OR ENGLISH UNITS MAYBE USED. ENTER:'
      WRITE (5,*) '       1   FOR SI UNITS'
      WRITE (5,*) '       2   FOR ENGLISH UNITS'
      READ (5,*) UNTS
      IF (UNTS.EQ.1) THEN
        K=60000.
        K1=1.
        LUNTS$='M'
        QUNTS$='L/MIN'
        VUNTS$='M/S'
      ELSE
        K=1.49*449
```

```
        K1=1.49
        LUNTS$='FT'
        QUNTS$='GPM'
        VUNTS$='FT/S'
      END IF
 10 WRITE (5,*) 'ENTER THE SLOPE (IN PERCENT) '
    READ (5,*) S
    WRITE (5,*) 'ENTER MANNINGS ROUGHNESS COEFFICIENT'
    READ (5,*) N
    WRITE (5,20) 'ENTER THE DESIRED CAPACITY OF THE CANAL IN ',QUNTS$
 20 FORMAT (A43,A5)
    READ(5,*) Q
    WRITE (5,30) 'ENTER THE DESIRED VELOCITY IN ',VUNTS$
 30 FORMAT (A30,A4)
    READ (5,*) V
    S=S/100.
    C=Q*N/(K*S**0.5)
200 WRITE (5,*) 'ENTER DESIRED CANAL OR DITCH SHAPE:      '
    WRITE (5,*) '  1  FOR RECTANGULAR '
    WRITE (5,*) '  2  FOR TRIANGULAR '
    WRITE (5,*) '  3  FOR TRAPEZOIDAL '
    READ (5,*) SHAPE
    IF ((SHAPE.LT.1) .OR. (SHAPE.GT.3)) THEN
      WRITE (5,*) 'YOU DID NOT ENTER EITHER 1, 2, OR 3. TRY AGAIN'
      GO TO 200
    END IF
    IF (SHAPE.EQ.1) CALL RECT(Q,N,S,V,D,C)
    IF (SHAPE.EQ.2) CALL TRI(Q,N,S,V,C)
    IF (SHAPE.EQ.3) CALL TRAP(Q,N,S,V,D,C)
    WRITE (5,*) 'DO YOU WISH TO CONSIDER ANOTHER CHANNEL? ENTER YES OR
   $ NO.'
    READ (5,5) A$
    IF (A$.EQ.'Y') GOTO 10
    WRITE (5,*) '*****  THE RUN IS COMPLETE  ******'
    WRITE (5,*) ' '
    STOP
    END
C
C
C    THIS SUBROUTINE CALCULATES DESIGN PARAMETERS FOR A
C    TRAPEZOIDAL SHAPED CHANNEL USING GIVEN INPUT.
C
C
    SUBROUTINE TRAP(Q,N,S,V,V1,C)
    COMMON LUNTS$,QUNTS$,VUNTS$
    CHARACTER LUNTS$*2,QUNTS$*5,VUNTS$*4
    CHARACTER*55 CHAR
```

```
      CHARACTER*1 A$,B$
      REAL N
      WRITE (6,*) ' '
      WRITE (6,*) ' '
      WRITE (6,*) ' '
      WRITE (6,40) 'TRAPEZOIDAL SHAPED CHANNEL WITH: '
  40  FORMAT (10X,A35)
      WRITE (6,*) ' '
      WRITE (6,50) 'Q =',Q,QUNTS$
      WRITE (6,50) 'S =',S*100,'% '
      WRITE (6,50) 'MANNINGS N =',N,' '
  50  FORMAT (10X,A21,F12.3,A6)
      WRITE (6,*) ' '
      CHAR = '      Z      B       Y      DESIRED    ACTUAL '
      WRITE (6,75) CHAR
      WRITE (6,70) LUNTS$,LUNTS$,'V (',VUNTS$,')    V (',VUNTS$,')'
  70  FORMAT (26X,A2,8X,A2,5X,A3,A4,A7,A4,A1)
      CHAR = '      -------------------------------------------------'
      WRITE (6,75) CHAR
  75  FORMAT (10X,A55)
  90  WRITE (5,100) ' ENTER BOTTOM WIDTH IN ',LUNTS$
 100  FORMAT (A23,A2)
      READ (5,*) B
      WRITE (5,*) 'ENTER SIDE SLOPE RATIO, Z '
      READ (5,*) Z
      Y=10
      DY=Y
      TOL=C/100.
      CHK2=5./3.
  68  C1=(((B+Z*Y)*Y)**(CHK2))
      C1=C1*(1./(B+2.*Y*(1.+Z**2)**0.5))**(2./3.)
      IF (ABS(C-C1).LT.TOL) THEN
        R=(B+Z*Y)*Y
        R=R/(B+2.*Y*(1.+Z**2)**0.5)
        V1=1./N*R**(2./3.)*S**0.5
        WRITE (6,170) Z,B,Y,V,V1
 170    FORMAT (10X,5(F10.3))
        WRITE (5,175) 'DESIRED VELOCITY = ',V,VUNTS$
        WRITE (5,175) 'ACTUAL VELOCITY  = ',V1,VUNTS$
 175    FORMAT (A19,F5.2,A5)
        WRITE (5,*) ' '
        WRITE (5,*) 'IS THIS SOLUTION SATISFACTORY? ENTER YES OR NO.'
        READ (5,180) A$
 180    FORMAT (A1)
        IF (A$.EQ.'Y') GOTO 80
        WRITE (5,190) ' CURRENT BOTTOM WIDTH = ',B,LUNTS$,'; CURRENT Z =
     $ ',Z
```

```
 190    FORMAT (A24,F5.2,A3,A14,F5.2)
          GO TO 90
        ELSE
          DY=DY/2.
        END IF
        IF (C1.GT.C) THEN
          Y=Y-DY
          GO TO 68
        ELSE IF (C1.LT.C) THEN
          Y=Y+DY
          GO TO 68
        END IF
  80 RETURN
        END
C
C
C     THIS SUBROUTINE CALCULATES DESIGN PARAMETERS FOR A TRIANGULAR
C     SHAPED CHANNEL USING GIVEN INPUT.
C
C
        SUBROUTINE TRI(Q,N,S,V,C)
        COMMON LUNTS$,QUNTS$,VUNTS$
        CHARACTER LUNTS$*2,QUNTS$*5,VUNTS$*4
        CHARACTER*55 CHAR
        CHARACTER*1 A$,B$
        REAL N
        WRITE (6,*) ' '
        WRITE (6,*) ' '
        WRITE (6,40) 'TRIANGULAR SHAPED CHANNEL WITH:  '
  40 FORMAT (10X,A35)
        WRITE (6,*) '   '
        WRITE (6,50) 'Q =',Q,QUNTS$
        WRITE (6,50) 'S =',S*100,'%    '
        WRITE (6,50) 'MANNINGS N =',N,'  '
  50 FORMAT (10X,A21,F12.3,A6)
        WRITE (6,*) ' '
        CHAR='                Z         Y        DESIRED    ACTUAL  '
        WRITE (6,60) CHAR
        WRITE (6,55) LUNTS$,'V (',VUNTS$,') V (',VUNTS$,')'
  55 FORMAT (36X,A2,6X,A3,A4,A6,A4,A1)
        CHAR='              ---------------------------------------'
        WRITE (6,60) CHAR
  60 FORMAT (10X,A55)
  65 WRITE (5,*) 'ENTER SIDE SLOPE RATIO, Z '
        READ( 5,*) Z
        TOL=C/100.
        R=Z/(2*(1+Z**2)**0.5)
```

```
      Y=(C/(Z*R**(2./3.)))**(3./8.)
      R=R*Y
      V1=1./N*R**(2./3.)*S**0.5
      WRITE (6,70) Z,Y,V,V1
   70 FORMAT (20X,4(F10.3))
      WRITE (5,71) 'DESIRED VELOCITY = ',V,VUNTS$
      WRITE (5,71) 'ACTUAL VELOCITY  = ',V1,VUNTS$
   71 FORMAT (A19,F5.2,A5)
      WRITE (5,*) ' '
      WRITE (5,*) 'IS THIS SOLUTION SATISFACTORY? ENTER YES OR NO.'
      READ (5,75) A$
   75 FORMAT (A1)
      IF (A$.EQ.'Y') GOTO 80
      WRITE (5,76) ' CURRENT SIDE SLOPE RATIO, Z = ',Z
   76 FORMAT (A31,F5.2)
      GO TO 65
   80 RETURN
      END
C
C
C     THIS SUBROUTINE CALCULATES DESIGN PARAMETERS FOR A
C     RECTANGULAR SHAPED CHANNEL USING GIVEN INPUT.
C
C
      SUBROUTINE RECT(Q,N,S,V,V1,C)
      COMMON LUNTS$,QUNTS$,VUNTS$
      CHARACTER LUNTS$*2,QUNTS$*5,VUNTS$*4
      CHARACTER*55 CHAR
      CHARACTER*1 A$,B$
      REAL N
      WRITE(6,*) ' '
      WRITE(6,*) ' '
      WRITE(6,*) ' '
      WRITE(6,40) 'RECTANGULAR SHAPED CHANNEL WITH: '
   40 FORMAT (10X,A35)
      WRITE(6,*) ' '
      WRITE (6,50) 'Q =',Q,QUNTS$
      WRITE (6,50) 'S =',S*100.,'%        '
      WRITE (6,50) 'MANNINGS N =',N,'  '
   50 FORMAT (10X,A21,F12.3,A6)
      WRITE (6,*) ' '
      CHAR='                 B         Y        DESIRED    ACTUAL  '
      WRITE (6,60) CHAR
      WRITE (6,55) LUNTS$,LUNTS$,'V (',VUNTS$,')   V (',VUNTS$,')'
   55 FORMAT (26X,A2,8X,A2,6X,A3,A4,A6,A4,A1)
      CHAR='                   -------------------------------------'
      WRITE (6,60) CHAR
```

```
   60 FORMAT (10X,A55)
   65 WRITE (5,72) 'ENTER BOTTOM WIDTH IN ',LUNTS$
   72 FORMAT (A22,A2)
      READ (5,*) B
      Y=10
      DY=Y
      TOL=C/100.
   68 C1=(B*Y)**(5./3.)*(1./(B+2*Y))**(2./3.)
      IF (ABS(C-C1).LT.TOL) THEN
        R=B*Y/(B+2*Y)
        V1=1./N*R**(2./3.)*S**0.5
        WRITE (6,70) B,Y,V,V1
   70    FORMAT (20X,4(F10.3))
        WRITE (5,71) 'DESIRED VELOCITY = ',V,VUNTS$
        WRITE (5,71) 'ACTUAL VELOCITY  = ',V1,VUNTS$
   71    FORMAT (A19,F5.2,A5)
        WRITE (5,*) ' '
        WRITE (5,*) 'IS THIS SOLUTION SATISFACTORY? ENTER YES OR NO.'
        READ (5,75) A$
   75    FORMAT (A1)
        IF (A$.EQ.'Y') GOTO 80
        WRITE (5,85) 'CURRENT BOTTOM WIDTH = ',B,LUNTS$
   85    FORMAT (A23,F5.2,A3)
        GO TO 65
      ELSE
        DY=DY/2.
      END IF
      IF (C1.GT.C) THEN
        Y=Y-DY
        GO TO 68
      ELSE IF (C1.LT.C) THEN
        Y=Y+DY
        GO TO 68
      END IF
   80 RETURN
      END
```

Appendix J

FORTRAN Code for Computing Uniformity, Efficiency, and Losses for a Surface Irrigation

```
      WRITE(5,*)'****************************************************'
      WRITE(5,*)'*   THIS PROGRAM CALCULATES THE UNIFORMITY AND      *'
      WRITE(5,*)'*   EFFICIENCY OF SURFACE IRRIGATION AS WELL AS     *'
      WRITE(5,*)'*   THE VOLUME OF INFILTRATION, DEEP PERCOLATION,   *'
      WRITE(5,*)'*   AND RUNOFF GIVEN ADVANCE, RECESSION, AND        *'
      WRITE(5,*)'*   INFILTRATION INFORMATION.            (LGJ 386)  *'
      WRITE(5,*)'****************************************************'
      WRITE(5,*)
      WRITE(5,30)
30    FORMAT(1X,'****************************************************',/
     &,1X,'*   YOU WILL BE ABLE TO CORRECT CERTAIN INPUTS      *',/
     &,1X,'*   (IDENTIFIED BY AN *) BEFORE THE CALCULATIONS    *',/
     &,1X,'*   BEGIN                                           *',/
     &,1X, '****************************************************')
C
C  IN THIS PROGRAM THE DEVICE NUMBERS FOR THE TERMINAL (SCREEN)
C  AND PRINTER ARE 5 AND 6, RESPECTIVELY
C
      REAL K,K1,K2,K3,K4,INFTM,IOT,INFIL,MANN,IRRTIM
      INTEGER TYPE,Z
      CHARACTER DUNTS$*3,QUNTS$*5,LUNTS$*2,Z$*1
      COMMON AT(0:25),S(0:25),RT(0:25),IOT(0:25),D(0:25),X(0:25)
      WRITE(5,*)'EITHER SI OR ENGLISH UNITS MAYBE USED. ENTER:'
      WRITE(5,*)'    1   FOR SI UNITS'
      WRITE(5,*)'    2   FOR ENGLISH UNITS'
      READ(5,50) IUNTS
50    FORMAT(I1)
      Z=0
C
C -----   UNIT CONVERSION FACTORS   ------
C
      IF (IUNTS.EQ.1) THEN
```

```
         K=1000.
         K1=1000.
         K2=1.
         K3=60000.
         K4=1000.
         DUNTS$='MM '
         LUNTS$='M '
         QUNTS$='L/MIN'
      ELSE
         K=43560.
         K1=27154.3
         K2=1.604
         K3=279.15
         K4=12.
         DUNTS$='INS'
         LUNTS$='FT'
         QUNTS$='GPM'
      END IF
      MANN=0
      WRITE(5,60) DUNTS$
60    FORMAT(1X,'ENTER THE DEPTH REQUIRED TO FILL THE SOIL',/
     &,1X,'TO FIELD CAPACITY IN ',A3,' *')
      READ(5,*) DD
70    FORMAT(F5.2)
      WRITE(5,*) 'EITHER OF THE FOLLOWING TWO INFILTRATION FUNCTIONS'
      WRITE(5,*) 'CAN BE USED:'
      WRITE(5,*) ' '
      WRITE(5,*) '     1   I=C*T**D        (EQ 7.11A)'
      WRITE(5,*) '     2   F=K*T**A+FO*T   (EQ 7.13)'
      WRITE(5,*) ' '
      WRITE(5,*) 'WHERE:'
      WRITE(5,80) DUNTS$,LUNTS$,QUNTS$,LUNTS$
80    FORMAT(1X,'          I = CUMULATIVE INFILTRATION DEPTH IN ',A3
     &    ,/,1X,'        C,K = COEFFIECIENTS'
     &    ,/,1X,'          T = TIME IN MINUTES'
     &    ,/,1X,'        A,D = EXPONENTS'
     &    ,/,1X,'          F = CUMULATIVE INFILTRATION IN SQ ',A2
     &    ,/,1X,'         FO = FINAL INFILTRATION RATE IN ',A5,'/',A2,
     &    //,1X,'ENTER 1 OR 2.')
      READ(5,50) INF
      WRITE(5,*) ' '
      IF (INF.EQ.1) THEN
         WRITE(5,*) 'ENTER C AND D.'
         READ (5,*) C,D1
         FO=0
      ELSE
         WRITE(5,*) 'ENTER K, A, AND FO.'
```

```
        READ(5,*) C,D1,FO
      END IF
      WRITE(5,100)
100   FORMAT(1X,'ENTER THE SYSTEM TYPE',/
     &,1X,'     1   FURROWS',/
     &,1X,'     2   BORDERS',/
     &,1X,'     3   BASINS',/)
      READ(5,50) TYPE
      IF (TYPE.EQ.1) THEN
        IF (IUNTS.EQ.1) THEN
          WRITE(5,*) 'ENTER THE FURROW SPACING IN M *'
        ELSE
          WRITE(5,*) 'ENTER THE FURROW SPACING IN INS *'
        END IF
        READ(5,*) W
        IF (INF.EQ.2) THEN
           C=C*K4/W
           FO=FO*K2/W
           IF (IUNTS.EQ.2) THEN
             C=C/12.
             FO=FO/12
           END IF
        END IF
        CALL FURROW(IUNTS,DD,C,D1,FO,TYPE,W,N,RTIM,Q,INFTM,J,Z,DUNTS$,
     &              LUNTS$,QUNTS$)
      ELSE IF (TYPE.EQ.2) THEN
        WRITE(5,318) LUNTS$
318   FORMAT(1X,'ENTER THE WIDTH OF THE BORDER IN ',A2,' *')
        READ(5,*) W
        IF (INF.EQ.2) THEN
           C=C*K4/W
           FO=FO*K2/W
        END IF
        CALL BORDER(IUNTS,K2,K3,K4,MANN,DD,C,D1,FO,TYPE,W,N,AVED,Q,INFTM
     &          ,J,Z,L,P,R,LAG,LAGTM,SO,IRRTIM,ADV,DUNTS$,LUNTS$,QUNTS$)
      ELSE IF (TYPE.EQ.3) THEN
        WRITE(5,426) LUNTS$
426   FORMAT(1X,'ENTER THE WIDTH OF THE BASIN IN ',A2,' *')
        READ(5,*) W
        IF (INF.EQ.2) THEN
           C=C*K4/W
           FO=FO*K2/W
        END IF
        CALL BASINS(IUNTS,K2,DD,C,D1,FO,TYPE,W,N,RTIM,Q,INFTM,Z,
     &              DUNTS$,LUNTS$,QUNTS$)
      END IF
C
```

```
C   DETERMINE IOT AND DEPTHS OF INFILTRATION
C
107    IF(TYPE.NE.2) THEN
          DO 110 I=0,N
          S(I)=RTIM
          IOT(I)=S(I)-AT(I)
          D(I)=C*IOT(I)**D1+FO*IOT(I)
110       CONTINUE
       END IF
C
C   DETERMINE THE AMOUNT OF INFILTRATION
C
       INFIL=D(0)*(X(1)-X(0))/2.+D(N)*(X(N)-X(N-1))/2.
       DO 120 I=1,N-1
          INFIL=INFIL+D(I)*(X(I+1)-X(I-1))/2.
120    CONTINUE
       AVED=INFIL/X(N)
       INFIL=AVED*W*X(N)/K
       IF (IUNTS.EQ.2) THEN
         IF (TYPE.EQ.1) THEN
           INFIL=INFIL/12.
         END IF
       END IF
       APPL=Q*INFTM/K1
       J=0
C
C   DETERMINE THE AMOUNT OF WATER STORED IN THE ROOT ZONE
C
       DO 130 I=0,N
          IF(D(I).LT.DD) THEN
          GOTO 140
          END IF
          J=I
130    CONTINUE
140    IF (J.EQ.N) THEN
          STORED=DD*W*X(N)/K
          GOTO 160
       ELSE IF (J.EQ.0) THEN
          STORED=INFIL
          GOTO 160
       END IF
C
C   DETERMINE LOCATION WHERE THE DEPTH OF INFILTRATION EQUALS THE
C   DESIRED DEPTH OF APPLICATION USING LINEAR INTERPOLATION
C
       X1=X(I-1)+(DD-D(I-1))*(X(I)-X(I-1))/(D(I)-D(I-1))
       STORED=DD*X1
```

```
      DUM=DD
      DO 150 J=I,N
          STORED=STORED+(D(J)+DUM)*(X(J)-X1)/2.
          X1=X(J)
          DUM=D(J)
150   CONTINUE
      STORED=STORED*W/K
      DUM=W/K
      IF (IUNTS.EQ.2) THEN
        IF (TYPE.EQ.1) THEN
          STORED=STORED/12.
          DUM=DUM/12.
        END IF
      END IF
160   DP=INFIL-STORED
      RO=APPL-INFIL
      EFF=100*STORED/APPL
      SEFF=100*STORED/(DD*X(N)*DUM)
C
C  DETERMINE CHRISTIANSEN COEFFICIENT OF UNIFORMITY (EQ 2.13A)
C
      DEV=ABS((D(0)-AVED)*(X(1)-X(0))/2.)+ABS((D(N)-AVED)*(X(N)-X(N-1))/
     &2.)
      DO 170 I=1,N-1
          DEV=DEV+ABS((D(I)-AVED)*(X(I+1)-X(I-1))/2.)
170   CONTINUE
      CU=100*(1.-DEV/(AVED*X(N)))
C
C  PRINT RESULTS
C
180   FORMAT(1X,'DISTANCE     ADVANCE      RECESSION    IOT       INFIL'
     &)
     . WRITE(6,180)
      IF (IUNTS.EQ.1) THEN
190     FORMAT(1X,'   M         TIME MINS    TIME MINS    MINS      MM')
        WRITE(6,190)
      ELSE IF (IUNTS.EQ.2) THEN
        WRITE(6,200)
200     FORMAT(1X,'   FT        TIME MINS    TIME MINS    MINS      INS')
      END IF
      WRITE(6,*) '---------------------------------------------------'
      DO 220 I=0,N
          WRITE(6,210) X(I),AT(I),S(I),IOT(I),D(I)
210       FORMAT(F7.2,F14.2,F13.2,F11.2,F9.2)
220   CONTINUE
      IF (IUNTS.EQ.1) THEN
        WRITE(6,230) Q,DD,APPL,INFTM,INFIL,STORED,DP,RO
```

```
230       FORMAT(//,1X,'Q               =',F8.2,' L/MIN',/,1X,'DESIRED D =',F
     &8.2,' MM',/,1X,'APPLIED      =',F8.2,' CUBIC M',/,1X,'INFL TIME  =',
     &  F8.2,' MINS',/,1X,'INFIL        =',F8.2,' CUBIC M',/,1X,'STORED
     &  =',F8.2,' CUBIC M',/,1X,'DEEP PERC =',F8.2,' CUBIC M',/,1X,
     &     'RUNOFF     =',F8.2,' CUBIC M')
          ELSE IF (IUNTS.EQ.2) THEN
            WRITE(6,240) Q,DD,APPL,INFTM,INFIL,STORED,DP,RO
240       FORMAT(//,1X,'Q               =',F8.2,' GPM',/,1X,'DESIRED D =',F8.2
     &,' INS',/,1X,'APPLIED      =',F8.2,' AC-INS',/,1X,'INFL TIME  =',
     &  F8.2,' MINS',/,1X,'INFIL        =',F8.2,' AC-INS',/,1X,'STORED
     & =',F8.2,' AC-INS',/,1X,'DEEP PERC =',F8.2,' AC-INS',/,1X,
     & 'RUNOFF     =',F8.2,' AC-INS')
          END IF
          WRITE(6,250) EFF,SEFF,CU
250   FORMAT(1X,'APPL EFF   =',F8.2,' PERCENT',/,1X,'STOR EFF   =',F8.2,
     &' PERCENT',/,1X,'CHR UNIF  =',F8.2)
        IF(TYPE.EQ.3) THEN
          IF (RO.GT.0) THEN
            WRITE(6,*) ' '
            WRITE(6,*) 'A POSITIVE VALUE FOR RUNOFF INDICATES THAT INFLOW'
            WRITE(6,*) 'EXCEEDS INFILTRATION. EITHER THE STREAMSIZE IS TOO'
            WRITE(6,*) 'LARGE OR THE INFILTRATION FUNCTION PARAMETERS ARE'
            WRITE(6,*) 'INCORRECT.'
          ELSE IF (RO.LT.0) THEN
            WRITE(6,*) ' '
            WRITE(6,*) 'A NEGATIVE VALUE FOR RUNOFF INDICATES THAT'
            WRITE(6,*) 'INFILTRATION EXCEEDS INFLOW. EITHER THE STREAMSIZE'
            WRITE(6,*) 'IS TOO SMALL OR THE INFILTRATION FUNCTION'
            WRITE(6,*) 'PARAMETERS ARE INCORRECT.'
          END IF
        END IF
C
C   ENTERING INFORMATION FOR ADDITIONAL RUNS
C
      WRITE(5,*) 'DO YOU WISH TO ENTER ANOTHER STREAMSIZE, YES OR NO?'
      READ(5,262) Z$
262   FORMAT(A1)
      IF (Z$.EQ.'N') GOTO 270
      Z=1
      IF (TYPE.EQ.1) THEN
        WRITE(5,264) QUNTS$
264     FORMAT(1X,'ENTER STREAMSIZE IN ',A5,' *')
        READ(5,*) Q
        GOTO 107
      ELSE IF (TYPE.EQ.3) THEN
        CALL BASINS(IUNTS,K2,DD,C,D1,FO,TYPE,W,N,RTIM,Q,INFTM,A,Z,
     &             DUNTS$,LUNTS$,QUNTS$)
```

```
            ELSE
                Z=1
                CALL BORDER(IUNTS,K2,K3,MANN,DD,C,D1,FO,TYPE,W,N,AVED,
        &       Q,INFTM,J,Z,L,P,R,LAG,LAGTM,SO,IRRTIM,ADV,DUNTS$,LUNTS$,QUNTS$)
            END IF
270     WRITE(5,275)
275     FORMAT(/,'*******   THE RUN IS COMPLETE ********')
        STOP
        END
C
C ********** FURROWS ********** FURROWS **********
C
        SUBROUTINE FURROW(IUNTS,DD,C,D1,FO,TYPE,W,N,RTIM,Q,INFTM,J,Z,
    &DUNTS$,LUNTS$,QUNTS$)
        REAL K,K1,K2,K3,INFTM,IOT,IRRTIM
        INTEGER TYPE,Z
        CHARACTER Z$*1,DUNTS$*3,QUNTS$*5
        COMMON AT(0:25),S(0:25),RT(0:25),IOT(0:25),D(0:25),X(0:25)
        WRITE(5,290) QUNTS$
290     FORMAT(1X,'ENTER STREAMSIZE IN ',A5,' *')
        READ(5,*) Q
        CALL ADVAN(IUNTS,DD,C,D1,FO,TYPE,W,N,Q,J,Z,L,P,R,LAGTM,DUNTS$,
        &           LUNTS$,QUNTS$,ADV,K4)
        IRRTIM=(DD/C)**(1/D1)
        IF (FO.GT.0.0001) THEN
            CALL IRTM (DD,IRRTIM,C,D1,FO)
        END IF
        INFTM=IRRTIM+AT(N)
        RTIM=INFTM
        WRITE(5,300)
300     FORMAT(1X,'THE RECESSION TIME IS ASSUMED TO EQUAL THE SUM OF ',/
    &,1X,'THE IRRIGATION AND ADVANCE TIMES UNLESS THE USER ',/
    &,1X,'SUPPLIES ANOTHER VALUE.  THE USER SUPPLIED RECESSION ',/
    &,1X,'TIME MUST EXCEED THE ADVANCE TIME.',//
    &,1X,'DO YOU WISH TO ENTER A RECESSION TIME, YES OR NO?')
        READ(5,303) Z$
303     FORMAT(A1)
        IF (Z$.EQ.'N') RETURN
        WRITE(5,*) 'ENTER A RECESSION TIME IN MINS'
306     READ(5,*) RTIM
        IF (RTIM.LT.AT(N)) THEN
            WRITE(5,307)
307         FORMAT(1X,'THE RECESSION TIME IS LESS THAN THE ADVANCE TIME',
    &'-- ENTER A LARGER RECESSION TIME IN MINS.')
            GOTO 306
        END IF
        WRITE(5,*) 'THE INFLOW TIME IS ASSUMED TO EQUAL THE SUM OF THE'
```

```
        WRITE(5,*) 'ADVANCE AND IRRIGATION TIMES UNLESS THE USER SUPPLIES'
        WRITE(5,*) 'ANOTHER VALUE.'
        WRITE(5,*) ' '
        WRITE(5,*) 'DO YOU WISH TO ENTER AN INFLOW TIME, YES OR NO?'
        READ(5,303) Z$
        IF (Z$.EQ.'N') THEN
            IF (RTIM.LT.INFTM) INFTM=RTIM
            RETURN
        END IF
        WRITE(5,310)
310     FORMAT(1X,'ENTER THE DESIRED INFLOW TIME IN MINS.  IF GREATER THAN
       & THE',/,1X,'RECESSION TIME THE INFLOW TIME WILL BE SET EQUAL TO TH
       &E',/,1X,'RECESSION TIME.')
        READ(5,*) INFTM
        IF (RTIM.LT.INFTM) THEN
            INFTM=RTIM
        END IF
        RETURN
        END
C
C ************ BORDERS ************* BORDERS ****************
C
        SUBROUTINE BORDER(IUNTS,K2,K3,K4,MANN,DD,C,D1,FO,TYPE,W,N,AVED,Q,
       &INFTM,J,Z,L,P,R,LAG,LAGTM,SO,IRRTIM,ADV,DUNTS$,LUNTS$,QUNTS$)
        REAL K,K1,K2,K3,K4,INFTM,MANN,L,LAGTM,IRRTIM,IOT
        INTEGER A,Z,ADV
        CHARACTER DUNTS$*3,QUNTS$*5,LUNTS$*2,Z$*1
        COMMON AT(0:25),S(0:25),RT(0:25),IOT(0:25),D(0:25),X(0:25)
315     FORMAT(I1)
316     FORMAT(F5.2)
317     FORMAT(2F5.2)
        IF (Z.EQ.1) GOTO 385
        WRITE(5,319) QUNTS$,LUNTS$
319     FORMAT(1X,'ENTER THE STREAMSIZE IN ',A5,'/',A2,' OF BORDER WIDTH')
        READ(5,*) Q
        Q=Q*W
        CALL ADVAN(IUNTS,DD,C,D1,FO,TYPE,W,N,Q,J,Z,L,P,R,LAGTM,DUNTS$,
       &LUNTS$,QUNTS$,ADV,K4)
          X(0)=0
          AT(0)=0
          RT(0)=0
        IF(ADV.EQ.1) THEN
          WRITE(5,320)
320       FORMAT(1X,'ENTER THE NUMBER OF STATIONS (UP TO 25) ACROSS THE',
       &  ' FIELD AT WHICH ADVANCE',/,1X,'AND RECESSION TIMES ARE KNOWN.',
       &  ' (DO NOT INCLUDE THE "0" STATION).')
322       READ(5,*) N
```

```
            IF (N.GT.25) THEN
              WRITE(5,*) 'N EXCEEDS 25, ENTER A SMALLER VALUE OF N.'
              GOTO 322
            END IF
            WRITE(5,323) LUNTS$
323         FORMAT(1X,'STARTING AT THE UPSTREAM END OF THE BORDER AND NOT ',
     &      'INCLUDING THE "0"',/,1X,'STATION, ENTER THE DISTANCE FROM THE',
     &      ' UPSTREAM END OF THE BORDER',/,1X,'IN ',A2,', THE ADVANCE T',
     &      'IME, AND THE RECESSION TIME (THE TIME SINCE WATER',/,1X,'STAR',
     &      'TED RECEEDING AT STATION "0") IN MINS FOR:')
            DO 325 I=1,N
              WRITE(5,350) I
              READ(5,*) X(I),AT(I),RT(I)
325         CONTINUE
      ELSE
            WRITE(5,330)
330         FORMAT(1X,'ENTER THE NUMBER OF STATIONS (UP TO 25) ACROSS THE',
     &      /,1X,'FIELD AT WHICH RECESSION DATA ARE AVAILABLE.')
338         READ(5,*) N
            IF (N.GT.25) THEN
              WRITE(5,*) 'N EXCEEDS 25, ENTER A SMALLER VALUE OF N.'
              GOTO 338
            END IF
            WRITE(5,340) LUNTS$
340         FORMAT(1X,'STARTING AT THE UPSTREAM END OF THE BORDER AND NOT ',
     &      'INCLUDING THE "0" ',/,1X,'STATION, ENTER THE DISTANCE FROM TH',
     &      'E UPSTREAM END OF THE',/,1X,'BORDER IN ',A2,' AND THE RECESSI',
     &      'ON TIME (THE TIME SINCE WATER',/,1X,'STARTED RECEEDING AT STA',
     &      'TION "0") IN MINS FOR:')
            DO 360 I=1,N
              WRITE(5,350) I
350           FORMAT(1X,'STATION# ',I3)
              READ(5,*) X(I),RT(I)
              AT(I)=(X(I)/P)**(1/R)
360         CONTINUE
      END IF
      IRRTIM=(DD/C)**(1/D1)
      IF (FO.GT.0.0001) THEN
         CALL IRTM (DD,IRRTIM,C,D1,FO)
      END IF
      WRITE(5,370)
370   FORMAT(1X,'YOU MAY ENTER A MEASURED RECESSION LAG TIME OR THE ',/
     &,1X,'PROGRAM WILL COMPUTE IT USING EITHER EQUATION 7.25A',/,
     &1X,'OR 7.25B. ENTER:',/,
     &1X,'                1  TO INPUT A MEASURED LAG TIME',/,
     &1X,'                2  TO HAVE THE PROGRAM COMPUTE ONE.')
      READ(5,315) LAG
```

```
          IF (LAG.EQ.1) THEN
            WRITE(5,375)
375         FORMAT(1X,'ENTER THE RECESSION LAG TIME IN MINS (THE TIME BETW',
     &      'EEN THE',/,1X,'END OF INFLOW AND THE START OF RECESSION AT ST',
     &      'ATION "0" *')
            READ(5,*) LAGTM
          ELSE
            IF (MANN.GT.0.0001) GOTO 385
            WRITE(5,380)
380         FORMAT (1X,'ENTER MANNINGS N AND THE SLOPE OF THE BORDER IN PE',
     &      'RCENT.')
            READ(5,*) MANN,SO
385         IF (SO.GE.0.4) THEN
               LAGTM=MANN**1.2*(Q/(K3*W))**.2/(120.*(SO/100.)**1.6)
            ELSE
               LAGTM=.0094*MANN*(Q/(K3*W))**.175
               LAGTM=LAGTM/(IRRTIM**.88*(SO/100.)**.5)
               LAGTM=(SO/100.+LAGTM)**1.6
               LAGTM=MANN**1.2*(Q/(K3*W))**.2/(120.*LAGTM)
            END IF
          END IF
          Z=1
          CALL ADVAN(IUNTS,DD,C,D1,FO,TYPE,W,N,Q,J,Z,L,P,R,LAGTM,DUNTS$,
     &               LUNTS$,QUNTS$,ADV,K4)
          INFTM=IRRTIM-LAGTM
          DUM=10.*IRRTIM
          DO 420 I=0,N
          S(I)=INFTM+LAGTM+RT(I)
          IOT(I)=S(I)-AT(I)
          IF (IOT(I).LT.DUM) THEN
            DUM=IOT(I)
          END IF
420       CONTINUE
            DUM=IRRTIM-DUM
            INFTM=INFTM+DUM
          DO 430 I=0,N
            IOT(I)=IOT(I)+DUM
            RT(I)=RT(I)+DUM
            D(I)=C*IOT(I)**D1+FO*IOT(I)
430       CONTINUE
          RETURN
          END
C
C************* BASINS ************** BASINS ***************
C
      SUBROUTINE BASINS(IUNTS,K2,DD,C,D1,FO,TYPE,W,N,RTIM,Q,INFTM,
     &Z,DUNTS$,LUNTS$,QUNTS$)
```

```
      COMMON AT(0:25),S(0:25),RT(0:25),IOT(0:25),D(0:25),X(0:25)
      REAL K,K1,K2,K3,INFTM,IRRTIM
      CHARACTER DUNTS$*3,LUNTS$*2,QUNTS$*5,Z$*1
425   FORMAT(F5.2)
      WRITE(5,430) QUNTS$,LUNTS$
430   FORMAT(1X,'ENTER THE STREAMSIZE IN ',A5,'/',A2,' OF BASIN WIDTH')
      READ(5,*) Q
      Q=Q*W
      CALL ADVAN(IUNTS,DD,C,D1,FO,TYPE,W,N,Q,J,Z,L,P,R,LAGTM,DUNTS$,
     &LUNTS$,QUNTS$,ADV,K4)
      IRRTIM=(DD/C)**(1/D1)
      IF (FO.GT.0.0001) THEN
         CALL IRTM (DD,IRRTIM,C,D1,FO)
      END IF
      RTIM=IRRTIM+AT(N)
      INFTM=DD*W*X(N)/(K2*Q)+AT(N)
      IF (INFTM.GT.RTIM) THEN
         RTIM=INFTM
      END IF
      RETURN
      END
C
C********** ADVANCE DATA ***** ADVANCE DATA ***************
C
      SUBROUTINE ADVAN(IUNTS,DD,C,D1,FO,TYPE,W,N,Q,J,Z,L,P,R,LAGTM,
     &DUNTS$,LUNTS$,QUNTS$,ADV,K4)
      COMMON AT(0:25),S(0:25),RT(0:25),IOT(0:25),D(0:25),X(0:25)
      REAL K,K1,K2,K3,INFTM,L,LAGTM,IOT,IRRTIM
      INTEGER TYPE,A,Z,CH,ADV
      CHARACTER LUNTS$*2,DUNTS$*3,QUNTS$*5,Z$*1
335   FORMAT(I1)
336   FORMAT(2F5.2)
337   FORMAT(F5.2)
      IF (Z.EQ.1) THEN
         GOTO 551
      END IF
      IF (TYPE.EQ.2) THEN
        WRITE(5,440)
440      FORMAT(1X,'A TABLE OF RECESSION DATA IS REQUIRED.  ADVANCE DATA'
     &,/,1X,'MAYBE ENTERED AS A TABLE OR AN EQUATION OF THE FORM'
     &,/,1X,'X=PT**R WHERE:')
      ELSE
        WRITE(5,450)
450      FORMAT(1X,'ADVANCE DATA MAYBE ENTERED AS EITHER A TABLE OF X,T'
     &,/,1X,'DATA PAIRS OR AN EQUATION OF THE FORM X=PT**R WHERE:')
      END IF
      WRITE(5,460) LUNTS$
```

```
460     FORMAT(1X,'                    X = DISTANCE FROM THE UPSTREAM END OF THE'
       &,/,1X,'                    FIELD IN ',A2
       &,/,1X,'                T = TIME FROM THE BEGINNING OF INFLOW TO WHEN'
       &,/,1X,'                    THE WATER FRONT REACHES DISTANCE X IN MINS'
       &,/,1X,'                P = COEFFICIENT'
       &,/,1X,'                R = EXPONENT'
       &,//,1X,'ENTER:'
       &,/,1X,'        1   TO INPUT ADVANCE DATA AS A TABLE'
       &,/,1X,'        2   TO INPUT P AND R IN THE ABOVE EQUATION')
        READ(5,*) ADV
        IF (ADV.EQ.1) THEN
           GOTO 510
        END IF
        N=20
470     FORMAT(1X,' ENTER P AND R IN THE ABOVE EQUATION.')
        WRITE(5,470)
        READ(5,*) P,R
        IF (TYPE.EQ.1) THEN
            WRITE(5,480) LUNTS$
480         FORMAT(1X,'ENTER THE LENGTH OF THE FIELD IN ',A2)
        ELSE IF (TYPE.EQ.2) THEN
            RETURN
        ELSE
            WRITE(5,490) LUNTS$
490         FORMAT(1X,'ENTER THE LENGTH OF THE BASIN IN ',A2)
        END IF
        READ(5,*) L
C
C   CREATING A TABLE OF ADVANCE DATA USING THE ADVANCE EQUATION
C
        DO 500 I=0,20
           X(I)=FLOAT(I)*L/20
           AT(I)=(X(I)/P)**(1/R)
500     CONTINUE
        GOTO 551
510     IF (TYPE.EQ.2) THEN
            RETURN
        END IF
C
C ENTERING A
C
520     FORMAT(1X,'ENTER THE NUMBER OF STATIONS ACROSS THE FIELD OF BASIN'
       &,/,1X,'AT WHICH ADVANCE TIMES ARE KNOWN (DO NOT INCLUDE THE'
       &,/,1X,'STATION 0)')
        WRITE(5,520)
        READ(5,*) N
        X(0)=0
```

```
      AT(0)=0
      WRITE(5,530)
530   FORMAT(1X,'ENTER THE DISTANCE FROM THE UPSTREAM END OF THE FIELD',
     &/,1X,'AND THE ADVANCE TIME (START AT THE FIRST STATION ',/,1X,
     &'DOWNSTREAM OF STATION 0)',/)
      DO 540 I=1,N
         WRITE(5,535) I
535      FORMAT(1X,'FOR STATION# ',I3,' *')
         READ(5,*) X(I),AT(I)
540   CONTINUE
C
C ----- ECHO CHECKING AND CORRECTING INPUT DATA ------
C
551   Y=0
      WRITE(5,552)
552   FORMAT(1X,'THE FOLLOWING DATA HAVE BEEN ENTERED:')
      IF (IUNTS.EQ.1) THEN
            WRITE(5,560) DD
560         FORMAT(1X,'    1.    DEPTH = ',F6.2,' MM')
      ELSE
            WRITE(5,570) DD
570         FORMAT(1X,'    1.    DEPTH = ',F6.2,' INS')
      END IF
      IF (FO.LT.001) THEN
         WRITE(5,580) C,D1
580      FORMAT(1X,'    2.    C = ',F10.2,/,
     &        1X,'    3.    D = ',F10.2)
      ELSE
         C2=C*W/K2
         F1=FO*W/K4
         WRITE(5,585) C2,F1,D1
585      FORMAT(1X,'    2.    K = ',F10.2,/,
     &        1X,'          FO = ',F10.2,/,
     &        1X,'    3.    A = ',F10.2)
      END IF
      IF(IUNTS.EQ.1) THEN
590         FORMAT(1X,'    4.    W = ',F10.2,' M')
            WRITE(5,590) W
            IF (TYPE.EQ.1) THEN
600               FORMAT(1X,'    5.    Q = ',F10.2,' L/MIN')
                  WRITE(5,600) Q
            ELSE
610               FORMAT(1X,'    5.    Q = ',F10.2,' L/MIN/M')
                  WRITE(5,610) Q/W
            END IF
      ELSE
            IF (TYPE.EQ.1) THEN
```

```
620             FORMAT(1X,'    4.     W = ',F10.2,' INS'
     &,/,1X,'    5.     Q = ',F10.2,' GPM')
                WRITE(5,620) W,Q
          ELSE
630             FORMAT(1X,'    4.     W = ',F10.2,' FT'
     &,/,1X,'    5.     Q = ',F10.2,' GPM/FT')
                WRITE(5,630) W,Q/W
          END IF
       END IF
       IF (ADV.EQ.2) THEN
640        FORMAT(1X,'    6.     P = ',F10.2,/,13X,'R = ',F10.2)
           WRITE(5,640) P,R
       ELSE
           IF (TYPE.NE.2) THEN
              WRITE(5,650)
650           FORMAT(1X,'    6.     DATA SET#',6X,'X',12X,'ADV TIME')
           ELSE
660           FORMAT(1X,'    6.     DATA SET#     X         ADV TIME',
     &'  REC TIME')
           END IF
           DO 690 I=1,N
              IF (TYPE.EQ.2) THEN
                 WRITE(5,670) I,X(I),AT(I),RT(I)
670              FORMAT(12X,I5,3F11.2)
              ELSE
                 WRITE(5,680) I,X(I),AT(I)
680              FORMAT(12X,I5,2F16.2)
              END IF
690        CONTINUE
       END IF
       WRITE(5,700)
700    FORMAT(/,1X,'ARE ALL THESE VALUES OKAY, YES OR NO')
       READ(5,825) Z$
       IF (Z$.EQ.'Y') THEN
          RETURN
       END IF
710    FORMAT(1X,'ENTER THE NUMBER OF THE VALUE YOU WISH TO CHANGE.')
       WRITE(5,710)
       READ(5,*) CH
       IF (CH.EQ.1) THEN
             WRITE(5,720) DUNTS$
720          FORMAT(1X,'ENTER THE NEW VALUE OF DEPTH TO FILL SOIL TO FIELD
     & CAPACITY IN ',A3)
             READ(5,*) DD
             GOTO 551
       ELSE IF(CH.EQ.2) THEN
          IF (FO.LT..001) THEN
```

```
              WRITE(5,730)
730           FORMAT(1X,'ENTER A NEW VALUE OF C')
              READ(5,*) C
          ELSE
              WRITE(5,735)
735           FORMAT(1X,'ENTER NEW VALUES OF K AND FO')
              READ(5,*) C,FO
              C=C*K2/W
              FO=FO*K4/W
              IF (TYPE.EQ.1) THEN
                C=C/12.
                FO=FO/12.
              END IF
          END IF
          IF (FO.GT..001) THEN
            C=C/W
            FO=FO/W
          END IF
          GOTO 551
      ELSE IF(CH.EQ.3) THEN
          WRITE(5,740)
740       FORMAT(1X,'ENTER A NEW VALUE OF D')
          READ(5,*) D1
          GOTO 551
      ELSE IF(CH.EQ.4) THEN
          IF (FO.GT..001) THEN
            C=C*W
            FO=FO*W
          END IF
          IF (IUNTS.EQ.1) THEN
            WRITE(5,*) ' ENTER A NEW VALUE OF W IN M'
          ELSE
            IF (TYPE.EQ.1) THEN
              WRITE(5,*) 'ENTER A NEW VALUE OF W IN INS'
            ELSE
              WRITE(5,*) 'ENTER A NEW VALUE OF W IN FT'
            END IF
          END IF
          READ(5,*) W
          GOTO 551
      ELSE IF (CH.EQ.5) THEN
          IF (TYPE.EQ.1) THEN
              WRITE(5,760) QUNTS$
760           FORMAT(1X,'ENTER A NEW VALUE OF Q IN ',A5)
          ELSE
              WRITE(5,770) QUNTS$,LUNTS$
770           FORMAT(1X,'ENTER A NEW VALUE OF Q IN ',A5,'/',A2)
```

```
            END IF
            READ(5,*)Q
            IF(TYPE.EQ.1) THEN
                GOTO 551
            END IF
            Q=Q*W
            GOTO 551
        END IF
        IF(ADV.EQ.2) THEN
780         FORMAT(1X,'ENTER NEW VALUES FOR P AND R')
            WRITE(5,780)
            READ(5,*) P,R
            GOTO 551
        END IF
        IF (CH.EQ.6) THEN
790         Y=0
791         FORMAT(1X,'ENTER THE NUMBER OF THE DATA SET TO BE CHANGED')
            WRITE(5,791)
            READ(5,*) J
            IF(TYPE.EQ.2) THEN
800             FORMAT(1X,'ENTER THE NEW VALUES OF X, ADV TIME (SINCE',
    &' THE START OF THE INFLOW)',/,1X,'AND REC TIME (SINCE WATER',
    &'STARTED RECEEDING AT STATION 0)')
                WRITE(5,800)
                READ(5,810) X(J),AT(J),RT(J)
810             FORMAT(3F5.2)
                WRITE(5,820)
820             FORMAT(1X,'MUST OTHER X S, ADV TIMES, AND/OR REC',
    &' TIMES BE CHANGED, YES OR NO')
            ELSE
                WRITE(5,830)
830             FORMAT(1X,'ENTER THE NEW VALUES OF X AND ADV TIME')
                READ(5,*) X(J),AT(J)
                WRITE(5,840)
840             FORMAT(1X,'MUST OTHER X S AND/OR ADV TIMES BE CHANGED',
    &' YES OR NO')
            END IF
            READ(5,825) Z$
825             FORMAT(A1)
            IF (Z$.EQ.'Y') GOTO 790
            IF (TYPE.NE.2) GOTO 551
            IF(CH.NE.7) GOTO 551
            WRITE(5,850)
850         FORMAT(1X,'ENTER A NEW VALUE OF LAG TIME IN MINS --THE TIME',
    &' BETWEEN THE END OF INFLOW AND THE START OF RECESSION AT',
    &' STATION 0--')
            READ(5,*) LAGTM
```

```
              GOTO 551
        END IF
        END
C
C    -------  LOOP FOR SOLVING EQ 7.13 FOR TIME  -------
C
        SUBROUTINE IRTM(DD,IRRTIM,C,D1,FO)
        REAL IRRTIM
        TOL=DD/1000.
        DUM=IRRTIM
        I=0
900     F=C*IRRTIM**D1+FO*IRRTIM
        I=I+1
        IF (I.GT.51) THEN
          WRITE(5,*) 'THE RUN HAS BEEN STOPPED BECAUSE A VALUE FOR IRRTIM'
          WRITE(5,*) 'HAS NOT BEEN OBTAINED IN 50 ITERATIONS IN'
          WRITE(5,*) 'SUBROUTINE IRTM. SEE THE INSTRUCTOR.'
          WRITE(5,*) ' '
          WRITE(5,*) '*******  THE RUN IS COMPLETE  ********'
          STOP
        END IF
        DIFF=F-DD
        IF (ABS(DIFF).LT.TOL) THEN
          RETURN
        ELSE
          DUM=DUM/2.
          IF (DIFF.GT.0) THEN
            IRRTIM=IRRTIM-DUM
            GOTO 900
          ELSE
            IRRTIM=IRRTIM+DUM
            GOTO 900
          END IF
        END IF
        STOP
        END
```

Appendix K

FORTRAN Code for Performing Land Smoothing Calculations

```
      WRITE(5,*) '**************************************************'
      WRITE(5,*) '*                                                *'
      WRITE(5,*) '*   THIS PROGRAM DETERMINES A LEAST SQUARES FIT  *'
      WRITE(5,*) '*   PLANE FOR AN M (HORIZONTAL) BY N (VERTICAL)  *'
      WRITE(5,*) '*   GRID OF LAND SURFACE ELEVATIONS. CUTS, FILLS,*'
      WRITE(5,*) '*   FINAL ELEVATIONS, AND VOLUME OF EXCAVATION   *'
      WRITE(5,*) '*   ARE COMPUTED.                   (LGJ - 1184) *'
      WRITE(5,*) '**************************************************'
C
C  IN THIS PROGRAM THE DEVICE NUMBERS FOR THE TERMINAL (SCREEN)
C  AND PRINTER ARE 5 AND 6, RESPECTIVELY
C
      DIMENSION  R4(30),R5(30),C4(30),C5(30),OP(30,30),E3(30,30)
      DIMENSION  E1(30,30)
      CHARACTER UNTS$*2,CUNTS$*9,Z$*3
      INTEGER M1,N1
      P=0
      FLAG=0
      WRITE (5,*) ' '
      WRITE(5,*) 'ENTER LOWEST ACCEPTABLE RATIO OF CUTS TO FILLS'
      WRITE(5,*) '1.2 IS MOST COMMON VALUE'
      READ(5,*) CFL
      WRITE(5,*) 'ENTER LARGEST RATIO OF CUTS TO FILLS FILLS, '
      WRITE(5,*) '1.5 IS MOST COMMON VALUE'
      READ(5,*) CFU
      WRITE(5,*) 'EITHER SI OR ENGLISH UNITS MAY BE USED. ENTER:'
      WRITE(5,*) '      1    FOR SI UNITS'
      WRITE(5,*) '      2    FOR ENGLISH UNITS'
      READ(5,*) UNTS
      IF (UNTS.EQ.1) THEN
        UNTS$='M'
        CUNTS$='CUBIC M  '
```

```
        ELSE
          UNTS$='FT'
          CUNTS$='CUBIC YDS'
        END IF
C
C     --------              GRID CORNER INPUTS          --------------  C
C
      WRITE(5,*) 'NUMBER OF GRID CORNERS IN M (HORZ) DIRECTION'
      WRITE (5,*) 'ENTER A VALUE FROM 1 TO 30  '
      READ(5,*) M1
      WRITE(5,*) 'NUMBER OF GRID CORNERS IN N (VERT) DIRECTION'
      WRITE(5,*) 'ENTER A VALUE FROM 1 TO 30'
      READ(5,*) N1
      WRITE(5,11) UNTS$
   11 FORMAT('ENTER THE DISTANCE IN ',A2,' BETWEEN GRID POINTS IN THE M
     $(HORZ) DIRECTION')
      READ(5,*) D2
      WRITE(5,12) UNTS$
   12 FORMAT('ENTER THE DISTANCE IN ',A2,' BETWEEN GRID POINTS IN THE N
     $(VERT) DIRECTION')
      READ(5,*) D1
C
C  -------          ENTER GRID POINT ELEVATIONS      ----------
C
    2 FORMAT('ENTER ELEVATION AT GRID POINT (',I3,',',I3,')')
      WRITE(5,*) ' '
      DO 10 I=1,N1
      DO 20 J=1,M1
        WRITE(5,2) I,J
        READ(5,*) E1(I,J)
        E3(I,J)=E1(I,J)
   20 CONTINUE
   10 CONTINUE
C
C    ------       CORRECTIONS TO CORNER ELEVATIONS    -------
C
      WRITE(5,*) 'WOULD YOU LIKE TO MAKE ANY CORRECTIONS, YES OR NO?'
      READ(5,21) Z$
   21 FORMAT(A3)
      IF (Z$.EQ.'NO') GOTO 103
   22 WRITE(5,*) 'ENTER M,N CORRDINATES FOR CORRECTION'
      READ(5,32) I,J
   32 FORMAT(I3,I3)
      WRITE(5,2) I,J
      READ(5,*) E1(I,J)
      WRITE(5,*) 'ANY MORE CORRECTIONS, YES OR NO ?'
       READ (5,21) Z$
```

```
          IF (Z$.EQ.'NO') GOTO 103
          GOTO 22
C
C   -----   COMPUTE SUM AND AVERAGE OF ELEVATIONS IN A ROW ----
C
   103 DO 30 I=1,N1
          R4(I)=0
          R5(I)=0
          DO 40 J=1,M1
          R4(I)=R4(I)+E1(I,J)
    40 CONTINUE
          R5(I)=R4(I)/FLOAT(M1)
    30 CONTINUE
C
C   ----    COMPUTE SUM AND AVERAGE FOR EACH COLUMN   ----
C
          DO 50 J=1,M1
          C4(J)=0
          C5(J)=0
          DO 60 I=1,N1
          C4(J)=C4(J)+E1(I,J)
    60 CONTINUE
          C5(J)=C4(J)/FLOAT(N1)
    50 CONTINUE
C
C   ----         COMPUTE CENTROID LOCATION AND ELEVATION    ----
C
          A=0.
          B=0.
          DO 70 I=1,N1
          B=B+R5(I)
          A=A+FLOAT(I)
    70 CONTINUE
          Y2=A/FLOAT(N1)
          A=0.
          DO 80 J=1,M1
          B=B+C5(J)
          A=A+FLOAT(J)
    80 CONTINUE
          X2=A/FLOAT(M1)
          E2=B/FLOAT(M1+N1)
C
C   ---- COMPUTING GRADE IN THE M (HORIZONTAL) DIRECTION   ----
C
          R1=0.
          R2=0.
          R3=0.
```

```
          R6=0.
        DO 90 I=1,N1
          R1=R1+FLOAT(I)*D2
          R2=R2+FLOAT(I)*FLOAT(I)*D2*D2
          R3=R3+R5(I)*D2*FLOAT(I)
          R6=R6+R5(I)
     90 CONTINUE
          A=R2
          B=R1**2
          C=R1*R6
          D=R3
          G1=(D-R1*R6/FLOAT(N1))/(A-B/FLOAT(N1))
C
C     ------ COMPUTE GRADE IN N (VERT) DIRECTION  ------
C
          C1=0.
          C2=0.
          C3=0.
          C6=0.
        DO 95 J=1,M1
          C1=C1+FLOAT(J)*D1
          C2=C2+FLOAT(J)*FLOAT(J)*D1*D1
          C3=C3+C5(J)*FLOAT(J)*D1
          C6=C6+C5(J)
     95 CONTINUE
          A=C2
          B=C1**2
          C=C1*C6
          D=C3
          G2=(D-C1*C6/FLOAT(M1))/(A-B/FLOAT(M1))
        WRITE(5,41) X2
     41 FORMAT(1X,'LOCATION OF CENTROID M  = ',F5.2)
        WRITE(5,42) Y2
     42 FORMAT(1X,'                      N  = ',F5.2)
        WRITE (5,43) E2,UNTS$
     43 FORMAT(1X,'CENTROID ELEVATION      = ',F5.2,1X,A2)
          G1T=G1*100
          G2T=G2*100
    555 WRITE(5,51) G1T,UNTS$,UNTS$
     51 FORMAT(1X,'GRADE IN M DIRECTION    = ',F5.2,1X,A2,'/100 ',A2)
        WRITE(5,52) G2T,UNTS$,UNTS$
     52 FORMAT(1X,'GRADE IN N DIRECTION    = ',F5.2,1X,A2,'/100 ',A2,/)
    105 WRITE(5,*) 'ARE THESE GRADES SATISFACTORY, YES OR NO?'
        READ(5,21) Z$
        IF (Z$.EQ.'YES') GOTO 104
    106 WRITE(5,53) UNTS$,UNTS$
     53 FORMAT ('ENTER THE DESIRED GRADE IN THE M (HORZ) DIRECTION IN ',A2
```

```
      $,'/100 ',A2)
        READ(5,*) G1T
        WRITE(5,54) UNTS$,UNTS$
   54 FORMAT ('ENTER THE DESIRED GRADE IN THE N (VERT) DIRECTION IN ',A2
      $,'/100 ',A2,/)
        READ(5,*) G2T
  104 G1=G1T/100
      G2=G2T/100
      WRITE(5,18) E2,UNTS$
   18 FORMAT(1X,'CENTROID ELEVATION IS ',F6.2,1X,A2,/)
      WRITE(5,*) ' '
      WRITE(5,*) 'IS CENTROID ELEVATION SATISFACTORY, YES OR NO?'
      READ(5,21) Z$
      IF (Z$.EQ.'YES') GOTO 194
  107 WRITE(5,19) UNTS$
   19 FORMAT (' ENTER THE DESIRED CENTROID ELEVATION IN ',A2)
      READ (5,*) E2
C
C   ------- ADJUSTING THE ELEVATION OF THE FIELD    -------
C
  194 C8=0.
      F8=0.
      A=E2-(G1*X2*D2)-(G2*Y2*D1)
      DO 25 I=1,N1
      DO 26 J=1,M1
       E3(I,J)=A+G1*FLOAT(J)*D2+G2*FLOAT(I)*D1
       B=E3(I,J)-E1(I,J)
       IF (E1(I,J) .GT. E3(I,J))   THEN
         B=ABS (B)
         C8=C8+B
         OP(I,J)=B
       ELSE
         B=ABS (B)
         OP(I,J)=B
         F8=F8+B
       END IF
   26 CONTINUE
   25 CONTINUE
      C9=D1*D2*C8
      F9=D1*D2*F8
      IF (UNTS .EQ. 1) GOTO 113
      C9=C9/27
      F9=F9/27
  113 IF (C9.GT.0.001 .AND. F9.GT.0.001) GOTO 110
      WRITE(5,*) 'THE LAND IS SMOOTH - THERE IS NO NEED TO CONTINUE'
      WRITE(5,*) 'UNLESS YOU WISH TO CHANGE THE SLOPE IN THE M'
      WRITE(5,*) '(HORZ) AND/OR N (VERT) DIRECTION.'
```

```
      WRITE(5,*) ' '
      G1T=G1*100
      G2T=G2*100
      FLAG=1
      GOTO 555
110 IF (ABS(C9-F9).GT..01) THEN
        R=C9/F9
      ELSE
        R=1
      END IF
      WRITE(5,231) C9,CUNTS$
      WRITE(5,232) F9,CUNTS$
      WRITE(5,233) R
      WRITE(5,234) G1T,UNTS$,UNTS$
      WRITE(5,235) G2T,UNTS$,UNTS$
      WRITE(5,236) E2,UNTS$
231 FORMAT(1X,'SUM OF CUTS          = ',F8.2,1X,A9)
232 FORMAT(1X,'SUM OF FILLS         = ',F8.2,1X,A9)
233 FORMAT(1X,'RATIO OF CUTS/FILLS  = ',F8.2)
234 FORMAT(1X,'GRADE IN M DIRECTION = ',F8.2,1X,A2,'/100 ',A2)
235 FORMAT(1X,'          N DIRECTION = ',F8.2,1X,A2,'/100 ',A2)
236 FORMAT(1X,'CENTROID ELEVATION   = ',F8.2,1X,A2,//)
      IF (R .GT. CFU) GOTO 501
      IF (R .LT. CFL) GOTO 502
      WRITE(5,*) 'THIS IS WITHIN THE ALLOWABLE RANGE'
      WRITE(5,*) 'ARE ALL PARAMETERS ACCEPTABLE, YES OR NO?'
      READ(5,21) Z$
      IF (Z$.EQ.'NO') GOTO 106
      GOTO 108
501 WRITE(5,*) 'RATIO IS TOO LARGE, RAISE CENTROID ELEVATION'
      GOTO 107
502 WRITE(5,*) 'RATIO IS TOO SMALL, LOWER CENTROID ELEVATION'
      GOTO 107
255 FORMAT(1X,'LOC M,N',6X,'ELEV.',7X,'GRD.EL.',5X,'OPERATION')
256 FORMAT(15X,A2,11X,A2,11X,A2)
257 FORMAT(1X,'-------',4X,'------',6X,'--------',5X,'---------'/)
258 FORMAT(1X,I2,2X,I2,4X,F8.2,3X,F10.2,7X,A1,F5.2)
108 WRITE(6,255)
      WRITE(6,256) UNTS$,UNTS$,UNTS$
      WRITE(6,257)
      DO 444 I=1,N1
        DO 443 J=1,M1
          IF (E1(I,J).LT.E3(I,J)) THEN
            Z$='F'
          ELSE
            Z$='C'
          END IF
```

```
          WRITE(6,258) I,J,E1(I,J),E3(I,J),Z$,OP(I,J)
443   CONTINUE
      WRITE(6,*) ' '
444 CONTINUE
      WRITE(6,*) ' '
      WRITE(6,*) ' '
      WRITE(6,334) G1T,UNTS$,UNTS$
334 FORMAT(1X,'FINAL GRADE IN M DIRECTION = ',F8.2,1X,A2,'/100 ',A2)
      WRITE(6,335) G2T,UNTS$,UNTS$
335 FORMAT(1X,'               N DIRECTION = ',F8.2,1X,A2,'/100 ',A2)
      WRITE(6,336) E2,UNTS$
336 FORMAT(1X,'FINAL CENTROID ELEVATION   = ',F8.2,1X,A2)
      WRITE(6,337) R
337 FORMAT(1X,'FINAL RATIO OF CUTS/FILLS  = ',F8.2)
      WRITE(6,338) C9,CUNTS$
338 FORMAT(1X,'FINAL VOLUME OF EXCAVATION = ',F8.2,1X,A9)
      WRITE (5,*) ' '
      WRITE (5,*) '**********  THE RUN IS COMPLETE  **********'
      STOP
      END
```

Appendix L

FORTRAN Code for Determining Infiltration Characteristics Using the Two-Point Method

```
      WRITE (5,*) '****************************************************'
      WRITE (5,*) '*                                                *'
      WRITE (5,*) '*   THIS PROGRAM DETERMINES "R" IN EQ 7.12 AND    *'
      WRITE (5,*) '*   "A" AND "K" IN EQ 7.13 USING THE TWO-POINT    *'
      WRITE (5,*) '*   METHOD OF ELLOITT AND WALKER, 1982. EITHER A  *'
      WRITE (5,*) '*   MEASURED AO CAN BE ENTERED OR AO CAN BE       *'
      WRITE (5,*) '*   ESTIMATED BY THE PROGRAM USING THE MANNING    *'
      WRITE (5,*) '*   EQUATION.                     (LGJ 287)       *'
      WRITE (5,*) '*                                                *'
      WRITE (5,*) '****************************************************'
C
C  IN THIS PROGRAM THE DEVICE NUMBERS FOR THE TERMINAL (SCREEN)
C  AND THE PRINTER ARE 5 AND 6, RESPECTIVELY
C
      CHARACTER   B$*1,C$*1,LUNTS$*2,QUNTS$*5
      REAL*4 TH,TF,K,K1,K2
      INTEGER UNTS
      WRITE (5,*) ' '
      WRITE (5,*) 'PRESS ANY CHARACTER KEY TO START THE PROGRAM'
      READ (5,5) B$
    5 FORMAT (A1)
      WRITE (5,*) 'EITHER SI OR ENGLISH UNITS MAYBE USED. ENTER:'
      WRITE (5,*) '     1   FOR SI UNITS'
      WRITE (5,*) '     2   FOR ENGLISH UNITS'
      READ (5,*) UNTS
      IF (UNTS.EQ.1) THEN
        K1=60000.
        K2=1000.
        LUNTS$='M'
        QUNTS$='L/MIN'
      ELSE
        K1=449*1.49
```

```
      K2=7.48
      LUNTS$='FT'
      QUNTS$='GPM'
      END IF
10 WRITE(5,*)   'T-HALF AND T-FULL ARE THE TIMES TO ADVANCE HALF THE
   $LENGTH AND THE FULL LENGTH OF THE BASIN, BORDER, OR FURROW, RESPEC
   $TIVELY.  ENTER T-HALF IN MINUTES.'
      READ (5,*) TH
      WRITE (5,*) 'ENTER T-FULL IN MINUTES'
      READ (5,*) TF
      WRITE (5,*) 'ENTER THE FINAL INFILTRATION RATE "FO" IN '
      WRITE (5,12) QUNTS$,'/',LUNTS$,' OF BASIN, BORDER, OF FURROW LENGT
   $H.'
12 FORMAT (1X,A5,A1,A2,A35)
      READ (5,*) FO
      WRITE (5,14) ' ENTER THE STREAMSIZE "Q" IN ',QUNTS$
14 FORMAT (A29,A5)
      READ(5,*) Q
      WRITE (5,*) 'ENTER THE LENGTH OF THE BASIN, BORDER,'
      WRITE (5,16) ' OR FURROW IN ',LUNTS$
16 FORMAT (A14,A2)
      READ (5,*) D
      WRITE(5,*) 'DO YOU WISH TO ENTER A MEASURED VALUE OF AO?  ENTER YE
   $S OR NO.'
      READ(5,5) B$
      IF (B$.EQ.'Y') THEN
         WRITE(5,18) ' ENTER THE MEASURED "AO" IN SQ ',LUNTS$
18       FORMAT (A31,A2)
         READ(5,*) AO
      ELSE
         CALL EST(Q,AO,UNTS,K1,LUNTS$)
      END IF
      R=(ALOG(2.)/(ALOG(TF)-ALOG(TH)))
      VFULL=Q*TF/(K2*D)-.77*AO-FO*TF/(K2*(1.+R))
      VHALF=2*Q*TH/(K2*D)-.77*AO-FO*TH/(K2*(1.+R))
      A=(ALOG(VFULL)-ALOG(VHALF))/(ALOG(TF)-ALOG(TH))
      SIGMA=(A+R*(1.-A)+1.)/((1.+A)*(1.+R))
      K=VFULL/(SIGMA*TF**A)
      WRITE (6,*) '      INPUT DATA'
      WRITE (6,*) '            '
      WRITE (6,50) 'Q      = ',Q,QUNTS$
      WRITE (6,51) 'LENGTH = ',D,LUNTS$
      WRITE (6,52) 'AO     = ',AO,' SQ ',LUNTS$
      WRITE (6,53) 'FO     = ',FO,QUNTS$,'/',LUNTS$
      WRITE (6,50) 'T-HALF = ',TH,'MINS'
      WRITE (6,50) 'T-FULL = ',TF,'MINS'
      WRITE (6,*) '       '
```

```
      WRITE (6,*) '    OUTPUT DATA'
      WRITE (6,*) ' '
      WRITE (6,50) ' R      = ',R,' '
      WRITE (6,50) ' K      = ',K,' '
      WRITE (6,50) ' A      = ',A,' '
      WRITE (6,*) '    '
   50 FORMAT (10X,A10,F9.3,A6)
   51 FORMAT (10X,A10,F9.3,A3)
   52 FORMAT (10X,A10,F9.3,A4,A2)
   53 FORMAT (10X,A10,F9.3,A6,A1,A2)
      WRITE(5,*) 'DO YOU WISH TO RUN THE PROGRAM AGAIN? ENTER YES OR NO'
      READ(5,5) C$
      IF (C$.EQ.'Y') GO TO 10
      WRITE (5,*) '********  THE RUN IS COMPLETE   *********'
      WRITE (5,*)
      STOP
      END
C
C
C -----     SUBROUTINE FOR ESTIMATING AO    -----
C
C
      SUBROUTINE EST(Q,AO,UNTS,K1,LUNTS$)
      REAL N,K1
      INTEGER UNTS
      CHARACTER LUNTS$*2
      WRITE (5,*) 'ENTER THE TYPE OF SURFACE SYSTEM BEING CONSIDERED: '
      WRITE (5,*) '    1   FOR A BASIN OR BORDER SYSTEM '
      WRITE (5,*) '    2   FOR A FURROW SYSTEM '
      READ (5,*) I
      WRITE (5,*) 'ENTER THE SLOPE IN PERCENT'
      READ (5,*) S
      WRITE (5,*) 'ENTER MANNING ROUGHNESS COEFFICIENT FROM TABLE 7.8.'
      READ (5,*) N
      IF (S.EQ.0.) THEN
        S=.001
      END IF
        S=S/100.
      IF (I.EQ.1) THEN
        C=Q*N/(K1*S**0.5)
        TOL=C/500.
        WRITE (5,75)' ENTER THE WIDTH OF THE BASIN OR BORDER IN ',LUNTS$
   75   FORMAT (A43,A2)
        READ(5,*) W
        Y=10
        DY=Y
   80   C1=(W*Y)**(5./3.)*(1/(W+2*Y))**(2./3.)
```

```fortran
      IF (ABS(C-C1).LT.TOL) THEN
        AO=Y*W
      ELSE
        DY=DY/2
        IF (C1.GT.C) THEN
          Y=Y-DY
          GO TO 80
        ELSE
          Y=Y+DY
          GO TO 80
        END IF
      END IF
   ELSE
C
C  FURROWS, USING EQUATIONS 15 AND 16 FROM WILKE AND SMERDON, 1965
C
      IF (UNTS.EQ.1) THEN
        Y=2.389*(Q/S**0.5)**0.4
        AO=0.000018507*Y**(5./3.)
      ELSE
        Y=0.013348*(Q/S**0.5)**0.4
        AO=2.75*Y**(5./3.)
      END IF
   END IF
   RETURN
   END
```

Appendix M

FORTRAN Code for Determining an Equation for Water Advance

```
      WRITE (5,*) '***************************************************'
      WRITE (5,*) '*                                                 *'
      WRITE (5,*) '*   THIS PROGRAM SOLVES EQS 7.12A AND 7.20        *'
      WRITE (5,*) '*   SIMULTANEOUSLY FOR R,TA,AND THALF. IT ALSO    *'
      WRITE (5,*) '*   COMPUTES P IN EQ 7.12. IT REQUIRES VALUES OF  *'
      WRITE (5,*) '*   Q, MANNING N, L, W AND K, A, AND FO IN EQ     *'
      WRITE (5,*) '*   7.13.                          (LGJ 187)      *'
      WRITE (5,*) '*                                                 *'
      WRITE (5,*) '***************************************************'
C
C  IN THIS PROGRAM THE DEVICE NUMBERS FOR THE TERMINAL (SCREEN)
C  AND PRINTER ARE 5 AND 6, RESPECTIVELY
C
      CHARACTER A$*1,DUNTS$*2,LUNTS$*2,QUNTS$*5
      REAL*4 TA,THALF,K,K1,K2,N
      WRITE (5,*) ' '
      WRITE (5,*) 'PRESS ANY CHARACTER KEY TO START THE PROGRAM'
      READ (5,5) A$
    5 FORMAT (A1)
      WRITE (5,*) ' '
      WRITE (5,*) 'EITHER SI OR ENGLISH UNITS MAYBE USED. ENTER:'
      WRITE (5,*) '   1   FOR SI UNITS'
      WRITE (5,*) '   2   FOR ENGLISH UNITS'
      READ (5,*) UNTS
      IF (UNTS.EQ.1) THEN
        K1=60000.
        K2=1000.
        DUNTS$='MM'
        LUNTS$='M'
        QUNTS$='L/MIN'
      ELSE
        K1=449*1.49
```

```
          K2=7.48
          DUNTS$='FT'
          LUNTS$='FT'
          QUNTS$='GPM'
        END IF
   10 WRITE (5,*) 'ENTER THE TYPE OF SURFACE SYSTEM TO BE CONSIDERED:'
      WRITE (5,*) '        1     FOR A BASIN OR BORDER SYSTEM'
      WRITE (5,*) '        2     FOR A FURROW SYSTEM'
      READ (5,*) ITYPE
      WRITE (5,22) ' ENTER THE STREAMSIZE Q IN ',QUNTS$
   22 FORMAT (A26,A6)
      READ( 5,*) Q
      WRITE (5,24) ' ENTER THE LENGTH OF THE FIELD L IN ',LUNTS$
   24 FORMAT (A36,A3)
      READ (5,*) D
      WRITE (5,*) 'ENTER MANNING ROUGHNESS COEFFICIENT FROM TABLE 7.8'
      READ (5,*) N
      WRITE (5,*) 'ENTER SLOPE OF FIELD IN PERCENT'
      READ (5,*) S
      IF (S.EQ.0.)THEN
        S=0.001
      ELSE
        S=S/100.
      END IF
      C=Q*N/(K1*S**0.5)
      TOL=C/500.
      IF (ITYPE.EQ.1) THEN
        WRITE (5,26) 'ENTER THE WIDTH OF THE BASIN OR BORDER IN ',LUNTS$
   26   FORMAT (A42,A2)
        READ (5,*) W
        Y=10
        DY=Y
   20   C1=(W*Y)**(5./3.)*(1./(W+2.*Y))**(2./3.)
        IF (ABS(C-C1).LT.TOL) THEN
          AO=Y*W
        ELSE
          DY=DY/2
          IF (C1.GT.C) THEN
            Y=Y-DY
          ELSE
            Y=Y+DY
          END IF
          GO TO 20
        END IF
      ELSE
C
C       FURROWS, USING EQUATIONS 15 AND 16 FROM WILKE AND SMERDON, 1965
```

```
C
          IF (UNTS.EQ.1) THEN
            Y=2.389*(Q/S**0.5)**0.4
            AO=0.000018507*Y**(5./3.)
          ELSE
            Y=.013348*(Q/S**0.5)**.4
            AO=2.75*Y**(5/3)
          END IF
        END IF
        WRITE (5,52)' AO = ',AO,' SQ ',LUNTS$,'  Y = ',Y,DUNTS$
     52 FORMAT (A6,F6.4,A4,A2,A6,F5.2,A3)
        WRITE (5,*)'ENTER K IN EQUATION 7.13'
        READ (5,*) K
        WRITE (5,*)'ENTER A IN EQUATION 7.13'
        READ (5,*) A
        WRITE (5,62) ' ENTER FO IN EQ 7.13 IN ',QUNTS$,'/',LUNTS$,' OF FIE
     $LD LENGTH.'
     62 FORMAT (A24,A5,A1,A2,A17)
        READ (5,*) FO
C
C       THE PROGRAM FOLLOWS THE SOLUTION STEPS IN EXAMPLE 7.7
C
        R=0.7
C
C       TA LOOP
C
     25 TA=5.*AO*D*K2/Q
        DEL=TA
        I=0
        SIGMAZ=(A+R*(1.-A)+1.)/((1.+A)*(1.+R))
        SIGMA=1./(1.+R)
        DO 30 I=1,51
          IF (I.GT.50) THEN
            WRITE (5,*) '50 ITERATIONS OF THE TA LOOP HAVE BEEN COMPLETED
     $ - SEE THE INSTRUCTOR'
            GO TO 80
          END IF
          ZERO=Q*TA/K2-0.77*AO*D-SIGMAZ*D*K*TA**A-SIGMA*FO*TA*D/K2
          IF (ABS(ZERO).LT.0.005) THEN
            GO TO 40
          ELSE
            DEL=DEL/2.
            IF (ZERO.GT.0.0) THEN
              TA=TA-DEL
            ELSE
              TA=TA+DEL
            END IF
```

```
           END IF
 30 CONTINUE
 40 I=0
    THALF=TA
    DEL=THALF
    DO 50 I=1,51
      IF (I.GT.50) THEN
         WRITE (5,*) '50 ITERATIONS OF THE THALF LOOP HAVE BEEN COMPLET
    $ED - SEE INSTRUCTOR'
           GO TO 80
      END IF
      ZERO=Q*THALF/K2-0.77*AO*D/2.-SIGMAZ*D/2.*K*THALF**A-SIGMA*FO*
    $THALF*D/(2*K2)
      IF (ABS(ZERO).LT.0.005) THEN
         GO TO 60
      ELSE
         DEL=DEL/2.
         IF (ZERO.GT.0.0) THEN
           THALF=THALF-DEL
         ELSE
           THALF=THALF+DEL
         END IF
      END IF
 50 CONTINUE
 60 R2=(ALOG(2.)/(ALOG(TA)-ALOG(THALF)))
    IF (ABS(R-R2).LT.0.001) THEN
      P=D/TA**R
    ELSE
      R=R2
      GO TO 25
    END IF
 80 WRITE (6,*) '    '
    WRITE (6,*) '    '
    WRITE (6,*) '         INPUT DATA: '
    WRITE (6,*) '    '
    WRITE (6,70) '   Q   = ',Q,QUNTS$
    WRITE (6,71) '   L   = ',D,LUNTS$
    WRITE (6,71) '   W   = ',W,LUNTS$
    WRITE (6,70) '   N   = ',N,' '
    WRITE (6,70) '   S   = ',S*100.,'%'
    WRITE (6,70) '   K   = ',K,' '
    WRITE (6,70) '   A   = ',A,' '
    WRITE (6,75) '   FO  = ',FO,QUNTS$,'/',LUNTS$
 75 FORMAT (11X,A9,F9.3,A6,A1,A2)
    WRITE (6,*) '      '
    WRITE (6,*) '      '
    WRITE (6,*) '         OUTPUT DATA: '
```

```
      WRITE (6,*) '        '
      WRITE (6,76) ' AO    = ',AO,' SQ ',LUNTS$
  76  FORMAT (11X,A9,F9.3,A4,A2)
      WRITE(6,70)' TA    = ',TA,'MINS'
      WRITE(6,70)' T-HALF= ',THALF,'MINS'
      WRITE(6,70)' R     = ',R,'       '
      WRITE(6,70)' P     = ',P,'       '
      WRITE(6,*)'       '
  70  FORMAT (11X,A9,F9.3,A6)
  71  FORMAT (11X,A9,F9.3,A3)
      WRITE (5,*) 'DO YOU WISH TO RUN THE PROGRAM AGAIN?  ENTER YES OR N
     $O'
      READ(5,5) A$
      IF (A$.EQ.'Y') GO TO 10
      WRITE (5,*) '********  THE RUN IS COMPLETE  ***********'
      WRITE (5,*)
      STOP
      END
```

Appendix N

Data for Determining Head-Discharge Relationships for Long-Throated Flumes with Trapezoidal and Circular Control Sections and for a Broad-Crested Weir in a Circular Pipe

Table N-1   Values of the Ratio y_c/H_1 as a function of z_c and H_1/b_c for Trapezoidal Control Sections

$\dfrac{H_1}{b_c}$	Side Slopes of Channel, Ratio of Horizontal to Vertical (z_c)									
	Vertical	0.25:1	0.50:1	0.75:1	1:1	1.5:1	2:1	2.5:1	3:1	4:1
0.00	0.667	0.667	0.667	0.667	0.667	0.677	0.667	0.667	0.667	0.667
0.01	0.667	0.667	0.667	0.668	0.668	0.669	0.670	0.670	0.671	0.672
0.02	0.667	0.667	0.668	0.669	0.670	0.671	0.672	0.674	0.675	0.678
0.03	0.667	0.668	0.669	0.670	0.671	0.673	0.675	0.677	0.679	0.683
0.04	0.667	0.668	0.670	0.671	0.672	0.675	0.677	0.680	0.683	0.687
0.05	0.667	0.668	0.670	0.672	0.674	0.677	0.680	0.683	0.686	0.692
0.06	0.667	0.669	0.671	0.673	0.675	0.679	0.683	0.686	0.690	0.696
0.07	0.667	0.669	0.672	0.674	0.676	0.681	0.685	0.689	0.693	0.699
0.08	0.667	0.670	0.672	0.675	0.678	0.683	0.687	0.692	0.696	0.703
0.09	0.667	0.670	0.673	0.676	0.679	0.684	0.690	0.695	0.698	0.706

0.10	0.667	0.670	0.674	0.677	0.680	0.686	0.692	0.697	0.701	0.709
0.12	0.667	0.671	0.675	0.679	0.684	0.690	0.696	0.701	0.706	0.715
0.14	0.667	0.672	0.676	0.681	0.686	0.693	0.699	0.705	0.711	0.720
0.16	0.667	0.672	0.678	0.683	0.687	0.696	0.703	0.709	0.715	0.725
0.18	0.667	0.673	0.679	0.684	0.690	0.698	0.706	0.713	0.719	0.729
0.20	0.667	0.674	0.680	0.686	0.692	0.701	0.709	0.717	0.723	0.733
0.22	0.667	0.674	0.681	0.688	0.694	0.704	0.712	0.720	0.726	0.736
0.24	0.667	0.675	0.683	0.689	0.696	0.706	0.715	0.723	0.729	0.739
0.26	0.667	0.676	0.684	0.691	0.698	0.709	0.718	0.725	0.732	0.742
0.28	0.667	0.676	0.685	0.693	0.699	0.711	0.720	0.728	0.734	0.744
0.30	0.667	0.677	0.686	0.694	0.701	0.713	0.723	0.730	0.737	0.747
0.32	0.667	0.678	0.687	0.696	0.703	0.715	0.725	0.733	0.739	0.749
0.34	0.667	0.678	0.689	0.697	0.705	0.717	0.727	0.735	0.741	0.751
0.36	0.667	0.679	0.690	0.699	0.706	0.719	0.729	0.737	0.743	0.752
0.38	0.667	0.680	0.691	0.700	0.708	0.721	0.731	0.738	0.745	0.754
0.40	0.667	0.680	0.692	0.701	0.709	0.723	0.733	0.740	0.747	0.756
0.42	0.667	0.681	0.693	0.703	0.711	0.725	0.734	0.742	0.748	0.757
0.44	0.667	0.681	0.694	0.704	0.712	0.727	0.736	0.744	0.750	0.759
0.46	0.667	0.682	0.695	0.705	0.714	0.728	0.737	0.745	0.751	0.760
0.48	0.667	0.683	0.696	0.706	0.715	0.729	0.739	0.747	0.752	0.761
0.5	0.667	0.683	0.697	0.708	0.717	0.730	0.740	0.748	0.754	0.762
0.6	0.667	0.686	0.701	0.713	0.723	0.737	0.747	0.754	0.759	0.767
0.7	0.667	0.688	0.706	0.718	0.728	0.742	0.752	0.758	0.764	0.771
0.8	0.667	0.692	0.709	0.723	0.732	0.746	0.756	0.762	0.767	0.774
0.9	0.667	0.694	0.713	0.727	0.737	0.750	0.759	0.766	0.770	0.776
1.0	0.667	0.697	0.717	0.730	0.740	0.754	0.762	0.768	0.773	0.778
1.2	0.667	0.701	0.723	0.737	0.747	0.759	0.767	0.772	0.776	0.782
1.4	0.667	0.706	0.729	0.742	0.752	0.764	0.771	0.776	0.779	0.784
1.6	0.667	0.709	0.733	0.747	0.756	0.767	0.774	0.778	0.781	0.786
1.8	0.667	0.713	0.737	0.750	0.759	0.770	0.776	0.781	0.783	0.787
2	0.667	0.717	0.740	0.754	0.762	0.773	0.778	0.782	0.785	0.788
3	0.667	0.730	0.753	0.766	0.773	0.781	0.785	0.787	0.790	0.792
4	0.667	0.740	0.762	0.773	0.778	0.785	0.788	0.790	0.792	0.794
5	0.667	0.748	0.768	0.777	0.782	0.788	0.791	0.792	0.794	0.795
10	0.667	0.768	0.782	0.788	0.791	0.794	0.795	0.796	0.797	0.798
∞		0.800	0.800	0.800	0.800	0.800	0.800	0.800	0.800	0.800

Source: M. G. Bos, J. A. Replogle, and A. J. Clemmens, *Flow Measuring Flumes for Open Channel Systems*, copyright © 1984 John Wiley and Sons, Inc., p. 215–216. Reprinted by permission of John Wiley and Sons, Inc.

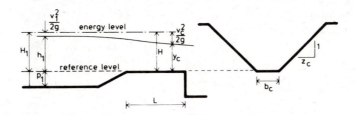

Table N-2 Ratios for Determining the Discharge Q of a Broad-Crested Weir and Long-Throated Flume with Circular Control Section[a]

$\dfrac{y_c}{d_c}$	$\dfrac{v_c^2}{2gd_c}$	$\dfrac{H_1}{d_c}$	$\dfrac{A_c}{d_c^2}$	$\dfrac{y_c}{H_1}$	$f(\theta)$	$\dfrac{y_c}{d_c}$	$\dfrac{v_c^2}{2gd_c}$	$\dfrac{H_1}{d_c}$	$\dfrac{A_c}{d_c^2}$	$\dfrac{y_c}{H_1}$	$f(\theta)$
0.01	0.0033	0.0133	0.0013	0.752	0.0001	0.46	0.1769	0.6369	0.3527	0.722	0.2098
0.02	0.0067	0.0267	0.0037	0.749	0.0004	0.47	0.1817	0.6517	0.3627	0.721	0.2186
0.03	0.0101	0.0401	0.0069	0.749	0.0010	0.48	0.1865	0.6665	0.3727	0.720	0.2276
0.04	0.0134	0.0534	0.0105	0.749	0.0017	0.49	0.1914	0.6814	0.3827	0.719	0.2368
0.05	0.0168	0.0668	0.0147	0.748	0.0027	0.50	0.1964	0.6964	0.3927	0.718	0.2461
0.06	0.0203	0.0803	0.0192	0.748	0.0039	0.51	0.2014	0.7114	0.4027	0.717	0.2556
0.07	0.0237	0.0937	0.0242	0.747	0.0053	0.52	0.2065	0.7265	0.4127	0.716	0.2652
0.08	0.0271	0.1071	0.0294	0.747	0.0068	0.53	0.2117	0.7417	0.4227	0.715	0.2750
0.09	0.0306	0.1206	0.0350	0.746	0.0087	0.54	0.2170	0.7570	0.4327	0.713	0.2851
0.10	0.0341	0.1341	0.0409	0.746	0.0107	0.55	0.2224	0.7724	0.4426	0.712	0.2952
0.11	0.0376	0.1476	0.0470	0.745	0.0129	0.56	0.2279	0.7879	0.4526	0.711	0.3056
0.12	0.0411	0.1611	0.0534	0.745	0.0153	0.57	0.2335	0.8035	0.4625	0.709	0.3161
0.13	0.0446	0.1746	0.0600	0.745	0.0179	0.58	0.2393	0.8193	0.4724	0.708	0.3268
0.14	0.0482	0.1882	0.0688	0.744	0.0214	0.59	0.2451	0.8351	0.4822	0.707	0.3376
0.15	0.0517	0.2017	0.0739	0.744	0.0238	0.60	0.2511	0.8511	0.4920	0.705	0.3487
0.16	0.0553	0.2153	0.0811	0.743	0.0270	0.61	0.2572	0.8672	0.5018	0.703	0.3599
0.17	0.0589	0.2289	0.0885	0.743	0.0304	0.62	0.2635	0.8835	0.5115	0.702	0.3713
0.18	0.0626	0.2426	0.0961	0.742	0.0340	0.63	0.2699	0.8999	0.5212	0.700	0.3829
0.19	0.0662	0.2562	0.1039	0.742	0.0378	0.64	0.2765	0.9165	0.5308	0.698	0.3947
0.20	0.0699	0.2699	0.1118	0.741	0.0418	0.65	0.2833	0.9333	0.5404	0.696	0.4068
0.21	0.0736	0.2836	0.1199	0.740	0.0460	0.66	0.2902	0.9502	0.5499	0.695	0.4189
0.22	0.0773	0.2973	0.1281	0.740	0.0504	0.67	0.2974	0.9674	0.5594	0.693	0.4314
0.23	0.0811	0.3111	0.1365	0.739	0.0550	0.68	0.3048	0.9848	0.5687	0.691	0.4440
0.24	0.0848	0.3248	0.1449	0.739	0.0597	0.69	0.3125	1.0025	0.5780	0.688	0.4569
0.25	0.0887	0.3387	0.1535	0.738	0.0647	0.70	0.3204	1.0204	0.5872	0.686	0.4701
0.26	0.0925	0.3525	0.1623	0.738	0.0698	0.71	0.3286	1.0386	0.5964	0.684	0.4835
0.27	0.0963	0.3663	0.1711	0.737	0.0751	0.72	0.3371	1.0571	0.6054	0.681	0.4971
0.28	0.1002	0.3802	0.1800	0.736	0.0806	0.73	0.3459	1.0759	0.6143	0.679	0.5109
0.29	0.1042	0.3942	0.1890	0.736	0.0863	0.74	0.3552	1.0952	0.6231	0.676	0.5252
0.30	0.1081	0.4081	0.1982	0.735	0.0922	0.75	0.3648	1.1148	0.6319	0.673	0.5397
0.31	0.1121	0.4221	0.2074	0.734	0.0982	0.76	0.3749	1.1349	0.6405	0.670	0.5546
0.32	0.1161	0.4361	0.2167	0.734	0.1044	0.77	0.3855	1.1555	0.6489	0.666	0.5698
0.33	0.1202	0.4502	0.2260	0.733	0.1108	0.78	0.3967	1.1767	0.6573	0.663	0.5855
0.34	0.1243	0.4643	0.2355	0.732	0.1174	0.79	0.4085	1.1985	0.6655	0.659	0.6015
0.35	0.1284	0.4784	0.2450	0.732	0.1289	0.80	0.4210	1.2210	0.6735	0.655	0.6180
0.36	0.1326	0.4926	0.2546	0.731	0.1311	0.81	0.4343	1.2443	0.6815	0.651	0.6351
0.37	0.1368	0.5068	0.2642	0.730	0.1382	0.82	0.4485	1.2685	0.6893	0.646	0.6528
0.38	0.1411	0.5211	0.2739	0.729	0.1455	0.83	0.4638	1.2938	0.6969	0.641	0.6712
0.39	0.1454	0.5354	0.2836	0.728	0.1529	0.84	0.4803	1.3203	0.7043	0.636	0.6903
0.40	0.1497	0.5497	0.2934	0.728	0.1605	0.85	0.4982	1.3482	0.7115	0.630	0.7102
0.41	0.1541	0.5641	0.3032	0.727	0.1683	0.86	0.5177	1.3777	0.7186	0.624	0.7312
0.42	0.1586	0.5786	0.3130	0.726	0.1763	0.87	0.5392	1.4092	0.7254	0.617	0.7533
0.43	0.1631	0.5931	0.3229	0.725	0.1844	0.88	0.5632	1.4432	0.7320	0.610	0.7769
0.44	0.1676	0.6076	0.3328	0.724	0.1927	0.89	0.5900	1.4800	0.7384	0.601	0.8021
0.45	0.1723	0.6223	0.3428	0.723	0.2012	0.90	0.6204	1.5204	0.7445	0.592	0.8293

0.91	0.6555	1.5655	0.7504	0.581	0.8592
0.92	0.6966	1.6166	0.7560	0.569	0.8923
0.93	0.7459	1.6759	0.7612	0.555	0.9297
0.94	0.8065	1.7465	0.7662	0.538	0.9731
0.95	0.8841	1.8341	0.7707	0.518	0.0248

[a] Note: $f(\theta) = (A_c/d_c^2)\{2(H_1/d_c - y_c/d_c)\}^{0.5}$

$\qquad\;\; = (\theta - \sin\theta)^{1.5}/[8(8 \sin\tfrac{1}{2}\theta)^{0.5}]$

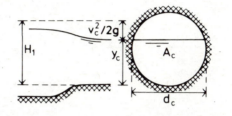

Source: M. G. Bos, J. A. Replogle, and A. J. Clemmens, *Flow Measuring Flumes for Open Channel Systems*, copyright © 1984 John Wiley and Sons, Inc., p. 216–217. Reprinted by permission of John Wiley and Sons, Inc.

Table N-3 Ratios for Determining the Discharge of a Broad-Crested Weir in a Circular Pipe[a]

$$f(\phi, \theta) = \frac{(\theta - \phi + \sin \phi - \sin \theta)^{1.5}}{8(8 \sin \tfrac{1}{2}\theta)^{0.5}}$$

$\dfrac{p_c + H_1}{d_c}$	$\dfrac{p_c}{d_c} = 0.15$	0.20	0.25	0.30	0.35	0.40	0.45	0.50
0.16	0.0004							
0.17	0.0011							
0.18	0.0021							
0.19	0.0032							
0.20	0.0045							
0.21	0.0060	0.0004						
0.22	0.0076	0.0012						
0.23	0.0094	0.0023						
0.24	0.0113	0.0036						
0.25	0.0133	0.0050						
0.26	0.0155	0.0066	0.0005					
0.27	0.0177	0.0084	0.0013					
0.28	0.0201	0.0103	0.0025					
0.29	0.0226	0.0124	0.0038					
0.30	0.0252	0.0145	0.0054					
0.31	0.0280	0.0169	0.0071	0.0005				
0.32	0.0308	0.0193	0.0090	0.0014				
0.33	0.0337	0.0219	0.0110	0.0026				
0.34	0.0368	0.0245	0.0132	0.0040				
0.35	0.0399	0.0273	0.0155	0.0057				
0.36	0.0432	0.0302	0.0179	0.0075	0.0005			
0.37	0.0465	0.0332	0.0205	0.0094	0.0015			
0.38	0.0500	0.0363	0.0232	0.0115	0.0027			
0.39	0.0535	0.0396	0.0260	0.0138	0.0042			
0.40	0.0571	0.0429	0.0289	0.0162	0.0059			
0.41	0.0609	0.0463	0.0320	0.0187	0.0077	0.0005		
0.42	0.0647	0.0498	0.0351	0.0214	0.0097	0.0015		
0.43	0.0686	0.0534	0.0383	0.0242	0.0119	0.0028		
0.44	0.0726	0.0571	0.0417	0.0271	0.0143	0.0043		
0.45	0.0767	0.0609	0.0451	0.0301	0.0167	0.0060		
0.46	0.0809	0.0648	0.0487	0.0332	0.0193	0.0079	0.0005	
0.47	0.0851	0.0688	0.0523	0.0365	0.0220	0.0100	0.0015	
0.48	0.0895	0.0729	0.0561	0.0398	0.0249	0.0122	0.0028	
0.49	0.0939	0.0770	0.0599	0.0432	0.0279	0.0145	0.0043	
0.50	0.0984	0.0813	0.0638	0.0468	0.0309	0.0170	0.0061	
0.51	0.1030	0.0856	0.0678	0.0504	0.0341	0.0197	0.0080	0.0005
0.52	0.1076	0.0900	0.0719	0.0541	0.0374	0.0224	0.0101	0.0015
0.53	0.1124	0.0945	0.0761	0.0579	0.0408	0.0253	0.0123	0.0028
0.54	0.1172	0.0990	0.0803	0.0618	0.0443	0.0283	0.0147	0.0044
0.55	0.1221	0.1037	0.0847	0.0658	0.0479	0.0314	0.0172	0.0061
0.56	0.1270	0.1084	0.0891	0.0699	0.0515	0.0346	0.0198	0.0080
0.57	0.1320	0.1132	0.0936	0.0741	0.0553	0.0379	0.0226	0.0101

0.58	0.1372	0.1180	0.0981	0.0783	0.0592	0.0413	0.0255	0.0123
0.59	0.1423	0.1230	0.1028	0.0826	0.0631	0.0448	0.0285	0.0147
0.60	0.1476	0.1280	0.1075	0.0870	0.0671	0.0484	0.0316	0.0172
0.62[b]		0.1382	0.1172	0.0960	0.0754	0.0559	0.0381	0.0225
0.64		0.1486	0.1271	0.1053	0.0840	0.0637	0.0449	0.0283
0.66		0.1593	0.1373	0.1149	0.0929	0.0718	0.0522	0.0346
0.68		0.1703	0.1477	0.1247	0.1020	0.0802	0.0597	0.0412
0.70		0.1815	0.1584	0.1348	0.1114	0.0888	0.0676	0.0481
0.72		0.1929	0.1692	0.1451	0.1211	0.0978	0.0757	0.0554
0.74		0.2045	0.1804	0.1556	0.1310	0.1070	0.0841	0.0629
0.76		0.2163	0.1917	0.1663	0.1411	0.1164	0.0928	0.0707
0.78		0.2283	0.2031	0.1773	0.1514	0.1260	0.1016	0.0788
0.80		0.2405	0.2148	0.1884	0.1618	0.1358	0.1107	0.0870
0.82		0.2528	0.2267	0.1997	0.1725	0.1458	0.1200	0.0955
0.84		0.2653	0.2386	0.2111	0.1833	0.1559	0.1294	0.1042
0.86		0.2780	0.2508	0.2227	0.1943	0.1662	0.1390	0.1130
0.88		0.2907	0.2630	0.2344	0.2054	0.1767	0.1487	0.1220
0.90		0.3036	0.2754	0.2462	0.2166	0.1872	0.1586	0.1311
0.92		0.3166	0.2879	0.2581	0.2279	0.1979	0.1686	0.1404
0.94		0.3297	0.3005	0.2701	0.2394	0.2087		
0.96		0.3428	0.3131	0.2823	0.2509			
0.98		0.3561	0.3259	0.2944				
1.00		0.3694	0.3387					
1.02		0.3827						
1.04		0.3961						

[a] $C_d = 1.0$; $\alpha_c = 1.0$; $H_1 = H_e$.
[b] Change in increment.

Source: M. G. Bos, J. A. Replogle, and A. J. Clemmens, *Flow Measuring Flumes for Open Channel Systems*, copyright © 1984 John Wiley and Sons, Inc., p. 218–219. Reprinted by permission of John Wiley and Sons, Inc.

Index

529